T0252151

Handbook of Learning and Approximate Dynamic Programming

Handbook of Learning and Approximate Dynamic Programming

Jennie Si
Andy Barto
Warren Powell
Donald Wunsch

IEEE Neural Networks Society, *Sponsor*

IEEE PRESS

A JOHN WILEY & SONS, INC., PUBLICATION

For general information on our other products and services please contact our Customer Care Department within the U.S. at 877-762-2974, outside the U.S. at 317-572-3993 or fax 317-572-4002.

Wiley also publishes its books in a variety of electronic formats. Some content that appears in print, however, may not be available in electronic format.

Library of Congress Cataloging-in-Publication Data is available.

ISBN 0-471-66054-X

10 9 8 7 6 5 4 3 2 1

Contents

23 Control, Optimization, Security, and Self-healing of Benchmark Power Systems   **599**

James A. Momoh and Edwin Zivi

Preface

Complex artificial systems have become integral and critical components of modern society. The unprecedented rate at which computers, networks, and other advanced technologies are being developed ensures that our dependence on such systems will continue to increase. Examples of such systems include computer and communication networks, transportation networks, banking and finance systems, electric power grid, oil and gas pipelines, manufacturing systems, and systems for national defense. These are usually multi-scale, multi-component, distributed, dynamic systems. While advances in science and engineering have enabled us to design and build complex systems, comprehensive understanding of how to control and optimize them is clearly lacking.

There is an enormous literature describing specific methods for controlling specific complex systems or instruments based on various simplifying assumptions and requiring a range of performance compromises. While much has been said about complex systems, these systems are usually too complex for the conventional mathematical methodologies that have proven to be successful in designing complex instruments. The mere existence of complex systems does not necessarily mean that they are operating under the most desirable conditions with enough robustness to withstand the kinds of disturbances that inevitably arise. This was made clear, for example, by the major power outage across dozens of cities in the Eastern United States and Canada in August of 2003.

Dynamic programming is a well known, general-purpose method to deal with complex systems, to find optimal control strategies for nonlinear and stochastic dynamic systems. It is based on the Bellman equation which suffers from a severe "curse of dimensionality" (for some problems, there can even be *three* curses of dimensionality). This has limited its applications to very small problems. The same may be said of the classical "min-max" algorithms for zero-sum games, which are closely related. Over the past two decades, substantial progress has been made through efforts in multiple disciplines such as adaptive/optimal/robust control, machine learning, neural networks, economics, and operations research. For the most part, these efforts have not been cohesively linked, with multiple parallel efforts sometimes being pursued without knowledge of what others have done. A major goal of the 2002 NSF workshop was to bring these parallel communities together to discuss progress and to share ideas. Through this process, we are hoping to better define a community with common interests and to help develop a common vocabulary to facilitate communication.

Despite the diversity in the tools and languages used, a common focus of these researchers has been to develop methods capable of finding high-quality approximate solutions to problems whose exact solutions via classical dynamic programming are not attainable in practice due to high computational complexity and lack of accurate knowledge of system dynamics. At the workshop, the phrase *approximate dynamic programming* (ADP) was identified to represent this stream of activities.

A number of important results were reported at the workshop, suggesting that these new approaches based on approximating dynamic programming can indeed scale up to the needs of large-scale problems that are important for our society. However, to translate these results into systems for the management and control of real-world complex systems will require substantial multi-disciplinary research directed toward integrating higher-level modules, extending multi-agent, hierarchical, and hybrid systems concepts. There is a lot left to be done!

This book is a summary of the results presented at the workshop, and is organized with several objectives in mind. First, it introduces the common theme of ADP to a large, interdisciplinary research community to raise awareness and to inspire more research results. Second, it provides readers with detailed coverage of some existing ADP approaches, both analytically and empirically, which may serve as a baseline to develop further results. Third, it demonstrates the successes that ADP methods have already achieved in furthering our ability to manage and optimize complex systems. The organization of the book is as follows. It starts with a strategic overview and future directions of the important field of ADP. The remainder contains three parts. Part One aims at providing readers a clear introduction of some existing ADP frameworks and details on how to implement such systems. Part Two presents important and advanced research results that are currently under development and that may lead to important discoveries in the future. Part Three is dedicated to applications of various ADP techniques. These applications demonstrate how ADP can be applied to large and realistic problems arising from many different fields, and they provide insights for guiding future applications.

Additional information about the 2002 NSF workshop can be found at http://www.eas.asu.edu/~nsfadp

JENNIE SI, TEMPE, AZ

ANDREW G. BARTO, AMHERST, MA

WARREN B. POWELL, PRINCETON, NJ

DONALD C. WUNSCH, ROLLA, MO

Acknowledgments

The contents of this book are based on the workshop: "Learning and Approximate Dynamic Programming," which was sponsored by the National Science Foundation (grant number ECS-0223696), and held in Playacar, Mexico in April of 2002. This book is a result of active participation and contribution from the workshop participants and the chapter contributors. Their names and addresses are listed below.

Charles W. Anderson
Department of Computer Science
Colorado State University
Fort Collins, CO 80523 USA

S. N. Balakrishnan
Department of Mechanical and Aerospace Engineering and Engineering Mechanics
University of Missouri-Rolla
Rolla, MO 65409 USA

Andrew G. Barto
Department of Computer Science
University Of Massachusetts
Amherst, MA 01003 USA

Dimitri P. Bertsekas
Laboratory for Information and Decision Systems
Massachusetts Institute of Technology
Cambridge, MA 02139 USA

Zeungnam Bien
Department of Electrical Engineering and Computer Science
Korea Advanced Institute of Science and Technology
Yuseong-gu, Daejeon 305-701 Republic of Korea

Vivek S. Borkar
School of Technology and Computer Science
Tata Institute of Fundamental Research
Mumbai, 400005 India

Xi-Ren Cao
Department of Electrical and Electronic Engineering
Hong Kong University of Science and Technology
Clear Water Bay, Kowloon , Hong Kong

Daniel Pucci de Farias
Department of Mechanical Engineering
Massachusetts Institute of Technology
Cambridge, MA 02139 USA

Thomas G. Dietterich
School of Electrical Engineering and Computer Science
Oregon State University
Corvallis, OR 97331 USA

Russell Enns
Boeing Company the Helicopter Systems
5000 East McDowell Road
Mesa, AZ 85215

Augustine O. Esogbue
Intelligent Systems and Controls Laboratory
Georgia Institute of Technology
Atlanta, GA 30332 USA

Silvia Ferrari
Department of Mechanical Engineering and Materials Science
Duke University
Durham, NC 27708 USA

Laurent El Ghaoui
Department of Electrical Engineering and Computer Science
University of California at Berkeley
Berkeley, CA 94720 USA

Mohammad Ghavamzadeh
Department of Computer Science
University of Massachusetts
Amherst, MA 01003 USA

Greg Grudic
Department of Computer Science
University of Colorado at Boulder
Boulder, CO 80309 USA

Dongchen Han
Department of Mechanical and Aerospace Engineering and Engineering Mechanics
University of Missouri-Rolla
Rolla, MO 65409 USA

Ronald G. Harley
School of Electrical and Computer Engineering
Georgia Institute of Technology
Atlanta, GA 30332 USA

Warren E. Hearnes II
Intelligent Systems and Controls Laboratory
Georgia Institute of Technology
Atlanta, GA 30332 USA

Douglas C. Hittle
Department of Mechanical Engineering
Colorado State University
Fort Collins, CO 80523 USA

Dong-Oh Kang
Department of Electrical Engineering and Computer Science
Korea Advanced Institute of Science and Technology
Yuseong-gu, Daejeon 305-701 Republic of Korea

Matthew Kretchmar
Department of Computer Science
Colorado State University
Fort Collins, CO 80523 USA

George G. Lendaris
Systems Science and Electrical Engineering
Portland State University
Portland, OR 97207 USA

Derong Liu
Department of Electrical and Computer Engineering
University of Illinois at Chicago
Chicago, IL 60612 USA

Sridhar Mahadevan
Department of Computer Science
University of Massachusetts
Amherst, MA 01003 USA

James A. Momoh
Center for Energy Systems and Control
Department of Electrical Engineering
Howard University
Washington, DC 20059 USA

Angelia Nedich
Alphatech, Inc.
Burlington, MA 01803 USA

James C. Neidhoefer
Accurate Automation Corporation
Chattanooga, TN 37421 USA

Arnab Nilim
Department of Electrical Engineering and Computer Sciences
University of California at Berkeley
Berkeley, CA 94720 USA

Warren B. Powell
Department of Operations Research and Financial Engineering
Princeton University
Princeton, NJ 08544 USA

Danil V. Prokhorov
Research and Advanced Engineering
Ford Motor Company
Dearborn, MI 48124 USA

Khashayar Rohanimanesh
Department of Computer Science
University of Massachusetts
Amherst, MA 01003 USA

Michael T. Rosenstein
Department of Computer Science
University of Massachusetts
Amherst, MA 01003 USA

Malcolm Ryan
School of Computer Science & Engineering
University of New South Wales
Sydney 2052 Australia

Shankar Sastry
Department of Electrical Engineering and Computer Sciences
University of California at Berkeley
Berkeley, CA 94720 USA

Jennie Si
Department of Electrical Engineering
Arizona State University
Tempe, AZ 85287 USA

Ronnie Sircar
Department of Operations Research and Financial Engineering
Princeton University
Princeton, NJ 08544 USA

Robert F. Stengel
Department of Mechanical and Aerospace Engineering
Princeton University
Princeton, NJ 08544 USA

Georgios Theocharous
Artificial Intelligence Laboratory
Massachusetts Institute of Technology
Cambridge, MA 02139 USA

Lyle Ungar
School of Engineering and Applied Science
University of Pennsylvania
Philadelphia, PA 19104 USA

Benjamin Van Roy
Departments of Management Science and Engineering and Electrical Engineering
Stanford University
Stanford, CA 94305 USA

Ganesh K. Venayagamoorthy
Department of Electrical and Computer Engineering
University of Missouri-Rolla
Rolla, MO 65409 USA

Paul Werbos
National Science Foundation
Arlington, VA 22230 USA

David White
Praesagus Corp.
San Jose, CA 95128 USA

Bernard Widrow
Department of Electrical Engineering
Stanford University
Stanford, CA 94305 USA

Donald C. Wunsch
Department of Electrical and Computer Engineering
University of Missouri-Rolla
Rolla, MO 65409 USA

Lei Yang
Department of Electrical Engineering
Arizona State University
Tempe, AZ 85287 USA

Peter Young
Department of Electrical and Computer Engineering
Colorado State University
Fort Collins, CO 80523 USA

Edwin Zivi
Weapons & Systems Engineering Department
U.S. Naval Academy
Annapolis, MD 21402 USA

The editors would like to acknowledge the efforts of Mr. Lei Yang and Mr. James Dankert of the Department of Electrical Engineering at Arizona State University, who assisted in many aspects during the preparation of this book and performed much of the electronic formatting of the book manuscript. The editors would also like to acknowledge the support from Ms. Catherine Faduska and Ms. Christina Kuhnen at IEEE Press. Finally, the editors would like to thank the reviewers and Dr. David Fogel for their encouragement and helpful inputs that have helped enhance the book.

Dedication

To those with an open mind and the courage to go beyond the ordinary.

Foreword

The NSF sponsored workshop on "Learning and Approximate Dynamic Programming" held in April 2002 at the magical resort town of Playa Del Carmen in Mexico brought together researchers from a number of different fields: control theory, operations research, artificial intelligence, robotics, and neural networks, to name a few. The workshop provided an environment to engage in thinking about common techniques and methodologies for the design of complex intelligent systems. While this endeavor has been long standing, it was gratifying to see the amount of progress that has been made in the development of scalable techniques for solving real world problems such as those in robotics, logistics, power systems, and aerospace engineering. It was also heartening to see the use of techniques that were hitherto considered to be computationally too demanding now being used in real time. This latter fact is attributable not only to the dramatic increases in computational power but also to the development of new conceptual frameworks, solution techniques and algorithms. A distinguishing feature of the workshop was diversity — a diversity of applications, a diversity of backgrounds, a diversity of vocabularies, and a diversity of algorithmic strategies. The workshop succeeded due to the common interest in a key set of unifying tools and methods which are in need of further conceptual development: the development of new approaches for approximate dynamic programming, approximate solutions to games, and learning. Thus the workshop was more than an interdisciplinary meeting; rather it was a dialog on issues of commonality spanning many areas of research and thought. The proceedings volume edited by Si, Barto, Powell and Wunsch captures the spirit of excitement that we all felt at this meeting.

With the spotlight on complex critical infrastructures and their robust performance in the face of both inadvertent mis-configuration or deliberate and malicious attack, I feel that the scalability of approximate dynamic programming and learning, real time solutions to multi-person dynamic games holds the key to the architecting of high confidence societal scale systems. While technological progress is on its way to enabling us to embed smart sensor networks in all of our infrastructures, the architecting of distributed control solutions which recognize and learn during operations is in its infancy. I feel that the issues raised and directions of work suggested in this volume will go a long way toward the development of robust, high confidence systems which can operate through attacks. Let me elaborate a little on the last point about operating through or degrading gracefully through attacks. The work in this area has been in the realm of cybersecurity research to date. However,

I feel that the key methodologies of approximate solutions of adversarial games and other (less malicious) sources of uncertainty developed in this volume are the right foundational and conceptual tools to bring rigor to the set of ad-hoc techniques that are currently the norm in the area of cybersecurity and critical infrastructure protection.

Of course there is far more detail on these points in the volume and especially in the papers of the organizers. I invite the reader to enjoy the papers and the richness of the potential of the methods in a wide variety of applications.

SHANKAR SASTRY, BERKELEY, CA

MAY 2004.

Handbook of Learning and Approximate Dynamic Programming
Edited by Jennie Si, Andy Barto, Warren Powell and Donald Wunsch
Copyright © 2004 The Institute of Electrical and Electronics Engineers, Inc.

1 ADP: Goals, Opportunities and Principles

PAUL WERBOS

National Science Foundation

1.1 GOALS OF THIS BOOK

Is it possible to build a general-purpose learning machine, which can learn to maximize *whatever* the user wants it to maximize, over time, in a strategic way, even when it starts out from zero knowledge of the external world? Can we develop general-purpose software or hardware to "solve" the Hamilton-Jacobi-Bellman equation for truly large-scale systems? Is it possible to build such a machine which works effectively in practice, even when the number of inputs and outputs is as large as what the mammal brain learns to cope with? Is it possible that the human brain itself is such a machine, in part? More precisely, could it be that the same mathematical design principles which we use to build rational learning-based decision-making machines are also the central organizing principles of the human mind itself? Is it possible to convert all the great rhetoric about "complex adaptive systems" into something that actually works, in maximizing performance?

Back in 1970, few informed scientists could imagine that the answers to these questions might ultimately turn out to be yes, if the questions are formulated carefully. But as of today, there is far more basis for optimism than there was back in 1970. Many different researchers, mostly unknown to each other, working in different fields, have independently found promising methods and strategies for overcoming obstacles that once seemed overwhelming.

This section will summarize the overall situation and the goals of this book briefly but precisely, in words. Section 1.2 will discuss the needs and opportunities for ADP systems across some important application areas. Section 1.3 will discuss the core mathematical principles which make all of this real.

Goal-oriented people may prefer to read Section 1.2 first, but bottom-up researchers may prefer to jump immediately to Section 1.3. The discussion of goals here and in Section 1.2 is really just a brief summary of some very complex lessons learned over time, which merit much more detailed discussion in another context.

This book will focus on the first three questions above. It will try to bring together the best of what is known across many disciplines relevant to the question: how can we develop *better* general-purpose tools for doing optimization over time, by using learning and approximation to allow us to handle *larger-scale, more difficult* problems? Many people have also studied the elusive subtleties of the follow-on questions about the human brain and the human mind [1]; however, we will never be able to construct a good straw-man model of how the mammalian brain achieves these kinds of capabilities, until we understand what these capabilities require of *any* information processing system on *any* hardware platform, wet, dry, hard or ethereal. The first three questions are a difficult enough challenge by themselves.

As we try to answer these three questions, we need to constantly re-examine what they actually mean. What are we trying to accomplish here? What is a "general-purpose" machine? Certainly no one will ever be able to build a system which is guaranteed to survive when it is given just a microsecond to adapt to a totally crazy and unprecedented kind of lethal shock from out of the blue. Mammal brains work amazingly well, but even they get eaten up sometimes in nature. But certainly, we need to look for some ability to learn or converge at reasonable speed (as fast as possible!) across a wide variety of complex tasks or environments. We need to focus on the long-term goal of building systems for the general case of nonlinear environments, subject to random disturbances, in which our intelligent system gets to observe only a narrow, partial window into the larger reality in which it is immersed.

In practical engineering, we, like nature, will usually not want to start our systems out in a state of zero knowledge. But even so, we can benefit from using learning systems powerful enough that they *could* converge to an optimum, even when starting from zero. Many of us consider it a gross abuse of the English language when people use the word "intelligent" to describe high-performance systems which are unable to learn anything fundamentally new, on-line or off-line.

In many practical applications today, we can use ADP as a kind of offline numerical method, which tries to "learn" or converge to an optimal adaptive control policy for a particular type of plant (like a car engine). But even in offline learning, convergence speed is a major issue, and similar mathematical challenges arise.

Is it *possible* to achieve the kind of general-purpose optimization capability we are focusing on here, as the long-term objective? Even today, there is only one exact method for solving problems of optimization over time, in the general case of nonlinearity with random disturbance: dynamic programming (DP). But exact dynamic programming is used only in niche applications today, because the "curse of dimensionality" limits the size of problems which can be handled. Thus in many engineering applications, basic stability is now assured via conservative design, but overall performance is far from optimal. (It is common to optimize or tweak a control parameter here and there, but that is not at all the same as solving the dynamic optimization problem.) Classical, deterministic design approaches cannot even address questions like: How can we minimize the *probability* of disaster, in cases where we cannot totally guarantee that a disaster is impossible?

There has been enormous progress over the past ten years in increasing the scale of what we can handle with *approximate* DP and learning; thus the goals of this book are two-fold: (1) to begin to unify and consolidate these recent gains, scattered across disciplines; and (2) to point the way to how we can bridge the gap between where we are now and *the next major watershed* in the basic science — the ability to handle tasks as large as what the mammal brain can handle, and even to connect to what we see in the brain.

The term "ADP" can be interpreted either as "Adaptive Dynamic Programming" (with apologies to Warren Powell) or as "Approximate Dynamic Programming" (as in much of my own earlier work). The long-term goal is to build systems which include *both* capabilities; therefore, I will simply use the acronym "ADP" itself. Various strands of the field have sometimes been called "reinforcement learning" or "adaptive critics" or "neurodynamic programming," but the term "reinforcement learning" has had many different meanings to many different people.

In order to reach the next major watershed here, we will need to pay more attention to two major areas: (1) advancing ADP as such, the main theme of this book; (2) advancing the critical subsystems we will need as components of ADP systems, such as systems that learn better and faster how to make predictions in complex environments.

1.2   FUNDING ISSUES, OPPORTUNITIES AND THE LARGER CONTEXT

This book is the primary product of an NSF-sponsored workshop held in Mexico in April 2002, jointly chaired by Jennie Si and Andrew Barto. The goals of the workshop were essentially the same as the goals of this book. The workshop covered some very important further information, beyond the scope of the book itself, posted at http://www.eas.asu.edu/ nsfadp

The potential of this newly emerging area looks far greater if we can *combine* what has been achieved across all the relevant disciplines, and put the pieces all together. There are substantial unmet opportunities here, including some "low lying fruit." One goal of this book is to help us see what these opportunities would look like, if one were to bring these strands together in a larger, more integrated effort. But as a practical matter, large funding in this area would require three things: (1) more proposal pressure in the area; (2) more unified and effective communication of the larger vision to the government and to the rest of the community; (3) more follow-through on the vision *within* the ADP community, including more cross-disciplinary research and more unified education.

Greater cross-disciplinary cooperation is needed for intellectual reasons, as you will see from this book. It is also needed for practical reasons. The engineering and operations research communities have done a good job on the applications side, on the whole, while the computer science communities have done a good job on educational outreach and infrastructure. We will need to combine both of these together, more effectively, to achieve the full potential in either area.

At a lower level, the Control, Networks and Computational Intelligence (CNCI) program in the Electrical and Communication Systems (ECS) Division of NSF has funded extensive work related to ADP, including earlier workshops which led to books of seminal importance [2, 3]. The center of gravity of CNCI has been the IEEE communities interested in these issues (especially in neural networks and in control theory), but ECS also provided some of the early critical funding for Barto, Bertsekas and other major players in this field. ADP remains a top priority in CNCI funding. It will benefit the entire community if more proposals are received in these areas. The Knowledge and Cognitive Systems program in Intelligent Information Systems (IIS) Division of NSF has also funded a great deal of work in related areas, and has joined with ECS in joint-funding many projects.

In the panels that I lead for CNCI, I ask them to think of the funding decision itself as a kind of long-term optimization problem under uncertainty. The utility function to be maximized is a kind of 50-50 sum of two major terms — the potential benefit to basic scientific understanding, and the potential benefit to humanity in general. These are also the main considerations to keep in mind when considering possible funding initiatives.

Occasionally, some researchers feel that these grand-sounding goals are too large and too fuzzy. But consider this analogy. When one tries to sell a research project to private industry, one *must* address the ultimate bottom line. Proposals which are not well-thought-out and specifically tuned to maximize the bottom line usually do not get funded by industry. The bottom line is different at NSF, but the same principle applies. Strategic thinking is essential in all of these sectors. All of us need to reassess our work, strategically, on a regular basis, to try to maximize our own impact on the larger picture.

The biggest single benefit of ADP research, in my personal view, is based on the hope that answers to the first three questions in Section 1.1 will be relevant to the further questions, involving the brain and the mind. This hope is a matter for debate in the larger world, where it contradicts many strands of inherited conventional wisdom going back for centuries. This book cannot do justice to those complex and serious debates, but the connection between those debates and what we are learning from ADP has been summarized at length elsewhere (e.g. [9, 52], with reference to further discussions).

1.2.1   Benefits to Fundamental Scientific Understanding

In practice, the fundamental scientific benefit of most of the proposals which I see comes down to the long-term watershed discussed above. There are few questions as fundamental in science as "What is Mind or Intelligence?" Thus I generally ask panelists to evaluate what the impact might be of a particular project in allowing us to get to the watershed earlier than we would without the project. Since we will clearly need some kind of ADP to get there, ADP becomes the top scientific priority. But, as this book will bring out, powerful ADP systems will also require powerful *components* or *subsystems*. This book will touch on some of the critical needs and opportunities for subsystems, such as subsystems which learn to predict or model the external world, subsystems for memory, subsystems for stochastic search and subsystems for more powerful function approximation. This book may be uniquely important in identifying what we need to get from such subsystems, but there are other books which discuss those areas in more detail. CNCI funds many projects in those areas from people who may not know what ADP is, but are providing capabilities important to the long-term goals. Likewise, there are many projects in control system design which can feed into the long-term goal here, with or without an explicit connection to ADP. In every one of these areas, greater unification of knowledge is needed, for the sake of deeper understanding, more effective education, and reducing the common tendencies toward reinventing or spinning wheels.

1.2.2   Broader Benefits to Humanity

Optimal performance turns out to be critical to a wide variety of engineering and management tasks important to the future of humanity — tasks which provide excellent test-beds or drivers for new intelligent designs. The ECS Division has discussed the practical needs of many major technology areas, directly relevant to the goals of achieving sustainable growth here on earth, of cost-effective settlement of space, and of fostering human potential in the broadest sense. ADP has important potential applications in many areas, such as manufacturing [3], communications, aerospace, Internet software and defense, all of interest to NSF; however, test-beds related to energy and the environment have the closest fit to current ECS activities.

This past year, ECS has played a major role in three cross-agency funding activities involving energy and the environment, where ADP could play a critical role in enabling things which could not be done without it. It looks to me as if all three will recur or grow, and all three are very open to funding partnerships between ADP researchers and domain experts. The three activities focus on: (1) new crossdisciplinary partnerships addressing electric power networks (EPNES); (2) space solar power (JIETSSP); and (3) sustainable technology (TSE). To learn about these activities in detail, search on these acronyms at http://www.nsf.gov. Here I will talk about the potential role of ADP in these areas.

1.2.2.1   Potential Benefits In Electric Power Grids   Computational intelligence is only one of the core areas in the CNCI program. CNCI is also the main funding program in the US government for support of electric utility grid research. CNCI has long-standing ties with the IEEE Power Engineering Society (PES) and the Electric Power Research Institute (EPRI). Transitions and flexibility in the electric power sector will be crucial to hopes of achieving a sustainable global energy system. Many people describe the electric power system of the Eastern United States as "the largest, most complicated single machine ever built by man."

The workshop led by Si and Barto was actually just one of *two* coordinated workshops held back-to-back in the same hotel in Mexico. In effect, the first workshop asked: "How can we develop the algorithms needed in order to better approximate optimal control over time of extremely large, noisy, nonlinear systems?" The second workshop, chaired by Ron Harley of the University of Natal (South Africa) and Georgia Tech, asked: "How can we develop the necessary tools in order to do integrated global optimization over time of the electric power grid as one single system?" It is somewhat frightening that some engineers could not see how there could be any connection between these two questions, when we were in the planning stages.

The Harley workshop — sponsored by James Momoh of NSF and Howard University, and by Massoud Amin of EPRI and the University of Minnesota — was actually conceived by Karl Stahlkopf, when he was a Vice-President of EPRI. Ironically, Stahlkopf urged NSF to co-sponsor a workshop on that topic as a kind of act of reciprocity for EPRI co-sponsoring a workshop I had proposed (when I was temporarily running the electric power area) on hard-core transmission technology and urgent problems in California. Stahlkopf was very concerned by the growing difficulty of getting better whole-system performance and flexibility in electric power grids, at a time when nonlinear systems-level challenges are becoming more difficult but control systems and designs are still mainly based on piecemeal, local, linear or static analysis. "Is it possible," he stressed, "to optimize the *whole* thing in an integrated way, as a single dynamical system?"

It will require an entire new book to cover all the many technologies and needs for optimization discussed at the second workshop. But there are five points of particular relevance here.

First, James Momoh showed how his new Optimal Power Flow (OPF) system — being widely distributed by EPRI — is much closer to Stahlkopf's grand vision than anything else now in existence.

Real-world suppliers to the electric utility industry have told me that OPF is far more important to their clients than all the other new algorithms put together. OPF does provide a way of coordinating and integrating many different decisions across many points of the larger electric power system. Momoh's version of OPF — using a combination of nonlinear interior point optimization methods and genetic algorithms — has become powerful enough to cope with a growing variety of variables, all across the system. But even so, the system is "static" (optimizes at a given time slice) and deterministic. Stahlkopf's vision could be interpreted *either* as: (1) extending OPF to the dynamic stochastic case — thus creating dynamic stochastic OPF (DSOPF); (2) applying ADP to the electric power grid. The chances of success here may be best if *both* interpretations are pursued together in a group which understands how they are equivalent, both in theory and in mathematical details. Howard University may be particularly well equipped to move toward DSOPF in this top-down, mathematically-based approach.

Some researchers in electric power are now working to extend OPF by adding the "S" but not the "D". They hope to learn from the emerging area of stochastic programming in operations research. This can improve the treatment of certain aspects of power systems, but it does provide a capability for anticipatory optimization. It does not provide a mechanism for accounting for the impact of present decisions on future situations. For example, it would not give credit to actions that a utility system could take at present in order to prevent a possible "traffic jam" or breakdown in the near future. Simplifications and other concepts from stochastic programming could be very useful within the context of a larger, unifying ADP system, but only ADP is general enough to allow a unified formulation of the entire decision problem.

Second, Venayagamoorthy of Missouri-Rolla presented important results which form a chapter of this book. He showed how DHP — a particular form of ADP design — results in an integrated control design for a system of turbogenerators which is able to keep the generators up, even in the face of disturbances three times as large as what can be handled by the best existing alternative methods. This was shown on an actual experimental electric power grid in South Africa. The neural network models of generator dynamics were trained in real time as part of this design. The ability to keep generators going is very useful in a world where containing and preventing system-wide blackouts is a major concern. This work opens the door to two possible extensions: (1) deployment on actual commercially running systems, in areas where blackouts are a concern or where people would like to run generators at a higher load (i.e. with lower stability margins); (2) research into building up to larger and larger systems, perhaps by incorporating power-switching components into the experimental networks and models. This may offer a "bottom-up" pathway toward DSOPF. This work was an outcome of a CNCI grant to fund a partnership between Ron Harley (a leading power engineer) and Don Wunsch (a leading researcher in ADP from the neural network community).

Third, the DHP system used in this example outputs value signals which may be considered as "price signals" or "shadow prices." They represent the *gradient* of the more conventional scalar value function used in the Bellman equation or in simpler ADP designs. They are generalizations of the Lagrange Multipliers used in Momoh's OPF system. They may also offer new opportunities for a better interface between the computer-based grid control systems and the money-based economic control systems.

Fourth, there is every reason to worry that the technology used by Venayagamoorthy — which works fine for a dozen to a few dozen continuous state variables — will start to break down as we try to scale up to thousands of variables, *unless* the components of the design are upgraded, as I will discuss later in this chapter. The effort to scale up is, once again, the core research challenge ahead of us here. One warning: some writers claim to handle a dozen state variables when they actually mean that they handle a dozen possible states in a finite-state Markhov chain system; however, in this case, I really am referring to a state space which is R12, a twelve-dimensional space defined by continuous variables, which is highly nonlinear, and not reducible to clusters around a few points in that space.

Finally, ADP researchers who are interested in exploring this area are strongly urged to look up "EPNES" at http://www.nsf.gov, and look in particular at the benchmark problems discussed there. The benchmark problems there, developed in collaboration between the two sponsors (NSF and the Navy), were designed to make it as easy as possible for non-power-engineers to try out new control approaches. As a practical matter, however, EPNES is more likely to fund partnerships which do include a collaborator from hard-core power engineering.

1.2.2.2   *Potential Benefits in Space Solar Power (SSP)*   The Millennium Project of the United Nations University (http://millennium-project.org) recently asked decision makers and science policy experts all over the world: "What challenges can science pursue whose resolution would significantly improve the human condition?" The leading response was: "Commercial availability of a cheap, efficient, environmentally benign non-nuclear fission and non-fossil fuel means of generating base-load electricity, competitive in price with today's fossil fuels." Space solar power is a high risk vision for future technology, but many serious experts believe that it is the most promising of the very few options for meeting this larger need. (Earth-based solar is another important option.) Thus in March 2002, NASA, NSF and EPRI issued a solicitation for proposals called the "Joint Investigation of Enabling Technologies for Space Solar Power (JIETSSP)." John Mankins of NASA and I served as co-chairs of the Working Group which managed this activity.

Reducing cost was the number one goal of this effort. For each proposal, we asked the reviewers to consider: What is the potential impact of funding this work on reducing the time we have to wait, before we know enough to build SSP systems which can safely beat the cost of nuclear power in developing regions like the Middle East? We also asked the usual questions NSF asks about the potential benefits to basic science and other impacts.

Earlier proposals for SSP, dating back to the 1960's and 1970's, could not meet this kind of cost target, for several reasons. One of the key problems was the enormous cost of sending up humans to do all of the assembly work in space. Consider the total costs of sending up six people to live in a space station, multiply by a thousand, and you begin to see how important it is to reduce this component of cost as much as possible.

JIETSSP was partly the outcome of a workshop organized by George Bekey of USC in April 2000, jointly sponsored by NSF and NASA. The original goal was to explore how radically new approaches to robotics, such as robots building robots or robots with real intelligence, might be used to reduce costs either in space solar power or in building earth-based solar power systems.

The most practical concept to emerge from that workshop was the concept of "teleautonomy" as described by Rhett Whittaker of Carnegie-Mellon. Truly autonomous, self-replicating robots may indeed become possible, using ADP and other new technologies, but we do not really need them for SSP, and we need to make SSP affordable as soon as possible. But by the same token, we cannot afford to wait until domain specialists in robotics develop complete three-dimensional physics models of every possible task and every possible scenario that might occur in space.

Whittaker proposed that we plan for a system in which hundreds of human operators (mostly on earth) control hundreds or thousands of robots in space, to perform assembly and maintenance. He proposed that we build up to this capability in an incremental way, by first demonstrating that we can handle the full range of required tasks in a more traditional telerobotics mode, and then focus on coordination and cost reduction. Ivory tower robotics research can easily become a gigantic, unfocused, nonproductive swamp; Whittaker's kind of strategy may be crucial to avoiding such pitfalls. Greg Baiden, of Penguin ASI in Canada and Laurentian University, has implemented real-world teleautonomy systems used in the mining industry which can provide an excellent starting point. Lessons from assembling the International Space Station need to be accounted for as well.

In the short term, the most promising role for ADP in SSP robotics may be mid-level work, analogous to the work of Venayagamoorthy in electric power. Venayagamoorthy has developed better sensor-based control policies, flexible enough to operate over a broader range of conditions, for a *component* of the larger power system. In the same way, ADP could be used to improve performance or robustness or autonomy for *components* of a larger teleautonomy system. Half of the intellectual challenge here is to scope out the testbed challenges which are really important within realistic, larger systems designs in the spirit of Whittaker or Baiden.

What kinds of robotics tasks could ADP really help with, either in space or on earth?

Practical robotics in the United States often takes a strictly domain-specific approach, in the spirit of expert systems. But the robotics field worldwide respects the achievements of Hirzinger (in Germany) and Fukuda (in Japan), who have carefully developed ways to use computational intelligence and learning to achieve better results in a wide variety of practical tasks. Fukuda's work on free-swinging, ballistic kinds of robots (motivated by the needs of the construction industry) mirrors the needs of space robotics far better than the domain-specific Japanese walking robots that have become so famous. Hirzinger has found ways to map major tasks in robotics into tasks for general-purpose learning systems [4], where ADP (and related advanced designs) may be used to achieve capabilities even greater than what Hirzinger himself has achieved so far.

It is also possible that new research, in the spirit of Warren Powell's chapter 10 in this book, could be critical to the *larger* organizational problems in moving thousands of robots and materials around a construction site in space. Powell's methods for approximating a value function have been crucial to his success so far in handling dynamic optimization problems with thousands of variables — but even more powerful function approximators might turn out to be important to SSP, if coupling effects and nonlinearities should make the SSP planning problems more difficult than the usual logistics problems on earth. We do not yet know.

In 2002, JIETSSP funded twelve or thirteen research projects (depending on how one counts). Five of these were on space robotics. These included an award to Singh,

Whittaker and others at Carnegie-Mellon, and other projects of a more immediate nature. They also included two more forwards-looking projects — one led by Shen at USC and one by Jennie Si — aiming at longer-term, more fundamental improvements in what can be done with robots. The Si project emphasizes ADP and what we can learn from biology and animal behavior. The Shen project emphasizes robots which can adapt their *physical form*, in the spirit of Transformers, in order to handle a wide range of tasks.

A deep challenge to NASA and NSF here is the need to adapt to changes in *high-level* design concepts for SSP. "What are we trying to build?" ask the roboticists. But we don't yet know. Will it be something like the Sun Tower design which Mankins' groups developed a few years ago? Or will the core element be a gigantic light-to-light laser, made up of pieces "floating" in a vacuum, held together by some kind of tensegrity principle with a bit of active feedback control? Could there even be some kind of safe nuclear component, such as a laser fusion system in space, using the new kinds of fuel pellets designed by Perkins of Lawrence Livermore which allow light-weight MHD energy extraction from the protons emerging from D-D fusion? One should not even underestimate ideas like the Ignatiev/Criswell schemes for using robotic systems to exploit materials on the moon. Given this huge spectrum of possibilities, we clearly need approaches to robotic assembly which are as flexible and adaptive as possible.

1.2.2.3   *Benefits to Technology for a Sustainable Environment (TSE)*   Complicated as they are, the electric power grid and the options for SSP are only two examples taken from a much larger set of energy/environment technologies of interest to the Engineering Directorate of NSF and to its partners, (e.g., see [5]).

Many of the other key opportunities for ADP could now fit into the recently enlarged scope of Technology for a Sustainable Environment (TSE), a joint initiative of EPA and NSF. Recent progress has been especially encouraging for the use of intelligent control in cars, buildings and boilers.

Building energy use is described at length in the chapter by Charles Anderson et al., based on a grant funded jointly by CNCI and by the CMS Division of NSF. Expert reviewers from that industry have verified that this new technology really does provide a unique opportunity to reduce energy use in buildings by a significant factor. Furthermore, use of ADP should make it possible to learn a *price-responsive* control policy, which would be very useful to the grid as a whole (and to saving money for the customer). There are important possibilities for improved performance through further research — but the biggest challenge for now is to transfer the success already achieved in the laboratory to the global consumer market. There are no insuperable barriers here, but it always takes a lot of time and energy to coordinate the many aspects of this kind of fundamental transition in technology.

ADP for cars has larger potential, but the issues with cars are more complicated. There are urgent needs for reduced air pollution and improved fuel flexibility and efficiency in conventional cars and trucks. Many experts believe that the emission of NOx compounds by cars and trucks is the main accessible cause of damage to human health and to nature today, from Los Angeles to the Black Forrest of Germany. Improved fuel flexibility could be critical to the ability of advanced economies to withstand sharp reductions in oil supply from the Middle East ten years in the future — an issue of growing concern. In the longer term, cost and efficiency concerns will be critical to the rate at which new kinds of cars — hybrid-electric, pure electric or fuel-cell cars — can actually penetrate the mass market.

Up to now, Feldkamp's group at Ford Research has demonstrated the biggest proven opportunity for learning systems to yield important benefits here. By 1998, they had demonstrated that a neural network learning system could meet three new stringent requirements of the latest Clean Air Act in an affordable way, far better than any other approach ever proven out: (1) on-board diagnostics for misfires, using time-lagged recurrent networks (TLRN); (2) idle speed control; (3) control of fuel/air ratios. In September of 1998, the President of Ford committed himself (in a major interview in *Business Week*) to deploying this system in every Ford car made in the world by 2001. An important element of this plan was a new neural network chip design from Mosaix LLC of California (led by Raoul Tawel of the Jet Propulsion Laboratory), funded by an SBIR grant from NSF, which would cost on the order of $1–$10 per chip. Ken Marko — who until recently ran a Division at Ford Research which collaborated closely with Feldkamp's Division — was asked to organize a major conference on clean air technology for the entire industry, based in part on this success. But there are many caveats here.

First, there have been changes in the Clean Air Act in the past few years, and major changes in Ford management.

Second, management commitment by itself is not sufficient to change the basic technology across a large world-wide enterprise, in the throes of other difficult changes. The deployment of the lead element of the clean air system — the misfire detector — has moved ahead very quickly, in actuality. Danil Prokhorov — who has an important chapter in this book — has played an important role in that activity, under Lee Feldkamp.

Third, the neural network control system was not based on ADP, and was not trained to reduce NOx as such. Statistics on NOx reduction were not available, at last check. The control system and the misfire detectors were *both* based on TLRNs trained by backpropagation through time (BPTT [7]). In effect, they used the kind of control design discussed in Section 2.10 of my 1974 Ph.D. thesis on backpropagation [6], which was later implemented by four different groups by 1988 [2] including Widrow's famous truck-backer-upper, and was later re-interpreted as "neural model predictive control" ([3, ch. 13], [8]). Some re-inventions of the method have described it as a kind of "direct policy reinforcement learning," but it is not a form of ADP. It does not use the time-forwards kind of learning system required of a plausible model of brain-like intelligence. Some control engineers would call it a kind of receding horizon method. Feldkamp's group has published numerous papers on the key ideas which made their success possible, summarized in part in [4].

In his chapter 15, Prokhorov raises a number of questions about ADP versus BPTT which are very important in practical engineering. Properly implemented (as at Ford),

BPTT tends to be far superior in practice to other methods commonly used. Stability is guaranteed under conditions much broader than what is required for stability with ordinary adaptive control [8, 9]. Whenever BPTT and ADP can both be applied to a task in control or optimization, a good evaluation study should use both on a regular basis, to provide a kind of cross-check. Powell's chapter gives one good example of a situation where ADP works far better than a deterministic optimization. (Unless one combines BPTT with special tricks developed by Jacobsen and Mayne [10], BPTT is a deterministic optimization method.) There are many connections between BPTT and ADP which turn out to be important in advanced research, as Prokhorov and I have discussed. Many of the techniques used by Ford with BPTT are essential to larger-scale applications of ADP as well.

More recently, Jagannathan Sarangapani of the University of Missouri-Rolla has demonstrated a 50 percent reduction in NOx (with a clear possibility for getting it lower) in tests of a simple neural network controller on a Ricardo Hydra research engine with cylinder geometry identical to that of the Ford Zetec engine [11], under a grant from CNCI. Sarangapani has drawn heavily on his domain knowledge from years of working at Caterpillar, before coming to the university. While BPTT and ADP both provide a way to maximize utility or minimize error *over the entire future*, Sarangapani has used a simpler design which tries to minimize error *one time step into the future*, using only one active "control knob" (fuel intake). But by using a good measure of error, and using *real-time* learning to update his neural networks, he was still able to achieve impressive initial results.

Saragapani explains the basic idea as follows. People have known for many years that better fuel efficiency and lower NOx could be achieved, if engines could be run in a "lean regime" (fuel air ratios about 0.7) and if exhaust gasses could be recycled into the engine at a higher level. But in the past, when people did this, it resulted in a problem called "cyclic dispersion" — a kind of instability leading to misfires and other problems. Physics-based approaches to modeling this problem and preventing the instabilities have not been successful, because of real-world variations and fluctuations in engine behavior. By using a simple neural network controller, inspired by ideas from his Ph.D. thesis adviser Frank Lewis [12], he was able to overcome this problem. This approach may lead to direct improvements in air quality from large vehicles which do not use catalytic converters (and are not likely to soon, for many reasons).

As with the Anderson work, it will be important to take full advantage of the initial breakthrough in performance here. But one can do better. One may expect better performance and extensions to the case of cars by incorporating greater foresight — by minimizing the same error or cost measure Sarangapani is already using, but minimizing it over the long-term, and multiplying it by a term to reflect the impact of fuel-oxygen ratios on the performance of the catalytic converter (or using a more direct measure or predictor of emissions); ADP provides methods necessary to that extension. ADP should also make it possible to account for additional "control knobs" in the system, such as spark plug advance and such, which will be very important to wringing out optimal performance from this system. It may also be important to use the approaches Ford has used to train TLRNs to *predict* the system, in order to obtain additional *inputs* to the control systems, in order to really maximize performance and adaptability of the system. (See 1.3.2.2.) It may even be useful to use something like the Ford system to *predict* the likelihood of misfire, to construct an error measure still better than what Saragapani is now using. In the end, theory tells us to expect improved performance if we take such measures, but we will not know how large such improvements can actually be until we try our best to achieve them.

ADP may actually be more important to *longer-term* needs in the world of cars and trucks.

For example, if mundane issues of mileage and pollution are enough to justify putting learning chips into all the cars and trucks in the world, then we can use those chips for additional purposes, without adding too much to the cost of a car. Years ago, the first Bush Administration made serious efforts to try to encourage *dual-fired* cars, able to flip between gasoline and methanol automatically. Roberta Nichols (now with University of California Riverside) once represented Ford in championing this idea at a national level. But at that time, it would have cost about $300 per car to add this flexibility, and oil prices seemed increasingly stable. In 2003, it seems increasingly clear that more "fuel insurance" would be a very good thing to have, and *also* an easier thing to accomplish. Materials are available for multi-fuel gas tanks. It does not have to be just gasoline *or* methanol; a wide spectrum can be accommodated. But more adaptive, efficient engine control becomes crucial. Learning-based control may provide a way to make that affordable. Fuel flexibility may also be important in solving the "chicken and egg" problems on the way to a more sustainable mix of energy sources.

Cost and efficiency of control are central issues as well in hybrid, electric and fuel cell vehicles. As an example, chemical engineers involved in fuel cells have often argued that today's fuel cell systems are far less efficient than they could be, if only they took better advantage of synergies like heat being generated in one place and needed in another, with the right timing. Optimization *across time* should be able to capture those kinds of synergies.

1.2.2.4 *Benefits in Other Application Domains*   For reasons of space, I cannot do real justice to the many other important applications of ADP in other areas. I will try to say a little, but apologize to those who have done important work I will not get to.

Certainly the broader funding for ADP, beyond ECS, would benefit from greater use of ADP in other well-funded areas such as counterterrorism, infrastructure protection above and beyond electric power, health care, transportation, other branches of engineering and software in general. Sometimes strategic partnerships with domain experts and a focus on benchmark challenges in those other sectors are the best way to get started. When ECS funding appears critical to opening the door to large funding from these other sources, many of us would bend over backwards to try to help, within the constraints of the funding process. In some areas — like homeland security and healthcare, in particular — the challenges to humanity are so large and complex that the funding opportunities available today may not always be a good predictor of what may become available tomorrow, as new strategies evolve to address the underlying problems more effectively.

Ferrari and Stengel in this book provide solid evidence for an astounding conclusion — namely, that significant improvements in fuel efficiency and maneuverability are possible even for conventional aircraft and conventional maneuvers, using ADP. Stengel's long track record in aerospace control and optimal control provides additional evidence that we should take these results very seriously.

ADP clearly has large potential in "reconfigurable flight control (RFC)," the effort to control aircraft so as to minimize the probability of losing the aircraft after damage so severe that an absolute guarantee of survival is not possible. The large current efforts in RFC across many agencies (most notably NASA Ames) can be traced back to successful simulation studies using ADP at McDonnell-Douglas, using McDonnell's in-house model of its F-15 [3]. Mark Motter of NASA organized a special session at the American Control Conference (ACC01 Proceedings) giving current status and possibilities. Lendaris' work in this book was partly funded by NSF, and partly funded under Motter's effort, in association with Accurate Automation Corporation of Tennessee. Jim Neidhoefer and Krishnakumar developed important ADP applications in the past, and have joined the NASA Ames effort, where Charles Jorgensen and Arthur Soloway made important earlier contributions, particularly in developing verification and validation procedures.

It is straightforward to extend ADP for use in strategic games. Some of the hybrid control designs described by Shastry and others are general-purpose ADP designs, with clear applicability to strategic games, autonomous vehicles and the like. Balakrishnan has demonstrated success in hit-to-kill simulations (on benchmark challenges provided by the Ballistic Missile Defense Organization) far beyond that of other competitors. As with electric power, however, upgraded *components* may be critical in making the transition from controlling one piece of the system, to optimal management of the larger theater.

There is also a strong parallel between the goal of "value-based management" of communication networks and the goal of DSOPF discussed in Section 1.2.2.1. In particular, preventing "traffic jams" is essentially a dynamic problem, which static optimization and static pricing schemes cannot address as well.

Helicopter control and semiconductor manufacturing with ADP have been addressed by Jennie Si in this book, and by David White in [3] and in other places. Unmet opportunities probably exist to substantially upgrade the manufacturing of high-quality carbon-carbon composite parts, based on the methods proven out by White and Sofge when they were at McDonnell-Douglas [3].

As this book was going to press, Nilesh Kulkarni of Princeton and Minh Phan of Dartmouth reported new simulations of the use of ADHDP in design of plasma hypersonic vehicles, a radically new approach to lower-cost reusable space transportation, where it is crucial to have a more adaptive nonlinear sort of control scheme like ADP. (This was jointly funded by CNCI and by the Air Force.) During the meeting at the Wright-Patterson Air Force Base, the key aerospace designers and funders were especially enthusiastic about the possible application of ADP to "Design for Optimal Dynamic Performance," which we had asked Princeton to investigate. In DODP, the simulator would contain truly vicious noise (reflecting parameter uncertainties), and ADP would be used to tune both the weights *and* the physical vehicle design parameters so as to optimize performance over time. Researchers such as Frank Lewis and Balakrishnan have demonstrated promising systems and opportunities with MicroElectroMechanical Systems (MEMS), a key stream of nanotechnology. Many forward-looking funding discussions stress how we are moving toward a new kind of global cyberinfrastructure, connecting in a more adaptive way from vast networks of sensors to vast networks of action and decision, with more and more fear that we cannot rely on Moore's Law alone to keep improving computing throughput per dollar beyond 2015. ADP may provide a kind of unifying framework for upgrading this new cyberinfrastructure, especially if we develop ADP algorithms suitable for implementation on parallel distributed hardware (such as cellular neural networks)with analog aspects (as in the work of Chris Diorio, funded by ECS).

Many researchers such as John Moody have begun to apply ADP or related methods in financial decision-making. Older uses of neural networks in finance have taken a "behaviorist" or "technical" approach, in which the financial markets are predicted as if they were stochastic weather systems without any kind of internal intelligence. George Soros, among others, has argued very persuasively and very concretely how limited and dangerous that approach can become (even though many players are said to be making a lot of money in proprietary systems). The financial system itself may be analyzed as a kind of value-calculation system or critic network. More concretely, the systems giving supply and demand (at a given price) may be compared to the Action networks or policy systems in an ADP design, and systems which adapt prices to reflect supply and demand and other variables may be compared to certain types of critic systems (mainly DHP). The usual pricing systems assumed in microeconomic optimality theorems tend to require perfect foresight, but ADP critics remain valid in the stochastic case; thus they might have some value in areas like auction system development, stability analysis, or other areas of economic analysis.

Again, of course, the work by Warren Powell on large-scale logistics problems is a major watershed in showing how ADP can already outperform other methods, and is doing so today on large important real-world management problems.

1.3   UNIFYING MATHEMATICAL PRINCIPLES AND ROADMAP OF THE FIELD

1.3.1   Definition of ADP, Notation and Schools of Thought

The key premise of this book is that many different schools of thought, in different disciplines, using different notation have made great progress in addressing the same underlying mathematical design challenge. Each discipline tends to be fiercely attached to its own notation, terminology and personalities, but here I will have to choose a notation "in the middle" in order to discuss how the various strands of research fit each other. I will try to give a relatively comprehensive review of the main ideas, but I hope that the important work which I do not know about will be discussed in the more specialized reviews in other chapters of this book.

Before I define ADP, I will first define a focused concept of reinforcement learning, as a starting point. The founders of artificial intelligence (AI)[13] once proposed that we build artificial brains based on *reinforcement learning* as defined by Figure 1.1:

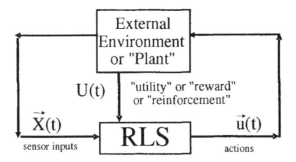

Fig. 1.1   Reinforcement Learning in AI. Reinforcement Learning Systems (RLS) choose actions u so as to maximize U, informed by raw sensor data X.

(See Figure 3.1 of [2] for a more intuitive cartoon version.). At each time t, a Reinforcement Learning System (RLS) outputs a list of numbers $u_1(1)$, $u_2(2) \ldots u_n(t)$, which form a vector $\vec{u}(t)$. These numbers form the "action vector" or "decisions" or "controls" which the RLS uses in order to influence its environment. The RLS is supposed to operate in a digital kind of fashion, from time tick to time tick; thus the time t is treated as an integer. I would define an RLS as a certain type of system which is designed to learn how to output actions, $\vec{u}(t)$, so as to maximize the expected value of the sum of future utility over all future time periods:

$$\text{MAXIMIZE} \left\langle \sum_{k=0}^{\infty} \left(\frac{1}{1+r} \right)^k U(t+k) \right\rangle. \tag{1.1}$$

Already some questions of notation arise here.

Here I am proposing that we should use the letter "U" for utility, because the goal of these systems is to maximize the expectation value of "cardinal utility," a concept rigorously developed by Von Neumann and Morgenstern [14] long before the less rigorous popular versions appeared elsewhere. The concept of cardinal utility is central to the field of decision analysis [15] (or "risk assessment") which still has a great deal to contribute to this area. Other authors have often used the letter "r" for "reinforcement" taken from animal learning theory. Animal learning theory also has a great deal to contribute here, but the concepts of "reinforcement learning" used in animal learning theory have been far broader and fuzzier than the specific definition here. The goal here is to focus on a mathematical problem, and mathematical specificity is crucial to that focus.

Likewise, I am using the letter "r" for the usual discount rate or interest rate as defined by economists for the past many centuries. It is often convenient, in pure mathematical work, to define a discount *factor* γ by:

$$\gamma = \frac{1}{1+r}. \tag{1.2}$$

However, the literature of economics and operations research has a lot to say about the meaning and use of r which is directly relevant to the use of ADP in engineering as well [16]. Some computer science papers discuss the choice of γ as if it were purely a matter of computational convenience — even in cases where the choice of γ makes a huge difference to deciding what problem is actually being solved. Values of γ much less than one can be a disaster for systems intended to survive for more than two or three time ticks.

From the viewpoint of economics, the control or optimization problems which we try to solve in engineering are usually just subproblems of a larger optimization problem — the problem of maximizing profits or maximizing value-added for the company which pays for the work. To do justice to the customer, one should use an interest rate r which is only a few percent *per year*. True, ethical foresight demands that we try to set $r = 0$, in some sense, for the utility function U which we use at the highest level of decision-making [16]. It is often more difficult to solve an optimization problem for the case where $r \to 0$; however, our ability to handle that case is just as important as our ability to scale up to larger plants and larger environments.

Finally, I am using angle brackets, taken from physics, to denote the expectation value.

There are many other notations for expectation value used in different disciplines; my choice here is just a matter of convenience and esthetics.

Over the years, the concept of an RLS in AI has become somewhat broader, fuzzier and not really agreed upon. I will define a more focused concept of RLS here, just for convenience.

Much of the research on RLS in AI considers the case where the time horizon is not *infinite* as in Eq. (1.1). That really is part of RLS research, and part of ADP as well, so long as people are using the kinds of designs which can address the case of an infinite time horizon. But when people address the classic problem of stochastic programming (i.e., when their future time horizon is just one period ahead), that is not ADP. There are important connections between stochastic programming and ADP, just as there are connections between nonlinear programming and ADP, but they are different mathematical tasks.

Likewise, in Figure 1.1, we are looking for an RLS design which marches forwards in time, and can be scaled up "linearly" as the complexity of the task grows. It is tricky to define what we mean by "linear scaling" in a precise way. (See [9] for a precise statement for the particular case of Multiple-Input Multiple-Output (MIMO) linear control.) But we certainly are not looking for designs whose foresight is totally based on an explicit prediction of every time period τ between the present t and some future ultimate horizon $t + H$. Designs of that sort are basically a competitor to ADP, as discussed in Section 1.2.2.3. We are not looking for designs which include explicit solution of the N^2-by-N^2 supraoperator [17] equations which allow explicit solution of matrix Riccati equations. Nevertheless, ADP research does include research which uses less scalable components, if it uses them *within* scalable designs and if it teaches us something important to our long-term goals here (e.g., [18, 19]).

ADP is almost the same as this narrow version of RLS, *except that* we assume that utility is specified as a function $U(\vec{\mathbf{X}}(t))$ instead of just a "reinforcement signal" $U(t)$. ADP is more general, because we could always append a "reinforcement signal" to the vector $\vec{\mathbf{X}}(t)$ of current inputs, and define $U(\vec{\mathbf{X}})$ as the function which picks out that component. ADP also includes the case where time t may be continuous.

In ADP, as in conventional control theory, we are interested in the general case of *partially observed* plants or environments. In my notation, I would say that the "environment" box in Figure 1.1 contains a "state vector," $\vec{\mathbf{R}}(t)$, which is not the same as $\vec{\mathbf{X}}(t)$. (In control theory [12, 20, 21], the state vector is usually denoted as $\vec{x}$, while the vector of observed or sensed data is called $\vec{y}$.) "R" stands for "reality." Within an ADP system, we do not know the true value of $\vec{\mathbf{R}}(t)$, the true state of objective reality; however we may use Recurrent systems to Reconstruct or update a Representation of Reality. (See Section 1.3.2.2.)

True ADP designs all include a component (or components) which estimates "value." There are fundamental mathematical reasons why it is impossible to achieve the goals described here, without using "value functions" somehow. Some work in ADP, in areas like hybrid control theory, does not use the phrase "value function," but these kinds of functions are present under different names. Value functions are related to utility functions U, but are not exactly the same thing. In order to explain why value functions are so important, we must move on to discuss the underlying mathematics.

1.3.2   The Bellman Equation, Dynamic Programming and Control Theory

Traditionally, there is only one exact and efficient way to solve problems in optimization over time, in the general case where noise and nonlinearity are present: dynamic programming. Dynamic programming was developed in the field of operations research, by Richard Bellman, who had close ties with John Von Neumann.

The basic idea of dynamic programming is illustrated in Figure 1.2. In dynamic programming, the user supplies both a utility function, the function to be maximized, and a stochastic model of the external plant or environment. One then *solves for* the unknown function $J(\vec{x}(t))$ which appears in the equation in Figure 1.2. This function may be viewed as a kind of secondary or strategic utility function. The key theorem in dynamic programming is roughly as follows: the strategy of action

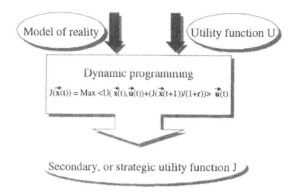

Fig. 1.2   The basic idea of dynamic programming: given U, solve the Bellman equation to get J; use J to calculate optimal actions.

$\vec{u}$ which maximizes J in the immediate future (time $t + 1$) *is* the strategy of action which maximizes the sum of U over the long-term future (as in Eq. (1.2)). Dynamic programming converts a problem in optimization over time into a "simple" problem in maximizing J just one step ahead in time.

1.3.2.1 Explanation of Basic Elements in Figure 1.2   Before going further, we need to explain Figure 1.2 very carefully. There are some hidden issues here which are important even to advanced research in ADP.

The equation in Figure 1.2 is a variant of the original equation developed by Bellman. The task of optimization in the deterministic case, where there is no noise, was studied centuries before Bellman. Bellman's equation was re-interpreted, years ago, as the stochastic generalization of the Hamilton-Jacobi equation of physics. Thus many control theorists now refer to it as the "Hamilton-Jacobi-Bellman" (HJB) Equation.

Many people have found that DP or ADP can be useful even in the deterministic case. The case of zero noise is just a special case of the general method. Many books start out by explaining DP in the case of zero noise, and then discussing the general case, which they call "stochastic dynamic programming." But for researchers in the field, "stochastic dynamic programming" is a redundant expression, like "mental telepathy."

In Figure 1.2, J is the value function. Many researchers use "J" [15, 25], but many others — especially in AI — use "V," in part because of the seminal paper by Barto, Sutton and Anderson [26]. The idea is that "V" stands for "value." But there are other types of value function which occur in ADP research.

The equation in Figure 1.2 deviates a little from the original formulation of Von Neumann, insofar as the utility function U is now shown as a function of the state $\vec{x}(t)$ *and of* the actions $\vec{u}(t)$. I did not include that extension, back in the first crude paper which proposed that we develop reinforcement learning systems by approximating Howard's version of dynamic programming [27, 22]. In that paper, I stressed the difference between primitive, unintelligent organisms — whose behavior is governed by in-born automatic stimulus-response action responses — versus intelligent organisms, which learn to choose actions based on the *results* which the actions lead to. There is no real need for the extension to $U(\vec{x}, \vec{u})$, since we can always account for variables like costs in the state vector. Nevertheless, the extension turns out to be convenient for everyday operations research. When we make this extension, the optimal $\vec{u}(t)$ is no longer the $\vec{u}(t)$ which maximizes $<J(\vec{x}(t+1))>$; rather, it is the $\vec{u}$ which maximizes $<U(t) + J(t+1)>$, as shown in Figure 1.2.

The reader who is new to this area may ask where the stochastic model fits into this equation. The stochastic model of reality is necessary in order to compute the expectation values in the equation.

1.3.2.2   *Partial Observability and Robust Control: Why "$\vec{x}$" in Figure 1.2 Is a Central Problem*   Note that I used "$\vec{x}$" here to describe the state of the plant or environment. I deliberately did not write "$\vec{X}$" or "$\vec{R}$," in Figure 1.2, because there is a very serious problem here, of central importance both in theory and in practice, in DP and in ADP. The problem is that classical DP assumes a *fully observed system*. In other words, it assumes that the system which chooses $\vec{u}(t)$ is able to make its decision based on *full knowledge* of $\vec{x}(t)$. More precisely, it assumes that we are trying to solve for an optimal strategy or policy of action, which may be written as a function $\vec{u}(\vec{x})$, and it assumes that the observed variables obey a Markhov process; in other words, $\vec{x}(t+1)$ is governed by a probability distribution which depends on $\vec{x}(t)$ and $\vec{u}(t)$, and is otherwise independent of anything which happened before time t. Failure to fully appreciate this problem and its implications has been a major obstacle to progress in some parts of ADP research.

In control theory and in operations research, there has long been active research into Partially Observed Markhov Decision Problems (POMDP). ADP, as defined in this book, is really an extension of the general case, POMDP, and not of DP in the narrowest sense.

Control theory has developed many insights into partial observability. I cannot summarize all of them here, but I can summarize a few highlights.

There are three very large branches of modern control theory: (1) robust control, (2) adaptive control and (3) optimal control. There are also some important emerging branches, like discrete event dynamical systems (DEDS) and hybrid control, also represented in this book. Even within control theory, there are big communication gaps between different schools of thought and different streams of practical applications.

Optimal control over time in the *linear case*, for a known plant subject to Gaussian noise, is a solved problem. There are several standard textbooks [20, 28]. For this kind of plant, an optimal controller can be designed by hooking together two components: (1) a Kalman filter, which estimates the state vector $\vec{R}(t)$, using a kind of recurrent update rule mainly based on the past $\vec{R}(t-1)$, the current observations $\vec{X}(t)$ and the past actions $\vec{u}(t - 1)$; (2) a "certainty equivalence" controller, which is essentially the same as what DP would ask for, *assuming that* the estimated state vector is the true state vector. This kind of "certainty equivalence" approach is not exactly correct in the nonlinear case, but it can be a useful approximation, far better than assuming that the state vector is just the same as the vector $\vec{X}$!

Robust and adaptive control are two different ways of responding to the question: "What can we do when the dynamics of the plant themselves are unknown?" Robust control and adaptive control were both developed around the goal of *stability*, first, rather than performance. In both cases, you usually assume that you have a mathematical model of the plant to be controlled, but there are parameters in that model whose values are unknown. In robust control, you try to find a control law or controller design $\vec{u}(\vec{X})$ which guarantees that the plant will not blow up, no matter what the unknown parameters are, over a very wide space of possibilities. There are computational tools now widely available for "mu synthesis" and such, which allow you to maximize the margin of safety (the acceptable range of parameters) for the usual linear case. In adaptive control, you try to *adapt* your control rule in real time, based on real-time observations of how the plant actually behaves. Note that adaptive dynamic programming (ADP) is not a special case of adaptive control in the usual sense!

Several advanced researchers in robust control, such as Athans and Barras, have addressed the task of robust control *in the general nonlinear case*. It turns out that the solution to that task reduces to the task of trying to solve the HJB equation! Thus the most accurate possible approximation to true nonlinear robust control, in the general case, can only be found by developing software which approximates the solution to the HJB equation. From the viewpoint of nonlinear robust control, this entire book can be seen as a book about the computational issues in implementing nonlinear robust control. But in actuality, when we use an HJB equation to derive a robust controller, we can add terms in the utility function to represent concepts like energy use and cost, to represent more accurately what we really want.

Barras [29] has done important fundamental work on partial observability in this case. Crudely, his work appears to say that we can solve for the true "optimal" nonlinear robust controller, if we take the certainty equivalence approach, but we use a different kind of filter instead of the Kalman filter. We must use a nonlinear function approximator, and we must tune or train it so as to minimize a special loss function which properly penalizes different kinds of prediction errors. The reader is urged to study the original paper for a more precise statement of the findings. (See [9, 20] and the chapter by Anderson et al. in this book for some other connections between robust control, optimal control, adaptive control and ADP.)

Both in nonlinear robust control *and* in other ADP applications, there will always be a certain amount of approximation error in solving the HJB equations, no matter how hard we work to minimize that error. This does have some implications for stability analysis, for real-world nonlinear robust control just as much as for other forms of ADP. As Balakrishnan has frequently argued, we know that the controller we get *after the fact* is stable, if we have found a way to converge to a reasonably accurate solution of the HJB equation in off-line learning, even if we had to try out different sets of initial weights to get there, and exploit tricks like "shaping" ([3], Foreword).

The problem of partial observability is crucial to practical applications, in many ways. For example, Section 1.2.2.3 mentions the work by Ford Research in developing a clean car chip. Both Ford and Sarangapani have stressed how variations between one car engine and another explain why classical types of control design (implemented at great detail and at great cost) failed to achieve the kind of success they were able to achieve. *No* fixed feedforward controller, conventional or neural, can be expected to perform well across such a wide range of plants. However, there are two approaches which have worked here: (1) as in Sarangapani, the approach of learning or adapting in real time to the individual plant; (2) as at Ford, the use of a kind of certainty equivalence approach.

In the simplest neural certainty equivalence approach (or "neural observer" approach, as described by Frank Lewis), we train a Time-Lagged Recurrent Network [3, 6] to predict the observed variables $\vec{X}(t)$ as a function of $\vec{X}(t-1)$, $\vec{R}(t-1)$ and $\vec{u}(t-1)$, where R now represents a set of recurrent neurons. In fact, the combination of $\vec{X}(t)$ and of this $\vec{R}(t)$ is basically just an estimated state vector! When we make this expanded vector of inputs available to our policy, $\vec{u}(\vec{X}, \vec{R})$, we arrive at a reasonable approximation to the exact results of Stengel, Barras and others. The recurrent tunable prediction structure does not have to be a neural network, but for the general nonlinear case, it should be some kind of universal nonlinear approximator.

Eventually, brain-like capability will require a *combination* of real-time learning (to learn the laws of nature, in effect) and time-lagged recurrence (to allow fast adaptation to changes in familiar but unobserved parameters like friction and how slippery a road is [4]). It will also require moving ahead from deterministic forecasting components, to more general stochastic components, unifying designs such as adaptive hidden Markhov models [30], Stochastic Encoder-Decoder-Predictors ([3, ch. 13]), and probability distribution estimators like Kohonen's self-organizing maps [31], etc. The development of better stochastic adaptive system identifiers is a major area for research in and of itself.

Finally, there may exist other ways to cope with partial observability which do not require use of a neural observer. Some kind of internal time-lagged recurrence or "memory" is necessary, because without memory our system cannot infer anything about the dynamics of the plant which it is trying to control. But the recurrence could be elsewhere. The recurrence could even be *within* the control rule or policy itself. In ([3, ch. 13]), I showed how the concept of "Error Critic" could be used to construct a model of the olive-cerebellum system in the brain, in which the "policy" is a time-lagged recurrent neural network. This approach results in a system which does not learn as fast, over time, as the neural observer approach, but it allows the resulting controller to function at a much higher sampling rate. One can easily imagine a kind of "master-slave" design in which the "master" ADP system learns quickly but has a lower sampling rate (say, 4 hertz or 10 hertz), while the "slave" system runs at a higher sampling rate (say, 200 hertz) and tries to maximize a utility function defined as something like $J^*(t+1) - J^*(t)$, where J^* is the value function computed by the "master" system.

Partial observability presents even larger challenges, in the longer term. In AI, there is a classic paper by Albus [32] which proposes a design for intelligent systems, in which there is *both* an estimated state vector *and* a "World Model" (a regularly updated map of the current state of the entire world of the organism). At first, this seems hard to understand, from an engineering point of view; after all, the estimated state vector is supposed to be a complete image of reality all by itself. But as the environment becomes more and more complex, something like this two-level system may turn out to be necessary after all. The Albus design may only be a kind of cartoon image of a truly optimizing learning system, but it may offer some real clues to the kinds of issues and patterns that we will need to address as we scale up to the level of complexity that the mammal brain can handle. Lokendra Shastri of Berkeley has addressed related issues as well, using ADP methods as part of a reasoning system.

In many engineering applications today, the easiest way to overcome the hurdles involving stability proof and verification is to combine off-line ADP learning with training sets based on very nasty "meta" simulation models like those used in the Ford work. It is essential that the ADP components contain the recurrence necessary to allow good performance for a fixed feedback controller over such nasty training sets. This kind of approach can be honestly described as an application of nonlinear robust control. It is necessary to achieve and verify convergence after the fact, but it is not necessary to provide an absolute guarantee that the offline learning would always converge without human assistance on all possible plants! In the case of true real-time adaptation, the adaptive control community has shown [21] that stability can be guaranteed only under very stringent conditions, even for multi-input multi-output (MIMO) linear systems. The usual methods for training critic or value-approximation networks do not possess guaranteed robust stability even in the linear case [9, sec. 7]. One may question whether even the brains of humans are universally stable under all possible streams of input experience. Nevertheless, I have proposed a suite of new ways to train value functions which do overcome these problems in the linear stochastic case [[9], section 9]. I conjecture that these methods provide the missing part of the long-sought design for true distributed MIMO adaptive control possessing total system stability for all controllable linear systems [9]. Proving such a theorem is one of the many important opportunities now at hand in this area.

1.3.3   ADP: Methods and Opportunities to Beat the Curse of Dimensionality

Exact dynamic programming, as illustrated in Figure 1.2, cannot be used in the general case. There is no general method to calculate an exact closed form solution to any arbitrary nonlinear differential equation! In fact, there is no such method even for simple algebraic equations, when they get beyond fourth-order or fifth-order polynomial equations. For the general case, *numerical methods* or *approximations* are the best that can be done. Even the human brain itself cannot exactly "solve"

NP-hard problems like playing a perfect game of chess or of Go; in reality, the goal is to do as well as possible, not more.

Exact solutions of the Bellman equation are possible in two important special cases: (1) when the dynamics of the environment are purely linear, with Gaussian noise and a quadratic utility function, the optimal control policy is well known [20, 28]; (2) when the plant or environment is a *finite-state* system, the mathematics become much easier.

A finite-state system is *not* a system which has a finite number of state variables; rather, it is a system which can only exist in a finite number of possible states. If a plant can only exist in N possible states, $i = 1$ to N, we only need to solve for N numbers $J(i)$. *After* we choose a specific policy or strategy of action, $\vec{u}(i)$, the stochastic model of the plant reduces to a simple Markhov chain, M_{ij}, a matrix containing the probabilities of transition from any state j at time t to state i at the next time tick. We can represent the state of such a system at any time as a simple integer i, but it turns out to be more convenient to use a more complicated representation, representing the state as a vector $\vec{x}$ in $\Re^N$. If the system is in state number i, we set $x_i = 1$ and $x_j = 0$ for all the other states j. In this representation, the function $J(\vec{x})$ is represented by a simple vector $\vec{J}$; in other words:

$$J(\vec{x}) = \vec{J}^T \vec{x}, \tag{1.3}$$

and likewise for $U(\vec{x})$, if we delete the unnecessary dependence on $\vec{u}$. The Bellman equation then reduces to:

$$\vec{J}^T \vec{x}(t) = (\vec{J}^T M^* + \vec{U}^T)\vec{x}(t) \text{ for all possible } \vec{x}(t), \tag{1.4}$$

where M^* is the transition matrix for the *optimal* policy. This is easily solved algebraically:

$$\vec{J} = (I - M^{*T})^{-1}\vec{U}. \tag{1.5}$$

Equation (1.5) is not a complete analytic solution of the Bellman equation in this case, because we still need to *find* the optimal $\vec{u}(i)$. Sometimes this can be done analytically. But often we can use the iterative dynamic programming methods of Howard [22], which could be viewed as the first ADP designs. Howard also developed methods to deal with "crossroads" problems which can occur sometimes, when the interest rate r is zero and the time horizon is infinite. (Crossroads as in "the crossroads of history.") These crossroads problems are a real substantive issue at times, and not just a mathematical anomaly; thus I deeply regret that I did not use the language of social science correctly in describing such problems, in the first published paper [33] which defined several of the basic ADP methods in relatively complete, mathematical fashion (stressing the use of backpropagation as a tool to enable scaling up to larger problems).

Very few decision or control problems in the real world fit exactly into one of these two special cases. Even for problems of inventory management and control, where the number of possible states N is often finite *in principle*, the number of possible states in the real world is often much too large to let us use Eq. (1.5) exactly.

Early attempts to approximate dynamic programming were mostly based on a very simple approximation to the Bellman equation. The true environment, however complex, was approximated as a finite state system. For example, if the state vector $\vec{x}$ is a bounded vector in $\Re^N$, we can represent $J(\vec{x})$ by using a kind of N-dimensional lookup table containing all possible values of $\vec{x}$. But the size of such a lookup table grows exponentially with N! This is the core of the "curse of dimensionality." Warren Powell, in his chapter, discusses the curse of dimensionality in more detail.

Strictly speaking, ADP does not throw out that old approach to approximation! It treats it as a special case of one of the **three more general principles which are the core of ADP**:

(1) **Value approximation:** Instead of solving for $J(\vec{x})$ *exactly*, we can use a universal approximation function $J(\vec{x}, W)$, containing a set of parameters W, and try to estimate parameters W which make $J(\vec{x}, W)$ a good approximation to the true function J;

(2) **Alternate starting points:** Instead of always starting from the Bellman equation directly, we can start from *related* recurrence equations (Section 1.3.3.2) which sometimes improve the approximation;

(3) **Hybrid design:** We can combine multiple ADP systems and other systems in more complex hybrid designs, without losing generality, in order to scale up to larger real-world problems, such as problems which involve multiple time scales and a mixture of continuous and discrete variables at multiple scales of space.

Dozens of different ADP designs or algorithms have been developed and used in different disciplines, based on one or more of these three general approaches. In order to build a *unified* ADP design, which can learn to handle the broadest possible variety of complex tasks (as well as the mammalian brain does), we will need to exploit all three approaches to the utmost, and learn how to combine them together more effectively.

The remainder of this section will discuss each of these three approaches, in turn. It will proceed systematically from simpler ADP designs — easier to implement but less able to cope with difficult tasks — all the way up to the larger research challenges critical to the long-term strategic goals discussed above. In so doing, it will try to provide a kind of global roadmap of this very complicated field.

1.3.3.1 Value Approximation and the Adaptive Critic   This subsection will mainly discuss how we choose and tune an approximate value function $J(\vec{x}, W)$ based on the Bellman equation proper. But the reader should be warned that the simplest reinforcement learning packages based on this approach often converge slowly when they are used on engineering tasks with a moderate number of continuous variables in them. Many critics of reinforcement learning have claimed that reinforcement learning could not be useful on large-scale problems — mainly because their students, using the easiest off-the-shelf tools in MatLab, were unable to tackle a large problem at first pass. The simplest designs are an important place to start, but one must build up beyond them (as in Sections 1.3.2.2, 1.3.3.2 and even 1.3.3.3) in order to handle larger problems. And of course we need further research and better software in order to capture the full power of the approach.

Even if we limit ourselves to ADP methods which tune $J(\vec{x}, W)$ based on the Bellman equation proper, there is still a huge variety of more-or-less general designs. There are four decisions you must make when you put together such a system:

(1) What function $J(\vec{x}, W)$ do you choose to approximate $J(\vec{x})$?

(2) **Value updates:** For any *given* strategy or policy or controller $\vec{u}(\vec{x})$, how do you adapt the weights W to "best fit the Bellman equation"?

(3) **Policy updates:** How do you represent and adapt $\vec{u}(\vec{x})$ itself as part of this process?

(4) How do you coordinate, manage and time the value updates and policy updates?

With regard to question 1, many researchers still use lookup tables or linear basis function approximators:

$$\hat{J}(\vec{x}, W) = \sum_{\alpha} W_{\alpha} f_{\alpha}(\vec{x}), \qquad (1.6)$$

where the functions f_{α} are preprogrammed basis functions or features chosen by the user. A truly general-purpose computer program to calculate value updates and policy updates should be able to accommodate these particular choices as a special case. These special cases can also yield mathematical insights [25] which might be useful as a stepping-stone to where we want to go. But our goal here is to get to the *general* case, where we need to be able to use a universal approximator able to approximate "any" nonlinear function. Powell, in particular, has stressed that his success in handling large problems is based on going beyond what he could get with linear basis functions as in [25].

But which universal approximator $J(\vec{x}, W)$ do we use? The choice of function approximator has been one of the key issues explaining why some researchers have done much better than others in real-world applications of ADP. We will also need to use approximators much more powerful than anyone has used in the past, in order to scale up to more brain-like performance.

Universal approximation theorems have been proven for many, many systems such as Taylor series, fuzzy logic systems, wavelet systems, simple "global" neural networks (like the Multilayer Perceptron, MLP [7, 34]), and "local" neural networks like the Radial Basis Function (RBF), the CMAC, the SOM, the ART, etc. These all have advantages in various application domains. Some people from statistics argue that there can never be any real science here, because "there is no free lunch." They argue that some approximators will be better in some domains, and others will be better in others, and there never can be a truly general-purpose system.

Certainly it makes sense for software designers to give their users a wide choice of options, but the "no free lunch" theory misses a few critical points. It misses a very large historical literature on the foundations of induction and learning, which is relevant here but is too complicated to discuss in detail (e.g., see ([4, ch. 10]) and sources cited in [34].) It misses the fact that approximators in family A can be dense in family B but not vice-versa. More concretely, it does not fully account for the important, practical theorems proven by researchers like Barron [35] and Sontag on function approximation.

Taylor series, lookup tables, gain schedulers, CMAC and RBF are all examples of the broad class of linear basis function approximators. Barron [35] proved that linear basis function systems and MLPs are both universal approximators, *but with different accuracy in approximation*. When we try to approximate smooth functions, and maintain a certain level of accuracy with more and more inputs (m), the *number of parameters* or nodes needed to maintain that accuracy rises at an exponential rate with m, for linear basis function approximators, while it only rises as a gentle polynomial function of m for MLPs. This makes intuitive sense, because we know that the required size of a lookup table grows exponentially with m, and we know that other linear basis function approximators are based on the same underlying principle.

Barron's result is only the tip of a very large iceberg, but it already conveys an important lesson. In practical work in the neural network field, it is well known that MLPs allow greater accuracy in approximation for *moderately large m* than the local approximators. (There are some exceptions, when the data is all clustered around a few points in state space, but that is not the general case.) A majority of the practical successes discussed in Section 1.2 were based on using MLPs as the value approximator(s). On the other hand, it usually takes more time and skill to train MLPs than to train linear basis function approximators. Thus one has to work harder if one wants to get better results; this is a recurrent theme across all parts of the ADP field. There is also still a need for more research on ways to blend global networks like MLPs and local networks together, to combine the advantages of both, *particularly* when the speed of learning is an urgent issue; this is one of the subsystem-level issues I referred to in Section 1.2.1.

I have speculated [36] that Elastic Fuzzy Logic systems (ELF) might be as powerful as MLPs in function approximation, but for now this is just a speculation, so far as I know.

MLPs can do well in approximating smooth functions with m on the order of $50 - 100$. That is a huge step up from the old lookup-table dynamic programming. *But it is still a far cry from a brain-like scale of performance.* It is good enough when the job is to control a single car engine, or airplane, or turbogenerator or router. It is *not good enough* to meet the needs of large-scale network control as discussed in Section 1.2.

Many people from AI and statistics have argued that we will never be able to scale up to brain-like complexity, without exploiting a whole lot of prior information. This is true, in principle. But researchers going back to the philosopher Emmanuel Kant and the statistician Brad Efron have stressed that we can use a particular *type* of prior information which does not block our ability to learn new things in an open-minded way. From a practical viewpoint, we can exploit tacit prior assumptions about symmetry in order to design whole new classes of neural network designs, which are capable of addressing large-scale network management, image processing and segmentation and the like. There is a ladder of designs which now exists here, rising up from the generalized MLP [7, 34], to the Simultaneous Recurrent Network (SRN), to the Cellular SRN [37, 38], to the ObjectNet [39]. These should not be confused with the simple recurrent networks which are widely discussed in psychology but are not so useful here.

A few quick comments about strategic games may be of interest to some readers. Fogel [18] has argued that the past success of reinforcement learning in playing games has been oversold. Reinforcement learning has generated human-class performance in backgammon and checkers, *but only* when the programmers spent enormous effort in crafting the features they use in a linear basis function value function. There was never any true learning of the game from scratch. With tic-tac-toe, he was able to train a perfect game player simply by using genetic algorithms to evolve an optimal game *player* $(\vec{u}(\vec{x}))$. But this did not really work for checkers, as he had hoped. He *was* able to achieve human-level performance for the first time in checkers without cheating, by evolving a value-approximating MLP network $J(\vec{x}, W)$, and using that network as a basis for choosing moves. This was an historic achievement. But no one has ever achieved real human-level performance in the game of Go, with or without learning; based on what we learned from earlier work on lattices similar to Go boards [37], I would conjecture that: (1) Fogel's approach might work with Go, if he replaced the MLP with a cellular SRN or ObjectNet; (2) even his earlier attempt to evolve a controller directly for checkers might have succeeded if he had used an SRN; (3) success in theater-like strategy games would also require Object Nets, but research on Go would be a good starting point to get there. I do not claim that this simple approach would do full justice to the game of Go. Rather, it offers a tangible hope to show that learning systems might outperform existing Go players. One could do better, to start, by using more advanced evolutionary computing methods, such as the use of particle swarm methods to adapt the continuous valued weights in the value-approximating network. In parallel, we need to develop better gradient-exploiting methods to train recurrent recurrent systems like ObjectNets, in order to train networks too large for evolutionary computing to handle, and in order to get closer to brain-like approaches. Go can provide a testbed for even more powerful general concepts, but it would be useful to get this stream of research extended and disseminated much further immediately, based on the best we can do today.

Again, this is all the tip of an iceberg, but the other three design questions above also merit serious discussion here.

First, the issue of how to do value updates. Beginners to ADP often suggest that we should try to train the weights or parameters W by minimizing the following measure of error as a function of W:

$$E_G = \sum_t \left(J(\vec{x}(t), W) - (U(\vec{x}(t), \vec{u}(t)) + \gamma J(\vec{x}(t+1), W) \right)^2, \qquad (1.7)$$

where the sum is taken over real or simulated pairs of data $(\vec{x}(t), \vec{x}(t+1))$. (The subscript "G" refers to Galerkin.) Years ago, I evaluated the performance of this value-updating method, for the case of linear multiple-input multiple-output (MIMO) plants with nonzero noise. I proved that it converges to the wrong estimate of W almost always [9, 40].

The connection to Galerkin is discussed at length in [9]. The procedure of minimizing E_G is essentially a special case of one of the more general methods proposed by Galerkin decades ago to approximate the solution to partial differential equations (PDE). But the Bellman equation is not strictly a PDE, because of its stochastic aspect. The stochastic aspect turns out to be the source of the problem in trying to apply Galerkin's approach here [9]. More recently, Balakrishnan has adapted a different method of Galerkin (specialized to time-forwards systems rather than PDE in general) to the challenge of optimal, ADP control of systems like fluid flows governed by PDE.

In 1977 [41], I proposed instead that we approximate Howard's iterative procedure directly, by *training* the network $J(\vec{x}, W)$ to try to match or predict the targets:

$$J^*(t) = U(t) + J(t+1)/(1+r). \qquad (1.8)$$

This training could be done by minimizing least square error and backpropagation, or we could use *any other* method used to train any kind of function approximator to match examples of input vectors and desired output vectors. I called this method "Heuristic Dynamic Programming" (HDP) [41]. This statement of the method is quite brief, but it is complete. The method really is quite simple, in the general case. (See [3] for elaborate discussions of how to embed this idea in simple, general form within a larger software system for ADP.)

Earlier, in 1973 [42], Widrow implemented an alternative training method for a neural network which he called a "critic", for use in playing blackjack. In effect, he trained $J(\vec{x}(t), W)$ directly to predict:

$$J^*(t) = \sum_{\tau=t}^{T-1} U(\tau) + U^*(T), \qquad (1.9)$$

where U^* is a kind of terminal payoff at a terminal time T. There was no connection made to utility functions or dynamic programming. In 1983, Barto, Sutton and Anderson [26] generalized Widrow's method to define the class of methods $TD(\lambda)$, such that $\lambda = 0$ and $\lambda = 1$ would yield a choice between Eq. (1.8) and Eq. (1.9). There was no indication of a connection to [41] or to any form of dynamic programming at that time; that connection was made later, in 1987, when Sutton read [43], and set up the discussions at GTE which later led to the 1988 NSF workshop reported in [2]. Since then, we have used the term "critic" or "critic network" to refer to any parameterized network used to approximate $J(\vec{x})$ *or to approximate* any of the other various value functions which emerge from relatives of the Bellman equation (Section 1.3.3.2). An "adaptive critic" system is any system which contains such a critic network, plus some mechanism to adapt or tune the critic based on off-line learning or on-line learning,

Many researchers such as Williams have been able to show how pathological training sets $\{\vec{x}\}$ can cause HDP itself to converge to the wrong answer or even to diverge. Van Roy [44] has proven that it will always converge to the right value, under reasonable conditions, *if* the training set is chosen from the normal stream of experience one would expect to see while controlling and observing the actual plant. There are several more complex variants of HDP, for which I have proven that the method will always converge to the right value function for linear MIMO plants (including stochastic versions), even when the training sets are chosen in a pathological manner [9]. There are also a wide variety of other methods available, some more truly learning-based and some based on linear programming, as described by Van Roy's chapter 10 in this book. It would be interesting to see whether Van Roy's approach could be made more powerful, by exploiting the nonlinear programming methods described in Momoh's chapter 22. It would also be interesting to see whether such methods could be restructured or adapted into true, scalable learning-based methods for adapting $J(\vec{x}, W)$, similar in spirit to some of the work by Trafalis in training simpler types of neural networks based on sophisticated ideas from operations research [45].

Policy updates are much more straightforward than value updates, in theory. Many researchers in AI address tasks where there is only a finite set of possible actions to take at any time t. In designing a system to play Go, for example, one might search over every possible move, and do the maximization on the fly by brute force. (There are many other similar choices familiar to AI researchers.) In engineering applications today, it is reasonable to train a controller $\vec{u}(\vec{x}, W_u)$, which may also be a universal approximator. For example, we may train the weights $W_{u\alpha}$ so as to maximize $<U(t) + \gamma J(t+1)>$, by adapting them in response to a stochastic gradient estimate:

$$\frac{\partial}{\partial W_{u\alpha}}\left(U(t) + \frac{J(t+1)}{1+r}\right)$$
$$\approx \sum_i \left(\left(\frac{\partial U}{\partial u_i(t)} + \gamma \sum_j \frac{\partial J(t+1)}{\partial x_j(t+1)} \cdot \frac{\partial x_j(t+1)}{\partial u_i(t)}\right) \cdot \frac{\partial u_i(t)}{\partial W_{u\alpha}}\right). \qquad (1.10)$$

The control theorist should immediately see that this is *exactly* the same as the procedure used to update or train a controller in Indirect Adaptive Control (IAC) [46], except that IAC eliminates "U" and it replaces $J(t+1)$ with a *fixed* measure of tracking error which is basically arbitrary. Both here and in IAC, we must train or obtain a model of the plant in order to estimate the required partial derivatives. In both cases, the derivative calculations and summation can be done more quickly and more efficiently in real time by using an algorithm which I call "backpropagation through a model." [3, 7]. This algorithm calculates the required derivatives efficiently through *any* differentiable model, neural, fuzzy, classical or other. In 1988 [2], I coined the term "Backpropagated Adaptive Critic" (BAC) [2] to name this way of training a controller or Action Net $u(\vec{x}, W_u)$. The stability guarantees for IAC depend on *highly* restrictive assumptions, even in the linear case [21]; however, there is good reason to believe that the *extension* of IAC provided by the use of new HDP variants can yield universal total system stability guarantees for *all* controllable linear plants [9].

More recently, some researchers have used the term "Policy Gradient Control Synthesis" (PGCS) for this approach to updating or training an Action Net. In certain designs, this may require the efficient real-time calculation of selected second derivatives. The form of second-order backpropagation required for such applications was first discussed in 1979 [33], but explained in more detail in [3].

High-level decision problems sometimes require more than just brute-force maximization or an Action network of this type. As we scale up to brain-like capabilities, we will need to design and train something like Option Networks, which generate reasonable *choices* for $\vec{u}(t)$ decisions, along with search and selection systems to go with them [47]. This involves some kind of stochastic search capability, related to evolutionary computing (EC), but not at all the same. In conventional EC, the user supplies a function $U(\vec{u})$, and a fixed algorithm searches through possible values of $\vec{u}$. But in brain-like stochastic search (BLiSS?), we try to develop learning systems which learn to do better and better stochastic search for *families* of optimization problems $U(\vec{u}, \vec{x})$. For example, one such function $U(\vec{u}, \vec{x})$ would be the *family* of all travelling salesman problems for 1,000 cities, where x is the vector of coordinates of those cities. There has been some preliminary research related to BLiSS by Wunsch and by Serpen, but much more will be needed.

Finally, the issue of how to manage, time and control value updates and policy updates is an extremely complex issue, very dependent on the specific choices one makes. Many chapters in this book will discuss various aspects of these issues. As a general rule, we do know that mammal brains somehow do concurrent real-time updates of *all* their components. But they also need periods of offline-learning and consolidation, in deep sleep and dreams. There are learning tasks in ADP which clearly relate to these biological functions [1, 2, 43], but more research will be needed to pin them down more completely.

1.3.3.2 *Alternative Starting Points: Beyond the Bellman Equation Proper*   The previous section discussed value approximation for ADP designs based *directly* on the Bellman equation — HDP, TD and variations of them. The Bellman equation itself may be viewed as a kind of recursion equation for the function $J(\vec{x})$.

There are two alternative starting points which are important to applications of ADP today. One of them is the recursion equation which underlies methods called Action-Dependent HDP (ADHDP), Q-learning, and related methods. The other is the recursion relation underlying Dual Heuristic Programming (DHP); that equation is essentially a stochastic generalization of the Pontryagin equation, which is as important and well known as the Hamilton-Jacobi equation in some fields of science.

Crudely speaking, the ADHDP/Q-recurrence leads to ADP designs which do not directly require a model of the plant or environment. (However, they do not escape the issues discussed in Section 1.3.3.2.) The DHP recurrence equation was developed in order to overcome scaling problems with all methods discussed so far in this section, for tasks involving continuous variables.

Q-learning and ADHDP are different methods, dating back to two independent sources in 1989 — the classic Ph.D. thesis of Watkins [23] and a far briefer paper given at the IEEE Conference on Decision and Control [24], respectively. But both can be derived from the same recurrence equation. Here I will derive that equation as in [24], but will now label the new value function as "Q" instead of "J."

If we write the Bellman equation as:

$$J(\vec{x}(t)) = \max_{\vec{u}(t)} < U(\vec{x}(t), \vec{u}(t)) + \gamma J(\vec{x}(t+1)) >, \qquad (1.11)$$

we may simply *define*:

$$Q(\vec{x}(t), \vec{u}(t)) = \; < U(\vec{x}(t), \vec{u}(t)) + \gamma J(\vec{x}(t+1)) > . \qquad (1.12)$$

Using this definition, Eq. (1.11) may be expressed equivalently as:

$$J(\vec{x}(t)) = \max_{\vec{u}(t)} Q(\vec{x}(t), \vec{u}(t)), \qquad (1.13)$$

substituting Eq. (1.13) into Eq. (1.12), we deduce:

$$Q(\vec{x}(t), \vec{u}(t)) = \; < U(\vec{x}(t), \vec{u}(t)) + \gamma \max_{\vec{u}(t+1)} Q(\vec{x}(t+1), \vec{u}(t+1)) > . \qquad (1.14)$$

Equation (1.14) is the recurrence equation underlying both ADHDP and Q-learning.

Q-learning generally uses lookup tables to represent or approximate the function Q. In the 1990 workshop which led to [3], Watkins did report an effort to approximate Q using a standard CMAC neural network, for a broom-balancing task; however, he reported that that variation did not work.

With Q-learning, as in TD learning, one can derive the policy $\vec{u}(\vec{x})$ simply by considering every possible choice of actions, in tasks where there is only a manageable, finite list of choices. However, in the TD case, dynamic programming clearly calls on us to try to *predict* $< J(t+1) >$ for each choice, which then requires some kind of model of the plant or environment. With Q-learning, you only need to know the Q function. You escape the need for a model. On the other hand, if $\vec{x}$ and $\vec{u}$ are actually continuous variables, and if you are using a lookup table approximation to Q, then the curse of dimensionality here is even worse than with conventional DP.

For Action-Dependent HDP (ADHDP) [3, 24], I proposed that we approximate $Q(\vec{x}(t), \vec{u}(t))$ and $\vec{u}(\vec{x})$ by using universal approximators, like neural networks. We can train a network $Q(\vec{x}, \vec{u}, W)$ to try to match targets Q^* based on the recurrence Eq. (1.14), using *exactly* the same procedure as in HDP. (See [3] for pseudo-code, flowcharts, examples and analysis.) We can train $\vec{u}(\vec{x}, W_u)$ to maximize Q, by tuning the weights of W_u in response to the derivatives of Q, backpropagated from Q to $\vec{u}$ to the weights.

Some neural network researchers would immediately ask: "Doesn't the use of backpropagation lead to slow learning here?" Not necessarily. When a traditional MLP is trained by backpropagation, the learning is indeed slower than what we see with radial basis functions (RBF), for example. But the problem lies with the MLP structure, more than with backpropagation. (It is also important to know modern techniques for adjusting learning rates and such.) More powerful function approximators (like humans!) are even trickier to train, because they are searching a space of possibilities which is even larger. But when backpropagation is used to train simpler, local networks, it can be reasonably fast.

As an example, David White — co-sponsor of the 1990 joint NSF/McDonnell-Douglas workshop which led to [3] — applied ADHDP in 1990 to control the McDonnell model of the F-15 aircraft in simulation. He and Urnes did a study in which the simulated aircraft was badly damaged in random ways, and the ADHDP system tried to relearn the control fast enough (2 seconds) to keep the craft from crashing. It succeeded about half the time, which was far better than conventional methods (2 percent survival or so). The neural networks used to approximate Q and $\vec{u}$ were a new, differentiable variant of the CMAC design. (See [3] for details.) White and Sofge also used ADHDP to produce high-quality carbon-carbon parts using a cost-effective process which had defied earlier, expensive approaches based on more traditional methods. (See [3] and Section 1.2 of this chapter.)

ADHDP in some form has been independently rediscovered by a number of researchers, who have reported good results. For example, Ford Research [48] reported a quick and easy solution of the bioreactor control challenge problem in [3], which, according to Ungar, defied the very best and most modern adaptive control methods. Many of Si's results reported in this book use a design in this category. Shibata's "fuzzy critic" for robot control [49], developed under Fukuda, also appears to be in this category. It is good news that we are beginning to integrate and consolidate more of this work.

ADHDP can handle larger-scale engineering problems than lookup-table Q or TD approaches. (From an engineering viewpoint, I have described the lookup-table and "associative trace" versions of Q and TD as "level 1," versus ADHDP as "level 2.") But it still encounters problems as one tries to scale up to even larger problems, and it also has problems related to "persistence of excitation" ([3, ch. 13]). Because of these issues, and because of the issues of partial observability, I have argued that the combination of HDP and BAC discussed in the previous section tends to be more powerful — "level 3" — on engineering problems which involve continuous variables. In order to scale up still further — "level 4" — many of us have gone on to use a different recurrence equation altogether, the DHP recurrence equation.

The correct recurrence equation for DHP may be derived in a straightforward way, by first defining:

$$\vec{\lambda}(t) = \nabla_{\vec{x}} J(\vec{x}(t)), \tag{1.15}$$

which is another way of writing:

$$\lambda_i(t) = \frac{\partial J(\vec{x}(t))}{\partial x_i(t)} \qquad (\forall i), \tag{1.16}$$

by performing the calculation for a particular policy $\vec{u}(\vec{x})$, and by differentiating Eq. (1.11) with respect to $x_i(t)$ to get:

$$
\begin{aligned}
\lambda_i(t) = \Bigg\langle & \frac{\partial U(\vec{x}(t),\vec{u}(t))}{\partial x_i} + \gamma \sum_j \frac{\partial J(\vec{x}(t+1))}{\partial x_j(t+1)} \cdot \frac{\partial x_j(t+1)}{\partial x_i(t)} \\
& + \sum_k \left(\frac{\partial U(\vec{x}(t),\vec{u}(t))}{\partial u_k(t)} + \gamma \sum_j \frac{\partial J(\vec{x}(t+1))}{\partial x_j(t+1)} \cdot \frac{\partial x_j(t+1)}{\partial u_k(t)} \right) \cdot \frac{\partial u_k(\vec{x}(t))}{\partial x_i(t)} \Bigg\rangle .
\end{aligned}
\tag{1.17}
$$

This general type of complex derivative calculation relies heavily on mathematical foundations discussed further in [6].

Some control theorists have asked at first: "Can you do this? Just differentiate the Bellman equation?" In fact, we *know* that Eq. (1.11) must hold for the optimal policy. So long as J and u and the plant are differentiable (within the Kolmogorov probability formalism [10]), we know that Eq. (1.17) *must hold* as well, for an optimal policy. It is a necessary condition for an optimum. Likewise, the policy update equivalent to Eq. (1.10) is trivial to deduce. (See [3].) But Eq. (1.10) is not a *sufficient* condition for a globally optimal policy; that is why I discussed stochastic search issues after the discussion of Eq. (1.10), and similar issues apply here. For complex nonlinear problems in general it is often possible to guarantee local optima, superiority to linear methods, and global optima in various convex cases in a rigorous way, but intelligent systems design can never guarantee "perfect" creativity in finding global optima. Stronger guarantees exist for GDHP (to be discussed below) than for DHP.

The recurrence equation for $\vec{\lambda}$ follows by using the definition of $\vec{\lambda}$ to simplify Eq. (1.17):

$$
\begin{aligned}
\lambda_i(t) = \Bigg\langle & \frac{\partial U}{\partial x_i} + \gamma \sum_j \lambda_j(t+1) \cdot \frac{\partial x_j(t+1)}{\partial x_i} \\
& + \sum_k \left(\frac{\partial U}{\partial u_k} + \gamma \sum_j \lambda_j(t+1) \cdot \frac{\partial x_j(t+1)}{\partial u_k} \right) \cdot \frac{\partial u_k}{\partial x_i} \Bigg\rangle ,
\end{aligned}
\tag{1.18}
$$

where variables without time arguments refer to values at time t.

The obvious way to approximate this recurrence relationship is to train an approximator, $\vec{\lambda}(\vec{x}, W)$, to meet the targets $\vec{\lambda}^*$ defined by the right-hand side of Eq. (1.18); more precisely, starting from any estimate of W, one can update the estimate of W to make $\vec{\lambda}(\vec{x}, W)$ better match $\vec{\lambda}^*(\vec{x}, W)$, where $\vec{\lambda}^*(\vec{x}, W)$ is defined by the right-hand side of Eq. (1.18), using the *current* estimates of W to estimate the values of $\vec{\lambda}(t+1)$ by $\vec{\lambda}(\vec{x}(t+1), W)$ as required to calculate to right-hand side of Eq. (1.18). Details, flow-charts, pseudo-code and some theoretical analysis (along with a couple of unfortunate typos) can be found in [3, ch. 13]. Using modular code design, and a basic understanding of backpropagation, it is not necessary to be an expert in control theory or on these equations in order to implement DHP.

Certain warnings are needed here. Although I discussed the general idea of DHP back in 1977 [41], I did not really specify the method then (unlike the case for HDP). In 1981 [50], I discussed a way to implement the idea, using backpropagation and an error function similar to Eq. (1.7); however, that kind of variation (which I now call DHPG) converges to the wrong answer almost always for linear dynamical systems with noise [9]. In order to converge to the correct answer, it is necessary in principle to include the $(\partial u_k / \partial x_i)$ terms, which were explained for the first time in [3]. (See [9] for extensions important to robust and adaptive control.) All of the actual implementations and applications of DHP occurred after the publication of [3]. Many of the most important ones appear in this book.

The vector $\vec{\lambda}$ of DHP has many connections to concepts in other fields of science and engineering. For example, to the economist, $\lambda_i(t)$ represents a kind of "marginal utility" or "shadow price" — the long-term market value of the commodity whose quantity or level is given by the variable x_i. DHP critics could actually be used as pricing systems. In the Pontryagin equation, λ_i is called a "costate variable" in modern optimal control [20, 28], the letter "λ" is still used for costate variables. The vector $\vec{\lambda}$ is also closely related to the gradients calculated in backpropagation through time, which creates many opportunities for seamless, integrated hybrid designs, discussed in part in the chapter 15 by Prokhorov.

DHP could have some interesting applications in network control, as discussed in Section 1.2.2. For a large network system, we would need something like an ObjectNet (Section 1.3.3.1) in order to input observations across an *entire electric power grid*, and output a vector of shadow prices or values for all of those observed values. *Physically*, an ObjectNet may be thought of as a specific kind of highly coordinated assembly of *component* networks, each of which refers to a particular object in the grid; thus we could choose to implement such a value-approximation network as a distributed system, with chips located near each object to implement the component network for that object. Such a system would output the estimate of values which apply to each object, from the node which is actually located at that object. At each object, we could perform a kind of local optimization over time, using a lower-level ADP system which responds to these global values, using a master-slave kind of arrangement similar to the one discussed in Section 1.3.2.2. Communication constraints could simply be represented as part of the topology specification of the ObjectNet to be trained.

DHP is not the only method in its class. There have also been a few simulations of Action-Dependent DHP (ADDHP), Globalized DHP (GDHP), ADGDHP and related Error Critic designs. (See [3, 9] and some of the work of Wunsch and Prokhorov.) In essence, GDHP is a hybrid of HDP and DHP. It trains a scalar critic which estimates the J function, but uses second derivatives in order to achieve the same effect as DHP. In effect, GDHP provides a way of performing DHP (as described above) while also guaranteeing strict adherence to the requirement that the vectors $\vec{\lambda}$ are the gradient of a function J. More generally [43], GDHP allows one to train the critic to minimize a weighted sum of the DHP second-order error measure and the usual HDP/TD first-order error measure; if *some* state variables are continuous while others are discrete, one can use GDHP on the entire J function by simply by not using the (undefined) second-order terms for the discrete state variables. (The discussion of [43] needs to be updated, to reflect the convergence results in [3] and [9].) Benchmark studies by Wunsch and Prokhorov show little difference in performance between the easier DHP method and GDHP in difficult engineering problems involving continuous variables. Thus it may be premature to say more about GDHP at the present time.

1.3.3.3   Hybrid and Large-Scale Designs: Closing the Gap to the Mammal Brain Level   This chapter started out by posing a question: when and how can we handle dynamic optimization problems as large and as complex as what the smallest mammal brain can learn to handle? (Of course, such systems could do far better than mice or humans on *some problems*, as a byproduct of this effort.)

Years ago, I believed that the kinds of designs discussed in Sections 1.3.3.1 and 1.3.3.2 might be enough to achieve that goal by themselves. After all, the famous neuropsychologist Hebb argued decades ago that all the complexities of higher-order intelligence might result as an *emergent property* of a much simpler kind of learning system [51]. Why could they not emerge from these kinds of more sophisticated and powerful learning design? The parallels to what was known about the brain in 1980 were also quite strong [43].

Recent research both in neuroscience and in technology has made it ever more clear that our earlier beliefs were mistaken.

For example, neuroscience has learned a great deal about the *basal ganglia* [52, 53]. Decision-making in the brain does not follow the highly rigid, pre-programmed kinds of hierarchies that were used in classical AI and in the old Red Army — but it clearly does have a mechanism for exploiting *multiple time scales*. Furthermore, it now seems more and more clear that critic methods can *converge faster* — by orders of magnitude at times — when they include a way to exploit such structure. The challenge lies in how to build systems which exploit such properties as effectively as possible, *without* using ad hoc patches and hierarchies that interfere with the flexibility of learning and the ability to converge in the end to a true multi-level optimum.

Three strands of research have demonstrated promising ideas to help us handle this issue. The field of *hybrid control*, represented here by Shankar Shastry, has studied decoupled methods for the HJB equation, particularly for problems involving a mix of continuous an discrete variables. AI researchers in reinforcement learning have also explored a number of ideas; the chapter 2 by Dietterich gives an excellent overview, and the chapter 14 by Barto describes further recent work. I myself have also developed some modified multilevel Bellman equations [47, 54] based on the concept of fuzzy or crisp partitions of the state space, which may have some role to play in this strand of research. (These were inspired in part by some earlier work by Sutton [55], and by some work on matrix decomposition theory which I did in 1978, but go substantially beyond the initial inspirations.)

Spatial structure is also very critical, as was already discussed in Section 1.3.3.1. Large-scale network control will be an important testbed in learning how to handle spatial structure, but concepts like the Object Net still leave open some key questions. For example, how do brains *learn* object types? How can systems like the brain implement such structures, using some kind of multiplexing or thalamic gating [56]? How does spatial complexity interface with temporal complexity?

As we move up to systems which truly make high-level decisions, in order to manage multiple levels of time, issues related to stochastic search and stochastic system identification become ever more important. This chapter has already discussed these issues, but we need to remember that they will become more important in the future.

None of this work would allow us to build an artificial human mind. The *human mind* involves a whole new set of issues [1] far beyond the scope of this chapter. But 99 percent of the human brain is more or less equivalent to structures which exist in the smallest mouse. A deeper, more functional understanding of the latter should be a big step forward, in *allowing* us someday to understand the former more deeply as well.

Bibliography

1. P. Werbos, What do neural nets and quantum theory tell us about mind and reality, in K. Yasue, M. Jibu and T Della Senta (eds.), *No Matter, Never Mind : Proc. of Toward a Science of Consciousness : Fundamental Approaches (Tokyo '99)*, John Benjamins Pub Co, 2002. See also P.Werbos, "Optimization: A Foundation for understanding consciousness," in D. Levine and W. Elsberry (eds.), *Optimality in Biological and Artificial Networks,* Erlbaum, 1997.

2. W. T. Miller, R. Sutton and P. Werbos (eds.), *Neural Networks for Control*, MIT Press, Cambridge, MA, 1990, now in paper.

3. D. White and D. Sofge (eds.), *Handbook of Intelligent Control*, Van Nostrand, New York, 1992.

4. P. Werbos, Neurocontrollers, in J. Webster (eds.), *Encyclopedia of Electrical and Electronics Engineering*, Wiley, New York, 1999.

5. M. Hoffert et al, Advanced Technology Paths to Global Climate Stability: Energy For a Greenhouse Planet, *Science*, 2002.

6. P. Werbos, *Beyond Regression: New Tools for Prediction and Analysis in the Behavioral Sciences*, Ph.D. Thesis, Committee on Applied Mathematics, Harvard U., 1974. Reprinted in its entirety in P. Werbos, *The Roots of Backpropagation: From Ordered Derivatives to Neural Networks and Political Forecasting*, Wiley, New York, 1994.

7. P. Werbos, Backpropagation through time: what it does and how to do it, *Proc. IEEE*, vol. 78, no. 10, 1990. Updated version reprinted as chapter 8 of [6].

8. J. A. Suykens, B. DeMoor and J. Vandewalle, Nlq theory: a neural control framework with global asymptotic stability criteria, *Neural Networks*, vol. 10, no. 4, pp. 615–637, 1997.

9. P. Werbos, *Stable Adaptive Control Using New Critic Designs*. ArXiv.org: adap-org/9810001 (1998)

10. D. Jacobson and D. Mayne, *Differential Dynamic Programming*, American Elsevier, 1970.

11. P. He and J. Sarangapani, Neuro Emission Controller for Minimizing Cyclic Dispersion in Spark Ignition Engines, TNN submitted, 2003. Condensed versions are in press in IJCNN 2003 Proc. (IEEE) and CCA 2003 Proc. (IEEE).

12. F. Lewis, J. Campos and R. Selmic, *Neuro-Fuzzy Control of Industrial Systems with Actuator Nonlinearities*, SIAM, Philadelphia, 2002.

13. E. A. Feigenbaum and J. Feldman, *Computers and Thought*, McGraw-Hill, 1963.

14. J. Von Neumann and O. Morgenstern, *The Theory of Games and Economic Behavior*, Princeton University Press, Princeton, NJ, 1953.

15. H. Raiffa, *Decision Analysis*, Addison-Wesley, Reading, MA 1968.

16. P. Werbos, Rational approaches to identifying policy objectives, *Energy: The International Journal*, vol. 15, no. 3/4, pp. 171–185, 1990.

17. D. F. Walls and G. F. Milburn, *Quantum Optics*, Springer, New York, 1994.

18. D. B. Fogel, *Blondie24: Playing at the Edge of AI*, Morgan-Kauffman, San Francisco, 2001.

19. T. Landelius, *Reinforcement Learning and Distributed Local Model Synthesis*, Ph.D. thesis and Report No. 469, Department of Electrical Engineering, Linkoping U., 58183, Linkoping, Sweden.

20. R. F. Stengel, *Optimal Control and Estimation*, Dover edition, 1994.

21. K. Narendra and A. Annaswamy, *Stable Adaptive Systems*, Prentice-Hall, Englewood Cliffs, NJ, 1989; Hemisphere, Washington, DC, 1982.

22. R. Howard, *Dynamic Programming and Markhov Processes*, MIT Press, Cambridge, MA 1960.

23. C. J. C. H. Watkins, *Learning From Delayed Rewards*, Ph.D. thesis, University of Cambridge, England, 1989. See also Watkins and Dayan, Technical note: Q-learning, *Machine Learning*, vol. 8, no. 3/4, pp. 279–292, 1992.

24. P. Werbos, Neural networks for control and system identification, *IEEE Proc. CDC89*, IEEE, 1989.

25. D. P. Bertsekas and J. N. Tsisiklis, *Neuro-Dynamic Programming*, Athena Scientific, Belmont, MA, 1996.

26. A. Barto, R. Sutton and C. Anderson, Neuronlike adaptive elements that can solve difficult learning control problems, *IEEE Trans. SMC*, vol. 13, no. 5, pp. 834–846, 1983.

27. P. Werbos, The elements of intelligence, *Cybernetica (Namur)*, no. 3, 1968.

28. A. Bryson and Y. C. Ho, *Applied Optimal Control*, Ginn, 1969.

29. J. S. Baras and N. S. Patel, Information state for robust control of set-valued discrete time systems, *Proc. 34th Conf. Decision and Control (CDC)*, IEEE, pp. 2302, 1995.

30. S. Mukhopadhyay and B. Jain, Multi-agent Markhov decision processes with limited agent communication, *Proc. of the Int'l Joint Conf. on Control Applications and Int'l Symposium on Intelligent Control (IEEE CCA/ISIC01)*, IEEE: 2001.

31. T. Kohonen, *Self-Organizing Maps*, New York: Spinger, 1997, Second Edition. Also see H. Ritter, T. Martinetz, and K. Schulten, *Neural Computation and Self-Organizing Maps*, Addison-Wesley, Reading, MA, 1992.

32. J. Albus, Outline of Intelligence, *IEEE Trans. Systems, Man and Cybernetics*, vol. 21, no. 2, 1991.

33. P. Werbos, Changes in global policy analysis procedures suggested by new methods of optimization, *Policy Analysis and Information Systems*, vol. 3, no. 1, 1979.

34. P. Werbos, Backpropagation: General Principles and Issues for Biology, in D. Fogel and C. Robinson (eds.), *Computational Intelligence: The Experts Speak*, IEEE, 2003.

35. A. R. Barron, Universal approximation bounds for superpositions of a sigmoidal function, *IEEE Trans. Info. Theory*, vol. 39, no. 3, pp. 930–945, 1993.

36. P. Werbos, Elastic fuzzy logic: a better fit to neurocontrol and true intelligence, *J. Intelligent & Fuzzy Systems*, vol. 1, no. 4, 1993.

37. X. Z. Pang and P. Werbos, Neural network design for J function approximation in dynamic programming, *Math. Modelling and Scientific Computing*, vol. 5, no. 2/3, 1996 (physically 1998). Available also as adap-org/9806001 at arXiv.org. See also P. Werbos and X. Z. Pang, "Generalized maze navigation: SRN critics solve what feedforward or Hebbian nets cannot," *Proc. Conf. Systems, Man and Cybernetics (SMC)*, Beijing, IEEE, 1996.

38. T. Yang and L. O. Chua, Implementing Back-Propagation-Through-Time Learning Algorithm Using Cellular Neural Networks, *Int'l J. Bifurcation and Chaos*, vol. 9, no. 9, pp. 1041–1074, 1999.

39. See P. Werbos posted at http://www.iamcm.org, and [47].

40. P. Werbos, Consistency of HDP applied to a simple reinforcement learning problem, *Neural Networks*, 1990.

41. P. Werbos, Advanced forecasting for global crisis warning and models of intelligence, *General Systems Yearbook*, 1977.

42. B. Widrow, N. Gupta and S. Maitra, Punish/reward: learning with a Critic in adaptive threshold systems, *IEEE Trans. SMC*, vol. 5, pp. 455–465, 1973.

43. P. Werbos, Building and understanding adaptive systems: A statistical/numerical approach to factory automation and brain research, *IEEE Trans. SMC*, 1987.

44. J. Tsitsiklis and B. Van Roy, An analysis of temporal-difference learning with function approximation, *IEEE Trans. Auto. Control*, vol. 42, no. 5, 1997.

45. T. B. Trafalis and S. Kasap, Artificial neural networks in optimization and applications, *Handbook of Applied Optimization*, in P. M. Pardalos and M. G. C. Resende (eds.), Cambridge University Press, 2000.

46. K. Narendra and S. Mukhopadhyay, Intelligent control using neural networks, in M. Gupta and N. Sinha (eds.), *Intelligent Control Systems*, IEEE Press, 1996.

47. P. Werbos, A Brain-Like Design To Learn Optimal Decision Strategies in Complex Environments, in M. Karny, K. Warwick and V. Kurkova (eds.), *Dealing with Complexity: A Neural Networks Approach*, Springer, London, 1998. Also in S. Amari and N. Kasabov, *Brain-Like Computing and Intelligent Information Systems*, Springer, 1998.

48. F. Yuan, L. Feldkamp, G. Puskorius and L. Davis, A simple solution to the bioreactor benchmark problem by application of *Q*-learning, *Proc. World Congress on Neural Networks*, Erlbaum, New York, 1995.

49. T. Shibata, *Hierarchical Intelligent Control of Robotic Motion*, Master's Thesis, chapter 5, Dept. of Electronic Mechanical Engineering, Nagoya University, Japan, 1992.

50. P. Werbos, Applications of advances in nonlinear sensitivity analysis, in R. Drenick and F. Kozin (eds.), *System Modeling and Optimization: Proc. IFIP Conf. (1981)*, Springer 1982; reprinted as chapter 7 in [6].

51. D. O. Hebb, *The Organization of Behavior*, Wiley, New York, 1949.

52. J. C. Houk, J. L. Davis and D. G. Beiser (eds.), *Models of Information Processing in the Basal Ganglia*, MIT Press, Cambridge, MA, 1995.

53. K. H. Pribram, (ed.), *Brain and Values,* Erlbaum: Hillsdale, NJ, 1998. (See also earlier books edited by Pribram in the same series from Erlbaum.)

54. P. Werbos, Multiple Models for Approximate Dynamic Programming and True Intelligent Control: Why and How, in K. Narendra (ed.), *Proc. 10th Yale Conf. on Learning and Adaptive Systems*, New Haven: K. Narendra, EE Dept., Yale University, 1998.

55. R. Sutton, TD Models: Modeling the World at a Mixture of Time Scales, *CMP-SCI Technical Report*, pp. 95–114, University of Massachussetts at Amherst,

December 1995, later published in *Proc. 12th Int. Conf. Macjine Learning*, pp. 531–539, Morgan Kaufmann, 1995.

56. C. H. Anderson, B. Olshausen and D. Van Essen, Routing networks in visual cortex, in M. Arbib (ed.), *The Handbook of Brain Theory and Neural Networks*, First Edition, pp. 823–826, MIT Press, Cambridge, MA, 1995.

Part I

Overview

2 Reinforcement Learning and Its Relationship to Supervised Learning

ANDREW G. BARTO   THOMAS G. DIETTERICH
University of Massachusetts   Oregon State University

Editor's Summary: This chapter focuses on presenting some key concepts of machine learning, approximate dynamic programming, and the relationships between them. Discussion and comparisons are made based on various aspects of the two fields such as training information, behavioral variety, problem conversion, applicable tasks, and so forth. This chapter mentions many real-world examples to illustrate some of the important distinctions being made. The primary focus of this chapter is a discussion of the concepts and strategies of machine learning, not necessarily algorithmic details. This chapter provides high-level perspective on machine learning and approximate dynamic programming.

2.1 INTRODUCTION

The modern study of approximate dynamic programming (ADP) combines ideas from several research traditions. Among these is the field of Artificial Intelligence, whose earliest period focused on creating artificial learning systems. Today, Machine Learning is an active branch of Artificial Intelligence (although it includes researchers from many other disciplines as well) devoted to continuing the development of artificial learning systems. Some of the problems studied in Machine Learning concern stochastic sequential decision processes, and some approaches to solving them are based on ADP. These problems and algorithms fall under the general heading of *reinforcement learning*. In this chapter, we discuss stochastic sequential decision processes from the perspective of Machine Learning, focusing on reinforcement learning and its relationship to the more commmonly studied supervised learning problems.

Machine Learning is the study of methods for constructing and improving software systems by analyzing examples of their behavior rather than by directly programming them. Machine Learning methods are appropriate in application settings where people are unable to provide precise specifications for desired program behavior, but where examples of desired behavior are available, or where it is possible to assign a measure of goodness to examples of behavior. Such situations include optical character recognition, handwriting recognition, speech recognition, automated steering of automobiles, and robot control and navigation. A key property of tasks in which examples of desired behavior are available is that people can perform them quite easily, but people cannot articulate exactly *how* they perform them. Hence, people can provide input–output examples, but they cannot provide precise specifications or algorithms. Other tasks have the property that people do *not* know how to perform them (or a few people can perform them only with great difficulty), but people are able to evaluate attempts to perform them, that is, to score behavior according to some performance criterion. Situations like this include playing master-level chess and controlling the nation-wide power grid in an efficient and fail-safe manner.

Machine Learning methods are also appropriate for situations where the task is changing with time or across different users, so that a programmer cannot anticipate exactly how the program should behave. For example, Machine Learning methods have been applied to assess credit-card risk, to filter news articles, to refine information retrieval queries, and to predict user browsing behavior in computer-based information systems such as the world-wide web.

Another area of application for Machine Learning algorithms is to the problem of finding interesting patterns in databases, sometimes called *data mining*. Many corporations gather information about the purchases of customers, the claims filed by medical providers, the insurance claims filed by drivers, the maintenance records of aircraft, and so forth. Machine Learning algorithms (and also, many traditional methods from statistics) can find important patterns in these data that can be applied to improve marketing, detect fraud, and predict future problems.

We begin by describing tasks in which examples of desired behavior are available.

2.2   SUPERVISED LEARNING

In *supervised learning*, the learner is given *training examples* of the form (x_i, y_i), where each input value x_i is usually an n-dimensional vector and each output value y_i is a scalar (either a discrete-valued quantity or a real-valued quantity). It is assumed that the input values are drawn from some fixed probability distribution $D(x)$ and then the output values y_i are assigned to them. The output values may be assigned by a fixed function f, so that $y_i = f(x_i)$, or they may be assigned stochastically by first computing $f(x_i)$ and then probabilistically perturbing this value with some random noise. This latter, stochastic view is appropriate when the output values are assigned by a noisy process (e.g., a human judge who makes occasional errors). In either case, the goal is to correctly predict the output values of new data points x *drawn from the same distribution* $D(x)$.

For example, in optical character recognition, each input value x_i might be a 256-bit vector giving the pixel values of an 8×8 input image, and each output value y_i might be one of the 95 printable ascii characters. When, as in this case, the output values are discrete, f is called a *classifier* and the discrete output values are called *classes*.

Alternatively, in credit card risk assessment, each input might be a vector of properties describing the age, income, and credit history of an applicant, and the output might be a real value predicting the expected profit (or loss) of giving a credit card to the applicant. In cases where the output is continuous, f is called a *predictor*.

A supervised learning algorithm takes a set of training examples as input and produces a classifier or predictor as output. The set of training examples provides two kinds of information. First, it tells the learning algorithm the observed output values y_i for various input values x_i. Second, it gives some information about the probability distribution $D(x)$. For example, in optical character recognition for ascii characters, the training data provides information about the distribution of images of ascii characters. Non-ascii characters, such as greek or hebrew letters, would not appear in the training data.

The best possible classifier/predictor for data point x would be the true function $f(x)$ that was used to assign the output value y to x. However, the learning algorithm only produces a "hypothesis" $h(x)$. The difference between y and $h(x)$ is measured by a *loss function*, $L(y, h(x))$. For discrete classification, the loss function is usually the 0/1 loss: $L(y, h(x))$ is 0 if $y = h(x)$ and 1 otherwise. For continuous prediction, the loss function is usually the squared error: $L(y, h(x)) = (y - h(x))^2$. The goal of supervised learning is to choose the hypothesis h that minimizes the expected loss: $\sum_x D(x) L(y, h(x))$. Hence, data points x that have high probability are more important for supervised learning than data points that have low or zero probability.

A good way to estimate the expected loss of a hypothesis h is to compute the average loss on the training data set: $1/N \sum_{i=1}^{N} L(y_i, h(x_i))$, where N is the number of training examples. Supervised learning algorithms typically work by considering a space of hypotheses, $\mathcal{H}$, that is chosen by the designer in the hopes that it contains a good approximation to the unknown function f. The algorithms search $\mathcal{H}$ (implicitly or explicitly) for the hypothesis h that minimizes the average loss on the training set.

However, if the training examples contain noise or if the training set is unrepresentative (particularly if it is small), then an h with zero expected loss on the training examples may still perform poorly on new examples. This is called the problem of *overfitting*, and it arises when h becomes overly complex and ad hoc as the learning algorithm tries to achieve perfect performance on the training set. To avoid overfitting, learning algorithms must seek a tradeoff between the simplicity of h (simpler hypotheses are less likely to be ad hoc) and accuracy on the training examples. A standard approach is to define a *complexity measure* for each hypothesis h and to search for the h that minimizes the sum of this complexity measure and the expected loss measured on the training data.

Learning algorithms have been developed for many function classes $\mathcal{H}$ including linear threshold functions (linear discriminant analysis, the naïve Bayes algorithm, the LMS algorithm, and the Winnow algorithm [13]), decision trees (the CART and C4.5 algorithms [9, 23, 26]), feed-forward neural networks (the backpropagation algorithm [6]), and various families of stochastic models (the EM algorithm [18]).

Theoretical analysis of supervised learning problems and learning algorithms is conducted by researchers in the area of computational learning theory [16]. One of the primary goals of research in this area is to characterize which function classes $\mathcal{H}$ have polynomial-time learning algorithms. Among the key results is a theorem showing that the number of training examples required to accurately learn a function f in a function class $\mathcal{H}$ grows linearly in a parameter known as the Vapnik-Chervonenkis dimension (VC-dimension) of $\mathcal{H}$. The VC-dimension of most commonly-used function classes has been computed. Another key result is a set of proofs showing that certain function classes (including small Boolean formulas and deterministic finite-state automata) cannot be learned in polynomial time by any algorithm. These results are based on showing that algorithms that could learn such function classes could also break several well known crytographic schemes which are believed to be unbreakable in polynomial time.

2.3   REINFORCEMENT LEARNING

Reinforcement learning comes into play when examples of desired behavior are not available but where it is possible to score examples of behavior according to some performance criterion. Consider a simple scenario. Mobile phone users sometimes resort to the following procedure to obtain good reception in a new locale where coverage is poor. We move around with the phone while monitoring its signal strength indicator or by repeating "Do you hear me now?" and carefully listening to the reply. We keep doing this until we either find a place with an adequate signal or until we find the best place we can under the circumstances, at which point we either try to complete the call or give up. Here, the information we receive is not directly telling us where we should go to obtain good reception. Nor is each reading telling us in which direction we should move next. Each reading simply allows us to evaluate the goodness of our current situation. We have to move around—explore—in order to decide where we should go. We are not given examples of correct behavior.

We can formalize this simple reinforcement learning problem as one of optimizing an unknown reward function R. Given a location x in the world, $R(x)$ is the reward (e.g., phone signal strength) that can be obtained at that location. The goal of reinforcement learning is to determine the location x^* that gives the maximum reward $R(x^*)$. A reinforcement learning system is not given R, nor is it given any training examples. Instead, it has the ability to take actions (i.e., choose values of x) and observe the resulting reward $R(x)$. The reward may be deterministic or stochastic.

We can see that there are two key differences from supervised learning. First, there is no fixed distribution $D(x)$ from which the data points x are drawn. Instead, the learner is in charge of choosing values of x. Second, the goal is not to predict the output values y for data points x, but instead to find a single value x^* that gives maximum reward. Hence, instead of minimizing expected loss over the entire space of x values (weighted according to $D(x)$), the goal is to maximize the reward at a single location x^*. If the reward $R(x)$ is stochastic, the goal is to maximize expected reward, but the expectation is taken with respect to the randomness in R at the single point x^*, and not with respect to some probability distribution $D(x)$.

We can formalize this simple form of reinforcement learning in terms of minimizing a loss function. The loss at a point x is the *regret* we have for choosing x instead of x^*: $L(x) = R(x^*) - R(x)$. This is the difference between the reward we could have received at x^* and the reward we actually received. However, this formulation is rarely used, because there is no way for the learner to measure the loss without knowing x^*—and once x^* is known, there is no need to measure the loss!

Given that this simple form of reinforcement learning can be viewed as optimizing a function, where does *learning* come in? The answer is long-term memory. Continuing the mobile phone example, after finding a place of good reception suppose we want to make another call from the same general area. *We go directly back to that same place, completely bypassing the exploratory search* (or we may not bother at all in the case when we were unsuccessful earlier). In fact, over time, we can build up a library of suitable spots in frequently visited locales where we go *first* when we want to make calls, and from which we possibly continue exploring. In a supervised version of this task, on the other hand, we would be directly told where the reception is best for a set of example locales. Reinforcement learning combines *search* and *long-term memory*. Search results are stored in such a way that search effort decreases—and possibly disappears—with continued experience.

Reinforcement learning has been elaborated in so many different ways that this core freature of combining search with long-term memory is sometimes obscured. This is especially true with respect to extensions of reinforcement learning that apply to sequential decision problems. In what follows, we first discuss in more detail several aspects of reinforcement learning and then specialize our comments to its application to sequential decision problems.

2.3.1 Why Call It Reinforcement Learning?

The term reinforcement comes from studies of animal learning in experimental psychology, where it refers to the occurrence of an event, in the proper relation to a response, that tends to increase the probability that the response will occur again in the same situation. The simplest reinforcement learning algorithms make use of the commonsense idea that if an action is followed by a satisfactory state of affairs, or an improvement in the state of affairs, then the tendency to produce that action is strengthened, that is, reinforced. This is the principle articulated by Thorndike in his famous "Law of Effect" [35]. Instead of the term reinforcement learning, however, psychologists use the terms *instrumental conditioning*, or *operant conditioning*, to refer to experimental situations in which what an animal actually does is a critical factor in determining the occurrence of subsequent events. These situations are said to include *response contingencies*, in contrast to Pavlovian, or classical, conditioning situations in which the animal's responses do not influence subsequent events, at least not those controlled by the experimenter. There are very many accounts of instrumental and classical conditioning in the literature, and the details of animal behavior in these experiments are surprisingly complex. See, for example, [15]. The basic principles of learning via reinforcement have had an influence on engineering for many decades (e.g., [19]) and on Artificial Intelligence since its very earliest days [21, 27, 36]. It was in these early studies of artificial learning systems that the term reinforcement learning seems to have originated. Sutton and Barto [29] provide an account of the history of reinforcement learning in Artificial Intelligence.

But the connection between reinforcement learning as developed in engineering and Artificial Intelligence and the actual details of animal learning behavior is far from straightforward. In prefacing an account of research attempting to capture more of the details of animal behavior in a computational model, Dayan [10] stated that "Reinforcement learning bears a tortuous relationship with historical and contemporary ideas in classical and instrumental conditioning." This is certainly true, as those interested in constructing artificial learning systems are motivated more by computational possibilities than by a desire to emulate the details of animal learning. This is evident in the view of reinforcement learning as a combination of search and long-term memory discussed above, which is an abstract computational view that does not attempt to do justice to all the subleties of real animal learning.

For our mobile phone example, the principle of learning by reinforcement is involved in several different ways depending on what grain size of behavior we consider. We could think of a move in a particular direction as a unit of behavior, being reinforced when reception improved, in which case we would tend to continue to move in the same direction. Another view, one that includes long-term memory, is that the tendency to make a call from a particular place is reinforced when a call from that place is successful, thus leading us to increase the probability that we will make a call from that place in the future. Here we see the reinforcement process manifested as the storing in long-term memory of the results of a successful search. Note that the principle of learning via reinforcement does not imply that only *gradual* or *incremental* changes in behavior are produced. It is possible for complete learning to occur on a single trial, although gradual changes in behavior make more sense when the contingencies are stochastic.

2.3.2   Behavioral Variety

Because it does not directly receive training examples or directional information, a reinforcement learning system has to actively try alternatives, process the resulting evaluations, and use some kind of selection mechanism to guide behavior toward the better alternatives. Many different terms have been used to describe this basic kind of process, which of course is also at the base of evolutionary processes: selectional (as opposed to instructional, which refers to processes like supervised learning), generate-and-test, variation and selection, blind variation and selection, and trial-and-error. These last two terms deserve discussion since there is some confusion about them. To many, blind variation and trial-and-error connote totally random, that is, uniformly distributed, behavior patterns. But this is not what those who have used these terms have meant (e.g., [12]). *Blind* variation refers to the need to sometimes take actions whose consequences cannot be foreseen, that is, which represent true leaps beyond the current knowledge base. This does not mean that these actions must be randomly chosen. They can be based on a large amount of accumulated knowledge, but they cannot have consequences that can be accurately deduced from the current knowledge base. Note that even a deterministic action selection process can satisfy this requirement. Similarly, trial-and-error learning has the same meaning. Trials do not have to be random.

As a result of the need for behavioral variety, reinforcement learning involves a conflict between *exploitation* and *exploration*. In deciding which action to take, the agent has to balance two conflicting objectives: it has to exploit what it has already learned in order to perform at a high level, and it has to behave in new ways—explore—to learn more. Because these needs ordinarily conflict, reinforcement learning systems have to somehow balance them. In control engineering, this is known as the conflict between control and identification, or the problem of dual control [14]. This conflict is not present in supervised learning tasks unless the learner can influence which training examples it processes, a setting that is known as *active learning*.

At the root of this conflict is that in a reinforcement learning task a search must be conducted for something that cannot be recognized based on its intrinsic properties. For example, the property of being the better of two alternatives depends on both alternatives—it is a relative property—and a search algorithm has to examine both alternatives to decide which is the better. It is logically necessary to examine the inferior alternative. In contrast, the objectives of other types of searches depend on intrinsic properties of the members of the search space. For example, in searching for the name "Adam Smith" in a telephone directory, one can recognize the target name when one sees it because being, or not being, the target name is an intrinsic property of individual names. When a name is found that satisfies the solution property, the search stops. In a search task involving intrinsic properties, such as a supervised learning task, it is conceivable that the search can be declared accomplished after examining a single element in the search space. If the search is based on relative properties, however, this is never possible. Note that since we are speaking about the logical properties of search processes, the distinction between searching defined by relative and intrinsic properties is somewhat different than the distinction between *satisficing* and *optimizing* [28], which refers to more practical issues in creating a stopping criterion.

2.3.3   Converting Reinforcement Learning to Supervised Learning

Having discussed key differences between reinforcement learning and supervised learning, the question arises as to whether these differences are fundamental or merely superficial differences that can be eliminated with suitable problem reformulation. In other words, are there ways of reducing one type of problem to the other? The first thing to note is that it is possible to convert any supervised learning task into a reinforcement learning task: the loss function of the supervised task can be used to define a reward function, with smaller losses mapping to larger rewards. (Although it is not clear why one would want to do this because it converts the supervised problem into a more difficult reinforcement learning problem.) But is it possible to do this the other way around: to convert a reinforcement learning task into a supervised learning task?

In general, there is no way to do this. The key difficulty is that whereas in supervised learning, the goal is to reconstruct the unknown function f that assigns output values y to data points x, in reinforcement learning, the goal is to find the input x^* that gives the maximum reward $R(x^*)$.

Nonetheless, is there a way that we could apply ideas from supervised learning to perform reinforcement learning? Suppose, for example, that we are given a set of training examples of the form $(x_i, R(x_i))$, where the x_i are points and the $R(x_i)$ are the corresponding observed rewards. In supervised learning, we would attempt to find a function h that approximates R well. If h were a perfect approximation of R, then we could find x^* by applying standard optimization algorithms to h. But notice that h could be a very poor approximation to R and still be very helpful for finding x^*. Indeed, h could be any function such that the maximum value of h is obtained at x^*. This means, for example, that we do not want to use the expected loss at each training point x_i as the goal of learning. Instead, we seek a function h whose maxima are good approximations of R's maxima. There are a variety of ways of formulating this problem as an optimization problem that can be solved. For example, we can require h be a good approximation of R but that it also satisfy the following constraint: for any two training examples $(x_1, R(x_1))$ and $(x_2, R(x_2))$, if $R(x_1) > R(x_2)$ then $h(x_1)$ must be greater than $h(x_2)$. Techniques of this kind are an area of active research [11]. However, note that these optimization problems are not equivalent to supervised learning problems.

2.4   SEQUENTIAL DECISION TASKS

The reinforcement learning tasks most relevant to approximate DP are sequential decision tasks. It is not an exaggeration to say that the application of reinforcement learning algorithms to these tasks accounts for nearly all of the current interest in reinforcement learning by Machine Learning researchers. In these problems, a computer program must make a sequence of decisions, where each decision changes the state of the program's environment and is followed by a numerical reward. The performance function is a measure of the total amount of reward received over a (possilby infinite) sequence of decisions. The case most commonly studied has the property that each immediate reward is zero until the end of the sequence, when it evaluates a final outcome. Imagine a computer playing the game of chess. The computer makes a long sequence of moves before it finds out whether it wins or loses the game. Similarly, in robot navigation, the robot must choose a sequence of actions in order to get from a starting location to some desired goal. We could train computers to play chess or control robots by telling them which move to make at each step. But this is difficult, tedious, and time-consuming. Furthermore, we may not know enough about the task to be able to give correct training information. It would be much nicer if computers could learn these tasks from only the final outcome—the win or loss in chess, the success or failure in robot navigation. Reinforcement learning algorithms are designed for this kind of *learning from delayed reward*.

Reinforcement learning researchers have widely adopted the framework of *Markov decision processes* (MDPs) to study sequential reinforcement learning. MDPs are discrete-time stochastic optimal control problems with a well-developed theory (see, e.g., [4]). A full specification of an MDP includes the probabilistic details of how state transitions and rewards are influenced by a set of actions. The objective is to compute an *optimal policy*, that is, a function from states to actions that maximizes the expected performance from each state, where a number of different performance measures are typically used. Given a full specification of an MDP with a finite number of states and actions, an optimal policy can be found using any of several stochastic DP algorithms, although their computational complexity makes them impractical for large-scale problems.

Reinforcement learning for sequential decision tasks consists of a collection of methods for approximating optimal policies of MDPs, usually under conditions in which a full specification of the MDP is unavailable. A typical reinforcement learning system learns a task by repeatedly performing it—that is, it makes moves in chess or issues control commands to a robot. Before each move, the algorithm can examine the current state, s, of the environment (i.e., the current board position in a chess game or the current sensor inputs from the robot together with any other relevant robot-internal information) and then choose and execute an action a (i.e., a chess move or a robot command). The action causes the environment to change to a new state, s'. After each state transition, the learning system receives a reward, $R(s, a, s')$. In chess, the reward is zero until the end of the game, where it is 1 (win), 0 (draw), or -1 (loss). In robot navigation, there is typically a large positive reward for reaching the goal position and a small negative reward for each step. There may also be negative rewards for using energy, taking time, or bumping into walls or other obstacles.

There are many alternative approaches to approximating optimal policies. The most direct approach is to directly learn a policy. In this approach, a space of possible policies is defined, usually by defining a parameterized family of policy functions that are continuously differentiable with respect to the parameters. Given a particular policy (corresponding to a particular parameter setting), there are algorithms that can estimate the gradient of the expected performance (i.e., the expected total reward, the expected discounted total reward, or the expected reward per step) of the policy with respect to its parameters by performing online trial executions of the policy (e.g., [2, 3]).

A somewhat less direct approach is to learn a *value function*, V, which assigns a real number $V(s)$ to each state s indicating how valuable it is for the system to be in that state. A closely-related method learns an *action-value function*, Q, where $Q(s, a)$ tells how valuable it is to do action a in state s. In either case, the value is an estimate of the total amount of reward that will accumulate over the future starting in the specfied state. (The counterpart of a value function in a cost-minimizing formulation is sometimes called the "cost-to-go" function.) Once the system has learned a good approximation of the value function, it can execute an improved policy by choosing actions that have higher values or lead to states that have higher values. For example, given the action-value function Q, the policy for state s can be computed as the action a that maximizes $Q(s, a)$. As this process continues, one expects the policy to improve toward optimality in an approximation of the policy improvement DP procedure. Reinforcement learning algorithms that learn value functions often update their policies before the value functions of their current policies have been fully learned. This allows them to decide on actions quickly enough to meet time constraints.

The most indirect approach to approximating an optimal policy is to learn a model of the MDP itself through a system identification procedure. For DPs this involves learning the reward function $R(s, a, s')$ and the transition probability function $P(s'|s, a)$ (i.e., the probability that the environment will move to state s' when action a is executed in state s). These two functions can be learned by interacting with the environment and using a system identification procedure. Each time the learner observes state s, performs action a, receives reward r and moves to the resulting state s', it obtains training examples for $P(s'|s, a)$ and $R(s, a, s')$. These examples can be given to supervised learning algorithms to learn P and R. Once these two functions have been learned with sufficient accuracy, DP algorithms can be applied to compute the optimal value function and optimal policy. Methods that learn a model are known as "model-based" methods. They typically require the fewest number of exploratory interactions with the environment, but they also typically do not scale well to very large problems.

These three approaches are not mutually exclusive. There are policy-search methods that learn partial models to help compute gradients [37] and value function methods that learn partial models and perform incremental DP [1, 24].

The algorithms that may scale best are model-free algorithms for estimating value functions, such as the Temporal Difference algorithm [30, 31]. Interestingly, the temporal difference family of algorithms can be viewed as supervised learning algorithms in which the training examples consist of $(s, \hat{V}(s))$ pairs, where s is a state and $\hat{V}(s)$ is an approximation of the value of state s. They are not true supervised learning algorithms, because the $\hat{V}(s)$ values are not provided by a teacher but instead are computed from the estimated values of future states. For this reason, they are sometimes called "bootstrapping" methods. The other reason that these algorithms are not true supervised learning algorithms is that the probability distribution over the states s is not a fixed distribution $D(s)$. Instead, the distribution depends on the current value function $\hat{V}$ and the current policy for choosing exploratory actions. Despite these differences, many algorithms from supervised learning can be applied to these temporal difference algorithms [8, 11, 17].

Reinforcement learning methods that use value functions are closely related to DP algorithms, which successively approximate optimal value functions and optimal policies for both deterministic and stochastic problems. Details are readily available elsewhere (see, e.g., [29]). Most reinforcement learning algorithms that estimate value functions share a few key features:

1. Because conventional DP algorithms require multiple exhaustive "sweeps" of the environment state set (or a discretized approximation of it), they are not practical for problems with very large finite state sets or high-dimensional continuous state spaces. Instead of requiring exhaustive sweeps, reinforcement learning algorithms operate on states as they occur in actual or simulated experiences in controlling the process. It is appropriate to view them as *Monte Carlo* DP algorithms.

2. Whereas conventional DP algorithms require a complete and accurate model of the process to be controlled, many reinforcement learning algorithms do not require such a model. Instead of computing the required quantities (such as state values) from a model, they estimate these quantities from experience. However, as described above, reinforcement learning methods can learn models in order to improve their efficiency.

3. Conventional DP algorithms use lookup-table storage of values for all states, which is impractical for large problems. Reinforcement learning algorithms often use more compact storage schemes in the form of parameterized function representations whose parameters are adjusted through adaptations of various function approximation methods.

2.4.1   Reinforcement Learning and Other Approximate DP Methods

While there is not a sharp distinction between reinforcement learning algorithms and other methods for approximating solutions to MDPs, several features tend to be associated with reinforcement learning. The most conspicuous one is that computation in reinforcement learning takes place during interaction between an active decision maker and its environment. The computational process occurs while the decision maker is engaged in making decisions as opposed to being an off-line batch process. Artificial Intelligence researchers say that the decision maker is "situated" in its environment. This feature arises from an underlying interest in *learning*, as we see it accomplished by ourselves and other animals, and not merely in general computational methods. Often this interaction is only virtual, as a simulated learning agent interacts with a simulated environment, but even in this case, the processes of *using* and *acquiring* knowledge are not separated into two distinct phases.

As a result of this emphasis, the objective of a reinforcement learning algorithm is not necessarily to approximate an optimal policy, at least not uniformly across the state space, as in conventional approaches to MDPs. It is more accurate to think of the objective from the active agent's point of view: it is to obtain as much reward over time as possible. Since an agent usually visits states non-uniformly while it is behaving, the approximation error is weighted by the agent's state-visitation distribution, a so-called *on-policy distribution* [29]. This is possible due to the situated nature of the reinforcement learning process. In some problems it confers significant advantages over conventional DP algorithms because large portions of the state space can be largely irrelevant for situated behavior.

A second feature that tends to be associated with reinforcement learning is the lack of complete knowledge of the MDP in question. The process of computing an optimal policy, or an approximately-optimal policy, given complete knowledge of an MDP's state transition and reward probabilities is considered to be more of a *planning* problem than a learning problem. However, there are many applications of reinforcement learning that make use of so-called *generative models*. These are simulation models of the MDP. Given a chosen state s and action a, a generative model can produce a next state s' sampled according to the probability transition function $P(s'|s, a)$ and an immediate reward generated according to $R(s, a, s')$. Some generative models permit the learner to jump around from one state to another arbitrarily, while other generative models can only simulate continuous trajectories through the state space. In principle, a generative model contains the same information as knowing P and R, but this information is not available in an explicit form and therefore cannot be used directly for DP. Instead, the generative model is applied to simulate the interaction of the learner with the environment, and reinforcement learning algorithms are applied to approximate an optimal policy. The advantage of generative models from an engineering perspective is that it is often much easier to design and implement a generative model of an application problem than it is to construct an explicit representation of P and R. In addition, learning from a generative model (e.g., of a robot aircraft) can be faster, safer, and cheaper than learning by interacting with the real MDP (e.g., a real robot). Using generative models in approximating solutions to MDPs is closely associated with reinforcement learning even though learning from on-line experience in the real world remains a goal of many reinforcement learning researchers.

2.5   SUPERVISED LEARNING FOR SEQUENTIAL DECISION TASKS

For the same reasons that non-sequential reinforcement learning cannot be reduced to supervised learning, sequential reinforcement learning can also not be reduced to supervised learning. First, the information provided by the environment (the next state and the reward) does not specify the correct action to perform. Second, the goal is to perform the optimal action in those states that are visited by the optimal policy. But this distribution of states is not a fixed distribution $D(s)$, but instead depends on the actions chosen by the learner. However, there is an additional reason why reinforcement learning in sequential decision tasks is different from supervised learning: In sequential decision tasks, *the agent must suffer the consequences of its own mistakes.*

Consider the problem of learning to steer a car down a highway. We can view this as a supervised learning problem in which an expert human teacher drives the car down the highway and the learner is given training examples of the form (s, a), where s is the current position of the car on the road (e.g., its position relative to the edges of the lane) and a is the steering action chosen by the teacher. We could now view this as a supervised learning problem where the goal is to learn the function $a = f(s)$. This approach has been termed "behavioral cloning" [20], because we wish to "clone" the behavior of a human expert.

Note that the distribution of states $D(s)$ contains only those states (i.e., those positions of the car) that are visited by a good human driver. If the function f can be learned completely and correctly and the car is started in a good state s, then there is no problem. But supervised learning is never perfect. Let h be the hypothesis output by the supervised learner, and let s_1 be a state where $h(s_1) \neq f(s_1)$. In this state, the learner will make a mistake, and the car will enter a new state s_2 chosen according to $P(s_2|s_1, h(s_1))$. For example, s_1 might be a state where the right wheels of the car are on the edge of the highway, and state s_2 might be a state where the right wheels are off the road. Now this state was never observed during training (because the human teacher would never make this mistake, so $D(s_2) = 0$). Consequently, the learner does not know how to act, and the car could easily leave the road completely and crash. The point is that even if h is 99.99% correct on the distribution $D(s)$, even a single state s_1 where h is wrong could lead to arbitrarily bad outcomes. Reinforcement learning cannot be solved by supervised learning, even with a perfect teacher. It is interesting to note that the ALVINN project [25] attempted to address exactly this problem by applying domain knowledge to generate synthetic training examples for a wide range of states including states where the car was far off the road.

Because reinforcement learning occurs online through interacting with the environment, the learner is forced to learn from its own mistakes. If the car leaves the road and crashes, the learner receives a large negative reward, and it learns to avoid those actions. Indeed, a reinforcement learning system can learn to avoid states that could potentially lead to dangerous states. Hence, the learner can learn to avoid states where the wheels get close to the edges of the lane, because those states are "risky." In this way, a reinforcement learning system can learn a policy that is better than the best human expert. This was observed in the TD-gammon system, where human experts changed the way they play certain backgammon positions after studying the policy learned by reinforcement learning [32, 33, 34].

2.6   CONCLUDING REMARKS

In this article, we have attempted to define supervised learning and reinforcement learning and clarify the relationship between these two learning problems. We have stressed the differences between these two problems, because this has often been a source of confusion. Nonetheless, there are many similarities. Both reinforcement learning and supervised learning are statistical processes in which a general function is learned from samples. In supervised learning, the function is a classifier or predictor; in reinforcement learning, the function is a value function or a policy.

A consequence of the statistical nature of reinforcement learning and supervised learning is that both approaches face a tradeoff between the amount of data and the complexity of the function that can be learned. If the space of functions being considered is too large, the data will be overfit, and both supervised and reinforcement learning will perform poorly. This is manifested by high error rates in supervised learning. In reinforcement learning, it is manifested by slow learning, because much more exploration is needed to gather enough data to eliminate overfitting.

Another similarity between reinforcement learning and supervised learning algorithms is that they both often make use of gradient search. However, in supervised learning, the gradient can be computed separately for each training example, whereas in reinforcement learning, the gradient depends on the relative rewards of two or more actions.

We note that the view we present of the key features distinguishing reinforcement learning from other related subjects leaves room for credible alternatives. Researchers do not thoroughly agree on these issues, and it is not clear that striving for definitive definitions serves a useful purpose. Modern problem formulations and algorithms can significantly blur some of these distinctions, or even render them irrelevant. Nevertheless, we hope that our discussion can serve as a useful guide to the core ideas behind reinforcement learning and their relationship to the fundamental ideas of supervised learning.

Acknowledgments

This material is based upon work supported by the National Science Foundation under Grant No. ECS-0218125. Any opinions, findings, and conclusions or recommendations expressed in this material are those of the authors and do not necessarily reflect the views of the National Science Foundation.

Bibliography

1. A. G. Barto, S. J Bradtke, and S. P. Singh, Learning to act using real-time dynamic programming, *Artificial Intelligence,* vol. 72, pp. 81–138, 1995.

2. J. Baxter and P. L. Bartlett, Infinite-horizon gradient-based policy search, *Journal of Artificial Intelligence Research,* vol. 15, pp. 319–350, 2001.

3. J. Baxter, P. L. Bartlett, and L. Weaver, Infinite-horizon gradient-based policy search: II. Gradient ascent algorithms and experiments, *Journal of Artificial Intelligence Research,* vol. 15, pp. 351–381, 2001.

4. D. P. Bertsekas, *Dynamic Programming: Deterministic and Stochastic Models,* Prentice-Hall, Englewood Cliffs, NJ, 1987.

5. D. P. Bertsekas and J. N. Tsitsiklis, *Neuro-Dynamic Programming,* Athena Scientific, Belmont, MA, 1996.

6. C. M. Bishop, *Neural Networks for Pattern Recognition,* Oxford University Press, Oxford, England, 1996.

7. J. A. Boyan, Least-squares temporal difference learning, in I. Bratko, and S. Dzeroski (eds.), *Machine Learning: Proc. of the 16th International Conference (ICML),* 1999.

8. S. J. Bradtke and A. G. Barto, Linear least–squares algorithms for temporal difference learning, *Machine Learning,* vol. 22, pp. 33–57, 1996.

9. L. Breiman, J. H. Friedman, R. A. Olshen, and C. J. Stone, *Classification and Regression Trees,* Wadsworth and Brooks, Monterey, CA, 1984.

10. P. Dayan, Motivated reinforcement learning, in T. G. Dietterich, S. Becker, and Z. Ghahramani (eds.), *Advances in Neural Information Processing Systems 14, Proc. of the 2002 Conference,* pp. 11–18, MIT Press, Cambridge, MA, 2003.

11. T. G. Dietterich and X. Wang, Batch value function approximation via support vectors, in T. G. Dietterich, S. Becker, and Z. Ghahramani (eds.), *Advances in Neural Information Processing Systems 14, Proc. of the 2002 Conference,* pp. 1491–1498, MIT Press, Cambridge, MA, 2003.

12. R. Dawkins, *The Blind Watchmaker*, Norton, New York, 1986.

13. R. O. Duda, P. E. Hart, and D. G. Stork, *Pattern Classification, Second Edition*, Wiley, New York, 2001.

14. A. A. Feldbaum, *Optimal Control Systems*, Academic Press, New York, 1965.

15. B. R. Hergenhahn and M. H. Olson, *An Introduction to Theories of Learning (Sixth Edition)*, Prentice-Hall, Upper Saddle River, NJ, 2001.

16. M. J. Kearns and U. V. Vazirani, *An Introduction to Computational Learning Theory*, MIT Press, Cambridge, MA, 1994.

17. M. G. Lagoudakis and R. Parr, Reinforcement learning as classification: leveraging modern classifiers, in T. G. Fawcett, N. Mishra (eds.), *Proc. 20th International Conference on Machine Learning*, pp. 424–431, AAAI Press, Menlo Park, CA, 2003.

18. G. J. McLachlan and T. Krishnan, *The EM Algorithms and Extensions*, John Wiley & Sons, Inc., New York, 1997.

19. J. M. Mendel and R. W. McLaren, Reinforcement learning control and pattern recognition systems, in J. M. Mendel and K. S. Fu (eds.), *Adaptive Learning and Pattern Recognition Systems: Theory and Applications*, pp. 287–318, Academic Press, New York, 1970.

20. D. Michie and C. Sammut, Behavioural clones and cognitive skill models, in K. Furukawa, D. Michie, and S. Muggleton (eds.), *Machine Intelligence 14: Applied Machine Intelligence*, pp. 387–395, Oxford University Press, New York, 1996.

21. M. L. Minsky, *Theory of Neural-Analog Reinforcement Systems and Its Application to the Brain-Model Problem*, Ph.D. dissertation, Princeton University, 1954.

22. M. L. Minsky, Steps toward artificial intelligence, *Proc. of the Institute of Radio Engineers*, vol. 49, pp. 8–30, 1961. Reprinted in E. A. Feigenbaum and J. Feldman (eds.), *Computers and Thought*, pp. 406–450, McGraw-Hill, New York, 1963.

23. T. Mitchell, *Machine Learning*, McGraw-Hill, New York, 1997.

24. A. W. Moore and C. G. Atkeson, Prioritized sweeping: reinforcement learning with less data and less real time, *Machine Learning*, vol. 13, pp. 103–130, 1993.

25. D. A. Pomerleau, Efficient training of artificial neural networks for autonomous navigation, *Neural Computation*, vol. 3, pp. 88–97, 1991.

26. J. R. Quinlan, *C4.5: Programs for Empirical Learning*, Morgan Kaufmann, San Francisco, 1993.

27. A. L. Samuel, Some studies in machine learning using the game of checkers, *IBM Journal on Research and Development,* vol. 3, pp. 211–229, 1959. Reprinted in E. A. Feigenbaum and J. Feldman (eds.), *Computers and Thought,* pp. 71–105, McGraw-Hill, New York, 1963.

28. H. A. Simon, *Administrative Behavior,* Macmillan, New York, 1947.

29. R. S. Sutton and A. G. Barto, *Reinforcement Learning: An Introduction,* MIT Press, Cambridge, MA, 1998.

30. R. S. Sutton, Learning to predict by the method of temporal differences, *Machine Learning,* vol. 3, pp. 9–44, 1988.

31. R. S. Sutton, Generalization in reinforcement learning: successful examples using coarse coding, in D. S. Touretzky, M. C. Moser and M. E. Hesselmo (eds.), *Advances in Neural Information Processing Systems, Proc. of the 1995 Conference,* pp. 1038–1044, MIT Press, Cambridge, MA, 1996.

32. G. J. Tesauro, Practical issues in temporal difference learning, *Machine Learning,* vol. 8, pp. 257–277, 1992.

33. G. J. Tesauro, TD–Gammon, A self-teaching backgammon program, achieves master-level play, *Neural Computation,* vol. 6, pp. 215–219, 1994.

34. G. Tesauro, Temporal Difference Learning and TD-Gammon, *Communications of the ACM,* vol. 28, pp. 58–68, 1995.

35. E. L. Thorndike, *Animal Intelligence,* Hafner, Darien, CT, 1911.

36. A. M. Turing, Computing machinery and intelligence, *Mind,* vol. 59, pp. 433–460, 1950. Reprinted in E. A. Feigenbaum and J. Feldman (eds.), *Computers and Thought,* pp. 11–15, McGraw-Hill, New York, 1963.

37. X. Wang and T. G. Dietterich, Model-based policy gradient reinforcement learning, in T. G. Fawcett, N. Mishra (eds.), *Proc. 20th International Conference on Machine Learning,* pp. 776–783, AAAI Press, Menlo Park, CA, 2003.

3 Model-Based Adaptive Critic Designs

SILVIA FERRARI   ROBERT F. STENGEL
Duke University   Princeton University

Editor's Summary: This chapter provides an overview of model-based adaptive critic designs including background, general algorithms, implementations, and comparisons. The authors begin by introducing the mathematical background of model-reference adaptive critic designs. Various ADP designs such as Heuristic Dynamic Programming (HDP), Dual HDP (DHP), Globalized DHP (GDHP), and Action-Dependent (AD) designs are examined from both a mathematical and implementation standpoint and put into perspective. Pseudocode is provided for many aspects of the algorithms. The chapter concludes with applications and examples. For another overview perspective that focuses more on implementation issues read Chapter 4: Guidance in the Use of Adaptive Critics for Control. Chapter 15 contains a comparison of DHP with back-propagation through time, building a common framework for comparing these methods.

3.1 INTRODUCTION

Under the best of circumstances, controlling a nonlinear dynamic system so as to minimize a cost function is a difficult process because it involves the solution of a two-point boundary value problem. The necessary conditions for optimality, referred to as *Euler-Lagrange equations*, include the final time. Hence, they cannot be accessed until the trajectory has been completed. The most that one can hope for is to solve the problem off line using an iterative process, such as steepest descent [1], before the trajectory begins. Then, the optimal control history can be applied to the actual system. If there are no errors in system description or control history implementation and if there are no disturbing forces, the system will execute the optimal trajectory in the specified time interval. For all other cases, optimality of the actual control system is, at best, approximate.

Of course, the real problem is more complex because there are non-deterministic (or stochastic) effects. The initial condition may differ from its assumed value, the system model may be imperfect, and there may be external disturbances to the dynamic process. At a minimum, closed-loop (feedback) control is required to account for perturbations from the open-loop-optimal trajectory prescribed by prior calculations. Because perturbations excite modes of motion that may be lightly damped or even unstable, the feedback control strategy must either assure satisfactory stability about the nominally optimal trajectory or produce a neighboring-optimal trajectory. If perturbations are small enough to be adequately described by a local linearization of the dynamic model, if random disturbances have zero mean, and if the model itself is not too imprecise, a *linear-quadratic neighboring-optimal controller* provides stability and generates the neighboring-optimal trajectory [1].

If, in addition, there are uncertain errors in the measurements required for feedback control, or if the full state is not measured, then some adjustment to control strategy must be made to minimize the degrading effect of imperfect information. For small perturbations and zero-mean measurement errors, the problem is solved by concatenating an optimal perturbation-state estimator with the feedback control law, forming a *time-varying, linear-quadratic-Gaussian regulator*. Inputs to the optimal estimator include a possibly reduced set of measurements and known control inputs. The algorithm makes predictions from the model of perturbation dynamics and, with knowledge of disturbance and measurement-error statistics, it corrects the predictions using the measurements to form a *least-squares state estimate*. In such case, the *separation* and *certainty-equivalence principles* apply to the perturbation system: the controller and estimator can be designed separately, and the controller itself is the same controller that would have been applied to the deterministic (i.e., certain) system state [2]. With perfect measurements, the neighboring-optimal controller is the same with or without zero-mean disturbance inputs: it is certainty-equivalent.

Thus, we see that real-time, exact optimal control of an actual system in an uncertain environment is not strictly possible, though there are approximating solutions that may be quite acceptable. In the remainder of this chapter, we present an approach to approximate optimal control that is based on dynamic programming, so we briefly relate the prior discussion to this alternative point of view [3, 4].

While there is no stochastic equivalent to the Euler-Lagrange equations, there is an equivalent stochastic dynamic programming formulation that extends to adaptive critic designs [5]. This approach seeks to minimize the *expected value* of the cost function with respect to the control, conditioned on knowledge of the system, its state, and the probability distributions of uncertainties [1]. For simplicity, this chapter only deals with adaptive critic designs that are deterministic.

Dual control is one method that offers a systematic solution to the problem of approximate optimization [6–8]. It optimizes a value function, or cost-to-go, composed of three parts that are associated with nominal optimal control, cautious feedback control, and probing feedback control. The first component represents the cost associated with optimization using present knowledge of the system and its trajectory. The second component represents cost associated with the effects of uncertain inputs and measurements. The third cost is associated with control inputs that improve knowledge of the system's unknown parameters. A numerical search over the present value of the control minimizes a stochastic *Hamilton-Jacobi-Bellman* (HJB) *equation* [6] providing a basis for real-time, approximate optimal control.

If the final time is finite and minimizing a terminal cost is important, the entire remaining trajectory must be evaluated to determine the future cost. If the final time is infinite, then the terminal cost is of no consequence, and the future trajectory is propagated far enough ahead that additional steps have negligible impact on the control policy. In some applications, it may be sufficient to propagate just one step ahead to approximate the future cost, greatly reducing the required computation. Thus, there is a close relationship between dual control and other *receding-horizon* or *predictive-adaptive* control approaches [9, 10].

Adaptive critic controllers provide an alternative, practical approach to achieving optimality in the most general case. A distinguishing feature of this approach is that the optimal control law and value function are modelled as parametric structures (e.g., computational neural networks), whose shapes are improved over time by solving the *Recurrence Relation of Dynamic Programming*. The methods use step-ahead projection to estimate the future cost. When the final time approaches infinity and the system dynamics are unchanging, the methods converge to the optimal control law. In practice, the parametric structures adapt to changing system parameters, including system failures, without explicit parameter identification. This requires an approximate model of the plant dynamics that admits satisfactory estimates of the future cost. In separate work [11, 12], we have shown how to pre-train neural networks to give these parametric structures good starting topologies (or *initializations*) prior to on-line learning by an adaptive critic approach.

3.2   MATHEMATICAL BACKGROUND AND FOUNDATIONS

This section introduces the foundations of model-reference adaptive critic designs, placing them within the framework of optimal control. The on-line solution of infinite-horizon problems, with information that becomes available incrementally over time, is emphasized. In optimal control problems the objective is to devise a strategy of action, or control law, that optimizes a desired performance metric or cost. Two well known solution approaches are the calculus of variations, involving the Euler-Lagrange equations, and backward dynamic programming. They are reviewed here because they both have strong ties with the adaptive critic approach.

Adaptive critic designs utilize two parametric structures called the *actor* and the *critic*. The actor consists of a parameterized control law. The critic approximates a value-related function and captures the effect that the control law will have on the future cost. At any given time the critic provides guidance on how to improve the control law. In return, the actor can be used to update the critic. An algorithm that successively iterates between these two operations converges to the optimal solution over time. More importantly, this adaptive critic algorithm can be used to design control systems that improve their performance on-line, subject to actual plant dynamics.

3.2.1   Solution of the Optimal Control Problem

The objective is to determine a stationary optimal control law that minimizes a performance measure J expressed by a scalar, time-invariant integral function of the state and controls and by a scalar terminal cost,

$$J = \varphi[\mathbf{x}(t_f)] + \sum_{t_k=t_0}^{t_f-1} \mathcal{L}[\mathbf{x}(t_k), \mathbf{u}(t_k)], \tag{3.1}$$

subject to the dynamic constraint imposed by plant dynamics. This *cost function* represents the cost of operation as it accrues from the initial time t_0 to the final time t_f. The integrand or *Lagrangian* $\mathcal{L}[\cdot]$ is the cost associated with one time increment; it also is referred to as utility or reward in the financial and operation research literature, where the objective typically is to maximize the overall performance (Eq. (3.1)). The plant dynamics are discrete, time-invariant, and deterministic, and they can be modelled by a difference equation of the form,

$$\mathbf{x}(t_{k+1}) = \mathbf{f}[\mathbf{x}(t_k), \mathbf{u}(t_k)], \tag{3.2}$$

with equally spaced time increments in the interval $t_0 \le t_k \le t_f$, and initial condition $\mathbf{x}(t_0)$. $\mathbf{x}$ is the $n \times 1$ plant state and $\mathbf{u}$ is the $m \times 1$ control vector. For simplicity, it also is assumed that the state is fully observable and that perfect output measurements are available.

The control law is assumed to be solely a function of the state

$$\mathbf{u} = \mathbf{c}(\mathbf{x}). \tag{3.3}$$

The control functional $\mathbf{c}(\cdot)$ may contain functions of its arguments such as integrals and derivatives, and the optimal form is denoted by $\mathbf{c}^*(\mathbf{x})$. At any moment in time, t_k, the cost that is going to accrue from that moment onward can be expressed by a *value function*,

$$V = \varphi[\mathbf{x}(t_f)] + \sum_{t_k}^{t_f-1} \mathcal{L}[\mathbf{x}(t_k), \mathbf{u}(t_k)] = V[\mathbf{x}(t_k), \mathbf{c}(\mathbf{x})], \tag{3.4}$$

which depends on the present value of the state, $\mathbf{x}(t_k)$, and on the chosen control law $\mathbf{c}(\mathbf{x})$. Therefore, $V[\mathbf{x}(t_k), \mathbf{c}(\mathbf{x})]$ also can be written in abbreviated form as $V[\mathbf{x}_k, \mathbf{c}]$, where $\mathbf{x}_k \equiv \mathbf{x}(t_k)$.

One approach to the optimization of Eq. (3.1) subject to Eq. (3.2) is to augment the cost function by the dynamic equation, recasting the problem in terms of the

Hamiltonian

$$\mathcal{H}(\mathbf{x}_k, \mathbf{u}_k, \lambda_k) = \mathcal{L}[\mathbf{x}(t_k), \mathbf{u}(t_k)] + \lambda^T(t_k)\mathbf{f}[\mathbf{x}(t_k), \mathbf{u}(t_k)]. \qquad (3.5)$$

λ is a *costate* or *adjoint* vector that contains Lagrange multipliers [1] and represents the cost sensitivity to state perturbations on the optimal trajectory; it can be shown that $\lambda(t_f) = \partial\varphi/\partial\mathbf{x}|_{t_f}$ (with the gradient defined as a column vector). When the final time is fixed, necessary conditions for optimality, the Euler-Lagrange equations, can be obtained by differentiating the Hamiltonian with respect to the state and the control [1]. Thus, the dynamic optimization problem is reduced to a *two-point boundary value problem*, where the state and the adjoint vector are specified at the initial and final time, respectively. Ultimately, necessary and sufficient conditions for optimality are provided by Pontryagin's Minimum Principle, stating that on the optimal trajectory $\mathcal{H}$ must be stationary and convex

$$\mathcal{H}^* = \mathcal{H}(\mathbf{x}_k^*, \mathbf{u}_k^*, \lambda_k^*) \leq \mathcal{H}(\mathbf{x}_k^*, \mathbf{u}_k, \lambda_k^*). \qquad (3.6)$$

Another approach to solving the optimal control problem consists of *imbedding* [13] the minimization of the cost function, Eq. (3.1), in the minimization of the value function, Eq. (3.4). When the system is in an admissible state $\mathbf{x}_k$, the value function, that is, the cost of operation from the instant t_k to the final time t_f, can be written as

$$V(\mathbf{x}_k, \mathbf{u}_k, \ldots, \mathbf{u}_{f-1}) = \mathcal{L}(\mathbf{x}_k, \mathbf{u}_k) + V(\mathbf{x}_{k+1}, \mathbf{u}_{k+1}, \ldots, \mathbf{u}_{f-1}). \qquad (3.7)$$

$\mathbf{x}_{k+1}$ depends on $\mathbf{x}_k$ and $\mathbf{u}_k$ through Eq. (3.2). All subsequent values of the state can be determined from $\mathbf{x}_k$ and from the chosen control history, $\mathbf{u}_k, \ldots, \mathbf{u}_{f-1}$. Therefore, the cost of operation from t_k onward can be minimized with respect to all future values of the control

$$V^*(\mathbf{x}_k^*) = \min_{\mathbf{u}_k, \ldots, \mathbf{u}_{f-1}} \{\mathcal{L}(\mathbf{x}_k^*, \mathbf{u}_k) + V(\mathbf{x}_{k+1}^*, \mathbf{u}_{k+1}, \ldots, \mathbf{u}_{f-1})\}. \qquad (3.8)$$

It follows that the optimal value function depends on the present state of the system and is independent of any prior history. Suppose a policy is optimal over the time interval $(t_f - t_k)$. Then, by the *Principle of Optimality* [14], whatever the initial state $(\mathbf{x}_k)$ and decision $(\mathbf{u}_k)$ are, the remaining decisions also must constitute an optimal policy with regard to the state $\mathbf{x}_{k+1}$, that is,

$$V^*(\mathbf{x}_k^*) = \min_{\mathbf{u}_k}\{\mathcal{L}(\mathbf{x}_k^*, \mathbf{u}_k) + V^*(\mathbf{x}_{k+1}^*)\}. \qquad (3.9)$$

The value function can be minimized solely with respect to the present value of the control, $\mathbf{u}_k$, provided the future cost of operation, $V(\mathbf{x}_{k+1}, \mathbf{u}_{k+1}, \ldots, \mathbf{u}_{f-1})$, is optimal (in which case, it only depends on the next value of the state on the optimal trajectory, that is, $\mathbf{x}_{k+1}^*$). Equation (3.9) constitutes the Recurrence Relation of Dynamic Programming. The optimal value function can be interpreted as the minimum cost for the time period that remains after a time t_k in a process that began

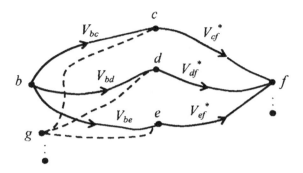

Fig. 3.1 Backward or discrete dynamic programming approach.

at t_0, that is, $(t_f - t_k)$. Alternatively, it can be viewed as the minimum cost for a process that begins at t_k and ends at t_f.

The recurrence relation can be used backwards in time, starting from t_f, to obtain an approximate solution to the exact optimal control history. This approach is referred to as *backward dynamic programming* (BDP). It discretizes the state space and makes a direct comparison of the cost associated with all feasible trajectories, guaranteeing a solution to the global optimal control problem. The space of admissible solutions is reduced by examining a multi-stage decision process as a sequence of one-stage processes. This approach typically is too computationally expensive for higher dimensional systems, with a large number of stages (or time increments). The required multiple generation and expansion of the state and the storage of all optimal costs lead to a number of computations that grows exponentially with the number of state variables. This phenomenon commonly is referred to as the "curse of dimensionality" or "expanding grid" [13].

Forward dynamic programming (FDP) and *temporal difference* methods use incremental optimization combined with a *parametric structure* to reduce the computational complexity associated with evaluating the cost [15–17]. A parametric structure consists of a functional relationship whose adjustable parameters allow it to approximate different mappings. Adaptive critic designs (ACD) derive from the forward-dynamic-programming approach, also called *approximate dynamic programming*.

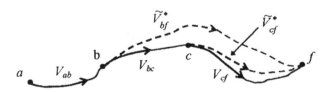

Fig. 3.2 Forward or approximate dynamic programming approach.

For comparison, the backward and forward DP approaches are illustrated for the last two stages of a hypothetical process in Figure 3.1 and Figure 3.2, respectively. The backward approach begins by considering the last stage. Since the optimal final state is not yet known, the following procedure has to be performed for all admissible values of the final state. The optimal paths are computed for all feasible intermediate-state values, c, d, and e, thereby producing the optimal costs V_{cf}^*, V_{df}^*, and V_{ef}^*, respectively. By the principle of optimality these paths also are optimal for the last stage of the optimal trajectories that go from b to f through the respective state values, for example, $V_{b(c)f}^* = V_{bc} + V_{cf}^*$. Thus, if the last-stage's costs, V_{cf}^*, V_{df}^*, and V_{ef}^*, are stored, then the total costs $V_{b(c)f}^*$, $V_{b(d)f}^*$, and $V_{b(e)f}^*$ can be compared to find the optimal path from b to f. For a process with more than two stages, the state b would be unknown. Hence, the same computation would be carried out for all possible state values, such as g in Figure 3.1, in order to determine V_{bf}^*, V_{gf}^*, and so on. The algorithm terminates when the initial stage is reached, where the state is known from the initial conditions.

Unlike backward DP, forward DP algorithms progress forward in time, and they approximate the optimal control and future cost by considering only the current value of the state. Suppose a is the initial state of a two-stage process or, equivalently, the state of an on-going process at the second-to-last stage, as outlined in Figure 3.2. Then, a is known and the cost V_{ab} can be computed from the Lagrangian for a chosen control policy. The optimal cost over all future stages, V_{bf}^*, is predicted by a function approximator or parametric structure. Hence, the sum $(V_{ab} + \tilde{V}_{bf}^*)$ can be minimized with respect to the present value of the control, according to the recurrence relation of dynamic programming (Eq. (3.9)). At the next stage b-c this procedure is repeated. The cost approximation, denoted by $\tilde{V}$, has been improved based on the information gathered during the first stage. Therefore, the next path, from c to f, is closer to the optimal trajectory.

Adaptive critic designs reproduce the most general solution of FDP by deriving recurrence relations for the optimal policy, the cost, and, possibly, their derivatives. The goal is to overcome the curse of dimensionality, while ensuring convergence to a near-optimal solution over time. The following section introduces the basic adaptive-critic algorithm and the related proof of convergence.

3.2.2 Adaptive Critic Algorithm

Adaptive critic designs are based on an algorithm that cycles between a *policy-improvement routine* and a *value-determination operation*. At each optimizing cycle, indexed by ℓ, the algorithm approximates the optimal control law (Eq. (3.3)) and value function (Eq. (3.8)) based on the state $\mathbf{x}_k$. We distinguish between iteration over k, which represents the passage of time in the dynamic process, and over ℓ, which relates to the search for an optimal solution. In off-line optimization, an ensemble of state vectors including all possible values of $\mathbf{x}_k$ is used during every cycle. In this case, the entire process can be simulated from t_0 to t_f. In on-line optimization, one cycle is carried out at each time step based on a single value of $\mathbf{x}_k$, that is, the actual state, such that $k = \ell$.

For simplicity, the following discussion is limited to the infinite-horizon optimal control problem, where the final time t_f tends to infinity. In this case, the cost function to be minimized takes the form,

$$J = \lim_{t_f \to \infty} \sum_{t_k=t_0}^{t_f-1} \mathcal{L}[\mathbf{x}(t_k), \mathbf{u}(t_k)], \tag{3.10}$$

and the terminal cost can be set equal to zero [1].

The projected cost $V(\mathbf{x}_{k+1}, \mathbf{c})$ that a sub-optimal control law $\mathbf{c}$ would incur from t_k to t_f can be estimated from Eq. (3.7). Using the projected cost as a basis, the control-law and value-function approximations can be improved during every optimization cycle, ℓ. Over several cycles, the policy-improvement routine generates a sequence of sub-optimal control laws $\{\mathbf{c}_\ell \mid \ell = 0, 1, 2, \ldots\}$; the value-determination operation produces a sequence of sub-optimal value functions $\{V_\ell \mid \ell = 0, 1, 2, \ldots\}$. The algorithm terminates when $\mathbf{c}_\ell$ and V_ℓ have converged to the optimal control law and value function, respectively. The proof of convergence [5] is outlined in the Appendix to this chapter.

Policy-improvement Routine   Given a value function $V(\cdot, \mathbf{c}_\ell)$ corresponding to a control law $\mathbf{c}_\ell$, an *improved* control law, $\mathbf{c}_{\ell+1}$, can be obtained as follows:

$$\mathbf{c}_{\ell+1}(\mathbf{x}_k) = \arg \min_{\mathbf{u}_k}\{\mathcal{L}(\mathbf{x}_k, \mathbf{u}_k) + V(\mathbf{x}_{k+1}, \mathbf{c}_\ell)\}, \tag{3.11}$$

such that $V(\mathbf{x}_k, \mathbf{c}_{\ell+1}) \leq V(\mathbf{x}_k, \mathbf{c}_\ell)$, for any value of $\mathbf{x}_k$. Furthermore, the sequence of functions $\{\mathbf{c}_\ell \mid \ell = 0, 1, 2, \ldots\}$ converges to the optimal control law, $\mathbf{c}^*(\mathbf{x})$.

Value-determination Operation   Given a control law $\mathbf{c}$, the value function can be updated according to the following rule:

$$V_{\ell+1}(\mathbf{x}_k, \mathbf{c}) = \mathcal{L}(\mathbf{x}_k, \mathbf{u}_k) + V_\ell(\mathbf{x}_{k+1}, \mathbf{c}), \tag{3.12}$$

such that the sequence $\{V_\ell \mid \ell = 0, 1, 2, \ldots\}$, in concert with the policy-improvement update, converges to the optimal value function $V^*(\mathbf{x})$.

The adaptive critic algorithm successively iterates between these two updates, as illustrated in Figure 3.3. The algorithm can begin in either box. The speed of convergence is dependent on the suitability of the initialized control law and value function $\mathbf{c}_0$ and V_0, as well as on the details of the adjustment rules. A simple example is given in Section 3.3.5. More detailed examples are given in [11, 18].

In the ℓth cycle, $\mathbf{c}_{\ell+1}$ is determined by the policy-improvement routine and can be used as the control law $\mathbf{c}$ in the value-determination operation. Then, in the $(\ell + 1)^{\text{th}}$ cycle, the updated value function $V_{\ell+1}$ can be used as $V(\cdot, \mathbf{c}_{\ell+1})$ in the policy-

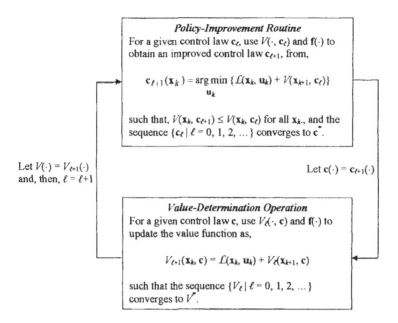

Fig. 3.3   Adaptive critic iteration cycle, after [5].

improvement routine. The algorithm terminates when two successive cycles produce the same control law, or a suitably close one. If the policy-improvement routine finds two control laws of equivalent performance, such that $[\mathcal{L}(\mathbf{x}_k, \mathbf{u}_k) + V(\mathbf{x}_{k+1}, \mathbf{c}_{\ell+1})] = [\mathcal{L}(\mathbf{x}_k, \mathbf{u}_k) + V(\mathbf{x}_{k+1}, \mathbf{c}_\ell)]$, but $\mathbf{c}_{\ell+1} \neq \mathbf{c}_\ell$, the old one ($\mathbf{c}_\ell$) always is chosen.

The adaptive critic algorithm has the following important properties [5]:

1. The algorithm improves its choice of control law during every cycle ℓ, such that each succeeding control functional has a better overall performance (lower cost) than the previous one.

2. The algorithm improves its choice of value function during every cycle ℓ, such that, for a given state, each succeeding value functional is smaller than the previous one.

3. The algorithm terminates on the optimal control law, provided there exists one. In other words, if there exists a superior control law it is found before the adaptive critic algorithm terminates.

4. The algorithm solves the optimal control problem without backward iteration in time.

Properties 1 through 4 are demonstrated in the Appendix to this chapter. Property 4 is fundamental to the solution approach. The following section discusses on-line implementations of the adaptive critic algorithm.

3.3   ADAPTIVE CRITIC DESIGN AND IMPLEMENTATION

The adaptive critic algorithm can be used to determine the optimal control law for a dynamic process on or off line. In the latter case, control design is based on the model of the system to be controlled, and the required values of the state are produced by simulation. The adaptive critic cycle generates a control law c_ℓ and a value function V_ℓ for an ensemble of state values. The procedure is repeated for $\ell = 0, 1, 2, \ldots$, until the algorithm terminates producing c^* and V^*. Subsequently, c^* can be implemented to operate the system in Eq. (3.2) in an optimal fashion.

The adaptive critic algorithm also can be used on-line. Unlike the Euler-Lagrange equations and backward DP, the computation required by each iteration depends only on the present state, x_k, and does not involve the final conditions (e.g., $\lambda(t_f)$). In on-line implementations state-vector values become available progressively in time as $\{x_k \mid k = 0, 1, 2, \ldots\}$. Then, the indices ℓ and k in Figure 3.3 coincide at every algorithmic cycle and moment in time. Although this procedure can be adopted for systems whose dynamic equation (Eq. (3.2)) is known precisely, it is particularly valuable when the system dynamics are not fully known or subject to change. Then, the adaptive critic algorithm can be implemented on-line, while the system is operating.

In on-line implementations, the control law c_ℓ, generated by the ℓth algorithmic cycle, is applied at time t_k. Assuming that the plant is fully observable, the *actual* value of the present state, x_k, is determined from available output measurements. The next value of the state, x_{k+1}, is *predicted* through the model in Eq. (3.2). Hence, the control law and value function can be improved at every moment in time t_k based on the observed value of x_k. The optimal control law can be determined on-line for systems whose true dynamic characteristics are revealed only during actual operation. Since the adaptive critic algorithm improves upon any sub-optimal control law (say $c_{\ell-1}$), it usually is convenient to implement c_ℓ as soon as it is generated. The system converges to optimal performance provided its dynamics remain unchanged. Otherwise, the algorithm continues to improve the control and value functions incrementally, subject to the varying dynamics. An underlying assumption is that the dynamic variations occur on a sufficiently large time scale with respect to the algorithmic cycles.

3.3.1   Overview of Adaptive Critic Designs

Adaptive critic designs implement approximate dynamic programming through recurrence relations for the control law, the value function and, possibly, their derivatives. The basic architectures can be classified in four categories [19, 20]:

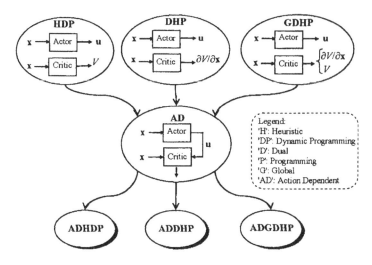

Fig. 3.4 Overview of well known adaptive critic designs.

1. Heuristic Dynamic Programming (HDP)

2. Dual Heuristic Dynamic Programming (DHP)

3. Globalized Dual Heuristic Dynamic Programming (GDHP)

4. Action-dependent (AD) designs

The distinguishing characteristics of each category are described in this section, and are summarized in Figure 3.4.

The salient feature of these designs is that they can be implemented on-line. The exact system dynamics need not be known *a priori* because the computations can be carried out while the system is operating, based on the observable state vector. Initially, the control law generated is sub-optimal. But, it is guaranteed to improve with every iteration cycle by the policy-improvement routine. The goal is to overcome the curse of dimensionality through function approximation, while approaching the optimal solution over time. The main challenge is to achieve satisfactory convergence to the optimal or *near*-optimal solution in a small number of cycles.

The four types of adaptive critic designs have been developed to accelerate convergence to the optimal solution. While they all are based on the adaptive critic algorithm (Section 3.2.2), they differ in what functionals they set out to approximate. The basic iteration cycle (Figure 3.3) involves the optimal control law and value function. Thus the most basic form of adaptive critic design, Heuristic Dynamic Programming (HDP), uses the parametric structure called actor (or action network) to approximate the control law, and another parametric structure called critic (or critic network) to approximate the value function. The role of the critic has been compared with that of "reinforcement" in psychology, and to the "utility function" in utility theory [21]. The critic provides the actor with a performance measure for its present actions by anticipating and evaluating future events. In practice, HDP converges slowly because the parameters of this architecture ultimately are updated based on a scalar evaluation function, that is, V.

An alternative approach referred to as Dual Heuristic Programming (DHP) was first proposed in [21, 22]. DHP uses the critic to approximate the derivatives of the value function with respect to the state, that is, $\partial V/\partial \mathbf{x}_k$. It correlates the parameters of the architecture to a larger number of dependent variables. The actor is used to approximate the control law, as in all other adaptive critic designs. An algorithm for updating the DHP functionals can be obtained from the policy-improvement routine and the value-determination operation, as explained in Section 3.3.3. In applications [23, 24], the DHP algorithm has been shown to find the optimal solution more rapidly (with less iteration cycles) than HDP. However, due to the use of derivative information, the relationships for updating the control and value-derivative functionals are more involved. Furthermore, since the DHP critic approximates a vector functional, the problem of function approximation is more challenging.

Many other methodologies have been proposed to alleviate some of the difficulties mentioned above. For example, Globalized Dual Heuristic Programming (GDHP) combines the advantages of both the HDP and DHP architectures [16, 25, 26]. In GDHP the critic approximates both the value function and its derivatives. Action-dependent (AD) versions of these approaches use a parametric structure, again referred to as critic, to approximate a value function that depends *explicitly* on the control (as shown in Figure 3.4). The motivation behind this [27] and other methodologies (e.g., [28]) is to achieve faster convergence to the optimal solution and to simplify the problem of function approximation.

In summary, an adaptive critic architecture consists of a parametric structure, the critic, which approximates the value function (Eq. (3.4)) or a closely related quantity, and another parametric structure, the actor, which approximates the control law. Once the functions to be approximated are chosen, an algorithm that updates them at every cycle can be obtained from the policy-improvement routine and the value-determination operation in Section 3.2.2. The objective is to produce two sequences of functions that eventually converge to the optimal solution.

3.3.2 Function Approximation

Three functionals play a key role in all adaptive critic designs. They are:

1. The control law

2. The value function

3. The model of the system to be controlled

The objective of the adaptive critic algorithm is to progressively improve its approximation of the optimal control law $\mathbf{c}^*(\mathbf{x}_k)$ and value function $V^*(\mathbf{x}_k)$. A model of the system (such as Eq. (3.2)) is required in order to predict the next value of the state x_{k+1} at every iteration cycle ℓ (Figure 3.3). Hence, designs based on this algorithm often are referred to as *model-based adaptive critics*. When the system's behavior is partly unanticipated the model provides an approximation to its dynamics [19].

The sequences of functions generated by the adaptive critic algorithm are synthesized by a convenient parametrization or function approximator that captures the relationship between the dependent and independent variables. Without the use of an appropriate parametric structure each function, say $u_k = c_\ell(x_k)$, would need to be represented by a lookup table. For large state and control spaces it is infeasible to carry out the required computation (Eqs. (3.11) and (3.12)) with lookup-table representations of c_ℓ, V_ℓ, $c_{\ell+1}$, and $V_{\ell+1}$. Moreover, it usually is desirable to obtain a functional representation of the optimal control and value functions once the algorithm converges. Therefore, a parametric structure that, ultimately, is capable of approximating the optimal solution must be chosen.

A *suitable function approximator* or parametric structure has the following characteristics:

- It is differentiable;

- It approximates the desired function with less complexity than a lookup-table representation;

- It is supplied with apposite algorithms for computing the parameter values;

- It is capable of approximating the optimal shape of the function within the desired accuracy.

There are many theoretical and practical issues associated with the problem of function approximation. The most important ones pertain to the effect that function approximation has on convergence. In order for the original proof (in the Appendix) to hold, this process must be consistent with the policy-improvement routine and value-determination operation in Section 3.2.2. Additional details can be found in [29–31].

We assume that the actor is a suitable parametric structure $\tilde{c}(x_k, a)$, where a is a vector of parameters to be optimized by the policy-improvement routine. The function approximator's parameters are updated at every cycle, ℓ, such that $\tilde{c}(x_k, a_\ell) \approx c_\ell(x_k)$. Then, the computational complexity of the adaptive critic algorithm is reduced at every cycle. Upon termination of the algorithm (when $c_\ell \rightarrow c^*$), the optimal control is readily approximated as $\tilde{c}(x_k, a^*) \approx c^*(x_k)$.

The parameters of the function approximator must be determined at every cycle of the algorithm without compromising convergence. This process, in itself, may require one or more iterations. The following sections describe how function approximation can be combined with the policy-improvement routine and the value-determination operation (Section 3.2.2) to produce heuristic and dual-heuristic adaptive critic designs.

3.3.3 Derivation of the Heuristic and Dual-Heuristic Dynamic Programming Algorithms

In on-line implementations of adaptive critic designs the control and value function approximations are updated incrementally over time. With one available value of the state, $\mathbf{x}_k$, the minimization problem in Eq. (3.11) can be solved by computing the stationary point of the scalar function $V(\mathbf{x}_k, \mathbf{c}_\ell) \equiv [\mathcal{L}(\mathbf{x}_k, \mathbf{u}_k) + V(\mathbf{x}_{k+1}, \mathbf{c}_\ell)]$. This point is defined as the value of $\mathbf{u}_k$ that satisfies the *optimality condition*,

$$\frac{\partial V(\mathbf{x}_k, \mathbf{c}_\ell)}{\partial \mathbf{u}_k} = \frac{\partial \mathcal{L}(\mathbf{x}_k, \mathbf{u}_k)}{\partial \mathbf{u}_k} + \left[\frac{\partial \mathbf{f}(\mathbf{x}_k, \mathbf{u}_k)}{\partial \mathbf{u}_k}\right]^T \frac{\partial V(\mathbf{x}_{k+1}, \mathbf{c}_\ell)}{\partial \mathbf{x}_{k+1}} = 0, \qquad (3.13)$$

along with the following *convexity condition*:

$$\frac{\partial^2 V(\mathbf{x}_k, \mathbf{c}_\ell)}{\partial \mathbf{u}_k^2} > 0. \qquad (3.14)$$

This condition requires a positive definite Hessian matrix [32]. The gradient of a scalar function [32], for example, $\partial V / \partial \mathbf{x}_k$, is defined as a column vector. The stationary control value can be used to improve upon the control law approximation $\mathbf{c}_\ell$ at the moment in time t_k. Rules for updating the critic on-line are similarly derived, as shown in the following sections.

Heuristic Dynamic Programming (HDP) Algorithm

In Heuristic Dynamic Programming, the actor is a parametric structure $\tilde{\mathbf{c}}$ with parameters a that is used to approximate the control law. The critic is a parametric structure $\tilde{V}$, with parameters $\mathbf{w}$, that is used to approximate the value function, that is:

$$\tilde{\mathbf{c}}(\mathbf{x}_k, \mathbf{a}_\ell) \approx \mathbf{c}_\ell(\mathbf{x}_k) \qquad (3.15)$$

$$\tilde{V}(\mathbf{x}_k, \mathbf{w}_\ell) \approx V_\ell(\mathbf{x}_k, \tilde{\mathbf{c}}) \qquad (3.16)$$

The value function estimate is assumed to be a *composite function* of $\mathbf{x}_k$, since $\mathbf{u}_k = \mathbf{c}(\mathbf{x}_k)$ and system dynamics are known (Eq. (3.2)). The state is its sole input, as this is the action-independent version of HDP. The optimal value function also is a function solely of the state, as shown in Eq. (3.8). When the chosen parametric structures are differentiable, the optimality condition (Eq. (3.13)) generates a sequence of successively improving control laws $\{\mathbf{c}_\ell \mid \ell = 0, 1, 2, \ldots\}$. Concurrently, the value-determination operation (Eq. 3.12) generates a sequence of successively improving value functions $\{V_\ell \mid \ell = 0, 1, 2, \ldots\}$. An HDP algorithm that iterates between these two operations is shown below. At every cycle ℓ, the actor parameters are updated based on the improved control law $\mathbf{c}_{\ell+1}$, and the critic parameters are updated based on the improved value function $V_{\ell+1}$.

Heuristic-Dynamic-Programming Actor Update Suppose a control-law approximator, $\tilde{c}(\cdot, a_\ell)$, and a corresponding value-function approximator, $\tilde{V}(\cdot, w_\ell)$, are given. Then, an improved control-law approximator $\tilde{c}(\cdot, a_{\ell+1})$ can be obtained by computing a desired control vector u_k^D, such that,

$$
\left. \frac{\partial \tilde{V}(\mathbf{x}_k, \mathbf{w}_\ell)}{\partial \mathbf{u}_k} \right|_{\mathbf{u}_k = \mathbf{u}_k^D} = \left. \left[\frac{\partial \mathcal{L}(\mathbf{x}_k, \mathbf{u}_k)}{\partial \mathbf{u}_k} + \left(\frac{\partial \mathbf{f}(\mathbf{x}_k, \mathbf{u}_k)}{\partial \mathbf{u}_k} \right)^T \frac{\partial \tilde{V}(\mathbf{x}_{k+1}, \mathbf{w}_\ell)}{\partial \mathbf{x}_{k+1}} \right] \right|_{\mathbf{u}_k = \mathbf{u}_k^D}
$$
$$
= 0, \tag{3.17}
$$

and the matrix $[\partial^2 \tilde{V}(\mathbf{x}_k, \mathbf{w}_\ell)/\partial \mathbf{u}_k^2]|_{\mathbf{u}_k = \mathbf{u}_k^D}$ is positive definite.

The improved actor parameters, $a_{\ell+1}$, can be computed as follows:

$$
\mathbf{a}_{\ell+1} = \arg \min_{\mathbf{a}} \{ \mathbf{u}_k^D - \tilde{\mathbf{c}}(\mathbf{x}_k, \mathbf{a}) \}, \tag{3.18}
$$

where $\mathbf{u}_k^D \equiv \mathbf{c}_{\ell+1}(\mathbf{x}_k)$, therefore $\tilde{\mathbf{c}}(\mathbf{x}_k, \mathbf{a}_{\ell+1}) \approx \mathbf{c}_{\ell+1}(\mathbf{x}_k)$.

Heuristic-Dynamic-Programming Critic Update Given the improved control-law approximator $\tilde{c}(\cdot, a_{\ell+1})$ and the value-function approximator $\tilde{V}(\cdot, w_\ell)$, an improved value-function approximator $\tilde{V}(\cdot, w_{\ell+1})$ can be obtained by computing its desired value,

$$
V_k^D \equiv V_{\ell+1}(\mathbf{x}_k, \tilde{\mathbf{c}}) = \mathcal{L}(\mathbf{x}_k, \mathbf{u}_k) + \tilde{V}(\mathbf{x}_{k+1}, \mathbf{w}_\ell), \tag{3.19}
$$

with $\mathbf{u}_k = \tilde{\mathbf{c}}(\mathbf{x}_k, \mathbf{a}_{\ell+1})$ and, subsequently, by determining the improved critic parameters,

$$
\mathbf{w}_{\ell+1} = \arg \min_{\mathbf{w}} \{ V_k^D - \tilde{V}(\mathbf{x}_k, \mathbf{w}) \}, \tag{3.20}
$$

such that, $\tilde{V}(\mathbf{x}_k, \mathbf{w}_{\ell+1}) \approx V_{\ell+1}(\mathbf{x}_k, \tilde{\mathbf{c}})$.

There exist other approaches for updating the actor based on the optimality condition. One possibility is to obtain the improved actor parameters by minimizing the right-hand side of Eq. (3.17) directly with respect to a:

$$
\mathbf{a}_{\ell+1} = \arg \min_{\mathbf{a}} \left\{ \left. \left[\frac{\partial \mathcal{L}(\mathbf{x}_k, \mathbf{u}_k)}{\partial \mathbf{u}_k} + \left(\frac{\partial \mathbf{f}(\mathbf{x}_k, \mathbf{u}_k)}{\partial \mathbf{u}_k} \right)^T \frac{\partial \tilde{V}(\mathbf{x}_{k+1}, \mathbf{w}_\ell)}{\partial \mathbf{x}_{k+1}} \right] \right|_{\mathbf{u}_k = \tilde{\mathbf{c}}(\mathbf{x}_k, \mathbf{a})} \right\}.
$$
$$
\tag{3.21}
$$

Although this method reduces the actor update to one step, it may be computationally less efficient and less reliable.

Dual Heuristic Dynamic Programming (DHP) Algorithm

In Dual Heuristic Dynamic Programming, the critic approximates a sequence of functionals, $\{\lambda_\ell \mid \ell = 0, 1, 2, \ldots\}$, which ultimately converges to the derivative of the optimal value function with respect to the state, defined as:

$$\lambda^*(\mathbf{x}_k^*) \equiv \frac{\partial V^*(\mathbf{x}_k^*)}{\partial \mathbf{x}_k^*}. \tag{3.22}$$

In essence, the role of the critic is to generate the costate or adjoint vector in the *Hamilton-Jacobi-Bellman* (HJB) equation, estimating the cost sensitivity to state perturbations. In the HJB approach, the Hamiltonian (Eq. (3.5)) is defined by adjoining the dynamic equation (Eq. (3.2)) to the Lagrangian by $\partial V / \partial \mathbf{x}_k$.

A parametric structure $\tilde{c}$ with parameters a, the actor, is used to approximate the control law. Another parametric structure $\tilde{\lambda}$ with parameters ω, the critic, is used to approximate the derivative of the value function with respect to the state:

$$\tilde{c}(\mathbf{x}_k, \mathbf{a}_\ell) \approx c_\ell(\mathbf{x}_k), \tag{3.15}$$

$$\tilde{\lambda}(\mathbf{x}_k, \omega_\ell) \approx \frac{\partial V_\ell(\mathbf{x}_k, \tilde{c})}{\partial \mathbf{x}_k} \equiv \lambda_\ell(\mathbf{x}_k, \tilde{c}). \tag{3.23}$$

When the chosen parametric structures are differentiable, the optimality condition (Eq. (3.13)) can be used to generate a sequence of successively-improving control laws $\{c_\ell \mid \ell = 0, 1, 2, \ldots\}$.

The value-determination operation (Eq. (3.12)) is differentiated with respect to the state to obtain a recurrence relation for the function λ:

$$
\begin{aligned}
\frac{\partial V_{\ell+1}(\mathbf{x}_k, \mathbf{c})}{\partial \mathbf{x}_k} \equiv \lambda_{\ell+1}(\mathbf{x}_k, \mathbf{c}) = {} & \frac{\partial \mathcal{L}(\mathbf{x}_k, \mathbf{u}_k)}{\partial \mathbf{x}_k} + \left[\frac{\partial c(\mathbf{x}_k)}{\partial \mathbf{x}_k}\right]^T \frac{\partial \mathcal{L}(\mathbf{x}_k, \mathbf{u}_k)}{\partial \mathbf{u}_k} \\
& + \left[\frac{\partial \mathbf{f}(\mathbf{x}_k, \mathbf{u}_k)}{\partial \mathbf{x}_k}\right]^T \frac{\partial V_\ell(\mathbf{x}_{k+1}, \mathbf{c})}{\partial \mathbf{x}_{k+1}} \\
& + \left[\frac{\partial c(\mathbf{x}_k)}{\partial \mathbf{x}_k}\right]^T \left[\frac{\partial \mathbf{f}(\mathbf{x}_k, \mathbf{u}_k)}{\partial \mathbf{u}_k}\right]^T \frac{\partial V_\ell(\mathbf{x}_{k+1}, \mathbf{c})}{\partial \mathbf{x}_{k+1}}.
\end{aligned}
\tag{3.24}
$$

Equation (3.24) generates the sequence of successively-improving value function derivatives $\{\lambda_\ell \mid \ell = 0, 1, 2, \ldots\}$, as the control law sequence is generated. A DHP algorithm that, at every step ℓ, updates the actor parameters based on the improved control law, $c_{\ell+1}$, and updates the critic parameters based on the improved value-function derivative, $\lambda_{\ell+1}$, is given below.

Dual-Heuristic-Programming Actor Update   Suppose a control-law approximator $\tilde{c}(\cdot, \mathbf{a}_\ell)$ and a corresponding value-function-derivative approximator $\tilde{\lambda}(\cdot, \omega_\ell)$ are given. Then, an improved control-law approximator $\tilde{c}(\cdot, \mathbf{a}_{\ell+1})$ can be obtained by computing a desired control vector u_k^D, such that,

$$\left[\frac{\partial \mathcal{L}(\mathbf{x}_k, \mathbf{u}_k)}{\partial \mathbf{u}_k} + \left(\frac{\partial \mathbf{f}(\mathbf{x}_k, \mathbf{u}_k)}{\partial \mathbf{u}_k}\right)^T \tilde{\lambda}(\mathbf{x}_{k+1}, \omega_\ell)\right]\Bigg|_{\mathbf{u}_k = \mathbf{u}_k^D} = 0 \qquad (3.25)$$

and the corresponding Hessian matrix is positive definite.

The improved actor parameters, $a_{\ell+1}$, are computed as follows:

$$a_{\ell+1} = \arg\min_{\mathbf{a}}\{\mathbf{u}_k^D - \tilde{c}(\mathbf{x}_k, \mathbf{a})\}, \qquad (3.26)$$

where $\mathbf{u}_k^D \equiv c_{\ell+1}(\mathbf{x}_k)$, therefore $\tilde{c}(\mathbf{x}_k, a_{\ell+1}) \approx c_{\ell+1}(\mathbf{x}_k)$.

Dual-Heuristic-Programming Critic Update   Given the improved control-law approximator $\tilde{c}(\cdot, a_{\ell+1})$ and the value-function-derivative approximator $\tilde{\lambda}(\cdot, \omega_\ell)$, an improved value-function-derivative approximator $\tilde{\lambda}(\cdot, \omega_{\ell+1})$ can be obtained by computing its desired value,

$$\lambda_k^D \equiv \lambda_{\ell+1}(\mathbf{x}_k, \tilde{c}) = \frac{\partial \mathcal{L}(\mathbf{x}_k, \mathbf{u}_k)}{\partial \mathbf{x}_k} + \left(\frac{\partial \tilde{c}(\mathbf{x}_k, a_{\ell+1})}{\partial \mathbf{x}_k}\right)^T \frac{\partial \mathcal{L}(\mathbf{x}_k, \mathbf{u}_k)}{\partial \mathbf{u}_k}$$
$$+ \left[\frac{\partial \mathbf{f}(\mathbf{x}_k, \mathbf{u}_k)}{\partial \mathbf{x}_k} + \frac{\partial \mathbf{f}(\mathbf{x}_k, \mathbf{u}_k)}{\partial \mathbf{u}_k}\frac{\partial \tilde{c}(\mathbf{x}_k, a_{\ell+1})}{\partial \mathbf{x}_k}\right]^T \tilde{\lambda}(\mathbf{x}_{k+1}, \omega_\ell), \quad (3.27)$$

with $\mathbf{u}_k = \tilde{c}(\mathbf{x}_k, a_{\ell+1})$. The improved critic parameters are determined by solving,

$$\omega_{\ell+1} = \arg\min_{\omega}\{\lambda_k^D - \tilde{\lambda}(\mathbf{x}_k, \omega)\}, \qquad (3.28)$$

such that, $\tilde{\lambda}(\mathbf{x}_k, \omega_{\ell+1}) \approx \lambda_{\ell+1}(\mathbf{x}_k, \tilde{c})$.

In the next section, a modular approach for implementing HDP and DHP is presented.

3.3.4 Implementation of the Heuristic and Dual-Heuristic Dynamic Programming Algorithms

Adaptive critic algorithms involve multiple computational levels that are conveniently interpreted by means of a modular approach. Individual modules can be modified independently of one another such that algorithmic changes and debugging can be performed quickly and reliably. The key modules and their characteristics summarized in Table 3.1, are of two types: *functional modules* and *algorithmic modules.*

A functional module is a parametric structure whose inputs and outputs are those of the mathematical function being represented. The module's structure is fixed, and the values of the parameters determine the shape of the function. Hence, each module inherits the subscript of the corresponding parameters (e.g., the actor inherits

Table 3.1 Functional and Algorithmic Modules and Their Characteristics within Heuristic Dynamic Programming (HDP) and Dual Heuristic Programming (DHP) Architectures

Module:	Implementation of:	Inputs:	Outputs:	Architecture:
Actor$_\ell$	$\widetilde{c}(\cdot, a_\ell)$	x_k	$\approx c_\ell(\cdot)$	Both
Critic$_\ell$	$\widetilde{V}(\cdot, w_\ell)$	x_k	$\approx V_\ell(\cdot)$	HDP
	$\widetilde{\lambda}(\cdot, \omega_\ell)$	x_k	$\approx \lambda_\ell(\cdot)$	DHP
Parameter update	Function approximation	e, p, and $\widetilde{g}(\cdot, v_\ell)$	$\widetilde{g}(\cdot, v_{\ell+1})$	Both
Actor target	Policy-improvement routine	$x_k, \widetilde{c}(\cdot, a_\ell)$, and $\widetilde{V}(\cdot, w_\ell)$	u_k^D	HDP
		$x_k, \widetilde{c}(\cdot, a_\ell)$, and $\widetilde{\lambda}(\cdot, \omega_\ell)$	u_k^D	DHP
Critic target	Value-determination operation	$x_k, \widetilde{c}(\cdot, a_{\ell+1})$, and $\widetilde{V}(\cdot, w_\ell)$	V_k^D	HDP
		$x_k, \widetilde{c}(\cdot, a_{\ell+1})$, and $\widetilde{\lambda}(\cdot, \omega_\ell)$	λ_k^D	DHP

the subscript of the a parameters). The features of the parametric structure can be made accessible to and modified by other modules. In most programming languages this is easily achieved, for example by making these features *global variables* [33]. The actor and the critic constitute the key functional modules.

An algorithmic module may involve one or more iterations whose purpose is to update parameters, solve the optimality condition, and compute the value function or its derivatives. The details of each module depend on the adaptive critic design and parametric structures. Table 3.1 shows how the properties of each module depend on the chosen architecture. Using this notation, both the HDP and DHP algorithms can be described by the diagram in Figure 3.5. The solid lines illustrate the flow of input and output information between the modules. The dashed lines in Figure 3.5 indicate that the updated functional modules replace the preceding ones. The modules' inputs and outputs are listed in Table 3.1.

3.3.4.1 *Parameter-Update Module for Function Approximation*

The actor and the critic functional modules represent the corresponding parametric structures, as illustrated in Table 3.1. These structures may consist of neural networks, polynomials, splines [34], or any other differentiable mappings with adjustable parameters. The

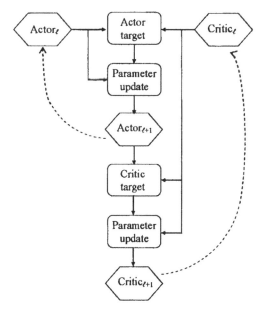

Fig. 3.5 Modular description of the $\ell'th$ iteration cycle in the HDP and DHP algorithms.

values of the parameters are adjusted by the parameter-update module during every cycle, ℓ, of the adaptive critic algorithm. For example, in the case of a neural network this module represents a training algorithm such as backpropagation [35]. In temporal-difference methods it represents the learning algorithm for updating the weight vector of the chosen basis functions.

In general, the functions in a sequence $\{g_\ell(\mathbf{p})\,|\,\ell = 0, 1, 2, \ldots\}$ must be successively approximated by a parametric structure, $\tilde{g}(\mathbf{p}, v)$, with input $\mathbf{p}$ and adjustable parameters v. The functions may comprise control laws or value-related functions, depending on which functional module is being updated. Eventually, the sequence converges to the optimal function $\mathbf{g}^*(\mathbf{p}) \approx \tilde{g}(\mathbf{p}, v)$. Prior to this, $\mathbf{g}_{\ell+1}(\mathbf{p})$ represents an estimate of $\mathbf{g}^*(\mathbf{p})$ that has been improved with respect to the preceding function $g_\ell(\mathbf{p})$ during the iteration cycle ℓ.

The parameter-update module determines the value of the parameters $v_{\ell+1}$ for which $\tilde{g}(\mathbf{p}, v_{\ell+1})$ most closely approximates $g_{\ell+1}(\mathbf{p})$. The parameters are initialized with either random variables or prior values. In subsequent cycles, an estimate of the parameters already is available in v_ℓ. This estimate can be improved by minimizing a specified error, denoted in general by e, with respect to the parameter values. The improved parameter values, $v_{\ell+1}$, replace v_ℓ in the updated structure $\tilde{g}(\mathbf{p}, v_{\ell+1})$ (as illustrated in Figure 3.6). Every error-minimization process may involve one or more iterations that are referred to as *epochs*. These are indexed by i to distinguish them from the adaptive critic iteration cycles, indexed by ℓ.

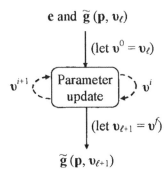

Fig. 3.6   Conceptual illustration of function approximation by the parameter-update module.

The quantities represented by e, g, and p depend on the context in which the parameter-update module is used. Therefore, the error to be minimized, e, and the parametric structure to be updated, $\tilde{g}(p, v_\ell)$, are inputs to the parameter-update module. For example, when the module is implemented to update the parameters a of the actor $\tilde{c}(x_k, a)$ according to Eq. (3.18), the error to be minimized is $\{u_k^D - \tilde{c}(x_k, a)\}$. If the actor parameters were updated according to Eq. (3.21), the error to be minimized would be given by $\partial \tilde{V} / \partial u_k$. When the module is used to update the parameters of an HDP critic, e consists of the term inside the curly brackets in Eq. (3.20).

3.3.4.2 Actor-Target Module for Policy Improvement   The actor-target module produces an improved control policy by solving the optimality condition for a desired control vector u_k^D. The calculations to be performed depend on the chosen architecture. They are illustrated here for the Heuristic Dynamic Programming (HDP) and the Dual Heuristic Dynamic Programming (DHP) algorithms derived in Section 3.3.3. Depending on the form of the dynamic equation (Eq. (3.2)) and on the chosen parametric structures, the optimality condition (Eq. (3.13)) leads to a set of simultaneous equations that are either linear or nonlinear. In the latter case, an iterative approach for solving nonlinear equations (e.g., a least-squares method) can be employed. The inputs to this module are the present state of the system and the latest actor and critic functionals (Table 3.1).

At every iteration ℓ, the HDP optimality condition reduces to Eq. (3.17) for the given control-law and value-function approximations, $\tilde{c}(\cdot, a_\ell)$ and $\tilde{V}(\cdot, w_\ell)$. These approximations are provided by the latest actor and critic functional modules, as shown in Figure 3.5. In on-line implementations, x_k is known from the actual state of the system observed at the present time, t_k. Then, the optimality condition can be implemented by the following actor-target module:

HDP actor-target module {

   Given $\mathbf{x}_k$, $\tilde{c}(\cdot, \mathbf{a}_\ell)$, and $\tilde{V}(\cdot, \mathbf{w}_\ell)$

   Initialize $\mathbf{u}_k^D$ by a guess or previous estimate

   while $\partial \tilde{V}/\partial \mathbf{u}_k|_{\mathbf{u}_k = \mathbf{u}_k^D} \neq 0$, update $\mathbf{u}_k^D$ and evaluate Eq. (3.17) by computing:

   $\quad \partial \mathcal{L}(\mathbf{x}_k, \mathbf{u}_k)/\partial \mathbf{u}_k|_{\mathbf{u}_k^D}$, from the derivative of the Lagrangian

   $\quad \mathbf{x}_{k+1} = \mathbf{f}(\mathbf{x}_k, \mathbf{u}_k)|_{\mathbf{u}_k^D}$

   $\quad \partial \tilde{V}(\mathbf{x}_{k+1}, \mathbf{w}_\ell)/\partial \mathbf{x}_{k+1}|_{\mathbf{u}_k^D}$, from the derivative of $\tilde{V}(\cdot, \mathbf{w}_\ell)$ and from $\mathbf{x}_{k+1}$

   $\quad \partial \mathbf{f}(\mathbf{x}_k, \mathbf{u}_k)/\partial \mathbf{u}_k|_{\mathbf{u}_k^D}$ (from the derivative of Eq. (3.2))

   end while

   Check that Hessian is positive definite at $\mathbf{u}_k^D$

}

The derivatives in the *while* loop can be obtained analytically or numerically, depending on the form of the governing equation (Eq. (3.2)), on the Lagrangian $\mathcal{L}(\cdot)$ (in Eq. (3.1)), and on the critic $\tilde{V}(\cdot)$. They are evaluated at $\mathbf{x}_k$ using the latest estimate $\mathbf{u}_k = \mathbf{u}_k^D$. Finally, $\mathbf{u}_k^D$ must satisfy the convexity condition in Eq. (3.14).

In the DHP algorithm the actor-target module implements the same policy-improvement routine, Eq. (3.13). However, the derivative of the value function with respect to the state is approximated by the critic, so the critic module need not be differentiated.

DHP actor-target module {

   Given $\mathbf{x}_k$, $\tilde{c}(\cdot, \mathbf{a}_\ell)$, and $\tilde{\lambda}(\cdot, \omega_\ell)$

   Initialize $\mathbf{u}_k^D$ by a guess or previous estimate

   while Eq. (3.25) is not satisfied, update $\mathbf{u}_k^D$ and re-evaluate Eq. (3.25) by computing:

   $\quad \partial \mathcal{L}(\mathbf{x}_k, \mathbf{u}_k)/\partial \mathbf{u}_k|_{\mathbf{u}_k^D}$, from the derivative of the Lagrangian

   $\quad \mathbf{x}_{k+1} = \mathbf{f}(\mathbf{x}_k, \mathbf{u}_k)|_{\mathbf{u}_k^D}$

   $\quad \tilde{\lambda}(\mathbf{x}_{k+1}, \omega_\ell)$, from the critic $\tilde{\lambda}(\cdot, \omega_\ell)$ and $\mathbf{x}_{k+1}$

   $\quad \partial \mathbf{f}(\mathbf{x}_k, \mathbf{u}_k)/\partial \mathbf{u}_k|_{\mathbf{u}_k^D}$ (from the derivative of Eq. (3.2))

   end while

   check that Hessian is positive definite at $\mathbf{u}_k^D$

}

The static minimization problem solved by the actor-target module (Eq. (3.13)) has the same dimension (m) as the control vector. Its computational difficulty is problem dependent. In the presence of multiple minima, an appropriate numerical search must be performed to ensure that $\mathbf{u}_k^D$ is the *global* minimum of $\tilde{V}(\cdot)$.

3.3.4.3 *Critic-Target Module for Value Determination*

The critic-target module computes an improved value function or value-function derivative for the present state, $\mathbf{x}_k$, according to the value-determination operation. The value of the improved function, denoted by the superscript $(\cdot)^{\mathrm{D}}$, can be used to update the critic parameters by means of the parameter-update module. Since the input/output characteristics of the critic depend on the chosen architecture, so do the computations performed by the critic-target module. In this section, the critic-target module is illustrated for the HDP and DHP algorithms derived in Section 3.3.3.

In the HDP algorithm, a critic-update routine is obtained by combining the original value-determination operation (Eq. (3.12)) with function approximation. The critic-target module computes the improved value V_k^{D} based on $\mathbf{x}_k$ and on the latest actor and critic approximations, $\tilde{c}(\cdot, a_{\ell+1})$ and $\tilde{V}(\cdot, \mathbf{w}_\ell)$.

HDP critic-target module {

    Given $\mathbf{x}_k$, $\tilde{c}(\cdot, a_{\ell+1})$, and $\tilde{V}(\cdot, \mathbf{w}_\ell)$

    Evaluate V_k^{D} from Eq. (3.19) by computing:

        $\mathcal{L}(\mathbf{x}_k, \mathbf{u}_k)$, using $\mathbf{u}_k = \tilde{c}(\mathbf{x}_k, a_{\ell+1})$

        $\mathbf{x}_{k+1} = \mathbf{f}(\mathbf{x}_k, \mathbf{u}_k)$, using $\mathbf{u}_k = \tilde{c}(\mathbf{x}_k, a_{\ell+1})$

        $\tilde{V}(\mathbf{x}_{k+1}, \mathbf{w}_\ell)$, from the critic $\tilde{V}(\cdot, \mathbf{w}_\ell)$ and $\mathbf{x}_{k+1}$

    }

The above algorithm implies that the actor parameters already have been updated from a_ℓ to $a_{\ell+1}$ according to the HDP actor update (Section (3.3.3)).

Similarly, in DHP the critic-target module computes the desired value-function derivative, λ_k^{D}, according to the critic-update operation. As before, λ_k^{D} is computed from the present state of the system, $\mathbf{x}_k$, and from the latest actor and critic structures, $\tilde{c}(\cdot, a_{\ell+1})$ and $\tilde{\lambda}(\cdot, \omega_\ell)$. Then, the critic parameters can be updated from ω_ℓ to $\omega_{\ell+1}$ by the parameter-update module. The DHP critic-target module implements a recurrence relation (Eq. (3.27)) that is obtained by differentiating the original value-determination operation (Eq. (3.12)).

DHP critic-target module {

    Given $\mathbf{x}_k$, $\tilde{c}(\cdot, a_{\ell+1})$, and $\tilde{\lambda}(\cdot, \omega_\ell)$

    Evaluate λ_k^{D} from Eq. (3.27) by computing:

        $\partial\mathcal{L}(\mathbf{x}_k, \mathbf{u}_k)/\partial\mathbf{u}_k$ and $\partial\mathcal{L}(\mathbf{x}_k, \mathbf{u}_k)/\partial\mathbf{x}_k$, from the derivatives of the Lagrangian

        Using $\mathbf{u}_k = \tilde{c}(\mathbf{x}_k, a_{\ell+1})$

        $\partial\mathbf{f}(\mathbf{x}_k, \mathbf{u}_k)/\partial\mathbf{u}_k$ and $\partial\mathbf{f}(\mathbf{x}_k, \mathbf{u}_k)/\partial\mathbf{x}_k$ (from the derivatives of Eq.(3.2)),

        Using $\mathbf{u}_k = \tilde{c}(\mathbf{x}_k, a_{\ell+1})$

        $\mathbf{x}_{k+1} = \mathbf{f}(\mathbf{x}_k, \mathbf{u}_k)$, using $\mathbf{u}_k = \tilde{c}(\mathbf{x}_k, a_{\ell+1})$

        $\tilde{\lambda}(\mathbf{x}_{k+1}, \omega_\ell)$, from the critic $\tilde{\lambda}(\cdot, \omega_\ell)$ and $\mathbf{x}_{k+1}$

        $\partial\tilde{c}(\mathbf{x}_k, a_{\ell+1})/\partial\mathbf{x}_k$, from the derivative of the actor $\tilde{c}(\cdot, a_{\ell+1})$

    }

The derivatives in the critic-target module can be obtained analytically or numerically, depending on the form of the governing equation (Eq. (3.2)), on the Lagrangian $\mathcal{L}(\cdot)$

(in Eq. (3.1)), and on the actor $\tilde{c}(\cdot)$. All of these quantities are evaluated at the present state x_k and at the control value produced by the updated actor, that is, $u_k = \tilde{c}(x_k, a_{\ell+1})$.

3.3.5   Example: Linear Quadratic Problem

Linear-quadratic (LQ) problems [1] involve a linear dynamic equation and a quadratic cost function. For this class of optimal control problems the solution can be computed by means of a *matrix Riccati Equation* [1]. Their solution by an adaptive critic approach is demonstrated to illustrate the concepts introduced in the previous sections.

A quadratic cost function of the form,

$$J = \lim_{t_f \to \infty} \frac{1}{2} \sum_{t_k = t_0}^{t_f - 1} [x_k^T Q x_k + u_k^T R u_k], \tag{3.29}$$

is to be minimized subject to the *linear time-invariant* (LTI) dynamic equation

$$x_{k+1} = F x_k + G u_k. \tag{3.30}$$

By assuming that the optimal value function is quadratic,

$$V^*(x_k^*) = \frac{1}{2} x_k^{*T} P x_k^*, \tag{3.31}$$

it can be proven [13] that the LQ optimal control law has the form,

$$u_k^* = -C(x_k^* - x_R), \tag{3.32}$$

where x_R represents the reference state value. The problem's objective is to determine the matrices C and P that bring the state to the value x_R, while optimizing Eq. (3.29) subject to Eq. (3.30).

Typically, the gain matrix C is obtained from the Riccati matrix P after solving a Riccati equation [1]. Here, we determine the optimal control and value function using the DHP approach. The functions to be calculated are the control law, c, and the value function derivative, λ. From Eqs. (3.31) and (3.32), it can be deduced that polynomials constitute suitable approximating structures for these functions. By considering the elements of C as the actor parameters (a) and the elements of P as the critic parameters (ω), the DHP actor and critic are given by:

$$\tilde{c}(x_k, a_\ell) = -C_\ell(x_k - x_R), \tag{3.33}$$

$$\hat{\lambda}(x_k, \omega_\ell) = P_\ell x_k. \tag{3.34}$$

Q, R, and P_ℓ are symmetric, positive-definite matrices. The DHP actor and critic updates (Section 3.3.3) are used to update the elements of C_ℓ and P_ℓ at every cycle, ℓ, of the adaptive critic algorithm.

The LQ problem can be solved on-line by using the observed value of the state x_k during the ℓth cycle of the DHP algorithm (letting $\ell = k$). After the actor and critic parameters are initialized to C_0 and P_0 the iterative process begins. Every cycle comprises an actor update followed by a critic update. During the ℓth cycle the DHP actor update is carried out for the given x_k, $\tilde{c}_\ell$, and $\tilde{\lambda}_\ell$. The DHP actor-target module (Section 3.3.4.2) computes u_k^D from Eq. (3.25), which takes the form

$$Ru_k^D + G^T P_\ell(Fx_k + Gu_k^D) = 0. \tag{3.35}$$

The new actor parameters $C_{\ell+1}$ are determined by the parameter-update module (Section 3.3.4.1). According to Eq. (3.26), this can be achieved by minimizing the error $\{u_k^D + C(x_k - x_R)\}$ with respect to C.

During the same cycle, ℓ, the DHP critic update is carried out for the given x_k, $\tilde{c}_{\ell+1}$, and $\tilde{\lambda}_\ell$. The DHP critic-target module (Section 3.3.4.3) computes λ_k^D from Eq. (3.27), which takes the form

$$\lambda_k^D = Qx_k + C_{\ell+1}^T RC_{\ell+1}x_k + [F + GC_{\ell+1}]^T P_\ell(F - GC_{\ell+1})x_k. \tag{3.36}$$

The new critic parameters $P_{\ell+1}$ are determined by the parameter-update module by minimizing the error $\{\lambda_k^D - Px_k\}$ with respect to P.

In this simple example the parameter-update module performs a linear minimization to adjust the actor and the critic functionals (Eqs. (3.33) and (3.34)). In both cases, an error with the form $\{b - Ax\}$ is to be minimized with respect to A. Since there are more unknowns (elements in A) than there are equations (error elements), the problem is underdetermined. Hence, even in this simple case there exist multiple alternatives for the design of the parameter-update module (including least-squares methods). This usually is the most challenging aspect of the implementation and is likely to affect convergence greatly.

LQ problems are a class of control problems for which there exist more effective methods of solution than adaptive critic designs. This example is offered mainly for illustrative purposes. Nevertheless, if the actual system dynamics were to differ from Eq. (3.30) while retaining this linear form, the adaptive critic approach described here could be implemented to perfect the control law (Eq. (3.33)) on-line. Thus, a similar method has been applied to aircraft optimal control in [40, 41].

3.4 DISCUSSION

This chapter focuses on on-line implementations of adaptive critic designs, where the actor and the critic are updated incrementally every time a state value is observed. Clearly, this approach is not always the most effective way to solve an optimal control problem. For example, when uncertainties and plant dynamics satisfy appropriate assumptions, other designs can be implemented more easily and effectively. Adaptive critic designs present many challenges, including intricate implementation details and lack of performance guarantees. Nevertheless, they remain an important area of research with many promises and accomplishments. Their most significant feature is that they can be used to develop control systems that can *learn* to improve their performance over time subject to the actual plant dynamics.

Because the behavior of most plants cannot be fully anticipated *a priori*, adaptive critic designs have been investigated for a wide range of applications. The DHP method has been successfully implemented to prevent cars from skidding when driving over unexpected patches of ice [36]. DHP-based controllers have been developed for missile interception [37] and for aircraft auto-landing [38]. Recently, adaptive critic designs have been used to replace the automatic voltage generator and the turbine governor of a turbogenerator connected to the power grid [24]. In this research, DHP is shown to converge faster than HDP, especially in cases where the system's dynamics and operating conditions change unexpectedly. Also, DHP has been used to improve the performance of an aircraft controller on-line in the presence of unmodelled dynamics, control failures, and parameter variations [39]. These numerical simulations show that learning can prevent loss of control during critical maneuvers while they are experienced by the controller for the first time. Another possible application of adaptive critic designs is on-line state estimation, which can be formulated as an optimal control problem (Section 3.2.1), where the control, u, is replaced by the estimate of the state.

3.5 SUMMARY

This chapter provides an introduction to the field of model-reference adaptive critic designs. These methods seek optimality in the most general case. Their use for the on-line approximate solution of infinite-horizon optimal control problems is emphasized here. Section 3.2 presents basic background material, including Howard's adaptive critic algorithm or iteration cycle, which forms the basis of most ACD. The cycle comprises a policy-improvement routine and a value-determination operation that can be used in sequence to improve the approximate solution over time. By combining this basic iteration cycle with function approximation a variety of adaptive critic designs can be obtained, as described in Section 3.3. The Heuristic and Dual-Heuristic Dynamic Programming (HDP and DHP) algorithms are derived in this section. Both algorithms can be implemented by a streamlined modular approach. This is based on two functional modules, the actor and the critic, and on three algorithmic modules, the parameter update, the actor target, and the critic target. Adaptive critic designs have been applied to a variety of problems, ranging from control of automobiles to regulation of turbogenerators. The DHP approach is demonstrated in Section 3.3.5 for a simple linear-quadratic problem. A purpose of this chapter is to motivate the further exploration of ACD for the control and estimation of complex systems.

Appendix: Proof of Convergence for the Adaptive Critic Algorithm

The following proof has been adapted from [5]. Given a generic control law $u_k = c(x_k)$, the following recurrence relation can be found by inspection from Eqs.

(3.4) and (3.7)

$$V(\mathbf{x}_k, \mathbf{c}) = \mathcal{L}(\mathbf{x}_k, \mathbf{u}_k) + V(\mathbf{x}_{k+1}, \mathbf{c}), \tag{3.37}$$

where the value function V is not necessarily optimal. For a process with a very large number of stages (or time increments), the total cost J in Eq. (3.1) can grow without limit as t_f tends to infinity. Therefore, it is convenient to define an *average total cost* corresponding to the control law c, that has the same optimality conditions as Eq. (3.1)

$$\bar{J}_\mathbf{c} = \lim_{t_f \to \infty} \frac{1}{t_f} \sum_{t_k=t_0}^{t_f-1} \mathcal{L}[\mathbf{x}(t_k), \mathbf{u}(t_k)]. \tag{3.38}$$

Also, as t_f approaches infinity in the limit, it is reasonable to let the terminal cost $\varphi[\mathbf{x}(t_f)]$ equal zero [1]. Subsequently, a *relative value function* that has value in proving convergence of the adaptive critic algorithm can be defined as

$$\upsilon_\mathbf{c}(\mathbf{x}_k) = \lim_{t_f \to \infty} \sum_{t_k=t_0}^{t_f-1} [\mathcal{L}(\mathbf{x}_k, \mathbf{c}) - \bar{J}_\mathbf{c}], \quad \mathbf{x}(t_0) = \mathbf{x}(t_k), \tag{3.39}$$

such that,

$$\bar{J}_\mathbf{c} + \upsilon_\mathbf{c}(\mathbf{x}_k) = \mathcal{L}(\mathbf{x}_k, \mathbf{u}_k) + V(\mathbf{x}_{k+1}, \mathbf{c}). \tag{3.40}$$

The relative value function differs from the original one by a constant that, under a given policy, can be eliminated by considering the difference in relative values, for example, $\upsilon_\mathbf{c}(\mathbf{x}_d) - \upsilon_\mathbf{c}(\mathbf{x}_e)$, representing the additional cost caused by starting at state d, rather than at state e.

Suppose that a control law $\mathbf{c}_A$ has been evaluated for the operation of the system, but the adaptive critic algorithm has produced a control law $\mathbf{c}_B \neq \mathbf{c}_A$. Then, property 1 of the algorithm can be proven by showing that the new control law has a smaller average cost, that is, $\bar{J}_B \leq \bar{J}_A$, or $J^\Delta \equiv (\bar{J}_A - \bar{J}_B) \geq 0$. Since, Eqs. (3.37) and (3.40) can be written solely in terms of $\mathbf{x}_k$ and $\mathbf{u}_k$ through Eq. 3.2, the subscript k can be omitted and the equations written with respect to $\mathbf{x}$, implying a generic moment in time. If the control law $\mathbf{c}_B$ was chosen over $\mathbf{c}_A$ by the policy-improvement routine, then the following relationship applies:

$$\mathcal{L}(\mathbf{x}, \mathbf{u}_B) + V[\mathbf{f}(\mathbf{x}, \mathbf{u}_B), \mathbf{c}_A] \leq \mathcal{L}(\mathbf{x}, \mathbf{u}_A) + V[\mathbf{f}(\mathbf{x}, \mathbf{u}_A), \mathbf{c}_A], \tag{3.41}$$

and the following positive function can be defined for all $\mathbf{x}$:

$$\gamma(\mathbf{x}) \equiv \mathcal{L}(\mathbf{x}, \mathbf{u}_A) + V[\mathbf{f}(\mathbf{x}, \mathbf{u}_A), \mathbf{c}_A] - \mathcal{L}(\mathbf{x}, \mathbf{u}_B) - V[\mathbf{f}(\mathbf{x}, \mathbf{u}_B), \mathbf{c}_A] \geq 0. \tag{3.42}$$

Equation (3.40) is applied to both control laws individually, as if they were each used to control the process from t_0 to t_f:

$$\bar{J}_A + \upsilon_A(\mathbf{x}) = \mathcal{L}(\mathbf{x}, \mathbf{u}_A) + V[\mathbf{f}(\mathbf{x}, \mathbf{u}_A), \mathbf{c}_A], \tag{3.43}$$
$$\bar{J}_B + \upsilon_B(\mathbf{x}) = \mathcal{L}(\mathbf{x}, \mathbf{u}_B) + V[\mathbf{f}(\mathbf{x}, \mathbf{u}_B), \mathbf{c}_B]. \tag{3.44}$$

When the above relations are subtracted from one another they produce

$$
\begin{aligned}
\bar{J}_A - \bar{J}_B + v_A(\mathbf{x}) - v_B(\mathbf{x}) =\ & \mathcal{L}(\mathbf{x}, \mathbf{u}_A) - \mathcal{L}(\mathbf{x}, \mathbf{u}_B) \\
& + V[\mathbf{f}(\mathbf{x}, \mathbf{u}_A), \mathbf{c}_A] - V[\mathbf{f}(\mathbf{x}, \mathbf{u}_B), \mathbf{c}_B].
\end{aligned}
\tag{3.45}
$$

which simplifies to

$$
\bar{J}_A - \bar{J}_B + v_A(\mathbf{x}) - v_B(\mathbf{x}) = \gamma(\mathbf{x}) + V[\mathbf{f}(\mathbf{x}, \mathbf{u}_B), \mathbf{c}_A] - V[\mathbf{f}(\mathbf{x}, \mathbf{u}_B), \mathbf{c}_B], \tag{3.46}
$$

by means of Eq. (3.42). By defining the quantities $v^\Delta(\mathbf{x}) \equiv v_A(\mathbf{x}) - v_B(\mathbf{x})$ and $V^\Delta[\mathbf{f}(\mathbf{x}, \mathbf{u}_B)] \equiv V[\mathbf{f}(\mathbf{x}, \mathbf{u}_B), \mathbf{c}_A] - V[\mathbf{f}(\mathbf{x}, \mathbf{u}_B), \mathbf{c}_B]$, Eq. (3.46) can be written as

$$
\bar{J}^\Delta + v^\Delta(\mathbf{x}) = \gamma(\mathbf{x}) + V^\Delta[\mathbf{f}(\mathbf{x}, \mathbf{u}_B)]. \tag{3.47}
$$

This relationship takes the same form as Eq. (3.40) with the average cost $\bar{J}_c$ given by Eq. (3.38). Thus, Eq. (3.47) can be solved similarly for $\bar{J}^\Delta$,

$$
\bar{J}^\Delta = \lim_{t_f \to \infty} \frac{1}{t_f} \sum_{t_k = t_0}^{t_f - 1} \gamma(\mathbf{x}_k), \tag{3.48}
$$

where $\gamma(\mathbf{x}_k) \geq 0$ for all $\mathbf{x}_k$. It follows that $\bar{J}^\Delta \geq 0$ and that the adaptive critic algorithm has improved the control law producing a functional that has better overall performance. This proves property 1 of the algorithm.

During every cycle, the value-determination operation updates the value functional from V_ℓ to $V_{\ell+1}$, for a given control law produced by the policy-improvement routine. Suppose during the last cycle the policy-improvement routine has chosen the control law $\mathbf{c}_B$ over $\mathbf{c}_A$, the control law from the previous cycle. Then, the following relationship must apply for all $\mathbf{x}$:

$$
V_\ell(\mathbf{x}, \mathbf{c}_B) \leq V_\ell(\mathbf{x}, \mathbf{c}_A). \tag{3.49}
$$

The value functional is updated by the value-determination operation according to the relationship:

$$
V_{\ell+1}(\mathbf{x}, \mathbf{c}_B) = \mathcal{L}(\mathbf{x}, \mathbf{u}_B) + V_\ell[\mathbf{f}(\mathbf{x}, \mathbf{u}_B), \mathbf{c}_B]. \tag{3.50}
$$

Thus the following inequality can be found to hold from Eqs. (3.37) and (3.49):

$$
V_{\ell+1}(\mathbf{x}, \mathbf{c}_B) = V_\ell(\mathbf{x}, \mathbf{c}_B) \leq V_\ell(\mathbf{x}, \mathbf{c}_A). \tag{3.51}
$$

The above equation shows that the value functional obtained during this last cycle, $V_{\ell+1}(\mathbf{x}, \mathbf{c}_B)$, is improved with respect to the one obtained during the previous cycle, $V_\ell(\mathbf{x}, \mathbf{c}_A)$. The equality represents the case in which the policy improvement routine has found two equally-good control laws, $\mathbf{c}_B$ and $\mathbf{c}_A$, such that $V_\ell(\mathbf{x}, \mathbf{c}_B) = V_\ell(\mathbf{x}, \mathbf{c}_A)$, but $\mathbf{c}_B \neq \mathbf{c}_A$. When this situation arises, the algorithm keeps the old control law, c_A, and value functional, $V_\ell(\mathbf{x}, c_A)$, as mentioned in Section 3.2.2. This proves property 2 of the adaptive critic algorithm. This property also implies that if the latest control law, c_B, were used for the remainder of the process, the cost that would accrue over all future times, $V_{\ell+1}(\mathbf{x}, c_B)$, would be lower than the cost that would have accrued by using the results from the previous cycle, that is, c_A and $V_\ell(\mathbf{x}, c_A)$.

Property 3 of the adaptive critic algorithm is proven by contradiction. Suppose a superior control law exists, call it c_A, but the algorithm has converged on a different functional, c_B, with a larger average cost, that is, $\bar{J}_A \leq \bar{J}_B$. It follows from Eq. (3.42) that $\gamma(\mathbf{x}) \leq 0$ for all $\mathbf{x}$. But, then, Eq. (3.48) implies that $\bar{J}^\Delta \leq 0$, which contradicts the assumption $\bar{J}_A \leq \bar{J}_B$. Thus, if the algorithm converges on a control law, it means that it is optimal, for if there were a superior control law it would be discovered before termination.

Property 4 can be deduced by observing that the converging sequences can be generated successively over time, by letting $\ell = k$.

Bibliography

1. R. F. Stengel, *Optimal Control and Estimation,* Dover Publications, New York, 1994.

2. Y. Bar-Shalom and E. Tse, Dual effect, certainty equivalence, and separation in stochastic control, *IEEE Trans. Automatic Control,* vol. 19, no. 5, pp. 494–500, 1974.

3. R. Bellman, *Dynamic Programming,* Princeton University Press, Princeton, NJ 1957.

4. R. Bellman and R. Kalaba, *Dynamic Programming and Modern Control Theory,* Academic Press, New York, 1965.

5. R. Howard, *Dynamic Programming and Markov Processes,* MIT Press, Cambridge, MA, 1960.

6. A. A. Feldbaum, *Optimal Control Systems,* Academic Press, New York, 1965.

7. E. Tse, Y. Bar-Shalom, and L. Meier, III, Wide-sense adaptive dual control for nonlinear stochastic systems, *IEEE Trans. Automatic Control,* vol. 18, no. 2, pp. 98–108, 1973.

8. Y. Bar-Shalom, Stochastic dynamic programming: caution and probing, *IEEE Trans. Automatic Control,* vol. 26, no. 5, pp. 1184–1195, 1981.

9. D. Q. Mayne and H. Michalska, Receding horizon control of non-linear systems, *IEEE Trans. Automatic Control,* vol. 35, no. 5, pp. 814–824, 1990.

10. G. C. Goodwin and K. S. Sin, *Adaptive Filtering Prediction and Control,* Prentice-Hall, Englewood Cliffs, NJ, 1984.

11. S. Ferrari and R. F. Stengel, Classical/neural synthesis of nonlinear control systems, *Journal of Guidance, Control and Dynamics,* vol. 25, no. 3, pp. 442–448, 2002.

12. S. Ferrari and R. F. Stengel, Algebraic training of a neural network, *Proc. 2001 American Control Conference,* Arlington, VA, pp. 1605–1610, 2001.

13. D. E. Kirk, *Optimal Control Theory: An Introduction*, Prentice-Hall, Englewood Cliffs, NJ, 1970.

14. R. E. Bellman and S. E. Dreyfus, *Applied Dynamic Programming*, Princeton University Press, Princeton, NJ, 1962.

15. R. Bellman, *Methods of Nonlinear Analysis: Volume II*, Academic Press, New York, 1973.

16. P. J. Werbos, Building and understanding adaptive systems: a statistical/numerical approach for factory automation and brain research, *IEEE Trans. Syst., Man, Cybern.*, vol. 17, no. 1, pp. 7–20, 1987.

17. D. P. Bertsekas, Distributed dynamic programming, *IEEE Trans. Automatic Control*, vol. 27, pp. 610–616, 1982.

18. S. Ferrari and R. F. Stengel, An adaptive critic global controller, *Proc. American Control Conference*, Anchorage, AK, 2002.

19. P. J. Werbos, Neurocontrol and Supervised Learning: An Overview and Evaluation, *Handbook of Intelligent Control*, D. A. White and D. A. Sofge (eds.), pp. 65–86, Van Nostrand Reinhold, New York, 1992.

20. V. Prokhorov and D. C. Wunsch, II, Adaptive critic designs, *IEEE Trans. Neural Networks*, vol. 8, no. 5, pp. 997–1007, 1997.

21. P. J. Werbos, A Menu of Designs for Reinforcement Learning Over Time, *Neural Networks for Control*, W. T. Miller, R. S. Sutton, and P. J. Werbos (eds.), pp. 67–96, MIT Press, Cambridge, MA, 1990.

22. P. J. Werbos, Advanced Forecasting Methods for Global Crisis Warning and Models of Intelligence, *General Systems Yearbook*, 1997.

23. G. G. Lendaris and T. Shannon, Application considerations for the DHP methodology, *Proc. International Joint Conference on Neural Networks*, Anchorage, AK, 1998.

24. G. K. Venayagamoorthy, R. G. Harley, and D. C. Wunsch, Comparison of heuristic dynamic programming and dual heuristic programming adaptive critics for neurocontrol of a turbogenerator, *IEEE Trans. Neural Networks*, vol. 13, no. 3, pp. 764–773, 2002.

25. A. Barto, R. Sutton, and C. Anderson, Neuronlike elements that can solve difficult learning control problems, *IEEE Trans. Systems, Man, and Cybernetics*, vol. 3, no. 5, pp. 834–846, 1983.

26. P. J. Werbos, Applications of advances in nonlinear sensitivity analysis, *System Modeling and Optimization: Proc. of the 10th IFIP Conference*, R. F. Drenick and F. Kozin (eds.), Springer-Verlag, New York, 1982.

27. C. Watkins, Learning from Delayed Rewards, Ph.D. Thesis, Cambridge University, Cambridge, England, 1989.

28. T. H. Wonnacott and R. Wonnacott, *Introductory Statistics for Business and Economics*, 2nd Ed., Wiley, New York, 1977.

29. R. E. Bellman and S. E. Dreyfus, Functional Approximation and Dynamic Programming, *Math. Tables and Other Aids Comp.*, Athena Scientific, Belmont, MA, 1995.

30. G. J. Gordon, Stable Function Approximation in Dynamic Programming, Technical Report CMU-CS-95-103, Carnegie Mellon University, Pittsburgh, 1995.

31. D. A. White and D. A. Sofge (eds.), *Handbook of Intelligent Control*, Van Nostrand Reinhold, New York, 1992.

32. G. Strang, *Linear Algebra and Its Applications*, 3rd Ed., Harcourt, Brace, Janovich, San Diego, 1988.

33. The MathWorks, Inc., *Getting Started with MATLAB*, http://www.mathworks.com, Version 5, September 1998.

34. J. N. Tsitsiklis and B. Van Roy, An analysis of temporal-difference learning with function approximation, *IEEE Trans. Automatic Control*, vol. 42, no. 5, pp. 674–690, 1997.

35. P. J. Werbos, Backpropagation through time: what it does and how to do it, *Proc. of the IEEE*, vol. 78, no. 10, pp. 1550–1560, 1990.

36. G. G. Lendaris, L. Schultz, and T. T. Shannon, Adaptive critic design for intelligent steering and speed control of a 2-axle vehicle, *Proc. International Joint Conference on Neural Networks*, Como, Italy, 2000.

37. D. Han and S. N. Balakrishnan, Adaptive critic based neural networks for control-constrained agile missile control, *Proc. American Control Conference*, San Diego, pp. 2600–2604, 1999.

38. G. Saini and S. N. Balakrishnan, Adaptive critic based neurocontroller for autolanding of aircraft, *Proc. American Control Conference*, Albuquerque, NM, pp. 1081–1085, 1997.

39. S. Ferrari, Algebraic and adaptive learning in neural control systems, Ph.D. Thesis, Princeton University, Princeton, NJ, 2002.

40. S. N. Balakrishnan and V. Biega, Adaptive-critic-based neural networks for aircraft optimal control, *Journal of Guidance, Control, and Dynamics*, vol. 19, no. 4, pp. 893–898, 1996.

41. K. KrishnaKumar and J. Neidhoefer, Immunized adaptive critics for level-2 intelligent control, *Proc. IEEE Int. Conf. Systems, Man and Cybernetics*, vol. 1, pp. 856–861, 1997.

Handbook of Learning and Approximate Dynamic Programming
Edited by Jennie Si, Andy Barto, Warren Powell and Donald Wunsch
Copyright © 2004 The Institute of Electrical and Electronics Engineers, Inc.

4 Guidance in the Use of Adaptive Critics for Control

GEORGE G. LENDARIS   JAMES C. NEIDHOEFER
Portland State University   Accurate Automation Corporation

Editor's Summary: This chapter, along with Chapter 3, provides an overview of several ADP design techniques. While Chapter 3 deals more with the theoretical foundations, Chapter 4 is more devoted to practical issues such as problem formulation and utility functions. The authors discuss issues associated with designing and training adaptive critics using the design techniques introduced in Chapter 3.

4.1 INTRODUCTION

The aim of this chapter is to provide guidance to the prospective user of Adaptive Critic / Approximate Dynamic Programming methods for designing the action device in certain kinds of control systems. While there are currently various different successful "camps" in the Adaptive Critic community spanning government, industry, and academia, and while the work of these independent groups may entail important differences, there are basic common threads. The latter include: Reinforcement Learning (RL), Dynamic Programming (DP), and basic Adaptive Critic (AC) concepts.

Describing and understanding the fundamental equations of DP is not difficult. Similarly, it is not difficult to show diagrams of different AC methodologies and understand conceptually how they work. However, understanding the wide variety of issues that crop up in actually applying the AC methodologies is both non-trivial and crucial to the success of the venture. Some of the important tasks include: formulating (appropriately) the problem-to-be-solved; defining a utility function that properly captures/embodies the problem-domain requirements; selecting the discount factor; designing the training "syllabus"; designing training strategies and selecting associated run-time parameters (epoch size, learning rates, etc.); deciding when to start and stop training; and, not the least, addressing stability issues. A brief overview of the three topics listed in the previous paragraph (RL, DP, and AC) is given first, followed by the main body of this chapter, in which selected issues important to the successful application of ACs are described, and some approaches to addressing them are presented.

Clearly, while much progress in the development and application of Adaptive Critics has already occurred, much remains to be done. The last section of the chapter describes some items the authors deem important be included in a future research agenda for the Adaptive Critic community.

4.2   REINFORCEMENT LEARNING

Reinforcement learning (RL) occurs when an agent learns behaviors through trial-and-error interactions with its environment, based on "reinforcement" signals from the environment. In the past ten to fifteen years, the potential of reinforcement learning has excited the imagination of researchers in the machine learning, intelligent systems, and artificial intelligence communities. Achievement of such potential, however, can be elusive, as formidable obstacles reside in the details of computational implementation.

In a general RL model, an agent interacts with its environment through sensors (perception) and actuators (actions) [6, 12]. Each interaction iteration typically includes the following: the agent receives inputs that indicate the state of the environment; the agent then selects and takes an action, which yields an output; this output changes the state of the environment, transitioning it to a "better" or a "worse" state; the latter are indicated to the agent by either a "reward" or a "penalty" from the environment, and the amount of such reward/penalty has the effect of a "reinforcement" signal to the agent. The behavior of a healthy agent tends to increase the reward part of the signal, over time, through a trial-and-error learning process. Thorndike's law of effect has been rephrased in [5, 56] to offer the following definition of reinforcement learning: "If an action taken by a learning system is followed by a satisfactory state of affairs, then the tendency of the system to produce that particular action is strengthened or reinforced. Otherwise, the tendency of the system to produce that action is weakened." Reinforcement learning differs from supervised-learning mainly in the kind of feedback received from the environment. In supervised-learning, the equivalent of a "teacher" function is available that knows the correct output, *a priori*, for each of the agent's outputs, and training/learning is based on output error data. In RL, on the other hand, the agent only receives a more general, composite reward/punish signal, and learns from this using an operating principle of increasing the amount of reward it receives over time. While RL has been implemented in a variety of different ways and has involved other related research areas (e.g., search and planning), in this chapter, we focus on application of the RL ideas to implementing approximate Dynamic Programming, often called Adaptive Critics. We comment that the phrase 'Adaptive Critic' was originally coined by Widrow, [61], in a manner that implies learning *with* a critic. The present authors, on the other hand, prefer that the term 'adaptive' in the phrase refer to the *critic's* learning attribute. We note that [59] also uses the term 'adaptive' in this latter sense.

4.3   DYNAMIC PROGRAMMING

Dynamic Programming (DP) [7] provides a principled method for determining optimal control policies for discrete-time dynamic systems whose states evolve according to given transition probabilities that depend on a decision/control u. Simultaneous with a transition from one state (call it $X(t)$) to the next $(X(t+1))$ under control u, a cost U is incurred [8]. Optimality is defined in terms of minimizing the sum of all the costs to be incurred while progressing from any state to the end state (both, finite and infinite cases are handled). This sum of costs is called 'cost-to-go,' and the objective of DP is to calculate numerically the optimal cost-to-go function J^*. An associated optimal control policy is also computed. Fundamental to this approach is Bellman's Principle of Optimality, which states that: "no matter how an intermediate point is reached in an optimal trajectory, the rest of the trajectory (from the intermediate point to the end) must be optimal." Unfortunately, the required DP calculations become cost-prohibitive as the number of states and controls become large (Bellman's "curse of dimensionality"); since most real-world problems fall into this category, approximating methods for DP have been explored since its inception (e.g., see [26]). The DP method entails the use of a Utility function, where the Utility function is crafted (by the user) to embody the design requirements of the given control problem. This function provides the above-mentioned 'cost' incurred while transitioning from a given state to the next one. A secondary utility function, known as the Value function (referred to above as the Cost-to-Go function), is defined in terms of the Utility function, and is used to perform the optimization process. (Bellman used the technically correct terminology Value *functional*, but much of our literature uses *function* instead; the latter is used in this chapter.) Once the 'optimal' version of the Value function has been determined, then the optimal controller may be designed, for example, via the Hamilton-Jacobi-Bellman equation.

4.4   ADAPTIVE CRITICS: "APPROXIMATE DYNAMIC PROGRAMMING"

The Adaptive Critic concept is essentially a juxtaposition of RL and DP ideas. It will be important to keep in mind, however, that whereas DP calculates the control via the optimal Value Function, the AC concept utilizes an approximation of the optimal Value Function to accomplish its controller design. For this reason, AC methods have been more properly referred to as implementing Approximate Dynamic Programming (ADP). A family (also called "ladder") of ADP structures was proposed by Werbos in the early 1990's [59, 60], and has been widely used by others [10, 11, 13, 14, 17, 18, 20–25, 31–33, 35–37, 40, 43, 45, 46]. While the original formulation was based on neural network implementations, it was noted that any learning structure capable of implementing the appropriate mathematics would work. Fuzzy Logic structures would be a case in point; recent examples may be found in [24, 44, 48, 53, 55]. This family of ADP structures includes: Heuristic Dynamic Programming (HDP), Dual Heuristic Programming (DHP), and Global Dual Heuristic Programming (GDHP). There are 'action dependent' (AD) versions of each, yielding the abbreviations: ADHDP, ADDHP, and ADGDHP. A detailed description of all these ADP structures is given in [43], called Adaptive Critic Designs there; additional details may also be found in [40].

The different ADP structures can be distinguished along three dimensions: (1) the inputs provided to the critic; (2) the outputs of the critic; and (3) the requirements for a plant model in the training process.

4.4.1   Critic Inputs

The critic typically receives information about the state of the plant (and of a reference model of the plant, where appropriate); in the action dependent structures, the critic is also provided the outputs of the action device (controller).

4.4.2   Critic Outputs

In the HDP structure, the critic outputs an approximation of the Value Function $J(t)$; in the DHP structure, it approximates the gradient of $J(t)$; and in the GDHP, it approximates both, $J(t)$ and its gradient.

4.4.3   Model Requirements

While there exist formulations that require only one training loop (e.g. [29]), the above ADP methods all entail the use of two training loops: one for the controller and one for the critic. There is an attendant requirement for two trainable function approximators, one for the controller and one for the critic. Depending on the ADP structure, one or both of the training loops will require a *model of the plant*. The controller training loop adapts the function approximator to be an approximately optimal controller (whose outputs are $u(t)$), via maximizing the secondary utility function $J(t)$. Since a gradient-based learning algorithm is typically used, derivatives' estimates are required for controller training (in the DHP version, these estimates are provided directly from the critic). Adaptation of the function approximation in the critic training loop is based on the consistency of its estimates through time, the exact implicit relationship being a function of the type of critic used and the structure of the primary utility function. In this chapter, we focus on the DHP structure; this structure requires a plant model for both loops. We mention that some view this model dependence to be an unnecessary "expense." The position of the authors, however, is that the expense is in many contexts more than compensated for by the additional information available to the learning/optimization process. We take further motivation for pursuing model-dependent versions from the biological exemplar: some explanations of the human brain developmental/learning process invoke the notion of 'model imperative' [38].

4.4.4 Model Use in Training Loops

Figure 4.1 provides a general diagrammatic layout for the ADP discussion. The base components are the action/controller and the plant; the controller receives measurement data about the plant's current state $X(t)$ and outputs the control $u(t)$; the plant receives the control $u(t)$, and moves to its next state $X(t + 1)$. The $X(t)$ data is provided to the critic and to the Utility function. In addition, the $X(t + 1)$ data is provided for a second pass through the critic. All of this data is needed in the calculations for performing the controller and critic training (the various dotted lines going into the 'calculate' boxes). This training is based on the Bellman Recursion:

$$J(t) = U(t) + \gamma J(t + 1). \tag{4.1}$$

We note that the term $J(t + 1)$ is an important component of this equation, and is the reason that $X(t + 1)$ is passed through the critic to get its estimate for time $(t + 1)$ (see [15, 16] for fuller expansion of the equations involved).

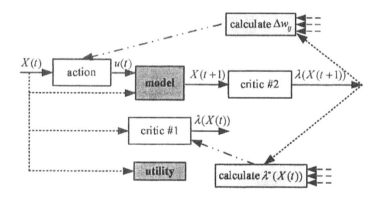

Fig. 4.1   General layout of Adaptive Critic structures.

The following is a verbal "walk through" of the six different AC structures, pointing out why and in which loop(s) of each structure a plant model is required. The results are tabulated in Table 4.1.

HDP: The critic estimates $J(t)$ based directly on the plant state $X(t)$; since this data is available directly from the plant, critic training does not need a plant model for its calculations. Controller training, on the other hand, requires finding the derivatives of $J(t)$ with respect to the control variables, obtained via the chain rule $\frac{\partial J(t)}{\partial u_i(t)} = \sum_{j=1}^{n} \frac{\partial J(t)}{\partial X_j(t)} \frac{\partial X_j(t)}{\partial u_i(t)}$. Estimates of the first term in this equation (derivatives of $J(t)$ with respect to the states) are obtained via Backpropagation through the critic network; estimates for the second term (derivatives of the states with respect to the controls) require a differentiable model of the plant (e.g., an explicit analytic model, a neural network model, etc). Thus HDP uses a plant model for the controller training but not the critic training.

ADHDP (Q-learning is in this category): Critic training is the same as for HDP. Controller training is simplified, in that since the control variables are inputs to the critic, the derivatives of $J(t)$ with respect to the controls, $\left(\frac{\partial X(t)}{\partial u(t)}\right)$, are obtained directly from Backpropagation through the critic. Thus ADHDP uses no plant models in the training process.

DHP: Recall that for this version, the critic directly estimates the derivatives of $J(t)$ with respect to the plant states, that is, $\lambda_i(t) = \frac{\partial J(t)}{\partial X_i(t)}$. The identity used for critic training is (in tensor notation):

$$\lambda_i(t) = \frac{\partial U(t)}{\partial X_i(t)} + \frac{\partial U(t)}{\partial u_i(t)}\frac{\partial u_j(t)}{\partial X_i(t)} + \lambda_k(t{+}1)\left[\frac{\partial X_k(t+1)}{\partial X_i(t)} + \frac{\partial X_k(t+1)}{\partial u_m(t)}\frac{\partial u_m(t)}{\partial X_i(t)}\right].$$

To evaluate the right-hand side of this equation, a full model of the plant dynamics is needed. This includes all the terms for the Jacobian matrix of the coupled plant-controller system, for example, $\frac{\partial X_j(t+1)}{\partial X_i(t)}$ and $\frac{\partial X_j(t+1)}{\partial u_i(t)}$. Controller training is much like that in HDP, except that the controller training loop directly utilizes the critic outputs along with the system model. So, DHP uses models for both critic and controller training.

ADDHP: ADDHP critics use both state and control variables as inputs, and output the gradient of $J(t)$ with respect to both states and controls, $\left(\frac{\partial J(t)}{\partial X(t)}\right)$ and $\left(\frac{\partial X(t)}{\partial u(t)}\right)$. This method utilizes the DHP critic training process, but gets the derivatives needed for controller training directly from the critic's output. Therefore ADDHP uses a plant model for critic training but not for controller training.

GDHP: GDHP critics have state variables as inputs, and they output both $J(t)$ and its gradient with respect to states. Critic training utilizes both the HDP and DHP recursions; controller training as in DHP. Therefore GDHP uses models for both critic and controller training.

ADGDHP: ADGDHP critics have both state and control variables as inputs, and they output both $J(t)$ and its gradient with respect to states and controls. As with GDHP, critic training utilizes both the HDP and DHP recursions, and controller training is as in ADDHP. Therefore ADGDHP uses a model for critic training but not for controller training.

Table 4.1   Summary of Requirement for Model in Training Loops

ADP STRUCTURE	Model NEEDED for training of	
	CRITIC	CONTROLLER
HDP		X
ADHDP		
DHP	X	X
ADDHP	X	
GDHP	X	X
ADGDHP	X	

4.5   SOME CURRENT RESEARCH ON ADAPTIVE CRITIC TECHNOLOGY

As part of providing guidance to prospective users of the ADP methods to develop controller designs, we sketch some of the work being done in the area, and provide citations that the reader may find useful.

Anderson and his colleagues at Colorado State University have been working on combining Robust control theory with reinforcement learning methodologies to develop proofs for both, static and dynamic stability, (e.g. [1, 2]). A reinforcement learning procedure has resulted which is guaranteed to remain stable even during training. In an attempt to speed up the process (which turns out being on the slow side), work is underway to use predicted reinforcement along with received reinforcement.

Balakrishnan and his colleagues at the University of Missouri—Rolla, have been working on applying adaptive critic based neurocontrol for distributed parameter systems (e.g., [10, 11, 25, 35–37]). The objectives of this research are to develop and demonstrate new adaptive critic designs, and to analyze the performance of these neurocontrollers in controlling parabolic, hyperbolic, and elliptic systems.

Barto and his colleagues at the University of Massachusetts, have been working on methods to allow an agent learning through reinforcement learning to automatically discover subgoals (e.g., [27, 39]). By creating and using subgoals, the agent is able to accelerate its learning on the current task, and to transfer its expertise to other related tasks. Discovery of subgoals is attained by examining commonalities across multiple paths to a solution. The task of finding these commonalities is cast as a multiple-instance learning problem, and the concept of diverse density is used to find the solution.

KrishnaKumar at the NASA Ames Intelligent Flight Controls Lab, and Neidhoefer, at Accurate Automation Coorporation, show an interesting implementation and application of adaptive critics ([13]). The basic idea is that if a nonlinear system can be linearized at representative points in the operational envelope, then the solution to the Ricatti equation at each point can be used as the Bellman Value function ("cost to go") for DP. If the Ricatti solutions also show a degree of statistical correlation, then an "Immunized" scheme (which mimics the building block scheme of biological immune systems) can be used with Ricatti solutions as "building blocks" to act as HDP Immunized Adaptive Critics (IAC).

Lendaris and his colleagues at the Portland State University NW Computational Intelligence Laboratory (NWCIL) have focused the past few years on exploring application issues related to ADP, in particular, the DHP version (e.g., [17, 18, 20–24, 48–53]). Much of the material reported in this chapter is an outgrowth of that work. A MATLAB based DHP computational platform has been developed, and is available for downloading and use from the NWCIL Web site: www.nwcil.pdx.edu. Key recent research and application results related to ADP involve the use of Fuzzy Logic structures for the controller, critic, and/or plant in the DHP ADP method (see Section 4.6.10). A recent application project is the design of a nonlinear controller for a hypersonic-shaped aircraft known as LoFlyte® [19]. Current work focuses on exploring methods of J^* surface generation for fast optimal decision/control design.

Prokhorov and his colleagues at the Ford Research Laboratory have done a significant amount of work in developing stability analysis techniques for neural networks (e.g., [3, 4]). An interesting application of the AC method was their experiment with a "real" ball-and-beam system. The benchmark ball-and-beam system was built in the lab, and different approaches were used to control the system. Neural networks were used in three roles: (1) to identify the system, (2) for the controller, and (3) for the critic. In one of their studies, they made the problem even more difficult by applying a sticky adhesive to the surface of the beam [9]; the ACs successfully handled the problem.

Saeks and his colleagues at Accurate Automation Corporation have been working with a variety of AC and adaptive dynamic programming implementations (e.g., [29, 45, 46]). Some of these implementations include an ADP algorithm based directly on the Hamilton-Jacobi-Bellman equation, and includes a continuous time stability proof that remains valid during on-line training. In [29], demonstrations of this algorithm are given for (i) the linear case, (ii) the nonlinear case using a locally quadratic approximation to the value functional, and (iii) the nonlinear case using a (potentially global) radial basis function approximation of the Value function. Another AC implementation has been developed suitable for real-time applications [31, 32]. This is a time-varying linear critic methodology based on LQR theory. Applications of these algorithms have included controlling the hybrid power system of a hybrid electric vehicle, pitch control in simulated aircraft problems, simulated X-43 autolanding analysis, path-planning for autonomous UAV flight, and the guidance module of a satellite formation flying algorithm.

Active work is also being performed by researchers such as Paul Werbos, Jennie Si, Sylvia Ferrari/Robert Stengel, and Ganesh Venayagamoorthy/Donald Wunsch. Please refer to their respective chapters in this book for a summary of their work.

4.6   APPLICATION ISSUES

In this section, we discuss various considerations that are important to the *application* of ADP. Before getting to the specifics, we comment that two major issues confronted in establishing practical training protocols are (1) the choice of control scenarios, and (2) the setting of values for the different parameters that govern the dynamics of the learning process. The control scenarios aspect includes the selection of regulation points, targets/target trajectories, initial plant states, noise regimes, and reset conditions (i.e., when should a training run be terminated). Training parameters of critical importance include the learning coefficients for both the critic and the controller, and the discount factor γ used in the Bellman recursion.

4.6.1   Problem Formulation

The mathematical formalism used in previous sections indicates that the plant's state vector X(t) is input to the critic and to the controller. An important pragmatic issue turns out being *what to include in the definition of* $X(t)$ for ADP computational purposes? The control engineer using this methodology must have deep understanding of the problem context and the physical plant to be controlled to successfully make the requisite choices for $X(t)$. A strong suggestion is to invoke your engineering intuition and whatever rigorous knowledge is available to satisfy yourself that the variables you select as inputs to the controller and critic are sufficient to assure that at every point in the corresponding state space, there will exist a unique action for the controller to take. If you decide such uniqueness is questionable, then you may have to estimate some (even a hybrid) variable that will make the state space unique. If this is not achieved, all is not lost, but more complex learning structures, for example, recurrent neural networks, may have to be used, and these are more difficult to train.

Not all mathematically describable states are observable; and even if they are in principle, there may be instrumentation constraints. Further, there are cases where we might be able to measure certain system variables (e.g., acceleration) whereas theory suggests fewer variables (e.g., only position and velocity) are required. But, experience informs us that in some situations, inclusion of the additional measurement could make the ADP process work better, for example, if the learning device has to infer the equivalent of acceleration to satisfy certain control objectives, providing acceleration directly might be beneficial. However, "more" is not always better, as more inputs potentially add to the computational and inferencing burden. In such a case, one could explore option(s) equivalent to providing position and acceleration instead of position and velocity.

When performing the problem-formulation task, it is useful to discern whether the plant is decomposable — that is, to determine whether certain aspects of the plant dynamics may be considered to be only loosely coupled. If so, this could be useful while crafting the Utility function (discussed below), and even provides the possibility that an equivalently loosely decoupled controller architecture might be appropriate.

While it may border on the obvious, another aspect of problem formulation that is critical to success is a clear statement of the control objectives. Only after clarity of the objectives is at hand, is one in a position to give explicit attention to the fundamental issue of how these objectives are to be represented for ADP application. The choice of this representation is a prerequisite to the next task, and is one of the key determinants of the eventual success or failure of the ADP design process.

4.6.2   Crafting the Utility Function

The Utility function is the only source of information the ADP process has about the task for which it is designing the controller. When the statement is made that Dynamic Programming designs an *optimal* controller, optimality is defined strictly in terms of the Utility function. It is important to recognize that a different Utility function will (typically) yield a different controller. The two key creative tasks performed by the user of ADP are:

1. Deciding what to include in the $X(t)$ vector, as discussed in the above sub-section;

2. *Crafting* the Utility function in a manner that properly captures/embodies the problem-domain requirements, and yields a desirable controller.

One mathematical formalism suggests designating the control task in terms of a reference trajectory, say $X^*(t)$ (which could in principle be obtained from a reference model), and defining the Utility function directly as $U(t) = \|X(t) - X^*(t)\|$ (e.g., see [40]). In practice, however, one finds that the ADP process can often be improved by treating some of the components of $X(t)$ in a non-uniform manner within $U(t)$. For example, different relative weightings might be used for various components, or more dramatically, some of the error components might use different powers, or alternatively, have nonlinear coefficients (e.g., see [20, 42]). Further, as suggested in [15] and [22], there is often substantial benefit to paring down $U(t)$ to contain the minimum number of terms necessary to accomplish the task (what these are, however, are not always easy to determine *a priori*).

We reproduce here a sequence of Utility functions reported in [20] that were crafted to represent an increasing set of constraints stipulated in the problem definition phase. The context was to design a steering and velocity controller for a 2-axle, terrestrial, autonomous vehicle; the task was to change lanes on a multi-lane road.

The first Utility function defined in that paper is an example of the suggestion above that it may be appropriate to have different weightings of the state variables:

$$U_1 = -\left(\frac{1}{2} y_{err}\right)^2 - \left(\frac{1}{8} v_{err}\right)^2 - \frac{1}{16}(\dot{v})^2.$$

The design objectives that motivated this Utility function definition were (a) reduce distance to centerline of adjacent lane (y-error) to zero, (b) reduce velocity error to zero and (c) don't be too aggressive in making the velocity corrections.

The second Utility function is an example where a non-linear rule is incorporated. To accommodate a stipulated requirement for handling a sudden change of friction between the tire and the road (e.g., hit an ice patch), an SI (sliding index) term was crafted to provide a proxy indication of where on the nonlinear tire-model curve (of tire side force vs. tire slip angle) the vehicle was operating in:

$$SI = \left(-10 \left(\frac{\frac{\partial a_y}{\partial \alpha_f} - \left(\frac{\partial a_y}{\partial \alpha_f} \right)_{base}}{\left(\frac{\partial a_y}{\partial \alpha_f} \right)_{base}} \right) \right),$$

where $\left(\frac{\partial a_y}{\partial \alpha_f} \right)_{base}$ is the slope at the linear portion of the curves. The terms in SI are calculated via (small) applied steering inputs and measured resulting side forces generated at the tire-road interface (via a lateral accelerometer on the vehicle). So defined, the sliding index approaches a value of 10 when sliding is occurring, and approaches zero for no sliding.

Then, a new Utility function was crafted as follows:

$$U_2 = \left[\begin{array}{ll} U_1 & \text{for SI} < 3 \\ U_1 - \frac{1}{4}(\text{SI})^2 & \text{for SI} \geq 3 \end{array} \right].$$

The SI value was input to the Critic and the Controller, and with this Utility function the DHP process developed a controller that successfully dealt with an ice patch in the road (and similarly, for a lateral wind gust), as described/discussed in [20].

In the third Utility function of the above reference, an additional term to limit lateral acceleration was included to accommodate a stipulation in the problem description concerning passenger "comfort" in automobiles, or for trucks, a "low tipping" requirement:

$$U_3 = U_2 - \frac{1}{8}(a_f)^2.$$

The reader may consult [20] to see the performances of the sequence of controllers generated by the DHP process using the above sequence of Utility functions.

Another kind of Utility function modification is to add time-lagged values of selected state variables, to help account for known delays in the plant being controlled. An example of this was used for one of the Narendra benchmark problems [30], presented in [23]:

$$U(t) = [x_1(t+1) - x_1'(t+1)]^2 + [x_2(t+2) - x_2'(t+2)]^2.$$

This Utility function did as well or better than more complex Utility functions previously reported in the literature for the same benchmark problem, and with substantially less computational overhead.

4.6.2.1 Decomposition of Utility Functions.   If during the Problem Formulation task it is determined that the plant is (even approximately) decomposable, then there is a potential for crafting separate Utility functions for each of the resulting "chunks."

In this case, it may be appropriate to define the overall Utility function as a sum of such component Utility functions, that is, $U(t) = U_1(t) + \ldots + U_p(t)$.

With such a formulation, a separate critic estimator could be used for each term. For HDP critics, one has

$$
\begin{aligned}
J(t) &= \sum_{i=0}^{\infty} \gamma^i U(t+i) \\
&= \sum_{i=0}^{\infty} \sum_{j=1}^{p} \gamma^i U_j(t+i) \\
&= \sum_{j=1}^{p} J_j(t),
\end{aligned}
$$

and for DHP

$$
\nabla J(t) = \sum_{j=1}^{p} \nabla J_j(t).
$$

In practice, this decomposition tends to speed up critic learning, as each sub-critic is estimating a simpler function.

In the case of multiple outputs from the controller, the controller learning process can also be simplified if the additive terms in the cost function correspond to separate modes, and the latter are dominated by distinct control variables. For example, consider a two dimensional nonlinear system:

$$
x_1 = f_1(x, u_1),
$$

$$
x_2 = f_2(x, u_2),
$$

with primary cost function

$$
U(t) = g_1(x, u_1, t) + g_2(x, u_2, t),
$$

and secondary cost function

$$
J(t) = J_1(t) + J_2(t).
$$

In DHP this could be approached using two critics, each estimating $\nabla \hat{J}_1(t)$ or $\nabla \hat{J}_2(t)$ respectively. The complete gradient for controller training would be

$$
\left(\frac{\partial}{\partial u_1} \hat{J}_1(t) + \frac{\partial}{\partial u_1} \hat{J}_2(t), \frac{\partial}{\partial u_2} \hat{J}_1(t) + \frac{\partial}{\partial u_2} \hat{J}_2(t) \right).
$$

For initial training, the cross terms could be dropped and the following approximation used: $\left(\frac{\partial}{\partial u_1}\hat{J}_1(t), \frac{\partial}{\partial u_2}\hat{J}_2(t)\right)$. This simplifies the learning of the dominant plant dynamics and control effects. It may be useful to include a subsequent fine tuning of the controller via inclusion of the cross terms, unless the interactions are *very weak*.

See [21] for an example of decomposed utility functions for the steering and speed control of a 2-axle terrestrial vehicle. Also, see [40] for a related kind of critic decomposition, one the author calls 'primitive adaptive critics.'

4.6.3 Scaling Variables

While theory does not speak to this issue directly, empirical evidence suggests that it is **eminently useful** to scale the components of $X(t)$ being fed into the controller and the critic (e.g., see [16]), such that each of the variable values are nominally in the range of ± 1, particularly when the critic and/or controller are implemented via neural networks (this recommendation is dependent on the approximating structure used). Further, as indicated above, it is important to pay attention to the relative scaling of the component terms in the Utility function. The latter may hinge solely on engineering intuition related to the problem domain and the control specifications.

4.6.4 Selecting the Discount Factor

The original equation defined by Bellman that led to Eq. (4.1) is as follows:

$$J(t) = U(t) + \sum_{k=1}^{\infty} \gamma^k U(t+k). \tag{4.2}$$

We notice that the Value function $J(t)$ is given in terms of the current $U(t)$, plus the sum of all future values of $U(\cdot)$, pre-multiplied by a discount factor γ, where $0 \leq \gamma \leq 1$. At $\gamma = 0$, only the present value of U is used, ignoring all future values. At $\gamma = 1$, all future values are deemed equally important ("infinite horizon" version).

In principle, we might expect the γ value to be governed by the requirements of the original problem formulation. In applying ADP, however, an important issue is how the value of γ influences the ADP convergence process (via its role in the Bellman recursion, Eq. (4.1)). The degree to which this is felt is different for the HDP, DHP, and GDHP structures. To help inform your intuition about this, note that the critic outputs values that are used to train itself, so at early stages of the process, the component in Eq. (4.1) contributed by the critic may be considered equivalent to 'noise.'

For the HDP structures, those that directly estimate $J(t)$ values, appropriately selecting γ is critical for convergence of critic training. Common practice (e.g., [9]) is to start training with low γ values and then anneal them up (progressively increment them). The low γ values represent a high discount rate that cancels out the right hand term of the Bellman recursion. This results in the critic learning to approximate (just) the primary utility function $U(t)$. Progressively incrementing γ then causes the critic to learn how the primary costs accumulate through time to form the long-term (secondary) value function, $J(t)$.

For the DHP structures, those that directly produce $\nabla J(t)$, this annealing process tends to be less necessary, often allowing large γ values to be used from the very beginning. For higher dimensional problems, however, even for the DHP structure, it has been found useful to "schedule" the γ values. A reasonable rule of thumb is that even if a larger value of γ is suggested by the problem formulation and/or the ADP structure type, use a small value at the early stages of the ADP process, while the state of knowledge in the critic is low, and as the critic's training proceeds, incrementally increase γ to higher levels.

4.6.5   Selecting Learning Rates

As mentioned earlier, the ADP structures addressed here all include two training loops: one for the critic and one for the controller. A *separate learning rate* (or learning-rate schedule) is associated with each training loop. In certain special cases, theory may be invoked to determine approximate desired values for the learning rates. For example, in Prokhorov et al. [9, 57], success was reported using extended Kalman filters to automatically adjust certain parameters, though this approach adds substantial computation to the process. More generally, however, "rule of thumb" is our primary guide, and even these are determined empirically within given problem contexts. Determining the values of these and other parameters turns out being the most labor-intensive aspect of employing the ADP methodology. In some cases, the user gets the feeling that the process requires an exhaustive search. The NWCIL DHP computational platform mentioned in Section 4.5 (available at www.nwcil.pdx.edu) provides a capability to experiment with lots of parameter values with minimal human intervention.

As is well known, the learning rate values mentioned above are important determinants of the training loop dynamics — and in particular, whether or not the process will converge, and if so, the convergence rate. Pragmatically, one determines useful ranges for the two learning rates empirically. Fortunately, once values are found that result in a convergent ADP process, these values are relatively robust to other process parameter changes the user may wish to explore. As with the γ values of the previous section, once useful ranges for learning rates are found, annealing (scheduling) the learning rates is also beneficial. The direction of annealing/scheduling in this case is the opposite: start with larger learning rates, and anneal downwards as learning progresses.

During the process of discovering useful values of the learning rates, if a selected set of rates results in the ADP process converging, but very slowly, then increase the learning rates incrementally — until one or both of the incremented values causes the process to diverge. Then just back down a notch or two. The more usual situation, however, is that early experiments result in a divergent process; in these cases, it is useful to observe which loop diverges first. The rate in this loop is adjusted first. Intuitively, since the controller is designed based on information acquired by the critic, it would make sense to use a larger learning-rate for the critic (to have it learn faster) than the controller. Indeed, there is empirical evidence for this. However, we have also seen counterexamples, where it worked better for the controller to have a higher learning rate. The rule-of-thumb we have developed is to start with a ratio of about 10:1 for the learning rates, the critic's being the larger one. Typical learning-rate values found useful in the problem domains explored by the first author in the past have been between 0.001 to 0.01, and sometimes up to 0.1.

General summary:

Guidance for selection of ADP process parameters is via "rules of thumb".

ADP parameter-value determination is the most labor-intensive aspect of employing the methodology.

Specific to this subsection:

Learning rate values determine training loop dynamics.

It is useful to use separate rates in each training loop.

One Rule of Thumb: Use a ratio of about 10:1 for the learning rates, the larger one for the critic loop (however, see caveat in the text).

To determine useful range(s) for learning rates, start exploration with (sometimes very) low values to find at least one set that will yield convergence; increase incrementally until process no longer converges; then back down a notch or two.

If no combination is found that yields convergence, see next subsection.

Learning-rate values found useful in applications to date (by first author and co-workers) for the critic loop are between $0.001 - 0.01$, and sometimes up to 0.1.

Once useful ranges of learning rates are determined, scheduling (annealing) the rates within these ranges may be beneficial during the design runs.

Scheduling of Learning Rate values goes from large to small (in contrast to scheduling gamma values of previous sub-section, which goes from small to large).

4.6.6   Convergence of the ADP (Controller Design) Process

The task of getting the ADP process to converge involves a carefully orchestrated selection of all of the above items. Experience indicates that there is strong interaction among their values in how they affect ADP convergence. If after scaling has been accomplished and exploration of learning rate and γ values has been performed with no successful convergence, we suggest reconsidering the Utility function formulation. We have examples of situations where seemingly minor changes in the formulation of the Utility function resulted in dramatically different ADP convergence behavior and resulting controller design. Associated with this, it may also be useful to reconsider the selection of variables being used as inputs to the controller and to the critic (cf. discussion in Section 4.6.1 as well).

4.6.7 Designing the Training "Syllabus"

Specific attention must be given to the design of the training regimen. Many issues need to be considered. In the control context, a key issue is *persistence of excitation*, which entails a requirement that the plant be stimulated such that all important modes are excited "sufficiently often" during the learning process. Additionally, it is also important that the full range of controller actions are experienced. A key rule-of-thumb in designing the regimen is to start the training with the simplest tasks first, and then build up the degree of difficulty. The truck backer-upper project of [34] provides an excellent example of this training principle, albeit in the context of a different learning methodology (Backpropagation through time).

The above rule-of-thumb includes considerations such as choosing initial plant states near regulation points or target states, selecting target trajectories that remain within a region of state space with homogenous dynamics, and initializing the controller with a stabilizing control law. This last approach falls under the topic of using *a priori* information to pre-structure either the controller or critic (more on this below). As the easier scenarios are successfully learned, harder scenarios are introduced in a manner that persistence of excitation across the entire desired operating region is achieved. In this stage, initial conditions for the plant are chosen farther and farther from regulation points, target trajectories are chosen so as to cross boundaries in qualitative dynamics, etc. The progression continues until the entire operating range of the controller is being exercised in the training runs.

A useful practice employed by the authors is to brainstorm how we would train animals or humans, including ourselves, to learn the given task. We then transform the insights gained into candidate training syllabi for the given ADP task.

4.6.8 Stopping/Reset Criteria

Another operational issue to consider is when to stop the learning process and start over again. In the well known cart-pole problem, there is a straightforward decision: When the pole drops, stop the process, reset the pole, and continue the training process (e.g. see [15, 16]). As another example, consider training a steering controller for a 4-wheeled terrestrial vehicle to change lanes on a highway: if the vehicle goes off the road, rather than continuing the training process to see if the controller can learn to get the vehicle back on the highway, instead, stop the process as soon as the vehicle goes "out of bounds," return to the starting point, and continue the training, starting the controller and critic weights (in the NN context) where they left off (e.g., see [20, 21]). The idea is to give an opportunity to improve the controller based on the design it had just before going out-of-bounds, rather than after it got "mired in the mud," as it might do in attempting to get back on the highway in the steering example. This idea may easily be generalized: Specify limits for each component of $X(t)$ (and $u(t)$ if appropriate) being used in the Utility function, create an out-of-bounds monitoring procedure, when an out-of-bound condition is detected (for one or more of the monitored variables), stop the process, return to an appropriate starting point, and continue the training.

This stop/reset strategy may also be usefully applied in those cases where the critic continues to diverge, no matter what choices are made with learning rate and/or other parameters. After a relatively "sweet" spot in the parameter values has been determined, even if the process does not converge by itself, the stop/reset strategy has been successfully employed to get the system to converge.

4.6.9 Simultaneous vs. Sequential Operation of Critic and Controller Training Loops

Once a forward computation is performed through the controller and plant, and a critic output is obtained (estimate of $J(t)$ or its derivatives), the ADP system is poised to perform a learning cycle in each of the two training loops. One strategy would be to simultaneously perform a learning cycle in both. This strategy works, and indeed, the authors routinely use it. However, experimentally determining values for the ADP process parameters discussed above is sometimes more difficult with this strategy than with other possibilities. In some early papers (e.g. [42, 43, 47, 57, 60], a "flip-flop" strategy was proposed wherein training was performed a number of times (called an epoch) in one loop while the training for the other loop was put on "hold", and then during the next epoch, the roles of being trained and being on hold were flipped. This flip-flop sequencing continued until the whole ADP process converged. While this strategy tends to be easier to get to converge, its convergence rate is slower than for other alternatives. This slower convergence is a consequence of losing information in those loops that are placed on hold. Additional strategies were subsequently developed (see [15, 16]) that also make use of the principle of separate (non-simultaneous) training, but in addition provide a means of preserving all the available information, thus avoiding the penalty of longer convergence times. The mechanism for preserving the information is called "shadow critic" in the critic training loop, and "shadow controller" in the controller training loop. The shadow concept entails performing the training updates in a COPY of the critic (rather than in the critic itself) and in a COPY of the controller during their respective "hold" epochs. Then at the end of the "hold" epoch, the design in the COPY (shadow version) is uploaded to the in-line version as a starting point for training during the next epoch. Various combinations are described: Shadow Critic Only; Shadow Controller Only; Shadow Critic and Shadow Controller. The motivating benefit for using these alternate strategies is their enhanced convergence performance. In addition, however, for some limited cases explored, the controller designs generated via the various strategies had some qualitative differences as well [23].

More recently, the Shadow Controller concept was incorporated in a proposed design of a method to deal with stability issues that arise when the ADP method is to be used in an on-line context [18]. See Section 4.6.11.

4.6.10   Embedding a priori Knowledge

If *a priori* knowledge is available about the problem domain that may be translated into a starting design of the controller and/or the critic, then it behooves us to use this knowledge as a starting point for the ADP procedures. While the ADP methods may be made to converge with random initializations of the controller and critic networks (usually only applicable in off-line situations), it is generally understood that the better the starting controller design, the "easier" it will be for the ADP process to converge. Another way to look at this is that if the starting controller design is "close" to an optimal design (e.g., the human designers already did a pretty good job), then the ADP system's task is one of *refining* a design — and this is intuitively easier than having to explore the design domain to even get to what is a *starting* point in the assumed context.

There are a variety of ways to obtain *a priori* information about the problem domain that can be used to initialize the trainable function approximator in the ADP process. For example, consider a system with an existing controller, but the controller's design is known to be non-optimal and it is desired to improve the design. One could train a neural network to copy this controller, substitute this NN in place of the controller, and implement an ADP process to optimize the controller design. If the ADP requires a differentiable plant model, then such a model will also have to be developed before starting the process. With this starting controller design, one would begin the ADP process with a long epoch to train just the critic, and then transition into one of the strategies described in the previous section to incrementally improve both, the critic's estimate of J^* and the corresponding controller design.

An alternate location to embed *a priori* knowledge is in the critic. For example, in the context of an LQR (linear quadratic regulator) problem, the J^* surface is known to be parabolic. While the various parameter values of the parabolic surface may not be known *a priori*, if the critic is pre-structured to just represent such surfaces, then ADP convergence is enhanced (e.g., see [54]). We think of this in terms of pre-biasing the critic's 'perception' of the problem.

Often times, the key source of *a priori* knowledge resides in the head of a human expert. There is little available in the neural network literature that provides guidance on how to embed such a priori knowledge into a neural network starting design. On the other hand, a large literature has developed in recent decades describing theory and methods for using Fuzzy Logic to capture such human expertise, and further, for using Fuzzy Systems in the controls context. Space limitations preclude surveying that literature here; a couple of accessible suggestions to the reader are [58, 62]. It is important to point out here that certain Fuzzy structures qualify as trainable universal function approximators, and thus, should in principle be usable in ADP processes. Indeed, successful use of Fuzzy structures for both controller and/or critic roles, and in fact, for the plant's differentiable model, have been accomplished (e.g., see [50, 51, 53]). We summarize below an example of such an application (taken from [50]), to convey the thought process one would use to employ such techniques. A summary of the results to be described in the following few paragraphs is given in Table 4.2.

A Fuzzy structure known as a first-order TSK model (e.g. see [62]) offers a direct approach for representing the relevant characteristics of the plant, and for prestructuring both the controller and critic. A very simple model of the well known cart-pole system was constructed using such a structure, and for DHP training, it was demonstrated that this model's effectiveness was comparable to the use of a full analytic model. This is especially interesting since no example-specific information (pole length or mass, cart mass, etc.) was included in the model.

The line of reasoning went as follows. First, it was noted that the six observable variables (related to pole angle θ and cart position x) constitute a coupled pair of second-order systems. It can be inferred that the derivatives $\frac{\partial\theta}{\partial\dot\theta}$, $\frac{\partial\theta}{\partial\ddot\theta}$, $\frac{\partial\theta}{\partial\ddot\theta}$, $\frac{\partial\dot\theta}{\partial\ddot\theta}$, $\frac{\partial\dot\theta}{\partial\ddot\theta}$, $\frac{\partial x}{\partial\dot x}$, $\frac{\partial x}{\partial\ddot x}$, $\frac{\partial x}{\partial\ddot x}$, $\frac{\partial\dot x}{\partial\ddot x}$ and $\frac{\partial\dot x}{\partial\ddot x}$ are all always positive. This observation constitutes a partial *qualitative model* of the plant's dynamics. An additional observation is that application of a positive control force to the cart tends to increase x, $\dot x$ and $\ddot x$, and decrease θ, $\dot\theta$ and $\ddot\theta$; this *a priori* knowledge allows setting $\frac{\partial x}{\partial u}$, $\frac{\partial\dot x}{\partial u}$, and $\frac{\partial\ddot x}{\partial u}$ positive, and $\frac{\partial\theta}{\partial u}$, $\frac{\partial\dot\theta}{\partial u}$ and $\frac{\partial\ddot\theta}{\partial u}$ negative. This collection of *assumptions* was defined as the Double Integrator Model (DIM). When the DIM was substituted into the baseline DHP training procedure in place of the true analytic plant model, the procedure successfully produced a controller 81 percent of the time (as compared to 99.99 percent with the analytic model).

Buoyed by this promising result, another piece of *a priori* knowledge was crafted out of the observable fact that when the pole is deflected from vertical, the force of gravity will tend to increase the angular acceleration of the pole in the same direction as the deflection while also imparting an acceleration to the cart in the opposite direction. This resulted in a new pair of rules:

$$\text{If }\theta\neq 0\text{ then }\frac{\partial\ddot x}{\partial\theta}\text{ is negative,}$$

and

$$\text{If }\theta\neq 0\text{ then }\frac{\partial\ddot\theta}{\partial\theta}\text{ is positive.}$$

The DIM augmented with these two rules was called the Crisp Rule Double Integrator Model (CRDIM). When used in the DHP training procedure this model turned out being only 76 percent effective.

While initially disappointing, this result provided the context for an important conclusion: while the linguistic description of the plant's behavior is correct, the *crisp implementation* of the rules that were used actually detracted from the effectiveness of the CRDIM for controller training. By moving to a Fuzzy framework for the *entire model*, substantially improved results were obtained.

To keep the fuzzy implementation simple, only three linguistic *values* were used for each variable: POSITIVE, ZERO and NEGATIVE. A triangular membership function was used for the ZERO linguistic value (with end points scaled consistent

Table 4.2   Effectiveness of Models in DHP Training (Cart-Pole Problem)

Type of Model	Effectiveness in Training
Analytic	99.99 %
Double Integrator	81 %
D. I. with Crisp Rules	76 %
Fuzzy Rules	99 %

with the expected range of the quantitative variable), and the membership functions for the POSITIVE and NEGATIVE values were defined so that the sum of membership values for any quantitative value sum to 1. The underlying observations included in the CRDIM were translated into fuzzy inference rules implemented using the sup-min operator for composition and the max operator for aggregation (cf. [62]). Height defuzzification was used, with centroid values of 1, 0 and −1 for POSITIVE, ZERO and NEGATIVE output values respectively. It should be clear that this Fuzzy Rule Model (FRM) would be a very poor numerical model of the system. Nevertheless it was 99 percent effective when used in DHP training, very close in performance to the true analytic model. 'Effectiveness' in this context is defined as the percentage of the trials in which the training procedure successfully produces a controller.

More advanced details about use of Fuzzy structures in DHP ADP systems are given in [52] and [53].

4.6.11   Stability issues

Direct application of Dynamic Programming (DP) as originally defined would be performed off-line, and would yield an optimal controller design that would then be implemented and inserted into the object system. The DP method guarantees that the resulting controller is a stabilizing one (entailed in the definition of *optimal*).

The *approximate* DP (ADP) methods considered in this book are also intended to yield optimal controllers, albeit only *approximately* optimal ones, after the underlying iterative approximation process has converged. Once the ADP process does converge, we can assume, with reasonable theoretical justification, that the resulting controller design is a stabilizing one (e.g., see [40]).

The stability story is more complicated, however, when the ADP methods are to be used *on-line* to modify the design of the controller to accommodate changes in the problem context (i.e., to be an *adaptive* controller in the traditional controls literature sense, or, to be a *reconfigurable* controller of the more recent literature). In this (on-line) case, the question of whether the controller design is a stabilizing one has to be asked *at each iteration* of the ADP design process. This is called *stepwise* stability in [29] or *static* stability in [2].

As in Section 4.5, we offer here a brief review of related research (here concerning stability issues) as part of providing guidance to prospective users of the ADP.

4.6.11.1   Recent Approaches to Stability Issues
The group at Colorado State University address the issue of stability in ADP systems in terms of what they call 'static' stability and 'dynamic' stability, [2]. Their static stability means that each time the controller's design is modified, it continues to be a stabilizing controller. This kind of stability is called step-wise stability in [29]. Their *dynamic* stability notion, on the other hand, refers to the dynamics introduced by the sequence of changed controller designs in the loop. They approach static stability with neural network controllers by first extracting the linear time-invariant (LTI) components of the neural network and representing the remaining parts as sector-bounded nonlinear uncertainties. Integral Quadratic Constraint (IQC) analysis [1] is then used to determine the stability of the system consisting of the plant, nominal controller, and the neural network with given weight values. The dynamic stability problem is addressed by once again treating the neural network's nonlinear components as sector bounded nonlinear uncertainties. In addition, uncertainty in the form of a slowly time-varying scalar is added to cover weight changes during learning. Finally, IQC analysis is applied to determine stability [28]. In this way, the network weight learning problem is transformed into one of network weight uncertainty; following this, a straightforward computation guarantees the stability of the network during training. The "down side" of this approach is its rather slow convergence to a design solution.

The group at the University of Massachusetts uses Lyapunov methods to successfully verify qualitative properties of controller designs, such as stability, or limiting behavior, [27]. Lyapunov-based methods are used to ensure that an agent learning through reinforcement learning exhibits behavior that satisfies qualitative properties relating to goal-achievement and safety.

The group at the Ford Research Laboratories has done a significant amount of work in analyzing the stability of recurrent neural networks (RNNs), [3, 4, 9]. Their work focuses on the global Lyapunov stability of multilayer perceptrons, where they assume the network weights are fixed. They perform a state space transformation to convert the original RNN equations to a form suitable for stability analysis. Then appropriate linear matrix inequalities (LMI) are solved to determine whether the system under study is globally exponentially stable. In [41], a technique to approximate Lyapunov functions using a form of Support Vector Machine (SVM) is developed. In [4], an algorithm to test whether a dynamic system has a convex Lyapunov function is proposed and evaluated.

The group at Portland State University's NW Computational Intelligence Laboratory has proposed a computation/simulation approach based on the Shadow Controller concept mentioned in Section 4.6.9 (e.g., see [17, 18]). This approach is predicted to become viable for on-line applications as computational power continues to increase. Since the issue during on-line training of a controller is to avoid instantiating into the control loop a controller that is not stabilizing, the (DHP, in their case) training is done only on the Shadow Controller. While many issues remain to be resolved for this proposed procedure, the idea is to determine the (local) stability of the current Shadow Controller design by performing a high-speed simulation of the closed loop (using the plant model already required for the DHP method), determine a local linearization, determine the s-plane pole locations, and from this test to determine whether the current Shadow Controller design meets minimum stability requirements; if so, upload the design to the on-line controller; if not, wait until another train/test cycle. The assumption is that stabilizing controller designs will occur sufficiently often to render the proposed procedure viable.

The group at Accurate Automation Corporation has developed an adaptive dynamic programming (incidentally, in [29], the acronym ADP is used for Adaptive Dynamic Programming, whereas in the present book ADP is used for Approximate Dynamic Programming) algorithm with a continuous time stability proof [29] . The algorithm is initialized with a (stabilizing) Value function, and the corresponding control law is computed via the Hamilton-Jacobi-Bellman Equation (which is thus guaranteed to be a stabilizing controller for this step), and the system is run; the resultant state trajectories are kept track of and used to update the Value function in a soft computing mode. The method is repeated to convergence. In [29], this method is shown to be globally convergent, with step-wise stability, to the optimal Value function / control law pair for an (unknown) input affine system with an input quadratic performance measure (modulo the appropriate technical conditions). This algorithm has been demonstrated on the example problems mentioned in Section 4.5 for Saeks and his colleagues.

4.7   ITEMS FOR FUTURE ADP RESEARCH

As mentioned in Section 4.1, much progress in the development and application of Adaptive Critics has already occurred, yet much remains to be done. We comment here on two topics that appear to us to have significant potential for expanded/enhanced application of the ADP methods. One relates to employment of Fuzzy communication among the actors in an ADP system, and the other relates to speeding up the process of J^* generation in selected problem domains.

The idea for the latter is to develop a computational intelligence methodology that efficiently designs optimal controllers for *additional problems* within an assumed problem domain, *based on knowledge of existing designs in that domain*. Research on this approach is just starting at the NWCIL, but promises to benefit from broader involvement. As currently envisioned, the key ingredient of this methodology will be a J^* Surface Generator (J^*SG). Fundamental tasks in this quest involve representation, representation, representation.

The idea for including Fuzzy communication among the actors in an ADP system (in contrast to within the various actors, as discussed in Section 4.6.10) is motivated by observing the process wherein, for example, a human athlete refines his/her performance based on verbal hints/instructions provided by an experienced coach (this may apply to many human activities, such as dancing, art, etc.). A potentially key location for receiving/embedding such knowledge communicated via a Fuzzy representation could be the Utility function. A hint that the Utility function could be the heart of such a refinement process may reside for us in the sequence of additions to the Utility functions described in Section 4.6.2, and the corresponding refinements in controller performance achieved. Each of us has many personal experiences of using verbally communicated guidance to enhance some kind of performance, and this could provide a rich source of intuition for such an approach. We encourage dialogue to begin within our research community.

Acknowledgments

Most of the work reported in Sections 4.4 and 4.6 of this chapter was supported by NSF Grant #ECS-9904378, G. G. Lendaris, PI.

Bibliography

1. C. Anderson, Approximating a policy can be easier than approximating a value function, Technical Report CS-00-101, Colorado State University, 2000.

2. C. Anderson, R. M. Kretchner, P. M. Young, and D. C. Hittle, Robust reinforcement learning control with static and dynamic stability, *International Journal of Robust and Nonlinear Control*, vol. 11, 2001.

3. N. Barabanov and D. Prokhorov, Stability analysis of discrete-time recurrent neural networks, *IEEE Trans. Neural Networks*, 2002.

4. N. Barabanov and D. Prokhorov, A new method for stability analyis of nonlinear discrete-time systems, *IEEE Trans. Automatic Control*, vol. 48, no. 12, 2003.

5. A. G. Barto, *Handbook of Intelligent Control*, chapter Reinforcement Learning and Adaptive Critic Methods, pp. 469–491, Van Nostrand-Reinhold, New York, 1992.

6. A. G. Barto and R. S. Sutton, *Reinforcement Learning: An Introduction*, MIT Press, Cambridge, MA, 1998.

7. R. E. Bellman, *Dynamic Programming*, Princeton University Press, Princeton, NJ, 1957.

8. D. P. Bertsekas and J. N. Tsitsiklis, *Neuro-Dynamic Programming*, Athena Scientific, Belmont, MA, 1996.

9. P. Eaton, D. Prokhorov, and D. Wunsch, Neurocontroller for fuzzy ball-and-beam systems with nonlinear, nonuniform friction, *IEEE Trans. Neural Networks*, pp. 423–435, 2000.

10. Z. Huang and S. N. Balakrishnan, Robust adaptive critic based neurocontrollers for missiles with model uncertainties, *2001 AAA Guidance, Navigation and Control Conference*, Montreal, Canada, 2001.

11. Z. Huang and S.N. Balakrishnan, Robust adaptive critic based neurocontrollers for systems with input uncertainties, *Proc. of IJCNN'2000*, pp. B–263, 2000.

12. L. P. Kaelbling, M. L. Littman, and A. W. Moore. Reinforcement learning: A survey, *Journal of Artificial Intelligence Research*, vol. 4, pp. 237–285, 1996.

13. K. KrishnaKumar and J. Neidhoefer, Immunized adaptive critics, invited session on *Adaptive Critics, ICNN '97*, Houston, 1997. A version of this was presented at ANNIE '96, November, St. Louis, MO.

14. K. KrishnaKumar and J. Neidhoefer, Immunized adaptive critic for an autonomous aircraft control application, *Artificial immune systems and their applications*, Springer-Verlag, New York, 1998.

15. G. G. Lendaris, T. T. Shannon, and C. Paintz, More on training strategies for critic and action neural networks in dual heuristic programming method (invited paper), *Proc. of Systems, Man and Cybernetics Society International Conference '97*, Orlando, FL, 1997.

16. G. G. Lendaris and C. Paintz, Training strategies for critic and action neural networks in dual heuristic programming method, *Proc. of International Conference on Neural Networks '97 (ICNN '97)*, Houston, 1997.

17. G. G. Lendaris, R. A. Santiago, and M. S. Carroll, Dual heuristic programming for fuzzy control, *Proceeedings of IFSA / NAFIPS Conference*, Vancouver, B.C., 2002.

18. G. G. Lendaris, R. A. Santiago, and M. S. Carroll, Proposed framework for applying adaptive critics in real-time realm, *Proc. of International Conference on Neural Networks '02 (IJCNN' 2002)*, Hawaii, 2002.

19. G. G. Lendaris, R. A. Santiago, J. McCarthy, and M. S. Carroll, Controller design via adaptive critic and model reference methods, *Proc. of International Conference on Neural Networks '03 (IJCNN' 2003)*, Portland, OR, 2003.

20. G. G. Lendaris and L. J. Schultz, Controller design (from scratch) using approximate dynamic programming, *Proc. of IEEE International Symposium on Intelligent Control '2000,(IEEE-ISIC' 2000)*, Patras, Greece, 2000.

21. G. G. Lendaris, L. J. Schultz, and T. T. Shannon, Adaptive critic design for intelligent steering and speed control of a 2-axle vehicle, *Proc. of International Conference on Neural Networks '00 (IJCNN '2000)*, Italy, 2000.

22. G. G. Lendaris and T. T. Shannon, Application considerations for the dhp methodology, *Proc. of the International Joint Conference on Neural Networks '98 (IJCNN '98)*, Anchorage, 1998.

23. G. G. Lendaris, T. T. Shannon, and A. Rustan, A comparison of training algorithms for dhp adaptive critic neuro-control, *Proc. of International Conference on Neural Networks '99 (IJCNN '99)*, Washington, DC, 1999.

24. G. G. Lendaris, T. T. Shannon, L. J. Schultz, S. Hutsell, and A. Rogers, Dual heuristic programming for fuzzy control, *Proceeedings of IFSA / NAFIPS Conference*, Vancouver, B.C., 2001.

25. X. Liu and S. N. Balakrishnan, Convergence analysis of adaptive critic based neural networks, *Proc. of 2000 American Control Conference,* Chicago, 2000.

26. R. Luus, *Iterative Dynamic Programmings,* CRC Press, Boca Raton, FL, 2000.

27. A. McGovern and A. G. Barto, Automatic discovery of subgoals in reinforcement learning using diverse density, *Proc. of the 18th International Conference on Machine Learning,* pp. 361–368, 2001.

28. A. Megretski and A. Rantzer, System analysis via integral quadratic constraints: Part II, Technical Report ISRN LUTFD2/TFRT-7559-SE, Lund Institute of Technology, 1997.

29. J. J. Murray, C. Cox, G.G. Lendaris, and R. Saeks, Adaptive dynamic programming, *IEEE Trans. on Systems, Man, and Cybernetics, Part C: Applications and Reviews,* vol. 32, no. 2, pp. 140–153, 2002.

30. K. S. Narendra and S. Mukhopadhyay, Adaptive control of nonlinear multivariable systems using neural networks, *Neural Networks,* vol. 7, no. 5, pp. 737–752, 1994.

31. J. C. Neidhoefer, Technical Report AAC-01-055, Accurate Automation Corp, 2001.

32. J. C. Neidhoefer, Technical Report AAC-02-016, Accurate Automation Corp, 2002.

33. J. C. Neidhoefer and K. Krishnakumar, Intelligent control for autonomous aircraft missions, *IEEE Trans. on Systems, Man, and Cybernetics, Part A,* 2001.

34. D. Nguyen and B. Widrow, The Truck Backer-Upper: An Example of Self Learning in Neural Networks, *Neural Networks for Control,* MIT Press, Cambridge, MA, 1957.

35. R. Padhi and S. N. Balakrishnan, Adaptive critic based optimal control for distributed parameter systems, *Proc. International Conference on Information, Communication and Signal Proc.,* 1999.

36. R. Padhi and S. N. Balakrishnan, A systematic synthesis of optimal process control with neural networks, *Proc. American Control Conference,* Washington, DC, 2001.

37. R. Padhi, S. N. Balakrishnan, and T. Randolph, Adaptive critic based optimal neuro control synthesis for distributed parameter systems, *Automatica,* vol. 37, pp. 1223–1234, 2001.

38. J. C. Pearce, *The Biology of Transcendence,* Park Street Press, Rochester, VT, 2002.

39. T. J. Perkins and A. G. Barto, Lyapunov design for safe reinforcement learning, AAAI Spring Symposium on Safe Learning Agents.

40. D. Prokhorov, *Adaptive Critic Designs and their Application*, Ph.D. Thesis, Texas Tech University, 1997.

41. D.V. Prokhorov and L.A. Feldkamp, Application of SVM to Lyapunov Function Approximation, *Proc. of International Conference on Neural Networks '99 (IJCNN'1999)*, Washington, DC, 1999.

42. D. Prokhorov, R. Santiago, and D. Wunsch, Adaptive critic designs: A case study for neurocontrol, *Neural Networks*, vol. 8, pp. 1367–1372, 1995.

43. D. Prokhorov and D. Wunsch, Adaptive critic designs, *IEEE Trans. Neural Networks*, vol. 8, no. 5, pp. 997–1007, 1997.

44. A. Rogers, T. T. Shannon, and G. G. Lendaris, A comparison of dhp based antecedent parameter tuning strategies for fuzzy control, *Proc. of IFSA/NAFIPS Conference*, Vancouver, B.C., 2001.

45. R. Saeks, C. Cox, J. Neidhoefer, and D. Escher, Adaptive critic control of the power train in a hybrid electric vehicle, *Proc. SMCia Workshop*, 1999.

46. R. Saeks, C. Cox, J. Neidhoefer, P. Mays, and J. Murray, Adaptive critic control of a hybrid electric vehicle, *IEEE Trans. on Intelligent Transportation Systems*, vol. 3, no. 4, 2002.

47. R. Santiago and P. Werbos, New progress towards truly brain-like intelligent control, *Proc. WCNN '94*, pp. 12–I33, Erlbaum, Hillsdale, NJ, 1994.

48. L. J. Schultz, T. T. Shannon, and G. G. Lendaris, Using dhp adaptive critic methods to tune a fuzzy automobile steering controller, *Proc. of IFSA/NAFIPS Conference*, Vancouver, B.C., 2001.

49. T. T. Shannon, Partial, noisy and qualitative models for adaptive critic based neuro-control, *Proc. of International Conference on Neural Networks '99 (IJCNN'99)*, Washington, DC, 1999.

50. T. T. Shannon and G. G. Lendaris, Qualitative models for adaptive critic neuro-control, *Proc. of IEEE SMC'99 Conference*, Tokyo, 1999.

51. T. T. Shannon and G. G. Lendaris, Adaptive critic based approximate dynamic programming for tuning fuzzy controllers, *Proc. of IEEE-FUZZ 2000*, 2000.

52. T. T. Shannon and G. G. Lendaris, A new hybrid critic-training method for approximate dynamic programming, *Proc. of International Society for the System Sciences, ISSS'2000*, Toronto, 2000.

53. T. T. Shannon and G. G. Lendaris, Adaptive critic based design of a fuzzy motor speed controller, *Proc. of ISIC2001*, Mexico City, 2001.

54. T. T. Shannon, R. A. Santiago, and G. G. Lendaris, Accelerated critic learning in approximate dynamic programming via value templates and perceptual learning, *Proc. of IJCNN'03*, Portland, OR, 2003.

55. S. Shervais and T. T. Shannon, Adaptive critic based adaptation of a fuzzy policy manager for a logistic system, *Proc. of IFSA /NAFIPS Conference*, Vancouver, B.C., 2001.

56. R. S. Sutton, A. G. Barto, and R. J. Williams, Reinforcement learning is direct adaptive optimal control, *Proc. of the American Control Conference*, Boston, pp. 2143–2146, 1991.

57. N. Visnevski and D. Prokhorov, Control of a nonlinear multivariable system with adaptive critic designs, *Proc. of Artificial Neural Networks in Engineering (ANNIE)*, vol. 6, pp. 559–565, 1996.

58. L. X. Wang, *A Course in Fuzzy Systems and Control*, Prentice-Hall, Englewood Cliffs, NJ, 1997.

59. P. J. Werbos, A Menu of Designs for Reinforcement Learning Over Time, *Neural Networks for Control*, pp. 67–95, MIT Press, Cambridge, MA, 1990.

60. P. J. Werbos, Approximate Dynamic Programming for Real-Time Control and Neural Modeling, *Handbook of Intelligent Control: Neural, Fuzzy, and Adaptive Approaches*, pp. 493–525, Van Nostrand Reinhold, New York, 1994.

61. B. Widrow, N. Gupta, and S. Maitra, Punish/reward: Learning with a critic in adaptive threshhold systems, *IEEE Trans. on Systems, Man and Cybernetics*, vol. 3, no. 5, pp. 455–465, 1973.

62. J. Yen and R. Langari, *Fuzzy Logic: Intelligence, Control and Information*, Prentice-Hall, Englewood Cliffs, NJ, 1999.

Handbook of Learning and Approximate Dynamic Programming
Edited by Jennie Si, Andy Barto, Warren Powell and Donald Wunsch
Copyright © 2004 The Institute of Electrical and Electronics Engineers, Inc.

5 Direct Neural Dynamic Programming

JENNIE SI   DERONG LIU
LEI YANG
Arizona State University   University of Illinois at Chicago

Editor's Summary: This chapter introduces direct neural dynamic programming (direct NDP), which belongs to the class of heuristic dynamic programming algorithms discussed in Chapters 3, 4, and 19. However, direct NDP is a model-independent approach to action-dependent heuristic dynamic programming. It is, therefore, an on-line learning control paradigm. This chapter contains a comparison study using other well-known algorithms to help readers gain quantitative insight on several ADP algorithms. It also contains results of direct NDP controlling a triple-link inverted pendulum using many continuous state variables and a continuous control, and direct NDP in a wireless network call admission control application. Furthermore, in Chapter 21 direct NDP is demonstrated on an industrial scale Apache helicopter model for stabilization, tracking control, and reconfiguration after component failure. Preliminary results indicate that direct NDP has the potential to address large-scale problems.

5.1 INTRODUCTION

The term reinforcement learning is often used when discussing approximate dynamic programming (ADP) in the computer science and machine learning community. Associated with this term is a wide range of solid results obtained both analytically and empirically by considering learning in the arena of Markov decision processes [5, 20]. In most of its analyses and applications, reinforcement learning results assume discrete cases, where the state and action spaces can be enumerated and stored in memory. Another class of systems, adaptive critic designs, address some very similar issues, namely, to develop learning systems that can improve their performance over time through interactions with the environment and through past experience. The foundation of this approach is built on the calculus of variations used in optimal control theory. The problem construct of adaptive critic designs introduces the possibility of designing learning systems that can handle continuous state space problems or the possibility of addressing generalization issues and design robustness issues.

This chapter introduces direct neural dynamic programming (direct NDP). The term "direct" is influenced by the adaptive control literature where "direct adaptive control" means no plant model, and thus no plant parameter estimation takes place but instead certain plant information is used directly to find appropriate and convergent control laws and control parameters. In direct NDP, two generic (neural) function approximators are used to represent both the value function (for control performance evaluation) and the action function (for control law generation). Furthermore, the state information is used directly in learning the control law where the controller parameters are the weight parameters in the action function implemented by a neural network. Therefore direct NDP is a model-independent approach to action-dependant heuristic programming [25].

The class of adaptive critic designs [22–26] encompasses several important ideas and implementations of model-based approximate dynamic programming including heuristic dynamic programming (HDP), dual heuristic dynamic programming (DHP), and globalized dual heuristic dynamic programming (GDHP). A distinguishing feature of the adaptive critic designs is the potential of addressing large scale problems including continuous state and control problems. Several applications and case studies have been developed to demonstrate the feasibility of some key adaptive critic design techniques where globalized dual dynamic programming has been found the most effective in the family as a nonlinear controller (refer to [14, 15] and results in Chapter 19). As a model-independent approach to the action-dependent heuristic dynamic programming, the direct NDP is expected to achieve computational efficiency and reliability due to its simple, completely gradient driven implementation in addressing large scale problems.

The chapter provides a detailed demonstration on how direct NDP works, some insights on why it is a relatively robust implementation of ADP, and how it generalizes. There are also discussions about important open issues regarding convergence, convergence speed, principle of optimality, controller performance guarantee, and so on. Here, non-trivial problems and benchmark studies are used to provide a quantitative assessment of the algorithm. A complex case study, namely helicopter stabilization, tracking, and reconfiguration control, will be provided in detail in Chapter 21 of this book to demonstrate the generalization capability of direct NDP, especially to large, complex, nonlinear continuous state space control problems, and to illustrate design issues associated with direct NDP.

5.2 PROBLEM FORMULATION

The focus of this chapter is to discuss the same design problem that has been addressed in reinforcement learning and adaptive critic designs: how to program a learning system by reward and punishment through trial and error to maximize its future performance. This problem setting has great intuitive appeal and has attracted considerable attention from different research fields. The authors choose not to use notational convention that is commonly used in reinforcement learning literature, but rather, the convention used in this chapter is rooted in classical control theory since the problem construct of this chapter can, in principle, deal with both discrete and continuous state and action problems. This convention is more in line with that used in adaptive critic designs.

To summarize the notation, $X(t)$ denotes a vector valued state variable at discretized time instance t, comparable to s modeled in a Markov decision process (MDP) frequently used in reinforcement learning literature; $u(t)$ denotes the control or action variable at time t, comparable to a in an MDP environment; $r(t)$ is the binary reinforcement signal provided from the external environment with "r_s" representing success and "r_f" for failure. As an example, we may choose $r_s = 0$ and $r_f = -1$, respectively. In the current problem setting, let the discounted total reward-to-go $R(t)$ at time t be given by

$$R(t) = r(t+1) + \alpha r(t+2) + \cdots = \sum_{k=1}^{\infty} \alpha^{k-1} r(t+k), \qquad (5.1)$$

where α is a discount factor. This total reward-to-go is also referred to as the value function. In the reinforcement learning literature, $R(t)$ or an approximation to $R(t)$, is usually represented by $V(t)$ or $Q(t)$, among others. In this chapter, $J(t)$ is adopted as an approximation to $R(t)$, which again is a notation from classic control theory.

The problem at hand is to consider a learning control process consisting of interactions between the learning system and the external environment. Specifically, for time step $t, t = 0, 1, \cdots$, the learning system receives some representation of the environment denoted as $X(t)$, which the learning system uses to determine a control action $u(t)$. After deciding on a value for the control variable $u(t)$, the learning system receives a reinforcement $r(t)$ which can be either a reward indicating success or a punishment indicating failure as a consequence of that control action. The goal of the learning system is to choose controls such that the overall reward over the long run, $R(t)$ in Eq. (5.1), is maximized.

5.3 IMPLEMENTATION OF DIRECT NDP

The approximate dynamic programming approach presented here is an on-line learning control scheme. The objective of the direct NDP controller is to optimize a desired performance measure by learning to choose appropriate control actions through interaction with the environment [19]. Learning is performed without requiring an explicit system model (e.g., $X(t+1) = f(X(t), u(t))$) as an (approximate) representation of the external environment prior to learning the controller. Instead, information about the system dynamics is directly "captured" by both the action and critic networks through learning. A general schematic diagram of the direct NDP is shown in Figure 5.1.

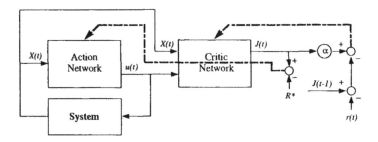

Fig. 5.1 Schematic diagram for implementation of direct neural dynamic programming. The solid lines denote system information flow, while the dashed lines represent error back-propagation paths to reduce the squared Bellman error: $([r(t) + \alpha J(t)] - J(t-1))^2$.

The direct NDP design structure in Figure 5.1 includes two networks, action and critic, as building blocks. The critic network is trained toward optimizing a total reward-to-go objective, namely to balance the Bellman equation. The action network is trained such that the critic output approaches an ultimate objective of success, $R^* = \frac{r_s}{1-\alpha}$, which is determined from Eq. (5.1). Due to this constraint a sufficient condition for the applicability of the learning structure proposed in Figure 5.1 is that $\alpha < 1$ to ensure a convergent sequence to the ultimate performance R^* derived from Eq. (5.1). During the learning process, the action network is constrained by the critic to generate controls that optimize the future reward-to-go instead of only temporarily optimal solutions. In contrast to usual supervised neural network learning applications, there are no readily available training sets of input-output pairs used for approximating the overall objective function $R(t)$ in terms of a least squares fit. Instead, both the control action u and the critic output J are updated according to an error function that changes from one time step to the next.

In the on-line learning control implementation, the controller is "naïve" when it starts to control. This is because, initially, both the action and critic networks possess random weights/parameters. Once a system state is observed, an action will be subsequently produced based on the parameters in the action network. A "better" control under the specific system state should result in a reduced Bellman error. This set of system operations will be reinforced through memory or association between states and control output in the action network. Otherwise, the control will be adjusted through tuning the weights in the action network to minimize the Bellman error.

5.3.1   Critic Network

The critic network is used to provide an output $J(t)$, which is an approximation for $R(t)$, the weighted total future reward-to-go. The reward function $R(t)$ at time t is given by Eq. (5.1).

We define the prediction error, and consequently the Bellman error, for the critic element as

$$e_c(t) = [r(t) + \alpha J(t)] - J(t-1), \qquad (5.2)$$

and the objective function to be minimized in the critic network is

$$E_c(t) = \frac{1}{2} e_c^2(t). \qquad (5.3)$$

Let w_c denote the set of weight parameters in the critic network. The critic network can then be represented by $J = \mathrm{nn}_c(X, u, w_c)$, where nn_c denotes the critic network. The weight update rule for the critic network is a gradient-based adaptation given by

$$w_c(t+1) \;=\; w_c(t) + \Delta w_c(t), \qquad (5.4)$$

$$\Delta w_c(t) \;=\; l_c(t)\left[-\frac{\partial E_c(t)}{\partial w_c(t)} \right], \qquad (5.5)$$

$$\frac{\partial E_c(t)}{\partial w_c(t)} \;=\; \frac{\partial E_c(t)}{\partial J(t)}\frac{\partial J(t)}{\partial w_c(t)}, \qquad (5.6)$$

where $l_c(t) > 0$ is the learning rate of the critic network at time t, which is typically selected to decrease with time to a small value. To re-iterate, the objective of the critic network learning is to minimize the Bellman error by properly adjusting the weight parameters in the critic network.

5.3.2   Action Network

The principle in adapting the action network is to back-propagate the error between the desired ultimate performance objective, denoted by R^*, and the approximate function J from the critic network. Since r_s has been defined as the reinforcement signal for "success," R^* is set to $\frac{r_s}{1-\alpha}$ in the direct NDP design paradigm and in subsequent case studies. In the action network, the state measurements are used as inputs to create a control as the output of the network. In turn, the action network can be implemented by either a linear or a nonlinear network, depending on the complexity of the problem. The weight update in the action network can be formulated as follows. Let

$$e_a(t) = J(t) - R^*. \qquad (5.7)$$

Let w_a denote the set of weight parameters in the action network, where the action network is represented by $u = \text{nn}_a(X, w_a)$. The weights in the action network are updated to minimize the following performance error measure:

$$E_a(t) = \frac{1}{2}e_a^2(t). \tag{5.8}$$

The update algorithm is then similar to the one in the critic network. By a gradient descent rule

$$w_a(t+1) = w_a(t) + \Delta w_a(t), \tag{5.9}$$

$$\Delta w_a(t) = l_a(t)\left[-\frac{\partial E_a(t)}{\partial w_a(t)}\right], \tag{5.10}$$

$$\frac{\partial E_a(t)}{\partial w_a(t)} = \frac{\partial E_a(t)}{\partial J(t)}\frac{\partial J(t)}{\partial w_a(t)}, \tag{5.11}$$

where $l_a(t) > 0$ is the learning rate of the action network at time t, which usually decreases with time to a small value.

5.3.3   On-line Learning Algorithms

To provide readers with concrete implementation details, for the rest of this chapter nonlinear multi-layer feed-forward networks are used as approximators for both the action and the critic networks. Specifically, in this design, one hidden layer is used in each network. As an example, and for the ease of introducing notation, a structure for the nonlinear, multi-layer critic network is shown in Figure 5.2.

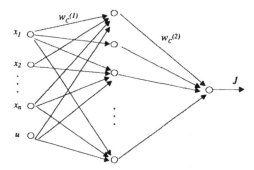

Fig. 5.2   Schematic diagram for the implementation of a nonlinear critic network using a feed-forward network with one hidden layer.

In the critic network, the output $J(t)$ is of the form,

$$J(t) = \sum_{i=1}^{N_{hc}} w_{c_i}^{(2)}(t)p_i(t), \tag{5.12}$$

$$p_i(t) = \frac{1 - e^{-q_i(t)}}{1 + e^{-q_i(t)}}, \; i = 1, \cdots, N_{hc}, \tag{5.13}$$

$$q_i(t) = \sum_{j=1}^{n+1} w_{c_{ij}}^{(1)}(t) x_j(t), \; i = 1, \cdots, N_{hc}, \tag{5.14}$$

where q_i is the ith hidden node input of the critic network, and p_i is the corresponding output of the hidden node. N_{hc} is the total number of hidden nodes in the critic network and $n + 1$ is the total number of inputs to the critic network including the analog action value $u(t)$ from the action network. By applying the chain rule, the adaptation of the critic network is summarized below.

(1) $\Delta \mathbf{w}_c^{(2)}$ (hidden to output layer):

$$\Delta w_{c_i}^{(2)}(t) = l_c(t) \left[-\frac{\partial E_c(t)}{\partial w_{c_i}^{(2)}(t)} \right], \tag{5.15}$$

$$\frac{\partial E_c(t)}{\partial w_{c_i}^{(2)}(t)} = \frac{\partial E_c(t)}{\partial J(t)} \frac{\partial J(t)}{\partial w_{c_i}^{(2)}(t)} = \alpha e_c(t) p_i(t). \tag{5.16}$$

(2) $\Delta \mathbf{w}_c^{(1)}$ (input to hidden layer):

$$\Delta w_{c_{ij}}^{(1)}(t) = l_c(t) \left[-\frac{\partial E_c(t)}{\partial w_{c_{ij}}^{(1)}(t)} \right], \tag{5.17}$$

$$\frac{\partial E_c(t)}{\partial w_{c_{ij}}^{(1)}(t)} = \frac{\partial E_c(t)}{\partial J(t)} \frac{\partial J(t)}{\partial p_i(t)} \frac{\partial p_i(t)}{\partial q_i(t)} \frac{\partial q_i(t)}{\partial w_{c_{ij}}^{(1)}(t)} \tag{5.18}$$

$$= \alpha e_c(t) w_{c_i}^{(2)}(t) \left[\frac{1}{2} \left(1 - p_i^2(t) \right) \right] x_j(t). \tag{5.19}$$

The adaptation in the action network is implemented by a feed-forward network similar to the one in Figure 5.2 except that the inputs are the n measured states and the output is the action $u(t)$. The associated equations for the action network are:

$$u(t) = \frac{1 - e^{-v(t)}}{1 + e^{-v(t)}}, \tag{5.20}$$

$$v(t) = \sum_{i=1}^{N_{ha}} w_{a_i}^{(2)}(t) g_i(t), \tag{5.21}$$

$$g_i(t) = \frac{1 - e^{-h_i(t)}}{1 + e^{-h_i(t)}}, \; i = 1, \cdots, N_{ha}, \tag{5.22}$$

$$h_i(t) = \sum_{j=1}^{n} w_{a_{ij}}^{(1)}(t) x_j(t), \; i = 1, \cdots, N_{ha}, \tag{5.23}$$

where v is the input to the action node, and g_i and h_i are the output and the input of the hidden nodes of the action network, respectively. Since the inputs to the action network only contain the state measurements, there is no $(n + 1)th$ term in (5.23) as in the critic network (see Eq. (5.14) for comparison). The update rule for the nonlinear multi-layer action network also contains two sets of equations.

(1) $\Delta \mathbf{w}_a^{(2)}$ (hidden to output layer):

$$\Delta w_{a_i}^{(2)}(t) = l_a(t) \left[-\frac{\partial E_a(t)}{\partial w_{a_i}^{(2)}(t)} \right], \tag{5.24}$$

$$\frac{\partial E_a(t)}{\partial w_{a_i}^{(2)}(t)} = \frac{\partial E_a(t)}{\partial J(t)} \frac{\partial J(t)}{\partial u(t)} \frac{\partial u(t)}{\partial v(t)} \frac{\partial v(t)}{\partial w_{a_i}^{(2)}(t)}, \tag{5.25}$$

$$= e_a(t) \sum_{i=1}^{N_{hc}} \left[w_{c_i}^{(2)}(t) \frac{1}{2} \left(1 - p_i^2(t) \right) w_{c_{i,n+1}}^{(1)}(t) \right] \cdot$$
$$\left[\frac{1}{2} \left(1 - u^2(t) \right) \right] g_i(t). \tag{5.26}$$

In the above equations, $\frac{\partial J(t)}{\partial u(t)}$ is obtained by changing variables and by chain rule. The result is the summation term. The term $w_{c_i,n+1}^{(1)}$ is the weight associated with the input element from the action network output.

(2) $\Delta \mathbf{w}_a^{(1)}$ (input to hidden layer):

$$\Delta w_{a_{ij}}^{(1)}(t) = l_a(t) \left[-\frac{\partial E_a(t)}{\partial w_{a_{ij}}^{(1)}(t)} \right], \tag{5.27}$$

$$\frac{\partial E_a(t)}{\partial w_{a_{ij}}^{(1)}(t)} = \frac{\partial E_a(t)}{\partial J(t)} \frac{\partial J(t)}{\partial u(t)} \frac{\partial u(t)}{\partial v(t)} \frac{\partial v(t)}{\partial g_i(t)} \frac{\partial g_i(t)}{\partial h_i(t)} \frac{\partial h_i(t)}{\partial w_{a_{ij}}^{(1)}(t)} \tag{5.28}$$

$$= e_a(t) \sum_{i=1}^{N_{hc}} \left[w_{c_i}^{(2)}(t) \frac{1}{2} \left(1 - p_i^2(t) \right) w_{c_{i,n+1}}^{(1)}(t) \right] \cdot$$
$$\left[\frac{1}{2} \left(1 - u^2(t) \right) \right] w_{a_i}^{(2)}(t) \left[\frac{1}{2} \left(1 - g_i^2(t) \right) \right] x_j(t). \tag{5.29}$$

In implementation, Eqs. (5.16) and (5.19) are used to update the weights in the critic network and Eqs. (5.26) and (5.29) are used to update the weights in the action network.

Pseudocode for implementing the direct NDP is summarized in Table 5.1.

Table 5.1   Pseudo Code for Direct NDP

Initialize w_a, w_c arbitrarily
Repeat (for each trial):
    Initialize state $X(t)$
    Repeat (for each step t of a trial):
        Choose action: $u(t) \leftarrow \mathbf{nn}_a(X(t))$
        Calculate estimated reward-to-go: $J(t) \leftarrow \mathbf{nn}_c(X(t), u(t))$
        Take action $u(t)$, observe $r(t)$, $X(t+1)$
        Repeat (Updating w_c):
$$e_c(t) \leftarrow \alpha J(t) - [J(t-1) - r(t)]$$
$$E_c(t) \leftarrow \tfrac{1}{2}e_c^2(t)$$
$$\Delta w_c(t) \leftarrow l_c(t)\left[-\frac{\partial E_c(t)}{\partial w_c(t)}\right]$$
$$w_c(t+1) \leftarrow w_c(t) + \Delta w_c(t)$$
        until $E_c(t) < T_c$ or maximum iteration number N_c reached
        Repeat (Updating w_a):
$$e_a(t) \leftarrow J(t) - R^*$$
$$E_a(t) \leftarrow \tfrac{1}{2}e_a^2(t)$$
$$\Delta w_a(t) \leftarrow l_a(t)\left[-\frac{\partial E_a(t)}{\partial w_a(t)}\right]$$
$$w_a(t+1) \leftarrow w_a(t) + \Delta w_a(t)$$
        until $E_a(t) < T_a$ or maximum iteration number N_a reached
        $X(t) \leftarrow X(t+1)$; $J(t-1) \leftarrow J(t)$
    until $X(t)$ is terminal
until maximum trial reached

5.4   COMPARISONS

In this section, the cart-pole balancing problem is used as a benchmark problem to compare four ADP algorithms. They are single-layer adaptive heuristic critic (1-AHC) [4], two-layer adaptive heuristic critic (2-AHC) [1, 2], the Q-learning method [21], and the direct NDP.

The cart-pole balancing problem is a classic control problem used in early reinforcement learning literature [1, 2, 4]. It has since been adopted frequently as a benchmark example to evaluate various learning algorithms [12, 13].

The system model under consideration was identical to that in [4]. This model provided four state variables: (1) $x(t)$, position of the cart on the track; (2) $\theta(t)$, angle of the pole with respect to the vertical position; (3) $\dot{x}(t)$, cart velocity; (4) $\dot{\theta}(t)$, angular velocity.

In this study a run consisted of a maximum of 50,000 consecutive trials. It was considered successful if the last trial (trial number less than 50,000) of the run lasted 600,000 time steps. Otherwise, if the controller was unable to learn to balance the cart-pole within 50,000 trials (i.e., none of the 50,000 trials lasted over 600,000 time steps), then the run was considered unsuccessful. In simulations, we used 0.02 seconds for each time step, and a trial was a complete process from start to fall. A pole was considered fallen when the pole was outside the range of $[-12°, 12°]$ and/or the cart was beyond the range of $[-2.4, 2.4]$ meters in reference to the central position on the track. Note that although the force F applied to the cart was binary, the control $u(t)$ fed into the critic network as shown in Figure 5.1 was continuous.

The direct NDP was implemented following the steps given in Table 5.1. Note that the weights in the action and the critic networks were trained using their internal cycles, N_a and N_c, respectively. That is, within each time step the weights of the two networks were updated for at most N_a and N_c times, respectively, or stopped once the internal training error threshold T_a and T_c were met. The parameters N_a, N_c, T_a, and T_c were later collectively referred to as stopping criteria.

Learning system performance was compared to evaluate the learning controller's overall ability to learn the task, to sustain noise, and to handle complex dynamics.

5.4.1   Algorithm Implementation

Here is a summary of implementation details of the four algorithms.

- Single-layer adaptive heuristic critic (1-AHC)

  The 1-AHC is one of the earliest reinforcement learning methods [4]. The learning controller implemented was from the original C source code off Richard Sutton's web site. The parameters were kept the same as follows: Action learning rate $\alpha = 1000$; Critic learning rate $\beta = 0.5$; TD discount factor $\gamma = 0.954$; Decay rate $\lambda = 0.8$; E trace decay rate $\delta = 0.9$. Quantized state variables were used as inputs. The state space was divided into 162 regions. A 162-component binary valued vector was a representation of one state input for the system. In this implementation, all components were zero except for one corresponding to the region where the current state resided.

- Two-layer adaptive heuristic critic (2-AHC)

  The 2-AHC is a direct extension of the 1-AHC by using two layer neural networks to present the actor and critic units, respectively [1, 2]. Therefore, continuous state inputs can be directly fed into each network after proper normalization. The algorithm implemented was from the original C source code off Charles Anderson's web site. The parameters were kept the same: Action learning rate $\alpha = 1.0$; Hidden action learning rate $\alpha_h = 0.2$; Critic learning rate $\beta = 0.2$; Hidden critic learning rate $\beta_h = 0.05$.

- Q-learning

  The one-step Q-learning with ϵ-greedy was implemented as described in [21]. The same input quantization as in 1-AHC was used for Q-learning. A Q-table with 162×2 entries were created and with initial Q values chosen to be 0. The ϵ-greedy policy allowed for an interleaving of "exploration" and "exploitation".

The learning parameters were chosen as follows: Q-learning rate $\beta = 0.5$; TD discount factor $\gamma = 0.999$; $\epsilon = 1 \times 10^{-5}$.

- Direct NDP

  The direct NDP was implemented per Table 5.1. The learning parameters were as follows: $l_c(0) = 0.3$; $l_a(0) = 0.3$; $l_c(f) = 0.005$; $l_a(f) = 0.005$, where $l_c(0)$ was the initial learning rate for the critic network which was later reduced to $l_c(f)$. The same can be said about $l_a(0)$ and $l_a(f)$. The stopping criteria used in training the neural networks were: $N_c = 50$; $N_a = 100$; $T_c = 0.05$; $T_a = 0.005$.

5.4.2 Simulation Implementation

All comparisons were based on 20 runs. A successful run was one where its last trial sustained 600,000 sequential control actions without failure (corresponding to 3 hours and 20 minutes in real time). Each implementation was written in C and ran on a Pentium IV PC.

The *first set* of experiments (with results shown in Table 5.2, Table 5.3, and Table 5.4) was to compare learning performance in a noisy cart-pole system. For each of the four algorithms, each run and each trial were conducted from the center start position. Three different noise levels (1%, 2%, and 5%) were evaluated. Uniform noise was added to the system states as follows. Let x be the current state variable. If the noise level was 5%, then the actual state values that fed into the learning controller was $x + random(-x * 0.05, x * 0.05)$. The same principle applied to other noise levels. In addition to sensor noise, a very small random disturbance with a magnitude less than ± 0.0001 was added into the system state measurements to ensure that the simulated system was subject to some level of noise even at $x = 0$. To provide a reference point, we also tabulated learning performance when no noise was added to the system. Results were compared for (1) number of trials needed to complete the learning control task; (2) learning time measured by CPU seconds; and (3) number of failed runs.

The *second set* of experiments (with results shown in Table 5.5) was to evaluate learning performance as in Anderson [1, 2] where both the cart and the pole started from random initial positions with random initial velocities. But no noise was added in this study. At the start of every trial, the positions and the velocities were set to new random values. Specifically, the four initial states were uniformly distributed within the following ranges: $x \in [-2.4, 2.4]$; $\dot{x} \in [-1.5, 1.5]$; $\theta \in [-12°, 12°]$; $\dot{\theta} \in [-1.5, 1.5]$.

5.4.3 Simulation Results

Statistical results for the first set of experiments are summarized in Tables 5.2 – 5.4. The results shown in Table 5.2 for 1-AHC, 2-AHC and Q-learning with 0% noise case were very close to those obtained in [12] when the authors compared the three algorithms to their SANE algorithm, using the identical cart-pole balancing system. The simulation results for 1-AHC, 2-AHC and Q-learning for other cases in Table 5.2 and Table 5.3 were comparable to the ones provided in the technical report [13]. The actual numbers may be a little different since a longer simulation time and a smaller number of allowed trials in each run were used to obtain the results.

Table 5.2   Comparison of Learning System Performance Measured by Trials Needed to Learn the Task under Different Levels of Noise From "0" Initial Condition

Noise Level	Mean Trials			
	1-AHC	2-AHC	Q-learning	Direct NDP
0%	320	5674	1815	10
1%	187	4890	3322	134
2%	2402	9887	9200	73
5%	N/A	14145	N/A	110

Table 5.3   Comparison of Learning System Performance Measured by Number of Failed Runs Out of 20 under Different Levels of Noise From "0" Initial Condition

Noise Level	Failed Runs			
	1-AHC	2-AHC	Q-learning	Direct NDP
0% & 1%	0	0	0	0
2%	0	1	0	0
5%	20	2	20	0

In obtaining the above results, it is noticed that direct NDP uses the least number of trials. But, that does not imply the shortest CPU time. In addition, Q-learning with a look-up table is the most efficient in terms of CPU time since it only updates a single table entry at every time step. However, Q-learning is efficient only if it is possible to explicitly create a table between the state/control and the Q values. As discussed earlier, this can pose serious constraints in real-world applications. Thus, in the following results, it is no longer taken into consideration since our focus is on examining scalability of the learning algorithms. The original 1-AHC is also omitted in the following results since it does not stand out using present learning performance measures.

Next in Table 5.4, direct NDP and 2-AHC were compared for CPU time. As shown in the table, the two algorithms were basically comparable considering the fact that the direct NDP implementation was not optimized for any learning parameters or the stopping criteria. It is to provide a realistic presentation to the readers of what direct NDP may be like without optimization. Our experience indicated that one could empiracally evaluate the stopping criteria and/or the learning parameters to result in more improved learning performance.

To illustrate the point, a linear action network and a linear critic network were used in the direct NDP scheme to measure the CPU time and the mean trial numbers. For the cases of adding 1%, 2% and 5% noise, the linear direct NDP successfully learned the task in under 70 trials and using less than 15 seconds of CPU time.

Some other considerations for direct NDP and 2-AHC are as follows. These algorithms involve a large number of free parameters during learning. In comparison, Q-learning only has three. This has created a dilemma or trade-off between consistent and repeatable learning results and the ability to generalize. The 2-AHC and the direct NDP are more noise tolerant and more likely to generalize than Q-learning; but Q-learning is more likely to generate consistant learning statistics due to its small number of free paremeters.

Table 5.4 Comparison of Learning System Performance Measured by CPU Time in Seconds under Different Levels of Noise from "0" Initial Condition

	CPU Time (s)			
	0% Noise	1% Noise	2% Noise	5% Noise
2-AHC	32	29	37	130
Direct NDP	134	92	77	90

As shown in Table 5.5, when the cart pole system was initialized to non-zero initial state conditions, Q-learning was still the fastest in CPU time (a few seconds). This is expected from its design principle. The learning speed was still comparable between the 2-AHC and the direct NDP. However, the 2-AHC failed at a considerably higher

Table 5.5 Comparison of Learning System Performance Measured by CPU Time, Mean Trials and Failed Runs for the Pole Balancing Task Starting from Random Initial Condition without Noise

	CPU Time (s)	Mean Trials	Failed Runs
1-AHC	162	2309	1
2-AHC	54	12310	10
Q-learning	4	1553	0
Direct NDP	41	64	0

rate now. This observation was quite similar to that reported in [12] for 2-AHC when dealing with complex system dynamics resulting from non-zero initial conditions. Note again that when two linear networks were used in place of the nonlinear neural networks, direct NDP learned the task from non-zero initial conditions in under 90 trials using less than 8 seconds of CPU time.

5.5   CONTINUOUS STATE CONTROL PROBLEM

The direct NDP design introduced in the previous sections was applied to a continuous state on-line learning control problem. The triple-link inverted pendulum problem with a single control input is illustrated in Figure 5.3.

5.5.1   Triple-Link Inverted Pendulum with Single Control Input

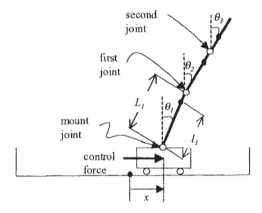

Fig. 5.3 Illustration of the triple-link inverted pendulum problem with state and control variables defined as depicted.

The system model for the triple-link problem was the same as that in [7]. The system parameters used in this study were identical to those in [7] as well.

The only control u (in volts) generated by the action network was converted into force by an analog amplifier through a conversion gain K_s (in Newtons/volt). In simulations, $K_s = 24.7125$ N/V. Each link could only rotate in the vertical plane about the axis of a position sensor fixed to the top of each link.

There were eight state variables in this model: (1) $x(t)$, position of the cart on the track; (2) $\theta_1(t)$, vertical angle of the first link joint to the cart; (3) $\theta_2(t)$, vertical angle of the second link joint to the first link; (4) $\theta_3(t)$, vertical angle of the third link joint to the second link; (5) $\dot{x}(t)$, cart velocity; (6) $\dot{\theta}_1(t)$, angular velocity of $\theta_1(t)$; (7) $\dot{\theta}_2(t)$, angular velocity of $\theta_2(t)$; (8) $\dot{\theta}_3(t)$, angular velocity of $\theta_3(t)$.

In the triple-link inverted pendulum problem, a run consisted of a maximum of 3,000 consecutive trials. A run was considered successful if the last trial of the run lasted 600,000 time steps. A unit time step was 5 ms. The constraints for the reinforcement learning were: (1) the cart track extended 1.0m to both ends from the center position; (2) the voltage applied to the motor was within $[-30, 30]$V range; (3) each link angle should be within the range of $[-20°, 20°]$ with respect to the vertical axis. In our simulations, condition (2) was assured by using a sigmoid function at the output of the action node. For conditions (1) and (3), if either one failed or both failed, the system provided an indicative signal $r = -1$ at the moment of failure, otherwise $r = 0$ all the time. Several experiments were conducted to evaluate the effectiveness of the direct NDP design. The results are reported in the following.

5.5.2 Simulation Results

Note that the triple-link system is highly unstable. To see this, observe that the positive eigenvalues of the linearized system model are far away from zero (the largest is around 10.0). In obtaining the linearized system model, the Coulomb friction coefficients are assumed to be negligible. Besides, the system dynamics change fast. It requires a sampling time below 10 ms.

Since the analog output from the action network was directly fed into the system, the controller was more sensitive to actuator noise than when using a binary control output. Experiments conducted included evaluations of direct NDP controller performance under uniform actuator noise, uniform or Gaussian sensor noise, and the case without noise. Specifically, the actuator noise was implemented through $u(t) = u(t) + \rho$, where ρ was a uniformly distributed random variable. For the sensor noise, both uniform and Gaussian random variables were added to the angle measurements θ. The uniform state sensor noise was implemented through $\theta = (1 +$ noise percentage$) \times \theta$. Gaussian sensor noise was zero mean with variance of either 0.1 or 0.2.

Before the presentation of our results, the learning parameters were summarized as follows: $l_c(0)$=0.8; $l_c(f)$=0.001; $l_a(0)$=0.8; $l_a(f)$=0.001. The stopping criteria in training the neural networks were: N_c=10; N_a=200; T_c=0.01; T_a=0.001. Simulation results are tabulated in Table 5.6.

Figure 5.4 is an example of typical angle trajectories of the triple-link under direct NDP control for a successful learning trial.

The results presented in this case study have demonstrated the applicability of the direct NDP designs to a non-trivial continuous state control problem. It is worth mentioning that the direct NDP controlled angle variations are significantly smaller than those obtained from nonlinear control system design [7].

Table 5.6   Performance Evaluation of Direct NDP Learning Controller When Balancing a Triple-Link Inverted Pendulum (The second column represents the percentage of successful runs out of 100. The third column depicts the average number of trials to learn the task. The average is taken over the successful runs.)

Noise Type	Success Rate	# of Trials
None	97%	1194
Uniform 5% actuator	92%	1239
Uniform 10% actuator	84%	1852
Uniform 5% sensor on θ_1	89%	1317
Uniform 10% sensor on θ_1	80%	1712
Gaussian sensor on θ_1 variance = 0.1	85%	1508
Gaussian sensor on θ_1 variance = 0.2	76%	1993

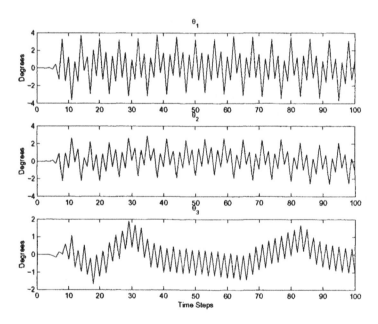

Fig. 5.4   Typical angle trajectories of the triple-link during a successful learning trial using on-line NDP control when the system is free of noise.

5.6   CALL ADMISSION CONTROL FOR CDMA CELLULAR NETWORKS

In the present example, direct NDP was applied to call admission control in SIR-based power-controlled direct-sequence code-division multiple access (DS-CDMA) cellular networks, where SIR denotes signal-to-interference ratio. For a given set of parameters including traffic statistics and mobility characteristics, fixed call admission control schemes can sometimes yield optimal solutions [16] using a grade of service measure. However, such schemes are not adaptable to changing network conditions [9, 16–18]. Therefore, we developed a direct NDP-based self-learning call admission control algorithm for the CDMA wireless network application. The performance of the present algorithm was studied through computer simulations and compared with existing call admission control algorithms. Results show that the direct NDP-based self-learning call admission control algorithm outperforms existing call admission control algorithms under the network conditions specified below.

Problem Definition

Consider a base station with N concurrent active connections. The power received from each connection is denoted by S_n, $n = 1, \cdots, N$. In an SIR-based power-controlled DS-CDMA network [3, 6, 10, 18], the desired value of the power level S_n is a function of the number of active connections in the home cell (home base station) and total other cell interference. A DS-CDMA cellular network with a single class of service (e.g., voice) is considered. The maximum received power at a base station for that class of service is denoted by H.

Two types of calls, new and handoff, are considered for admission controls. For an arriving call requesting admission, one of the two control actions, accept or reject, is selected. Typically handoff calls are given higher priority than new calls [9, 16–18]. This is accomplished by using different thresholds for new calls and handoff calls as shown below.

The criteria for accepting new calls can be expressed as

$$u(t) = \begin{cases} 1 \text{ (accept)}: & \text{if } 0 < S_{new} \leq T \text{ and} \\ & \qquad 0 < S_n < H \quad \text{for} \quad n = 1, 2, \cdots, N, \\ -1 \text{ (reject)}: & \text{else,} \end{cases}$$

where $T \leq H$ is the threshold for new calls.

For handoff calls, a more relaxed threshold is given as

$$u(t) = \begin{cases} 1 \text{ (accept)}: & \text{if } 0 < S_{handoff} \leq H \text{ and} \\ & \qquad 0 < S_n \leq H \quad \text{for} \quad n = 1, 2, \cdots, N, \\ -1 \text{ (reject)}: & \text{else.} \end{cases}$$

The objective of this study is to optimize the grade of service (GoS) measure under various, possibly changing environmental conditions and user behaviors. The GoS measure [28] is defined as

$$\text{GoS} = P(\text{call blocking}) + w \times P(\text{handoff failure}), \tag{5.30}$$

where the first term is the probability of call blocking and the second term is the probability of handoff blocking with the weighting factor w (typically chosen as 10) signifying a relative emphasis on the two terms.

Algorithm Implementation

Since only binary control, accept or reject, was needed in call admission, only the critic network shown in Figure 5.1 was utilized where the control input to the critic network was $u(t) = +1$ for accept and $u(t) = -1$ for reject. A correct admission decision was associated with a reward signal of "0" otherwise a penalty signal of a positive value. The inputs to the critic network were the system states, the admission control decision (accept or reject) and the call type (new call or handoff call). The system state was the total interference received at the base station. In DS-CDMA cellular systems, interference increases as traffic load increases.

To formulate a proper value function for the learning controller, a cost is first associated with each active connection n as follows:

- For new call arrival

$$E_n = \begin{cases} \xi \cdot \max\left\{\left[\frac{S_n}{T} - 1\right], 0\right\}, & \text{when } u(t) = 1, \\ \xi \cdot \max\left\{\left[1 - \frac{S_n}{H}\right], 0\right\}, & \text{when } u(t) = -1, \end{cases} \tag{5.31}$$

- For handoff call arrival

$$E_n = \xi \cdot \max\left\{u(t) \cdot \left[\frac{S_n}{H} - 1\right], 0\right\}, \tag{5.32}$$

where $\xi > 0$ is a coefficient, $T \leq H$, and $E_n \geq 0$ for all $n = 0, 1, \cdots, N$ (S_0 represents S_{new} or $S_{handoff}$).

A collective cost is then chosen from all N connections as

$$E = \begin{cases} \max_{0 \leq n \leq N}(E_n), & \text{if } u(t) = 1, \\ \min_{0 \leq n \leq N}(E_n), & \text{if } u(t) = -1. \end{cases} \tag{5.33}$$

We can now define an instantaneous reinforcement signal r as

$$r = \frac{E}{1 + E}. \tag{5.34}$$

The instantaneous reinforcement signal r defined in (5.34) has the following properties: (1) $r \approx 0$ when $E \ll 1$, and (2) $r \approx 1$ when $E \gg 1$. The choice of the present reinforcement signal in (5.34) clearly shows minimum points (the flat areas) that our call admission control tries to reach and the points with high penalty to avoid. The objective of the learning controller is to make the correct admission control decisions such that the total reward-to-go $R(t)$ in Eq. (5.1) is optimized. We have thus formulated an approximate dynamic programming problem. Next, we use the learning architecture shown in Figure 5.1 to obtain an approximate solution. Note again that in this example, the action network is omitted since the binary control decision is well defined.

With the reinforcement signal given by (5.33) and (5.34), we have

$$0 < J(t) < \frac{1}{1 - \alpha},$$

where α is the discount factor in Eq. (5.1).

Alternatively, a different cost and consequently instantaneous reinforcement can be defined in place of (5.31) as follows:

$$E_n = \begin{cases} \xi \cdot \max \left\{ \left[\frac{S_n}{H} - 1 \right], 0 \right\}, & \text{when } u(t) = 1, n_a \leq N_h, \\ \xi \cdot \max \left\{ \left[\frac{S_n}{T} - 1 \right], 0 \right\}, & \text{when } u(t) = 1, n_a > N_h, \\ \xi \cdot \max \left\{ \left[1 - \frac{S_n}{H} \right], 0 \right\}, & \text{when } u(t) = -1, \end{cases} \tag{5.35}$$

where n_a is the number of handoff calls in the cell that are accepted and are still active, and N_h is a fixed parameter indicating the threshold for low traffic load. It is expected that different cost measure will lead to different system performance.

Data are collected prior to training the learning controller. When a call arrives, one can accept or reject the call with any scheme and calculate the instantaneous reinforcement signal r for the system as presented above. For instance, a call can be accepted and rejected simply by using a equal probability of 0.5. In the mean time, state and environmental data are collected corresponding to each action. These include total interference, call type (new call or handoff call), and traffic load indication (high or low).

Simulation Results

To obtain simulation results, similar network parameters were used as those in [18, 10]. The arrival rate consisted of the new call attempt rate λ_c and the handoff call attempt rate λ_h. The parameter λ_c depended on the expected number of subscribers per cell while λ_h depended on traffic load, user velocity, and cell coverage areas [8, 9]. In our simulations, we assumed that $\lambda_c : \lambda_h = 5 : 1$ [9]. A channel was released upon call completion or handoff to a neighboring cell. The channel occupancy time was assumed to be exponentially distributed [8, 9] with identical mean value of $1/\mu = 3$ minutes.

We first conducted a comparison study between the present direct NDP call admission control algorithm and a static call admission control algorithm developed in [11] with fixed thresholds for new calls given by $T = H, T = 0.8H$, and $T = 0.5H$, respectively. The arrival rate in all neighboring cells was fixed at 18 calls/minute. The training data were collected as mentioned in the previous section. Parameters $\xi = 10$ and $T = 0.5H$ were chosen for Eq. (5.31).

The critic network was implemented as a multilayer feed-forward neural network with 3–6–1 structure, that is, 3 neurons at the input layer, 6 neurons at the hidden layer, and 1 neuron at the output layer. Both the hidden and output layers used the hyperbolic tangent function as the activation function. The three inputs were the total interference, the action and the call type.

Simulation results in Figure 5.5 show that the performance of the direct NDP algorithm is similar to the case of static algorithm with $T = 0.5H$, since we chose $T = 0.5H$ in Eq. (5.31) for our learning control algorithm. The direct NDP algorithm performs worse than the other two cases of static algorithms ($T = 1.0H$ and $T = 0.8H$) at low call arrival rates, since it reserved too much capacity for handoff calls and rejected too many new calls.

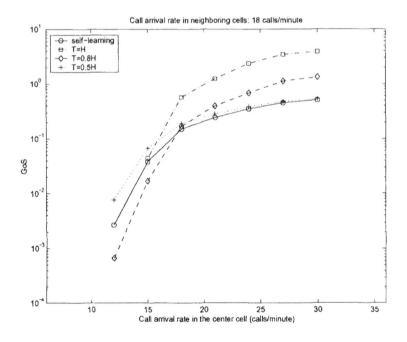

Fig. 5.5 Comparison result using cost function defined in (5.31).

To improve performance when the call arrival rate was low, the alternative cost was used for new calls as in Eq. (5.35), where $N_h = 15$ in our simulation. The training data were collected using one of the static algorithms with fixed threshold or the previous critic network. Four inputs were provided to the critic network: the same three as before and a new input of either "1" when $n_a \leq N_h$ or "−1" for otherwise. A neural network of the structure 4–8–1 was used for the critic network.

Figure 5.6 shows the result of applying the new critic network to the same traffic pattern as in the previous case. The self-learning controller using the new critic network has rendered the best performance by simply changing the cost function from (5.31) to (5.35).

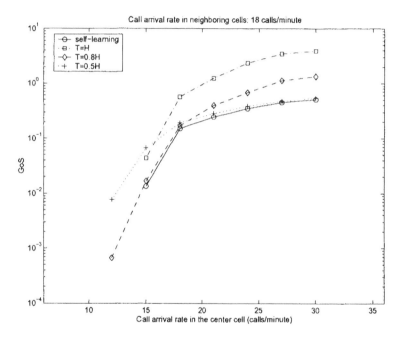

Fig. 5.6 Comparison study using cost function defined in (5.35).

The next comparison study is between the present self-learning call admission control algorithm and that of [18]. The algorithm of [18] was implemented as follows. First, the base station controller read the current interference from the power strength measurer. The current interference margin (CIM) and handoff interference margin (HIM) were then estimated, where CIM < HIM. A total interference margin (TIM) was set according to the quality of service target. If CIM > TIM, reject the call admission request. If HIM < TIM, accept the call request. If CIM < TIM < HIM, then only handoff calls were accepted.

Simulation results in Figure 5.7 compares the self-learning algorithm with the one in [18] that reserved 1, 2, 3 channels for handoff calls, respectively. The arrival rate in all neighboring cells was fixed at 18 calls/minute. The self-learning algorithm performs the best in Figure 5.7. Since the algorithm in [18] used a guard channel policy in CDMA systems, GC = 1 performed the best when the load was low, and GC = 3 performed the best when the load was high. However, our proposed algorithm can adapt to varying traffic load conditions. It has the best overall performance under various traffic loads being considered.

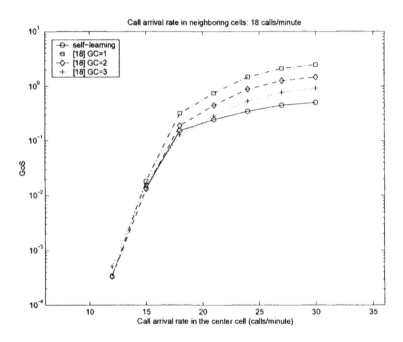

Fig. 5.7   Comparison studies with the algorithm in [18].

5.7   CONCLUSIONS AND DISCUSSIONS

This chapter aims at providing a clear introduction of the direct NDP design and its implementation. Detailed derivations of the algorithm are provided along with the pseudocode. With this information, together with the on-line Matlab code at the first author's web site on the benchmark cart-pole balancing example, readers should be able to evaluate the results using direct NDP.

This chapter has also provided a broad perspective on how the direct NDP fits in the large existing ADP literature. Empirical results indicate that direct NDP is capable of producing successful learning results under a wide range of parameterization and system scenarios. However, it is not as consistent as Q-learning in terms of arriving at similar controllers after learning. Q-learning has very few free parameters, which is the key to producing consistent learning results or solutions. However, as discussed earlier in the chapter, it is difficult to scale or generalize for large problems.

As an alternative to a tabulated value function for each possible state and control action, a neural network based value function approximation has the advantage of generalization over the input space and being more resistant to noise. However a trade-off, observed empirically, is that on-line training of the neural value function may not always be accurate when compared to table-based approaches. This is mainly due to the randomness in parameter initialization, and the possibility that the parameter set converges to a local minimum in the error function space. With this said, our experience indicate that direct NDP is in general insensitive to parameters such as initial weights in the action and/or critic network and the learning rate coefficients as long as they satisfy some general guidelines such as those specified in stochastic approximation. Simulation results in the present chapter and Chapter 21 have demonstrated that the direct NDP mechanism is capable of handling complex system dynamics, and it is robust in the presence of noise and disturbance.

The examples used in this chapter are continuous state problems. The triple-link pendulum is a multi-state-single-control problem. Even though still a toy problem, it is not an easy problem for a typical classic controller. The three controlled pendulum angles shown in Figure 5.4 are much tighter than those in [7]. Later in Chapter 21, direct NDP is used to control a helicopter under realistic flying conditions for stabilization, maneuvering, and reconfiguration after a component failure. All these results point to the possibility that direct NDP type of designs do have the potential to generalize. However many challenges remain to be addressed.

Direct NDP is a model-independent approach to ADP. As seen from comparisons to Q-learning and actor-critic methods, even though clearly defined gradient information may have provided a clear parameter search direction, it still contains certain levels of randomness due to the very nature of neural networks as approximators. A recent study has demonstrated that this randomness can result in large variances in controller bandwidth and other control performance measures [27]. This is not surprising since unlike many classic control design methods where specific system performance measures are explicitly taken into design consideration, in the ADP setting, one is only given a delayed, qualitative measure of the system's well being. The immediate issue is whether or not one can take advantage of a system model and a priori system knowledge and use them in the design process.

As discussed earlier, a good learning controller should be one that learns to perform a task quickly and also with a guarantee to learn to perform the task successfully through trial and error. Another factor that may not be explicitly present in the reinforcement signal is the degree of meeting the performance requirement. In both pendulum case studies, however, it is quite intriguing to see that the learning controllers not only balanced the poles, but also maintained the poles as centered as possible.

In summary, our results have demonstrated the potential of direct NDP in solving large scale problems guided by delayed, high-level feedbacks despite the open issues discussed earlier.

Acknowledgments

The first and second authors' research was supported by the National Science Foundation under grants ECS-0002098 and ECS-0233529. The third author's research was supported by the National Science Foundation under grants ECS-9996428 and ANI-0203063.

Bibliography

1. C. Anderson, Strategy learning with multi-layer connectionist representations, *Proc. Fourth International Workshop on Machine Learning,* pp. 103–114, 1987.

2. C. Anderson, Learning to control an inverted pendulum using neural networks, *IEEE Control Systems Magazine,* vol. 9, no. 3, pp. 31–37, 1989.

3. S. Ariyavisitakul, Signal and interference statistics of a CDMA system with feedback power control – Part II, *IEEE Trans. Communications,* vol. 42, pp. 597–605, 1994.

4. A. G. Barto, R. S. Sutton, and C. W. Anderson, Neuron like adaptive elements that can solve difficult learning control problems, *IEEE Trans. Systems, Man, and Cybernetics,* vol. 13, pp. 834–847, 1983.

5. D. P. Bertsekas and J. N. Tsitsiklis, *Neuro-Dynamic Programming,* Athena Scientific, Belmont, MA, 1996.

6. Z. Dziong, M. Jia, and P. Mermelstein, Adaptive traffic admission for integrated services in CDMA wireless-access networks, *IEEE Journal on Selected Areas in Communications,* vol. 14, pp. 1737–1747, 1996.

7. K. D. Eltohamy and C. Y. Kuo, Nonlinear optimal control of a triple link inverted pendulum with single control input, *International Journal of Control,* vol. 69, no. 2, pp. 239–256, 1998.

8. R. A. Guerin, Channel occupancy time distribution in a cellular radio system, *IEEE Trans. Vehicular Technology,* vol. 35, pp. 89–99, 1987.

9. D. Hong and S. S. Rappaport, Traffic model and performance analysis for cellular mobile radio telephone systems with prioritized and nonprioritized handoff procedures, *IEEE Trans Vehicular Technology,* vol. 35, pp. 77–92, 1986.

10. D. K. Kim and D. K. Sung, Capacity estimation for an SIR-based power-controlled CDMA system supporting ON-OFF traffic, *IEEE Trans. Vehicular Technology,* vol. 49, pp. 1094–1100, 2000.

11. D. Liu and Y. Zhang, An adaptive call admission control algorithm for SIR-based power-controlled integrated services CDMA cellular networks, *Proc. International Conference on Telecommunications,* pp. 610–638, 2002.

12. D. E. Moriarty and R. Miikkulainen, Efficient reinforcement learning through symbiotic evolution, *Machine Learning,* vol. 22, no. 11-32, pp. 11–32, 1996.

13. M. Pendrith, On reinforcement learning of control actions in noisy and non-Markovian domains, *Technical Report UNSW-CSE-TR-9410,* 1994.

14. D. V. Prokhorov, R. A. Santiago, and D. C. Wunsch II, Adaptive critic designs: a case study for neuro-control, *Neural Networks,* vol. 8, pp. 1367–1372, 1995.

15. D. V. Prokhorov and D. C. Wunsch II, Adaptive critic designs, *IEEE Trans. Neural Networks,* vol. 8, no. 5, pp. 997–1007, 1997.

16. R. Ramjee, D. Towsley, and R. Nagarajan, On optimal call admission control in cellular networks, *Wireless Networks,* vol. 3, no. 1, pp. 29–41, 1997.

17. S. S. Rappaport and C. Purzynski, Prioritized resource assignment for mobile cellular communication systems with mixed services and platform types, *IEEE Trans. Vehicular Technology,* vol. 45, pp. 443–458, 1996.

18. S. M. Shin, C. H. Cho, and D. K. Sung, Interference-based channel assignment for DS-CDMA cellular systems, *IEEE Trans. Vehicular Technology,* vol. 48, pp. 233–239, 1999.

19. J. Si and Y. Wang, Online learning control by association and reinforcement, *IEEE Trans. Neural Networks,* vol. 12, no. 2, pp. 264–276, 2001.

20. R. S. Sutton and A. G. Barto, *Reinforcement Learning: An Introduction,* MIT Press, Cambridge, MA, 1998.

21. C. J. C. H. Watkins, *Learning from Delayed Rewards,* Ph.D. Thesis, University of Cambridge, UK, 1989.

22. P. Werbos, Advanced forecasting methods for global crisis warning and models of intelligence, *General System Yearbook,* vol. 22, pp. 25-38, 1977.

23. P. Werbos, A menu of design for reinforcement learning over time, *Neural Networks for Control,* in W. T. Miller III, R. S. Sutton, and P. J. Werbos (eds.), MIT Press, Cambridge, MA, 1990.

24. P. Werbos, Neuro-control and supervised learning: An overview and valuation, *Handbook of Intelligent Control,* in D. White and D. Sofge (eds.), Van Nostrand Reinhold, New York, 1992.

25. P. Werbos, Approximate dynamic programming for real-time control and neural modeling, *Handbook of Intelligent Control,* in D. White and D. Sofge (eds.), Van Nostrand Reinhold, New York, 1992.

26. P. Werbos, Tutorial on neurocontrol, control theory and related techniques: from back propagation to brain-like intelligent systems, *12th Intl Conference*

on Math. & Computer Modeling & Scientific Computing, 1999. Also available at http://www.iamcm.org/pwerbos/.

27. L. Yang, J. Si, K. Tsakalis, and A. Rodriguez, Analyzing and enhancing direct NDP designs using a control-theoretic approach, *IEEE International Symposium on Intelligent Control,* Houston, TX, pp. 529-532, 2003.

28. A guide to DECT features that influence the traffic capacity and the maintenance of high radio link transmission quality, including the results of simulations, *ETSI Technical Report: ETR 042,* 1992.

6 The Linear Programming Approach to Approximate Dynamic Programming

DANIELA PUCCI DE FARIAS

Massachusetts Institute of Technology

Editor's Summary: This chapter addresses the issue of the "curse of dimensionality" by treating ADP as the "dual" of the linear programming problem and introduces the concept of approximate linear programming (ALP). It provides a brief introduction to the use of Markov Decision Process models. For a more comprehensive study of MDP models, and the techniques that can be used with them, read Chapters 11 and 12. This chapter discusses the performance of approximate LP policies, approximation error bounds, and provides an application to queueing networks. Another queueing network example can be found in Chapter 12. The chapter finishes with an efficient constraint sampling scheme.

6.1 INTRODUCTION

Dynamic programming offers a unified approach to solving problems of stochastic control. Central to the methodology is the cost-to-go function, which is obtained via solving Bellman's equation. The domain of the cost-to-go function is the state space of the system to be controlled, and dynamic programming algorithms compute and store a table consisting of one cost-to-go value per state. Unfortunately, the size of a state space typically grows exponentially in the number of state variables. Known as the *curse of dimensionality*, this phenomenon renders dynamic programming intractable in the face of problems of practical scale.

One approach to dealing with the curse of dimensionality is to generate an approximation within a parameterized class of functions, in a spirit similar to that of statistical regression. In particular, to approximate a cost-to-go function J^* map-

153

ping a state space S to reals, one would design a parameterized class of functions $\tilde{J} : S \times \Re^K \mapsto \Re$, and then compute a parameter vector $r \in \Re^K$ to "fit" the cost-to-go function, so that

$$\tilde{J}(\cdot, r) \approx J^*.$$

Note that there are two important preconditions to the development of an effective approximation. First, it is necessary to choose a parameterization $\tilde{J}$ that can closely approximate the desired cost-to-go function, that is, the parameterization must be such that $\inf_r \|J^* - \tilde{J}(\cdot, r)\|$ is small. Second, it is necessary to have an algorithm that identifies a parameter vector $\bar{r}$ such that $\|J^* - \tilde{J}(\cdot, \bar{r})\|$ is small.

At this stage, choosing appropriate classes of parametric functions is an art. One must balance issues such as efficiency of computation and storage of $\tilde{h}$, difficulty in computing appropriate parameters r, and ability of the parametric class to produce reasonable approximations. In practice, the choice of a suitable class typically requires analysis and/or experimentation with the specific problem at hand. "Regularities" associated with the function, for example, can guide the choice of representation.

In this chapter, we assume that a linearly parameterized class $\tilde{J}$ is given. Such a class can be represented by

$$\tilde{J}(\cdot, r) = \sum_{k=1}^{K} r_k \phi_k,$$

where each ϕ_k is a "basis function" mapping S to $\Re$ and the parameters $r_1, \ldots, r_K$ represent basis function weights. We consider the problem of generating an appropriate parameter r through *approximate linear programming* (ALP) [8, 25], an algorithm based on a formulation that generalizes the linear programming approach to exact dynamic programming [5, 11–13, 18, 20]. The development in this chapter is twofold. First, we analyze the performance of approximate linear programming. We address questions such as:

- If the set of basis functions has the potential to produce a close approximation to the cost-to-go function, is approximate linear programming also able to produce a close approximation?

- What guarantees can we offer for the expected costs induced by policies produced by approximate linear programming? How close are they to the optimal costs?

- How do the approximation errors and performance losses scale as we increase the size of the state space of the system? What about increasing the dimension of the state space?

Second, we develop some analysis leading to streamlined guidelines for implementation of approximate linear programming. We investigate issues such as how different choices for the algorithm's free parameters affect the quality of the approximation being generated, and how we can deal with the large number of constraints involved in the linear programs that have to be solved by the algorithm.

6.1.1 Performance Analysis for Approximate Linear Programming

Over the years, interest in approximate dynamic programming has been fuelled to a large extent by stories of empirical success in applications such as backgammon [26], job shop scheduling [31], elevator scheduling [6] and pricing of American options [19, 29]. These case studies point to approximate dynamic programming as a potentially powerful tool for large-scale stochastic control. However, significant trial and error is involved in most success stories found in the literature, and duplication of the same success in other applications has proven difficult. This is partly due to a relatively poor understanding of how, why and when approximate dynamic programming algorithms work, and poses a barrier to the use of such algorithms in industry. Hence the analysis of performance and errors associated with approximate dynamic programming algorithms is not only an interesting theoretical challenge but also a practical necessity, and the recent advancements in the theoretical analysis of approximate linear programming can be viewed as steps toward the ultimate goal of developing approximate dynamic programming into a truly practical, streamlined methodology for aiding decision-making in large-scale problems.

The version of approximate linear programming described here was first introduced in [8] and represents a variant of the algorithm originally proposed by Schweitzer and Seidman [25]. While the original ALP algorithm may exhibit poor scaling properties, the version introduced in [8] enjoys strong theoretical guarantees.

A central question about any approximate dynamic programming algorithm is how well it approximates the cost-to-go function. The quality of the approximation depends on the choice of basis functions, hence the error induced by an approximate dynamic programming algorithm is usually characterized in relative terms — we compare it against the best error that can be achieved given the selection of basis functions.

Section 6.5 presents an error bound establishing that approximate linear programming yields an approximation error that is comparable to the best that can be achieved given the selection of basis functions. Besides offering guarantees about the behavior of the algorithm, the bound provides insight about the impact of the algorithm's free parameters onto the final approximation, thus leading to some guidance in their selection. It also provides information about how well the algorithm scales as the problem dimensions increase.

Comparison of approximations to the cost-to-go function involve the choice of a metric on the space of functions over the state space that determines how good different approximations are perceived to be. Of course, the ultimate comparison is the expected cost of the policy induced by the approximation, and this should be reflected in the metric being chosen. Section 6.4 includes a bound on the expected cost increase due to using policies generated by approximate linear programming instead of the optimal policy. The bound suggests a natural metric for comparison of different approximations to the cost-to-go function and gives reassurance that the approximation error bounds under consideration are meaningful, as they are stated in terms of a metric that is compatible with the natural performance-based metric. The performance bound also provides guidance on how to set the free parameters in the approximate linear programming algorithm so as to indirectly optimize performance of the policy being obtained.

6.1.2   Constraint Sampling

The approximate linear programming method, as the name suggests, relies on a linear program — the *approximate LP* — to compute an approximation to the cost-to-go function. The number of variables involved in the approximate LP is relatively small — one per basis function. However, the number of constraints — one per state-action pair — is generally intractable, and this presents an obstacle.

Section 6.7 describes an approximation scheme based on constraint sampling that replaces the approximate LP with an auxiliary linear program — the *reduced LP* — with the same variables and objective as the approximate LP, but with only a sampled subset of constraints. The reduced LP is motivated by the fact that, because there is a relatively small number of variables in the approximate LP, many of the constraints should have a minor impact on the feasible region and do not need to be considered; in particular, it can be shown that, by sampling a tractable number of constraints according to any given distribution, the solution of the resulting reduced LP is guaranteed to be *near-feasible* with high probability. In other words, with high probability, the set of constraints violated by the solution of the reduced LP has small probability under the sampling distribution.

The fact that constraint sampling yields near-feasible solutions is a general property of LP's with a large number of constraints and relatively few variables. Properties specific to approximate linear programming imply that, under certain conditions, the error in approximating the cost-to-go function yielded by the reduced LP should be close to that yielded by the approximate LP, with high probability.

6.1.3   Literature Review

Much of the effort in the literature on approximate linear programming has been directed toward efficient implementation of the algorithm. Some general analysis of the algorithm can be found in [15, 27, 28]. Trick and Zin [27, 28] develop heuristics for combining the linear programming approach with successive state aggregation/grid refinement in two-dimensional problems. Some of their grid generation techniques are based on stationary state distributions, which also appear in the analysis of state-relevance weights presented here. An important feature of the linear programming approach is that it generates *lower bounds* as approximations to the cost-to-go function; Gordon [15] discusses problems that may arise from this fact and suggests constraint relaxation heuristics. One of these problems is that the linear program used in the approximate linear programming algorithm may be overly constrained, which may lead to poor approximations or even infeasibility. The approach taken here prevents this — part of the different between the variant of approximate linear programming presented here and the original one proposed by Schweitzer and Seidmann is that the former involves certain basis functions that guarantee feasibility and also lead to guaranteed bounds on the approximation error.

The constraint sampling scheme presented here does not exploit any possible regularity associated with structure in the constraints arising in specific problems, or a particular choice of basis functions, which might lead to much tighter bounds or even methods for exact solution of the approximate LP. Results of this nature can be found in the literature. Morrison and Kumar [21] formulate approximate linear programming algorithms for queueing problems with a certain choice of basis functions that renders all but a relatively small number of constraints redundant. Guestrin et al. [17] exploit the structure arising when factored linear architectures are used for approximating the cost-to-go function in factored MDP's to efficiently implement approximate linear programming with a tractable number of constraints. Traditional methods designed to deal with linear programs with large numbers of constraints, such as cutting planes and column generation, may also be useful in specific settings. Schuurmans and Patrascu [24] devise a constraint generation scheme for factored MDP's with factored linear architectures that lacks the guarantees of the algorithm presented in [17], but requires a smaller amount of time on average. Grötschel [16] presents an efficient cutting-plane method for the traveling salesman problem.

Approximate linear programming also appears in the literature as a tool for analysis and solution of stochastic control problems in various application areas. Paschalidis and Tsitsiklis [22] apply the algorithm to two-dimensional problems arising in pricing of network services. Morrison and Kumar [21] use approximate linear programming to derive bounds on the performance of queueing networks. Adelman [1, 2] uses approximate linear programming as an analytical tool aiding in the development of policies for inventory control. De Farias et al. [7] use approximate linear programming to derive policies for optimal resource allocation in web server farms.

The results presented here appeared originally in articles [8, 9] by de Farias and Van Roy. Proofs are omitted and can be found in these references.

6.1.4   Chapter Organization

This chapter is organized as follows. Sections 6.2 and 6.3 provide some background, including a formulation of the stochastic control problem under consideration and discussion of linear programming approaches to exact and approximate dynamic programming. Section 6.4 presents bounds on the performance of policies generated by ALP, and a discussion on how performance optimization can be incorporated in the algorithm by appropriate choice of certain parameters in the approximate LP. Section 6.5 presents bounds on the error in the approximation to the cost-to-go function yielded by ALP. The error bounds involve problem-dependent terms, and Section 6.6 focuses on characteristics of these terms in examples involving queueing networks. Section 6.7 describes a constraint sampling scheme for dealing with the large number of constraints involved in the approximate LP. Finally, Section 6.8 discusses directions for future research on approximate linear programming.

6.2 MARKOV DECISION PROCESSES

A Markov decision process (MDP) in discrete-time is characterized by a tuple $(\mathcal{S}, \mathcal{A}, P.(\cdot, \cdot), g.(\cdot))$, with the following interpretation. $\mathcal{S}$ represents a finite state space $\mathcal{S}$ of cardinality $|\mathcal{S}| = N$. For each state $x \in \mathcal{S}$, there is a finite set of available actions $\mathcal{A}_x$. Taking action $a \in \mathcal{A}_x$ when the current state is x incurs cost $g_a(x)$. State transition probabilities $P_a(x, y)$ represent, for each pair (x, y) of states and each action $a \in \mathcal{A}_x$, the probability that the next state will be y given that the current state is x and the current action is a.

A *policy* u is a mapping from states to actions. Given a policy u, the dynamics of the system follow a Markov chain with transition probabilities $P_{u(x)}(x, y)$. For each policy u, we define a transition matrix P_u whose (x, y)th entry is $P_{u(x)}(x, y)$.

We let $x_t, t = 0, 1, \ldots$ denote the state of the MDP at time t.

The problem of stochastic control amounts to selection of a policy that optimizes a given criterion. We will employ as the optimality criterion the infinite-horizon discounted cost

$$J_u(x) = \mathrm{E}\left[\sum_{t=0}^{\infty} \alpha^t g_u(x_t) | x_0 = x\right],\tag{6.1}$$

where $g_u(x)$ is used as shorthand for $g_{u(x)}(x)$ and the discount factor $\alpha \in (0, 1)$ reflects inter–temporal preferences. Throughout the paper, we let $\mathrm{E}[\cdot]$ denote expected value. It is well known that there exists a single policy u that minimizes $J_u(x)$ simultaneously for all x, and the goal is to identify that policy.

Let us define operators T_u and T by

$$T_u J = g_u + \alpha P_u J$$

and

$$T J = \min_u T_u J,$$

where the minimization is carried out component–wise. For any vector J, we also define the *greedy* policy associated with J given by

$$u_J(x) = \operatorname*{argmin}_{a \in \mathcal{A}_x}(g_a(x) + \alpha \sum_{y \in \mathcal{S}} P_a(x, y) J(y)).$$

Dynamic programming involves solution of Bellman's equation

$$J = T J.$$

The unique solution J^* of this equation is the optimal cost–to–go function

$$J^* = \min_u J_u,$$

and a policy is optimal if and only if it is greedy with respect to J^*.

Dynamic programming offers a number of approaches to solving Bellman's equation. One of particular relevance to our chapter makes use of linear programming, as we will now discuss. Consider the problem

$$\max_J \quad c^T J \tag{6.2}$$
$$\text{s.t.} \quad TJ \geq J,$$

where $c : S \mapsto \Re$ is a vector with positive components, which we will refer to as *state–relevance weights*, and c^T denotes the transpose of c. It can be shown that any feasible J satisfies $J \leq J^*$. It follows that, for any set of positive weights c, J^* is the unique solution to (6.2).

Note that T is a nonlinear operator, thus the constrained optimization problem (6.2) is not a linear program. However, it is easy to reformulate the problem as a linear program, by noting that each nonlinear constraint $(TJ)(x) \geq J(x)$ is equivalent to a set of linear constraints

$$g_a(x) + \alpha \sum_{y \in S} P_a(x, y) J(y) \geq J(x), \; \forall a \in \mathcal{A}_x.$$

We refer to problem (6.2) as the *exact LP*.

As mentioned in the introduction, state spaces for practical problems are huge due to the curse of dimensionality. Hence the linear program of interest involves prohibitively large numbers of variables and constraints. The approximation algorithm discussed in the sequel reduces dramatically the number of variables.

6.3   APPROXIMATE LINEAR PROGRAMMING

An approach to alleviating the curse of dimensionality is to generate approximations to the cost-to go function of the form

$$J^*(x) \approx \tilde{J}(x, r) = \sum_{i=1}^{K} r_i \phi_i(x), \tag{6.3}$$

where $\phi_i : S \mapsto \Re, i = 1, \ldots, K$ are prespecified basis functions. Define a matrix $\Phi \in \Re^{|S| \times K}$ given by

$$\Phi = \begin{bmatrix} | & & | \\ \phi_1 & \cdots & \phi_K \\ | & & | \end{bmatrix}, \tag{6.4}$$

that is, each of the basis functions is stored as a column of Φ, and each row corresponds to a vector $\phi(x) \in \Re^K$ of the basis functions evaluated at a state x. Then $\tilde{J}(\cdot, r)$ can be represented in matrix notation as Φr, where r is a K-dimensional vector with ith entry corresponding to r_i. The following optimization problem might be used for computing a weight vector $\tilde{r} \in \Re^K$ such that $\Phi\tilde{r}$ is a close approximation to J^*:

$$\max_r \quad c^T \Phi r \tag{6.5}$$
$$\text{s.t.} \quad T\Phi r \geq \Phi r.$$

Given a solution $\tilde{r}$, one might hope that taking the greedy policy with respect to the *approximate cost-to-go function* $\Phi\tilde{r}$ would lead to near-optimal decisions. Like the exact LP, the optimization problem (6.5) can be recast as a linear program. We will refer to this problem as the *approximate LP*.

Note that, although the number of variables in the approximate LP is reduced to K, the number of constraints remains as large as in the exact LP. Fortunately, most of the constraints become inactive, and solutions to the approximate LP can be approximated efficiently. Linear programs involving few variables and a large number of constraints are often tractable via constraint generation. In the specific case of the approximate LP, Section 6.7 describes how the special structure of dynamic programming can be exploited in an efficient constraint sampling algorithm that leads to good approximations to the approximate LP solution.

6.4   STATE-RELEVANCE WEIGHTS AND THE PERFORMANCE OF ALP POLICIES

With the limited approximation capacity offered by the selection of basis functions, an approximation that is uniformly good throughout the state space; in particular, the maximum error over all states can become arbitrarily large as the problem dimensions increase. This raises the question of how to balance the accuracy of the approximation over different portions of the state space. In this section, it is shown that approximate linear programming allows for assigning different weights to approximation errors over different portions of the state space, thereby allowing for emphasis of regions of greater importance. Moreover, a bound on the expected cost increase resulting from using a policy generated by ALP instead of the optimal cost-to-go function provides guidance on which portions of the state space should be emphasized.

The next lemma shows that tradeoffs between approximation errors over different states can be controlled via the state-relevance weights.

Henceforth we will use norms $\| \cdot \|_{p,\gamma}$, defined by

$$\|J\|_{p,\gamma} = \left(\sum_{x \in \mathcal{S}} \gamma(x)|J(x)|^p \right)^{1/p}$$

for each $p \geq 1$ and weight vector $\gamma : \mathcal{S} \mapsto \Re^{+}$.

Lemma 6.4.1 *[8] A vector $\tilde{r}$ solves the approximate LP (6.5) if and only if it solves*

$$\min_r \quad \|J^* - \Phi r\|_{1,c}$$
$$\text{s.t.} \quad T\Phi r \geq \Phi r.$$

The approximate LP can be viewed as minimizing a certain weighted norm of the approximation error, with weights equal to the state-relevance weights. Therefore the vector c specifies the tradeoff in the quality of the approximation across different states, and better approximations may be generated in a region of the state space by assigning relatively larger weight to that region.

The next result provides some guidance on which regions of the state space should be emphasized. Underlying the choice of state-relevance weights is the question of how to measure quality of the approximate cost-to-go function. A possible measure of quality is the distance to the optimal cost-to-go function. A more direct measure is a comparison between the actual costs incurred by using the greedy policy associated with the approximate cost-to-go function and those incurred by an optimal policy. Theorem 6.4.1 provides a bound on the cost increase incurred by using approximations generated by approximated linear programming.

We consider as a measure of the quality of policy u the expected increase in the infinite-horizon discounted cost, conditioned on the initial state of the system being distributed according to a probability distribution ν, that is,

$$E_{X \sim \nu}\left[J_u(X) - J^*(X)\right] = \sum_{x \in \mathcal{S}} \nu(x)(J_u(x) - J^*(x)) = \|J_u - J^*\|_{1,\nu}.$$

It will be useful to define a measure $\mu_{u,\nu}$ over the state space associated with each policy u and probability distribution ν, given by

$$\mu_{u,\nu}^T = (1 - \alpha)\nu^T \sum_{t=0}^{\infty} \alpha^t P_u^t. \tag{6.6}$$

It is easy to show that it is a probability distribution over the state space $\mathcal{S}$.

The measure $\mu_{u,\nu}$ captures the expected frequency of visits to each state when the system runs under policy u, conditioned on the initial state being distributed according to ν. Future visits are discounted according to the discount factor α.

Theorem 6.4.1 *[8] Let $J : \mathcal{S} \mapsto \Re$ be such that $TJ \geq J$. Then*

$$\|J_{u_J} - J^*\|_{1,\nu} \leq \frac{1}{1-\alpha}\|J - J^*\|_{1,\mu_{u_J,\nu}}. \tag{6.7}$$

Theorem 6.4.1 has some interesting implications. Recall from Lemma 6.4.1 that the approximate LP generates an approximate cost-to-go function $\Phi\tilde{r}$ minimizing $\|\Phi r - J^*\|_{1,c}$ over the feasible region; contrasting this result with the bound on the

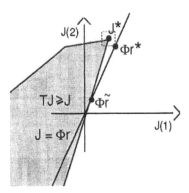

Fig. 6.1   Graphical interpretation of approximate linear programming

increase in costs (6.7), it makes sense to choose state-relevance weights c that capture the (discounted) frequency with which different states are expected to be visited. The theorem also sheds light on how beliefs about the initial state of the system can be factored into the approximate LP.

Note that the frequency with which different states are visited in general depends on the policy being used. This poses a difficulty in the choice of state-relevance weights, since the "ideal" distribution $\mu_{u,\nu}$ depends on the policy u that is induced by the approximate LP, which is not known *a priori*. One possibility is to have an iterative scheme, where the approximate LP is solved multiple times with state-relevance weights adjusted according to the intermediate policies being generated. Another possibility is to determine, via analysis of the problem being solved, the general structure of the stationary distributions, and experiment with state-relevance weights c having that structure. This approach is illustrated in Section 6.6, where approximate linear programming is applied to problems involving queueing networks.

6.5   APPROXIMATION ERROR BOUNDS

The central question in this section is whether having a good selection of basis functions is sufficient for the approximate LP to produce good approximations. Figure 6.1 illustrates the issue. Consider an MDP with states 1 and 2. The plane represented in the figure corresponds to the space of all functions over the state space. The shaded area is the feasible region of the exact LP, and J^* is the pointwise maximum over that region. In the approximate LP, solutions are constrained to lie in the subspace $J = \Phi r$. Note that the span of the basis functions comes relatively close to the optimal cost-to-go function J^*; for instance, a maximum-norm projection of J^* onto the subspace $J = \Phi r$ yields the reasonably good approximation Φr^*. At the same time, the approximate LP yields the approximate cost-to-go function $\Phi \tilde{r}$.

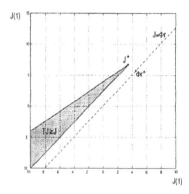

Fig. 6.2 Graphical representation of a 2-state problem with $\Phi r^* \approx J^*$ and an infeasible approximate LP.

This section demonstrates that the approximate LP can be set up so that $\Phi\tilde{r}$ is not too much farther from J^* than Φr^* is, and gives an explicit bound relating these distances. Note that without further assumptions such a result cannot be established; it is easy to construct examples where Φr^* is arbitrarily close to J^* and yet $\Phi\tilde{r}$ is arbitrarily far, or the approximate LP is even infeasible, as illustrated in Figure 6.2.

Performance guarantees for the approximate LP given in this section involve two different bounds on the approximation error $\|J^* - \Phi\tilde{r}\|_{1,c}$. The first bound, presented in Theorem 6.5.1, has poor scaling properties, and it helps explain the relevance of the assumptions involved in the main result, presented in Theorem 6.5.2

Henceforth we let $\| \cdot \|_{\infty,\gamma}$ denote the weighted maximum norm, defined for all $\gamma : S \mapsto \Re^+$ by

$$\|J\|_\infty = \max_{x \in S} \gamma(x)|J(x)|.$$

We also let e denote the vector of ones, and define $\| \cdot \|_\infty = \| \cdot \|_{\infty,e}$.

Theorem 6.5.1 *[8] Let e be in the span of the columns of Φ and c be a probability distribution. Then, if $\tilde{r}$ is an optimal solution to the approximate LP,*

$$\|J^* - \Phi\tilde{r}\|_{1,c} \le \frac{2}{1-\alpha} \min_r \|J^* - \Phi r\|_\infty.$$

This bound implies that when the optimal cost-to-go function lies close to the span of the basis functions, the approximate LP generates a good approximation. In particular, if the error $\min_r \|J^* - \Phi r\|_\infty$ goes to zero (e.g., as we make use of more and more basis functions) the ALP error $\|J^* - \Phi\tilde{r}\|_{1,c}$ also goes to zero.

Although the bound above offers some support for the linear programming approach, there are some significant weaknesses:

1. The bound calls for an element of the span of the basis functions to exhibit uniformly low error over all states. In practice, however, $\min_r \|J^* - \Phi r\|_\infty$ is typically huge, especially for large-scale problems.

2. The bound does not take into account the choice of state-relevance weights. As demonstrated in the previous section, these weights can significantly impact the quality of the approximate cost-to-go function. A more meaningful bound should take them into account.

To set the stage for the development of an improved bound, let us establish some notation. Let H be the operator given by

$$(HV)(x) = \max_{a \in \mathcal{A}_x} \sum_{y \in S} P_a(x, y)V(y),$$

for all $V : S \mapsto \Re$. For any V, $(HV)(x)$ represents the maximum expected value of $V(Y)$ if the current state is x and Y is a random variable representing the next state.

For each $V : S \mapsto \Re$, we define a scalar β_V given by

$$\beta_V = \max_x \frac{\alpha(HV)(x)}{V(x)}. \tag{6.8}$$

Definition 6.5.1 (Lyapunov function) *We call $V : S \mapsto \Re^+$ a Lyapunov function if $\beta_V < 1$.*

Existence of a Lyapunov function translates into the condition that there exist $V > 0$ and $\beta < 1$ such that $\alpha(HV)(x) \leq \beta V(x)$, $\forall x \in S$. If α were equal to 1, this would look like a Lyapunov stability condition: the maximum expected value $(HV)(x)$ at the next time step must be less than the current value $V(x)$. In general, α is less than 1, and this introduces some slack in the condition.

We are now ready to state the main result of this section. For any given function V mapping S to positive reals, we use $1/V$ as shorthand for a function $x \mapsto 1/V(x)$.

Theorem 6.5.2 *[8] Let $\tilde{r}$ be a solution of the approximate LP. Then, for any $v \in \Re^K$ such that Φv is a Lyapunov function, we have*

$$\|J^* - \Phi\tilde{r}\|_{1,c} \leq \frac{2c^T \Phi v}{1 - \beta_{\Phi v}} \min_r \|J^* - \Phi r\|_{\infty, 1/\Phi v}. \tag{6.9}$$

Let us now discuss how Theorem 6.5.2 addresses the shortcomings of Theorem 6.5.1.

1. The norm $\| \cdot \|_\infty$ appearing in Theorem 6.5.1 is undesirable largely because it does not scale well with problem size. In particular, for large problems, the optimal cost-to-go function can take on huge values over some (possibly infrequently visited) regions of the state space, and so can approximation errors

in such regions. Observe that the maximum norm of Theorem 6.5.1 has been replaced in Theorem 6.5.2 by $\| \cdot \|_{\infty, 1/\Phi v}$. Hence, the error at each state is now weighted by the reciprocal of the Lyapunov function value. This should to some extent alleviate difficulties arising in large problems. In particular, the Lyapunov function should take on large values in undesirable regions of the state space — regions where J^* is large. Hence, division by the Lyapunov function acts as a normalizing procedure that scales down errors in such regions.

2. As opposed to the bound of Theorem 6.5.1, the state-relevance weights do appear in the new bound. In particular, there is a coefficient $c^T \Phi v$ scaling the right-hand side. In general, if the state-relevance weights are chosen appropriately, we expect that $c^T \Phi v$ will be reasonably small and independent of problem size. Section 6.6 provides further qualification of this statement and a discussion of approaches to choosing c in the context of a concrete example.

It is important to note that the Lyapunov function Φv plays a central role in the bound of Theorem 6.5.2. Its choice influences three terms on the right-hand side of the bound: the error $\min_r \| J^* - \Phi r \|_{\infty, 1/\Phi v}$, the term $1/(1 - \beta_{\Phi v})$, and the inner product $c^T \Phi v$ with the state-relevance weights. An appropriately chosen Lyapunov function should make all three of these terms relatively small. Furthermore, for the bound to be useful in practical contexts, these terms should not grow much with problem size. We illustrate choices of Lyapunov functions via an application to a multiclass queueing problem, in the next section.

6.6   APPLICATION TO QUEUEING NETWORKS

This section presents an example involving choices of state-relevance weights and Lyapunov functions in a multiclass queueing problem. The intention is to illustrate more concretely how state-relevance weights and Lyapunov functions might be chosen and show that reasonable choices lead to practical error bounds that are independent of the number of states and of the number of state variables.

Consider a queueing network with d queues and finite buffers of size B. We assume that the number of exogenous arrivals occuring in any time step has expected value less than or equal to Ad, for a finite A. The state $x \in \Re^d$ indicates the number of jobs in each queue. The cost per stage incurred at state x is given by $g(x) = \frac{1}{d} \sum_{i=1}^d x_i$, the average number of jobs per queue.

Let us first consider the optimal cost-to-go function J^* and its dependency on the number of state variables d. The goal is to establish bounds on J^* that will offer some guidance on the choice of a Lyapunov function V that keeps the error $\min_r \| J^* - \Phi r \|_{\infty, 1/V}$ small. Since $J^* \geq 0$ only upper bounds will be derived.

Instead of carrying the buffer size B throughout calculations, attention is restricted to the infinite buffer case. The optimal cost-to-go function for the finite buffer case should be bounded above by that of the infinite buffer case, as having finite buffers corresponds to having jobs arriving at a full queue discarded at no additional cost.

We have

$$E_x\left[|x_t|\right] \leq |x| + Adt,$$

since the expected total number of jobs at time t cannot exceed the total number of jobs at time 0 plus the expected number of arrivals between 0 and t, which is less than or equal to Adt. E_x stands for the expected value conditioned on $x_0 = x$, and $|x| = \sum_{i=1}^{d} x_i$. It follows from this inequality that

$$
\begin{aligned}
E_x\left[\sum_{t=0}^{\infty} \alpha^t |x_t|\right] &= \sum_{t=0}^{\infty} \alpha^t E_x\left[|x_t|\right] \\
&\leq \sum_{t=0}^{\infty} \alpha^t (|x| + Adt) \\
&= \frac{|x|}{1-\alpha} + \frac{Ad}{(1-\alpha)^2}.
\end{aligned}
\tag{6.10}
$$

The first equality holds because $|x_t| \geq 0$ for all t; by the monotone convergence theorem, we can interchange the expectation and the summation. We conclude from (6.10) that the optimal cost-to-go function in the infinite buffer case should be bounded above by a linear function of the state; in particular,

$$0 \leq J^*(x) \leq \frac{\rho_1}{d}|x| + \rho_0,$$

for some positive scalars ρ_0 and ρ_1 independent of the number of queues d.

As discussed before, the optimal cost-to-go function in the infinite buffer case provides an upper bound for the optimal cost-to-go function in the case of finite buffers of size B. Therefore, the optimal cost-to-go function in the finite buffer case should be bounded above by the same linear function regardless of the buffer size B.

We will establish bounds on the terms involved in the error bound of Theorem 6.5.2. Consider a Lyapunov function $V(x) = \frac{1}{d}|x| + C$ for some constant $C > 0$. Then

$$
\begin{aligned}
\min_r \|J^* - \Phi r\|_{\infty,1/V} &\leq \|J^*\|_{\infty,1/V} \\
&\leq \max_{x \geq 0} \frac{\rho_1|x| + d\rho_0}{|x| + dC} \\
&\leq \rho_1 + \frac{\rho_0}{C},
\end{aligned}
$$

and the bound above is independent of buffer size and number of queues.

Consider the term β_V. We have

$$
\begin{aligned}
\alpha(HV)(x) &\leq \alpha\left(\frac{|x| + Ad}{d} + C\right) \\
&\leq V(x)\left(\alpha + \frac{\alpha A}{\frac{|x|}{d} + C}\right) \\
&\leq V(x)\left(\alpha + \frac{\alpha A}{C}\right),
\end{aligned}
$$

and it is clear that, for C sufficiently large and independent of d, there is a scalar $\beta < 1$ independent of d such that $\alpha HV \leq \beta V$, and therefore $\frac{1}{1-\beta_V}$ is uniformly bounded on B and d.

Finally, consider the term $c^T V$. Under some stability assumptions, the tail of the steady-state distribution will have an upper bound with geometric decay [4]. This motivates choosing $c(x) = \left(\frac{1-\xi}{1-\xi^{B+1}}\right)^d \xi^{|x|}$. The state-relevance weights c are equivalent to the conditional joint distribution of d independent and identically distributed geometric random variables conditioned on the event that they are all less than $B + 1$. Therefore,

$$
\begin{aligned}
c^T V &= E\left[\frac{1}{d}\sum_{i=1}^{d} X_i + C \mid X_i < B + 1, i = 1, ..., d\right] \\
&< E[X_1] + C \\
&= \frac{\xi}{1 - \xi} + C,
\end{aligned}
$$

where $X_i, i = 1, ..., d$ are identically distributed geometric random variables with parameter $1-\xi$. It follows that $c^T V$ is uniformly bounded over buffer size and number of queues. We conclude that the error produced by approximate linear programming in this problem is uniformly bounded on the size and dimension of the state space.

6.7   EFFICIENT CONSTRAINT SAMPLING SCHEME

While the approximate LP may involve only a small number of variables, there is a potentially intractable number of constraints — one per state-action pair. This section describes a tractable approximation to the approximate LP: the reduced linear program (reduced LP).

Generation of a reduced LP relies on three objects: (1) a constraint sample size m, (2) a probability measure ψ over the set of state-action pairs, and (3) a bounding set $\mathcal{N} \subseteq \Re^K$. The probability measure ψ represents a distribution from which constraints are sampled. In particular, we consider a set $\mathcal{X}$ of m state-action pairs, each independently sampled according to ψ. The set $\mathcal{N}$ is a parameter that restricts the magnitude of the reduced LP solution. This set should be chosen such that it contains $\Phi\hat{r}$. The reduced LP is defined by

$$\begin{aligned}
\max \quad & c^T\Phi r \\
\text{s.t.} \quad & g_a(x) + \alpha\sum_y P_a(x,y)(\Phi r)(y) \geq (\Phi r)(x), \ \forall(x,a) \in \mathcal{X}. \quad (6.11) \\
& r \in \mathcal{N}
\end{aligned}$$

Let $\tilde{r}$ be an optimal solution of the approximate LP and let $\hat{r}$ be an optimal solution of the reduced LP. In order for the solution of the reduced LP to be meaningful, $\|J^* - \Phi\hat{r}\|_{1,c}$ should close to $\|J^* - \Phi\tilde{r}\|_{1,c}$. This is formalized as a requirement that

$$\text{Prob}\left(\|J^* - \Phi\hat{r}\|_{1,c} \leq \|J^* - \Phi\tilde{r}\|_{1,c} + \epsilon\|J^*\|_{1,c}\right) \geq 1 - \delta,$$

where $\epsilon > 0$ is an error tolerance parameter and $\delta > 0$ parameterizes a level of confidence $1 - \delta$. The main result of this section is the characterization of a tractable sample size m that is sufficient for meeting such a requirement. In particular, it is shown that, if ψ and $\mathcal{N}$ are well-chosen, an error tolerance of ϵ can be accommodated with confidence $1 - \delta$ given a sample size m that grows as a polynomial in K, $1/\epsilon$, and $\log 1/\delta$, and is *independent* of the total number of approximate LP constraints.

The analysis is based on the notion of near-feasibility. Given a probability distribution ψ over state-action pairs and a scalar $\epsilon > 0$, r is said to be (ψ, ϵ)-near-feasible if

$$\psi\left(\{(x,a) : (T_a\Phi r)(x) < (\Phi r)(x)\}\right) < \epsilon,$$

that is, the set of constraints violated by r has measure less than or equal to ϵ.

The following theorem is a direct application of results on uniform convergence of empirical probabilities [14, 30], and establishes that only a tractable number of constraints needs to be considered in order to ensure near-feasibility.

Theorem 6.7.1 *[9] Let $0 < \delta < 1$, $0 < \epsilon < 1$ be given constants and let $\mathcal{W}$ be a set formed by sampling m i.i.d. state-action pairs sampled from ψ. Then if*

$$m \geq \frac{32}{\epsilon^2}\left[\ln 8 + \ln\frac{1}{\delta} + k\left(\ln\frac{16e}{\epsilon} + \ln\ln\frac{16e}{\epsilon}\right)\right], \quad (6.12)$$

we have

$$\sup_{\substack{r:(T_a\Phi r)(x)\geq(\Phi r)(x) \\ \forall(x,a)\in\mathcal{W}}} \psi\left(\{(x,a) : (T_a\Phi r)(x) < (\Phi r)(x)\}\right) \leq \epsilon, \quad (6.13)$$

with probability at least $1 - \delta$.

It is easy to see that near-feasibility is not enough to ensure a small error if the solution of the reduced LP involves arbitrarily large violations of the non-sampled constraints. The next result presents conditions that enable a graceful bound on the error in the approximation of the cost-to-go function incurred by solving the reduced LP instead of the approximate LP.

Let us first introduce certain constants and functions involved in our error bound.

For any Lyapunov function V, we define a family of probability distributions on the state space S, given by

$$\mu_{u,c,V}(x) = \frac{\mu_{u,c}(x)V(x)}{\mu_{u,c}^T V}.$$

We also define a distribution over state-action pairs

$$\psi_{u,c,V}(x,a) = \frac{\mu_{u,c,V}(x)}{|\mathcal{A}_x|}, \forall a \in \mathcal{A}_x.$$

Finally, let the constants A and θ be given by

$$A = \max_x |\mathcal{A}_x|$$

and

$$\theta = \frac{\mu_{u^*}^T V}{c^T J^*} \sup_{r \in \mathcal{N}} \|J^* - \Phi r\|_{\infty, 1/V}.$$

Theorem 6.7.2 is the main result of this section — a bound on the approximation error introduced by constraint sampling.

Theorem 6.7.2 *[9] Let u^* be an optimal policy and $\mathcal{X}$ be a (random) set of m state-action pairs sampled independently according to the distribution $\psi_{u^*,c,V}(x,a)$, for some Lyapunov function V, where*

$$m \geq \frac{512 A^2 \theta^2}{(1-\alpha)^2 \epsilon^2} \left[\ln \frac{8}{\delta} + K \ln \left(\frac{64 A \theta e}{(1-\alpha)\epsilon} \ln \frac{64 A \theta e}{(1-\alpha)\epsilon} \right) \right]. \tag{6.14}$$

Let $\tilde{r}$ be an optimal solution of the approximate LP that is in $\mathcal{N}$, and let $\hat{r}$ be an optimal solution of the corresponding reduced LP. If $\tilde{r} \in \mathcal{N}$ then, with probability at least $1 - \delta$, we have

$$\|J^* - \Phi\hat{r}\|_{1,c} \leq \|J^* - \Phi\tilde{r}\|_{1,c} + \epsilon\|J^*\|_{1,c}. \tag{6.15}$$

Certain aspects of Theorem 6.14 deserve further consideration. A major weakness of Theorem 6.14 is that it relies on an idealized choice of ψ. In particular, the choice assumes knowledge of an optimal policy. Alas, we typically do not know an optimal policy — which is what we are after in the first place. Nevertheless, the result provides guidance on what makes a desirable choice of distribution. The spirit here is analogous to one present in the importance sampling literature. In that context, the goal is to reduce variance in Monte Carlo simulation through intelligent choice of a sampling distribution and appropriate distortion of the function being integrated. Characterizations of idealized sampling distributions guide the design of heuristics that are ultimately implemented.

The set $\mathcal{N}$ also plays a critical role in the bound. It influences the value of θ, and an appropriate choice is necessary in order for this term to scale gracefully with problem size. Ideally, given a class of problems, there should be a mechanism for generating $\mathcal{N}$ such that θ grows no faster than a low-order polynomial function of the number of basis functions and the number of state variables. As illustrated later through an example involving controlled queueing networks, we expect that it will be possible to design effective mechanisms for selecting $\mathcal{N}$ for practical classes of problems.

Finally, the number of sampled constraints grows polynomially with the maximum number of actions available per state A, which makes the proposed approach inapplicable to problems with a large number of actions per state. In [9], it is shown that complexity in the action space can be exchanged for complexity in the state space, so that such problems can be recast in an amenable format.

Note that the sample complexity bounds in Theorems 6.7.1 and 6.7.2 are loose. The emphasis is on showing that the number of required samples can be independent of the total number of constraints and can scale gracefully with respect to the number of variables. Furthermore, the emphasis is on a general result that holds for a broad class of MDPs, therefore special regularities associated with particular choices of basis functions or specific problems are not being exploited. In the presence of such special structure, it may be possible to provide much tighter bounds or even methods for exact solution of the approximate LP. Examples of how problem-specific structure may lead to tractable solution of the approximate LP can be found in the literature in applications such as queueing networks [21], factored MDP's [17, 24] or the traveling salesman problem [16]. Exploitation of particular properties and regularities of the problem at hand is obviously useful and the constraint sampling scheme is not meant as a substitute for that. The significance of the results presented here is that they suggest viability of the linear programming approach to approximate dynamic programming even in the absence of such favorable special structure.

6.7.1 Example: Controlled Queueing Networks

In order for the error bound (6.15) to be useful, the parameter

$$\theta = \frac{\mu_{u_*}^T V}{c^T J^*} \sup_{r \in \mathcal{N}} \|J^* - \Phi r\|_{\infty, 1/V},$$

should scale gracefully with problem size. We anticipate that for many relevant classes of MDPs, natural choices of V and $\mathcal{N}$ will ensure this. This section illustrates this point through an example involving controlled queueing networks. The key result is Theorem 6.7.3, which establishes that — given certain reasonable choices of Φ, $\mathcal{N}$, and V — θ grows at most linearly with the number of queues.

We consider classes of problems denoted by $\mathcal{Q}(\xi, \alpha, \lambda)$, with the following interpretation. Each problem instance $Q \in \mathcal{Q}(\xi, \alpha, \lambda)$ corresponds to a queueing control problem of the form described in Section 6.6 and identified by a quadruple:

- number of queues $d_Q \geq 1$;

- buffer size $B_Q \geq d_Q \xi/(1 - \xi)$;

- action sets $\mathcal{A}^Q$;

- transition probabilities $P^Q(\cdot, \cdot)$.

Each $Q \in \mathcal{Q}(\xi, \alpha, \lambda)$ involves a discount factor α, and the expected number of arrivals in any time step is less than or equal to λd.

Let u_Q^* and J_Q^* denote an optimal policy and the optimal cost-to-go function for a problem instance Q. We consider approximating J_Q^* by a linear combination of basis functions $\phi_k^Q(x) = x_k, k = 1, \ldots, d_Q$ and $\phi_{d_Q+1}^Q(x) = 1$, using the following approximate LP:

$$
\begin{aligned}
\text{maximize} \quad & \sum_{x \in S} c_Q(x) \left(\sum_{k=1}^{d_Q+1} r_k x_k + r_{d+1} \right) \\
\text{subject to} \quad & \frac{1}{d_Q} \sum_{i=1}^{d_Q} x_i + \alpha \sum_{y \in S} P_a^Q(x, y) \left(\sum_{k=1}^{d_Q} r_k y_k + r_{d_Q+1} \right) \\
& \geq \sum_{k=1}^{d_Q} r_k x_k + r_{d_Q+1}, \quad \forall x \in S_Q, a \in \mathcal{A}_x^Q,
\end{aligned}
\tag{6.16}
$$

where $S_Q = \{0, \ldots, B_Q\}^{d_Q}$ and the state-relevance weights are given by

$$
c_Q(x) = \frac{\xi^{-\sum_{i=1}^{d_Q} x_i}}{\sum_{y \in S_Q} \xi^{-\sum_{i=1}^{d_Q} y_i}}.
$$

The number of constraints imposed by the approximate LP (6.16) grows exponentially with the number of queues d_Q. For even a moderate number of queues (e.g., ten), the number of constraints becomes unmanageable. Constraint sampling offers an approach to alleviating this computational burden.

Formulating a reduced LP, given a problem instance Q, requires defining a constraint set $\mathcal{N}_Q$ and a sampling distribution ψ_Q. Let $\mathcal{N}_Q$ to be the set of vectors $r \in \Re^{d+1}$ that satisfy the following linear constraints:

$$
r_{d_Q+1} \leq \frac{\lambda}{(1 - \alpha)^2},
\tag{6.17}
$$

$$
B_Q r_k + r_{d_Q+1} \leq \frac{B_Q}{(1 - \alpha)d_Q} + \frac{\lambda}{(1 - \alpha)^2} \quad \forall k = 1, \ldots, d_Q,
\tag{6.18}
$$

$$
\left(\frac{\xi}{1 - \xi} - \frac{\xi^{B_Q+1}(B_Q + 1)}{1 - \xi^{B_Q+1}} \right) \sum_{k=1}^{d_Q} r_k + r_{d_Q+1} \geq 0.
\tag{6.19}
$$

Note that the resulting reduced LP is a linear program with $m + d_Q + 2$ constraints, where m is the number of sampled ALP constraints.

It turns out that $\mathcal{N}_Q$ has two desirable properties: it contains optimal solutions of the ALP (6.16) and is uniformly bounded over $\mathcal{Q}(\xi, \alpha, \lambda)$.

Lemma 6.7.1 *[9] For each $\xi \in (0,1)$, $\alpha \in (0,1)$, and $\lambda \in (0,\infty)$ and each $Q \in \mathcal{Q}(\xi, \alpha, \lambda)$, $\mathcal{N}_Q$ contains every optimal solution of the ALP (6.16).*

Lemma 6.7.2 *[9] For each $\xi \in (0,1)$, $\alpha \in (0,1)$, and $\lambda \in (0,\infty)$, there exists a scalar $C_{\xi,\alpha,\lambda}$ such that*

$$\sup_{r \in \mathcal{N}_Q} \|r\|_\infty \leq C_{\xi,\alpha,\lambda},$$

for all $Q \in \mathcal{Q}(\xi, \alpha, \lambda)$.

Now consider the distribution $\psi_Q = \psi_{u_Q^*, V_Q}$, where

$$V_Q(x) = \frac{1}{d_Q(1-\alpha)} \sum_{i=1}^{d_Q} x_i + \frac{2\lambda}{(1-\alpha)^2}. \tag{6.20}$$

The following lemma establishes that V_Q is an optimal-policy Lyapunov function.

Lemma 6.7.3 *[9] For each $\xi \in (0,1)$, $\alpha \in (0,1)$ and $\lambda \in (0,\infty)$ and each $Q \in \mathcal{Q}(\xi, \alpha, \lambda)$, V_Q is an optimal-policy Lyapunov function.*

Recall that the bound on sample complexity for the reduced LP, given by Eq. (6.14), is affected by a parameter θ. In the context of controlled queueing networks, we have a parameter θ_Q for each problem instance $Q \in \mathcal{Q}(\xi, \alpha, \lambda)$:

$$\theta_Q = \frac{\mu_{u_Q^*}^T V_Q}{c_Q^T J_Q^*} \sup_{r \in \mathcal{N}_Q} \|J_Q^* - \Phi_Q r\|_{\infty, 1/V_Q}.$$

The final result presented here establishes that θ_Q can be bounded above by a linear function of the number of queues.

Theorem 6.7.3 *For each $\xi \in (0,1)$, $\alpha \in (0,1)$, and $\lambda \in (0,\infty)$, there exists a scalar $C_{\xi,\alpha,\lambda}$ such that $\theta_Q \leq C_{\xi,\alpha,\lambda} d_Q$.*

Combining this theorem with the sample complexity bound of Theorem 6.7.2, it follows that for any $Q \in \mathcal{Q}(\xi, \alpha, \lambda)$, a number of samples

$$m = O\left(\frac{A_Q d_Q}{(1-\alpha)\epsilon}\left(d_Q \ln \frac{A_Q d_Q}{(1-\alpha)\epsilon} + \ln \frac{1}{\delta}\right)\right),$$

where $A_Q = \max_{x \in S_Q} |\mathcal{A}_x^Q|$, suffices to guarantee that

$$\|J_Q^* - \Phi^Q \hat{r}\|_{1,c_Q} \leq \|J_Q^* - \Phi^Q \tilde{r}\|_{1,c_Q} + \epsilon \|J_Q^*\|_{1,c_Q},$$

with probability $1 - \delta$. Hence, the number of samples grows at most quadratically in the number of queues.

6.8   DISCUSSION

The present analysis represents a step toward making approximate linear programming a streamlined approach to large-scale dynamic programming. It offers a better understanding of the role of state-relevance weights and insights on how they can be chosen so as to indirectly optimize performance of the policy being generated. The analysis also offers guarantees for the error in the approximation of the optimal cost-to-go function — approximate LP produces approximations that are comparable to the best that could have been achieved with the given selection of basis functions. The performance and approximation error bounds enable us to predict the scaling properties of approximate LP for broadly defined classes of problems, as illustrated by the study of multiclass queueing networks. Finally, this chapter presents general approach to dealing with the large number of constraints involved in the approximate LP, which one can resort to in the absence of problem-specific structure allowing for exact solution of the approximate LP.

There are a number of directions to be pursued in the further development of approximate linear programming. We mention a few:

Large discount factors and average-cost criterion. This chapter investigated how approximate linear programming scales with the size and dimension of the state space of the system, but there is no consideration of how the algorithm scales as the discount factor increases. The current performance and error bounds do not scale well with discount factor; in particular, the error bound (6.9) should scale with $1/(1-\alpha)^2$. The Lyapunov function analysis used to develop the bound also seems overly restrictive in the average-cost setting, as in this case it requires stability of all policies. Initial exploration of the average-cost case can be found in [10]. It is shown that a naive approximate LP formulation of the average-cost problem, which follows directly from the discounted-cost version, may lead to undesirable behavior of approximate LP, and therefore a more sophisticated version of the algorithm is required.

Adaptive choice of state-relevance weights. The performance bound in Theorem 6.4.1 suggests finding a weight vector r such that Φr minimizes

$$\|\Phi r - J^*\|_{1,\mu_{\nu,u_r}},$$

where ν is the estimate for the initial state distribution. By contrast, approximate linear programming minimizes

$$\|\Phi r - J^*\|_{1,c},$$

with state-relevance weights c determined by the user, and we have strong guarantees for the error $\|\Phi\tilde{r} - J^*\|_{1,c}$ at the optimal solution $\tilde{r}$ of the approximate LP. If c is close to μ_{ν,u_r}, combining the performance and error bounds leads to a strong guarantee on the performance of the policy generated by approximate LP. These observations suggest an iterative scheme for adaptive choices of state-relevance weights:

1. Start with an arbitrary policy u_0, let $k = 0$;

2. Solve the approximate LP with state-relevance weights μ_{ν,u_k} to determine a new policy u_{k+1};

3. Let $k = k + 1$ and go back to step 2.

Questions regarding the convergence of such a scheme and how much it would improve performance must be addressed. Note that the above algorithm is based on the assumption that a initial distribution ν over the states is given. Running approximate linear programming online also raises the possibility of adapting the initial distribution ν to reflect the current state of the system, so that if at the kth of the approximate LP the system is at state x, we may set $\nu = 1_x$.

Robustness to model uncertainty. In principle, approximate linear programming requires explicit knowledge of the costs and transition probabilities in the system. This requirement can be relaxed by noting that, in fact, transition probabilities appear in the approximate LP only through the expected values

$$\sum_y P_a(x, y)\Phi$$

of the basis functions evaluated at the next state X_{k+1} given the current state $X_k = x$ of the system. Hence the approximate LP can be implemented without major changes if there is access to a system simulator that takes as input any state x and produces a next state y distributed according to $P_a(x, \cdot)$. A more interesting situation arises in the case where such a simulator is not available, but rather one must learn from online observations of the system, and possibly adapt to changes in the system behavior over time. In this case, it would be interesting to identify whether it is possible to continually update the approximate LP parameters, easily determine when it is necessary to re-solve the LP, and exploit knowledge of previous solutions to speed up computations in the approximate LP.

Basis function generation. A central issue in any approximate dynamic programming algorithm that has not been addressed in this chapter regards the choice of basis functions. Note that including additional basis functions in the approximate LP corresponds to adding extra columns, hence it may be possible to apply general linear programming column generation techniques to approximate linear programming in order to generate new basis functions leading to improvement in the approximation error. For an initial exploration of this idea, see [23], where it is shown how the Lagrange multipliers of the approximate LP can be used to assess whether adding any given new basis functions should lead to improvement in the approximation.

Appropriate distributions for constraint sampling. Theorem 6.7.2, involves an assumption that states are sampled according to distribution μ_{u^*} for some optimal policy u^*. In general it is not possible to compute μ_{u^*} in practice, without knowledge of the optimal policy u^*; hence to constraints have to be sampled according to an alternative distribution $\bar\mu$. Intuitively, if $\bar\mu$ is reasonably close to μ_{u^*}, the reduced LP should still produce meaningful answers. How close $\bar\mu$ and μ_{u^*} have to be and how to choose an appropriate $\bar{\mu}$ are still open questions. As a simple heuristic, noting that $\mu^*(x) \to c(x)$ as $\alpha \to 0$, one might choose $\bar{\mu} = c$. This choice is also justified by the realization that, in many applications, c is an estimate for π_{u_*}, and if $c = \pi_{u_*}$, it is also true that $\mu_{u_*} = c = \pi_{\mu_*}$.

Bibliography

1. D. Adelman, *A price-directed approach to stochastic inventory/routing,* Preprint, 2002.

2. D. Adelman, *Price-directed replenishment of subsets: Methodology and its application to inventory routing,* Preprint, 2002.

3. D. Bertsekas and J. N. Tsitsiklis, *Neuro-Dynamic Programming,* Athena Scientific, Belmont, MA, 1996.

4. D. Bertsimas, D. Gamarnik, and J.N. Tsitsiklis, Performance of multiclass Markovian queueing networks via piecewise linear Lyapunov functions, *Annals of Applied Probability,* vol. 11, no. 4, pp. 1384–1428, 2001.

5. V. Borkar, A convex analytic approach to Markov decision processes, *Probability Theory and Related Fields,* vol. 78, no. 4, pp. 583–602, 1988.

6. R. H. Crites and A. G. Barto, Improving elevator performance using reinforcement learning, *Advances in Neural Information Processing Systems,* vol. 8, 1996.

7. D. P. de Farias, *The Linear Programming Approach to Approximate Dynamic Programming,* Ph.D. Thesis, Stanford University, 2002.

8. D. P. de Farias and B. Van Roy, The linear programming approach to approximate dynamic programming, *Operations Research,* vol. 51, no. 6, pp. 850-865, 2003.

9. D. P. de Farias and B. Van Roy, The linear programming approach to approximate dynamic programming, *Mathematics of Operations Research,* vol. 51, no. 6, pp. 850–865, 2003.

10. D. P. de Farias and B. Van Roy, Approximate linear programming for average-cost approximate dynamic programming, *Advances in Neural Information Processing Systems,* vol. 15, 2003.

11. G. de Ghellinck, Les problèmes de décisions séquentielles, *Cahiers du Centre d'Etudes de Recherche Opérationnelle,* vol. 2, pp. 161–179, 1960.

12. E. V. Denardo, On linear programming in a Markov decision problem, *Management Science,* vol. 16, no. 5, pp. 282–288, 1970.

13. F. D'Epenoux, A probabilistic production and inventory problem, *Management Science*, vol. 10, no. 1, pp. 98–108, 1963.

14. R. M. Dudley, *Uniform Central Limit Theorems*, Cambridge University Press, Cambridge, MA, 1998.

15. G. Gordon, *Approximate Solutions to Markov Decision Processess*, Ph.D. Thesis, Carnegie Mellon University, Pittsburgh, 1999.

16. M. Grötschel and O. Holland, Solution of large-scale symmetric travelling salesman problems, *Mathematical Programming*, vol. 51, pp. 141–202, 1991.

17. C. Guestrin, D. Koller, and R. Parr, Efficient solution algorithms for factored MDPs, *Journal of Artificial Intelligence Research*, vol. 19, pp. 399–468, 2003.

18. A. Hordijk and L. C. M. Kallenberg, Linear programming and Markov decision chains, *Management Science*, vol. 25, pp. 352–362, 1979.

19. F. Longstaff and E.S. Schwartz, Valuing American options by simulation: A simple least squares approach, *Review of Financial Studies*, vol. 14, pp. 113–147, 2001.

20. A. S. Manne, Linear programming and sequential decisions, *Management Science*, vol. 6, no. 3, pp. 259–267, 1960.

21. J. R. Morrison and P. R. Kumar, New linear program performance bounds for queueing networks, *Journal of Optimization Theory and Applications*, vol. 100, no. 3, pp. 575–597, 1999.

22. I. C. Paschalidis and J. N. Tsitsiklis, Congestion-dependent pricing of network services, *IEEE/ACM Trans. Networking*, vol. 8, no. 2, pp. 171–184, 2000.

23. P. Poupart, R. Patrascu, D. Schuurmans, C. Boutilier, and C. Guestrin, Greedy linear value function approximation for factored Markov decision processes, *Eighteenth National Conference on Artificial Intelligence*, Edmonton, Alberta, 2002.

24. D. Schuurmans and R. Patrascu, Direct value-approximation for factored MDPs, *Advances in Neural Information Processing Systems*, vol. 14, 2001.

25. P. Schweitzer and A. Seidmann, Generalized polynomial approximations in Markovian decision processes, *Journal of Mathematical Analysis and Applications*, vol. 110, pp. 568–582, 1985.

26. G.J. Tesauro, Temporal difference learning and TD-gammon, *Communications of the ACM*, vol. 38, pp. 58–68, 1995.

27. M. Trick and S. Zin, A linear programming approach to solving dynamic programs, Unpublished manuscript, 1993.

28. M. Trick and S. Zin, Spline approximations to value functions: A linear programming approach, *Macroeconomic Dynamics*, vol. 1, 1997.

29. J.N. Tsitsiklis and B. Van Roy, Regression methods for pricing complex American-style options, *IEEE Trans. Neural Networks*, vol. 12, no. 4, pp. 694–703, 2001.

30. M. Vidyasagar, *A Theory of Learning and Generalization*, Springer, London, 1997.

31. W. Zhang and T.G. Dietterich, High-performance job-shop scheduling with a time-delay TD(λ) network, *Advances in Neural Information Processing Systems*, vol. 8, 1996.

Handbook of Learning and Approximate Dynamic Programming
Edited by Jennie Si, Andy Barto, Warren Powell and Donald Wunsch
Copyright © 2004 The Institute of Electrical and Electronics Engineers, Inc.

7 Reinforcement Learning in Large, High-Dimensional State Spaces

GREG GRUDIC   LYLE UNGAR
University of Colorado at Boulder   University of Pennsylvania

Editor's Summary: The previous chapter addresses the "curse of dimensionality" by treating ADP as the dual of the linear programming problem and introduces the method known as approximate linear programming. This chapter presents another method for dealing with the "curse of dimensionality," the policy gradient reinforcement learning framework. The Action Transition Policy Gradient (ATPG) algorithm presented here estimates a gradient in the policy space that increases reward. Following a brief motivation the authors present their algorithm in detail and discuss its properties. Finally, detailed experimental results are presented to show the types of problems that the algorithm can be applied to and what type of performance can be expected. Another algorithm, Boundary Localized Reinforcement Learning, is also discussed in this chapter. This is a mode switching controller that can be used to increase the rate of convergence.

7.1 INTRODUCTION

In Reinforcement Learning (RL) an agent uses a trial-and-error strategy to explore its environment with the goal of learning to maximize some (often infrequent) reward. This trial-and-error learning process is governed by stochastic search, which defines a probability distribution of actions taken during exploration. This type of stochastic search strategy has proven effective in many RL applications that intrinsically have low-dimensional state spaces [14].

However, applying a stochastic search strategy (or any poorly directed search strategy) to higher-dimensional problems is problematic because, in general, the search space grows exponentially with the number of state variables. As a consequence, the computational cost and the time to convergence of reinforcement learning can quickly become impractical as the dimension of the problem increases. Function approximation techniques have been proposed for learning generalizations across large

179

state spaces, as one possible solution to this curse of dimensionality problem in RL. However, even when function approximation techniques successfully generalize, the dimension of the search remains unchanged, and its computational cost and rate of convergence can still be impractical.

The goal of this chapter is to present a detailed summary of our recent efforts in addressing the convergence problems of RL in large state spaces. The main theme of our efforts has been to systematically direct search in ways that allow quick convergence to *locally* optimal policies. We specifically focus our theory at convergence to *local* optimums because, in the large problem domains we are interested in, convergence to globally optimal policies is simply not possible. In support of this claim, we pose the following question: can any human or animal claim to have achieved a globally optimal solution for how to live life? The answer to this question obviously is no, and although our systems do not approach the complexity of living animals, they nevertheless cannot hope to achieve globally optimal policies. One example of this is a real mail delivery robot, where obstacles it must avoid continually move around (e.g. humans), sensors are noisy, actuators don't always perform the same, task parameters change, and the size of the state space can be thousands of real-valued variables (vision sensors, sonar sensors, collision sensors, etc). Given all these uncertainties and the large problem space, globally optimal control polices are simply not reasonable.

The approach to RL advocated here has three specific properties. First we formulate our solution within the Policy Gradient Reinforcement Learning (PGRL) framework, where the agent estimates a gradient in policy space which increases reward, and incrementally updates its policy along this gradient. There are a number of formulations of PGRL, all of which have theoretical guarantees to locally optimal policies [2, 5, 6, 12, 26]. In Section 7.1.1, we present our motivation for using this framework over the more traditional value function RL framework [21].

The key to effective learning in PGRL is efficiently obtaining accurate estimates of the performance gradient. Therefore, the second property of our RL formulation is that we propose a PGRL algorithm, called Action Transition Policy Gradient (ATPG), which restricts estimates of the performance gradient to coincide with states where the agent changes from executing one action to another [2]. As argued below (see Section 7.1.2), this formulation reduces the variance in policy gradient estimates and greatly improves the rate of convergence of PGRL.

The third, and final, key property of our RL formulation is that we further reduce the computational cost of search in high-dimensional spaces by restricting the agent's policy to a mode switching controller, which restricts the search to very limited regions of the state space [1]. The size of the search region bounds the computational cost of RL. Intuitively, the smaller the search region, the lower the computational cost of learning, making it possible to apply RL to very-high-dimensional problems. This formulation of RL is called Boundary Localized Reinforcement Learning (BLRL) and is motivated in Section 7.1.3.

7.1.1   Motivation for Policy Gradient Reinforcement Learning (PGRL)

There has recently been a renewed interest in the Policy Gradient formulation of Reinforcement Learning (PGRL) [4–7]. There are three main motivations behind the PGRL framework. First, PGRL uses a function approximation (FA) representation (e.g., Neural Networks, decision trees, etc.) of the agent's policy (i.e., mapping between state and action), which directly addresses the need for generalization in large RL problems [14]. Second, PG algorithms learn by estimating the gradient of the agent's reward function with respect to the parametrization of the agent's policy. The computational cost of this estimate is linear in the number of parameters describing the policy, which is in stark contrast to the exponential growth associated with traditional value function RL algorithms [14], and makes the PG formulation very attractive for high-dimensional problems [1].

A third reason for renewed interest in PGRL is that they are provably convergent. Specifically, one of the earliest examples of PGRL, REINFORCE [12], as well as more resent examples [5, 6], are all theoretically guaranteed of converging to locally optimal policies.

In PGRL, the agent's policy is represented as $\pi(s, a;\ \theta)$, which denotes the probability that the agent chooses action a in state s. The key difference between the value function formulation of RL and PGRL is the parameter vector θ, which represents all of the modifiable parameters of the agent's policy. PGRL algorithms directly modify these parameters as follows. A reward function $\rho(\theta)$ is defined and learning is done by estimating the *performance gradient* $\partial\rho/\partial\theta$, which modifies the policy in a direction that increases reward. The agent learns by first estimating this gradient $\widehat{\partial\rho/\partial\theta}$ for the current policy π, and then updating the policy parameters θ as follows:

$$\theta_{t+1} = \theta_t + \alpha\frac{\widehat{\partial\rho}}{\partial\theta}, \tag{7.1}$$

where α is a small positive step size. The performance gradient $\widehat{\partial\rho/\partial\theta}$ is estimated using the state action value function $Q^\pi(s, a)$, which is the value of executing action a in state s, under the current policy π.

7.1.2   Motivation for Action Transition Policy Gradient Algorithm (ATPG)

Looking at Eq. (7.1), it is not difficult to observe that the efficacy of a PGRL formulation is directly affected by how easily the agent can estimate the performance gradient $\partial\rho/\partial\theta$. Typically, the problem with estimating the performance gradient is one of high variance. For example, although REINFORCE is known to give an unbiased estimate of $\partial\rho/\partial\theta$, the variance in this estimate is very large, resulting in very slow convergence [4, 5]. As argued below, the main reason for this high variance is that REINFORCE uses a stochastic sampling technique that does not account for the fact that a key requirement of PGRL is that *relative* estimates of the value of executing two or more actions in each state are necessary for convergence. Therefore, an agent using REINFORCE must pass through the same set of states many times, and hope that its stochastic policy executes a sufficient number of different actions in each state to obtain this relative value estimate.

Function approximation representations of $Q^\pi(s, a)$ (i.e., knowing the value of executing every action in every state under the current policy) have been proposed as one way of decreasing the variance in estimating $\partial\rho/\partial\theta$. Given $Q^\pi(s, a)$, the relative value of executing each action in each state is explicitly known [5, 6]. This approach seeks to build function approximation representations $f(s, a) \approx Q^\pi(s, a)$, and then use these approximations instead of direct samples of $Q^\pi(s, a)$ to estimate $\widehat{\partial\rho/\partial\theta}$. In [5, 6] it is shown that if the form of $f(s, a)$ is restricted appropriately, then the resulting PG algorithm is guaranteed to converge to a locally optimal policy. However, as argued below, function approximation approaches to policy gradients are not necessarily optimal.

How PGRL algorithms should be formulated to give fast convergence to locally optimal solutions is an important ongoing area of research [3]. In PGRL formulations performance gradient estimates typically have the following form:

$$\widehat{\frac{\partial\rho}{\partial\theta}} = f\left(\left[\hat{Q}(s_1, a_1) - b(s_1)\right], ..., \left[\hat{Q}(s_T, a_T) - b(s_T)\right]\right), \qquad (7.2)$$

where $\hat{Q}(s_i, a_i)$ is the estimate of the value of executing action a_i in state s_i (i.e. the state action value function), $b(s_i)$ the bias subtracted from $\hat{Q}(s_i, a_i)$ in state s_i, T is the number of steps the agent takes before estimating $\partial\rho/\partial\theta$, and the form of the function $f(.)$ depends on the PGRL algorithm being used (see Section 7.2.2, Eq. (7.5) for the form being considered here). The effectiveness of PGRL algorithms strongly depends on how $\hat{Q}(s_i, a_i)$ is obtained and the form of $b(s_i)$.

This chapter summarizes two recent theoretical results which shed new light on how PGRL algorithms should be formulated [3]. The first addresses the issue of whether function approximation representations of the state action value function Q can improve the convergence of the performance gradient. It has been proven that specific linear FA formulations can be incorporated into PGRL algorithms, while still guaranteeing convergence to locally optimal solutions [5, 6]. However, whether linear FA representations actually improve the convergence properties of PGRL is an open question. To shed light on this, we present theory showing that using linear basis function representations of Q, rather than direct observations of it, can *slow* the rate of convergence of PG estimates by a factor of $O(ML)$ (see **Theorem 7.2.1** in Section 7.2.3.1). This result suggests that PGRL formulations should avoid the use of linear FA techniques to represent Q. In Section 7.3, experimental evidence is presented supporting this conjecture.

The second theoretical result is concerned with the bias term. Specifically, can a non-zero bias term $b(s)$ in (7.2) improve the convergence properties of PG estimates? There has been speculation that an appropriate choice of $b(s)$ can improve convergence properties [5, 12], but theoretical support has been lacking. We present theory showing that if $b(s) = (1/M) \sum_a Q(s, a)$, where M is the number actions, then the rate of convergence of the PG estimate is *improved* by $O(1 - (1/M))$ (see **Theorem**

7.2.2 in Section 7.2.3.2). This suggests that the convergence properties of PGRL algorithms can be improved by using a bias term that is the average of Q values in each state. Section 7.3 gives experimental evidence supporting this conjecture.

This second theoretical result motivates the Action Transition Policy Gradient (ATPG) algorithm, which has the unique property that it estimates the performance gradient using direct estimates of the *relative* value of each action with respect to the average value of all the actions [2]. The ATPG performance gradient estimate is:

$$\frac{\widehat{\partial \rho}}{\partial \theta} = \sum_s \sum_a g_\pi (s,a) \left(\widehat{Q^\pi} (s,a) - \widehat{\overline{V^\pi}}(s) \right), \tag{7.3}$$

where $g_\pi(s,a)$ is analytically derived from the policy (see Section 7.2.2, eq. (7.5), and $\widehat{\overline{V^\pi}}(s)$ is an estimate of $\overline{V^\pi}(s)$, which is the bias term in **Theorem 7.2.2** (see Section 7.2.3.2). For a policy of M possible actions, the exact expression for $\overline{V^\pi}(s)$ is defined by:

$$\overline{V^\pi}(s) = \frac{1}{M} \sum_{j=1}^{M} Q^\pi (s, a_j). \tag{7.4}$$

The intuition behind ATPG RL is as follows: If the execution in state s of action a_j has better than average reward under π, then the policy gradient in (7.3) will update the parameters θ such that the probability of executing a_j in s is increased. Conversely, if the execution in state s of action a_j has worse than average reward under π, then the policy gradient in (7.3) will update the parameters θ such that the probability of executing a_j in s is decreased.

Theory shows that, under appropriate piece-wise continuity conditions on the policy $\pi(s, a; \theta)$ and the state-action value function $Q^\pi(s, a)$, the ATPG algorithm converges to a locally optimal policy (see **Theorem 7.2.3** in Section 7.2.4). These theoretical results are supported by an experimental comparison of the ATPG algorithm with REINFORCE and with the Policy Iteration and Function Approximation PIFA algorithm proposed by [5]. Our experiments indicate that ATPG consistently outperforms PIFA, giving an order of magnitude faster convergence when the value function is highly nonlinear or discontinuous. Furthermore, PIFA consistently outperformed REINFORCE by at least an order of magnitude in all experiments. These experimental results are summarized in Section 7.3.

7.1.3 Motivation for Boundary Localized Reinforcement Learning (BLRL)

To further bias search in PGRL to improve convergence, we consider the class of deterministic mode switching controllers, where the agents deterministically choose which action (or mode) is executed in each region of the state space (see Figure 7.1). Mode switching controllers are commonly used in many control applications and allow complex control systems to be developed using many simple controllers, each operating in different regions of the state space. Early examples of this are aircraft controllers, where a different control system is used when an aircraft is climbing steeply vs. cruising at constant elevation [18]. Mode switching controllers are also commonly used in Robotics where robots are required to switch between different modes such as avoid obstacle, follow leader, find target, etc, depending on the robot's state [30].

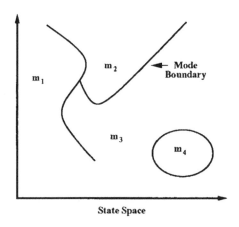

State Space

Fig. 7.1   A Mode Switching Controller consists of a finite number of modes $m_1, m_2, \ldots$ or actions, which are deterministically applied in specific regions of the workspace. The state space is therefore divided into regions specified by Mode Boundaries.

Boundary Localized Reinforcement Learning (BLRL) [1] is a formulation of reinforcement learning for agents that use a mode switching controller. In BLRL trial-and-error search is limited to regions near mode boundaries, which greatly improves convergence. Because BLRL is concerned only with updating the boundary locations between modes, these boundaries can be implicitly parameterized, and PGRL is used to update these parameter values. Thus learning takes place by shifting mode switching boundaries in directions that increase reward.

We summarize two theoretical results on the BLRL formulation. The first shows that any stochastic policy (i.e., stochastic control strategy) can be transformed into a mode switching policy, which localizes search to near mode boundaries (see Section 7.2.5). The practical consequence of this result is that any RL problem can be converted to a BLRL problem, thus taking advantage of the convergence properties of BLRL in high-dimensional state spaces. The second shows that convergence to a locally optimal mode switching policy is obtained when stochastic search is limited to near mode boundaries (see **Theorem 7.2.4** in Section 7.2.6). Therefore most of the agent's state space can be ignored, while still guaranteeing convergence to a locally optimal solution. These theoretical results are supported experimentally via simulation studies in Section 7.3.

7.2   THEORETICAL RESULTS AND ALGORITHM SPECIFICATIONS

7.2.1   Reinforcement Learning as a MDP

We model the RL problem as a Markov Decision Process (MDP) [14]. The agent's state at time $t \in \{1, 2, ...\}$ is given by $s_t \in S$, $S \subseteq \Re^D$. At each time step the agent chooses from a finite set of $M > 1$ actions $a_t \in A = a_1, ..., a_M$ and receives a reward $r_t \in \Re$. The dynamics of the environment are characterized by transition probabilities $P_{ss'}^a = Pr\{s_{t+1} = s' | s_t = s, a_t = a\}$ and expected rewards $R_s^a = E\{r_{t+1} | s_t = s, a_t = a\}, \forall s, s' \in S, a \in A$. The policy followed by the agent is characterized by a parameter vector $\theta \in \Re^d$, and is defined by the probability distribution $\pi(s, a; \theta) = Pr\{a_t = a | s_t = s; \theta\}, \forall s \in S, a \in A$. We assume that $\pi(s, a; \theta)$ is differentiable with respect to θ.

7.2.2   Policy Gradient Formulation

We line out analysis to the start state discount reward formulation of the reward function, where the reward function $\rho(\pi)$ and state action value function $Q^\pi(s, a)$ are defined as:

$$\rho(\pi) = E\left\{\sum_{t=1}^{\infty} \gamma^t r_t \middle| s_0, \pi\right\}, \quad Q^\pi(s, a) = E\left\{\sum_{k=1}^{\infty} \gamma^{k-1} r_{t+k} \middle| s_t = s, a_t = a, \pi\right\},$$

where $0 < \gamma \leq 1$. Then, as shown in [5], the exact expression for the performance gradient is:

$$\frac{\partial \rho}{\partial \theta} = \sum_s d^\pi(s) \sum_{i=1}^{M} \frac{\partial \pi(s, a_i; \theta)}{\partial \theta} (Q^\pi(s, a_i) - b(s)), \qquad (7.5)$$

where $d^\pi(s) = \sum_{t=0}^{\infty} \gamma^t Pr\{s_t = s | s_0, \pi\}$ and $b(s) \in \Re$.

In order to implement this PGRL algorithm, we require that the state-action value function, Q^π, under the current policy be estimated. This estimate, $\hat{Q}^\pi$, is derived using the observed value $Q_{obs}^\pi(s, a_i)$. We assume that $Q_{obs}^\pi(s, a_i)$ has the following form:

$$Q_{obs}^\pi(s, a_i) = Q^\pi(s, a_i) + \varepsilon(s, a_i),$$

where $\varepsilon(s, a_i)$ has zero mean and finite variance σ_{s, a_i}^2. Therefore, if $\hat{Q}^\pi(s, a_i)$ is an estimate of $Q^\pi(s, a_i)$ obtained by averaging N observations of $Q_{obs}^\pi(s, a_i)$, then the mean and variance are given by:

$$E\left[\hat{Q}^\pi(s, a_i)\right] = Q^\pi(s, a_i), \quad V\left[\hat{Q}^\pi(s, a_i)\right] = \frac{\sigma_{s, a_i}^2}{N}. \qquad (7.6)$$

In addition, we assume that $Q_{obs}^{\pi}(s, a_i)$ are independently distributed. This is consistent with the MDP assumption.

7.2.3   Rate of Convergence Results

We first state the following definitions:

$$\sigma_{max}^2 = \max_{s \in S, i \in \{1, ..., M\}} \sigma_{s, a_i}^2, \quad \sigma_{min}^2 = \min_{s \in S, i \in \{1, ..., M\}} \sigma_{s, a_i}^2, \qquad (7.7)$$

where σ_{s, a_i}^2 is defined in (7.6) and

$$
\begin{aligned}
C_{min} &= \left[\sum_s (d^{\pi}(s))^2 \sum_{i=1}^{M} \left(\frac{\partial \pi(s, a_i; \theta)}{\partial \theta} \right)^2 \right] \sigma_{min}^2, \\
C_{max} &= \left[\sum_s (d^{\pi}(s))^2 \sum_{i=1}^{M} \left(\frac{\partial \pi(s, a_i; \theta)}{\partial \theta} \right)^2 \right] \sigma_{max}^2.
\end{aligned}
\qquad (7.8)
$$

7.2.3.1   Rate of Convergence of PIFA Algorithms
The PIFA algorithm [5] which uses the following basis function representation for estimating the state action value function, $\hat{Q}^{\pi}$:

$$\hat{Q}^{\pi}(s, a_i) = f_{a_i}^{\pi}(s) = \sum_{l=1}^{L} w_{a_i, l} \phi_{a_i, l}(s), \qquad (7.9)$$

where $w_{a_i, l} \in \Re$ are weights and $\phi_{a_i, l}(s)$ are basis functions defined in $s \in \Re^D$. The weights $w_{a_i, l}$ are chosen using the observed $Q_{obs}^{\pi}(s, a_i)$, and the basis functions, $\phi_{a_i, l}(s)$, satisfy the conditions defined in [5, 6], then the performance gradient is given by:

$$\frac{\partial \rho}{\partial \theta}_F = \sum_s d^{\pi}(s) \sum_{i=1}^{M} \frac{\partial \pi(s, a_i; \theta)}{\partial \theta} f_{a_i}^{\pi}(s). \qquad (7.10)$$

Then, following theorem, first presented in [3], establishes bounds on the rate of convergence for this representation of the performance gradient.

Theorem 7.2.1 PIFA Algorithms. *Let* $\widetilde{\frac{\partial \rho}{\partial \theta}}_F$ *be an estimate of (7.10) obtained using the PIFA algorithm and the basis function representation (7.9). Then, given the above and Eqs. (7.7) and (7.8), the rate of convergence of a PIFA algorithm is bounded below and above by:*

$$C_{min} \frac{ML}{N} \leq V \left[\widehat{\frac{\partial \rho}{\partial \theta}}_F \right] \leq C_{max} \frac{ML}{N}, \qquad (7.11)$$

where L is the number of basis functions, M is the number of possible actions, and N is the number of independent estimates of the performance gradient.

Proof: See [3].

7.2.3.2   Rate of Convergence of Direct Sampling Algorithms   Next we establish rate of convergence bounds for performance gradient estimates that directly use the observed $Q_{obs}^{\pi}(s, a_i)$ without the intermediate step of building the FA representation. These bounds are established for the conditions $b(s) = (1/M) \sum_a Q(s,a)$ and $b(s) = 0$ in (7.5).

Theorem 7.2.2 Direct Sampling Algorithms. *Let $\widehat{\frac{\partial \rho}{\partial \theta}}$ be an estimate of (7.5), be obtained using direct samples of Q^{π}. Then, if $b(s) = 0$, and given the above assumptions and Eqs. (7.7) and (7.8), the rate of convergence of $\widehat{\frac{\partial \rho}{\partial \theta}}$ is bounded by:*

$$C_{\min} \frac{1}{N} \le V \left[\widehat{\frac{\partial \rho}{\partial \theta}} \right] \le C_{\max} \frac{1}{N}, \tag{7.12}$$

where M is the number of possible actions, and N is the number of independent estimates of the performance gradient. If $b(s) \neq 0$ is defined as:

$$b(s) = \frac{1}{M} \sum_{j=1}^{M} Q^{\pi}(s, a_j), \tag{7.13}$$

then the rate of convergence of the performance gradient $\widehat{\frac{\partial \rho}{\partial \theta}}_b$ is bounded by:

$$C_{\min} \frac{1}{N} \left(1 - \frac{1}{M} \right) \le V \left[\widehat{\frac{\partial \rho}{\partial \theta}}_b \right] \le C_{\max} \frac{1}{N} \left(1 - \frac{1}{M} \right). \tag{7.14}$$

Proof: See [3].

Comparing (7.14) and (7.12) to (7.11) we see that policy gradient algorithms such as PIFA (which build FA representations of Q) converge by a factor of $O(ML)$ slower than algorithms which directly sample Q. Furthermore, if the bias term is as defined in (7.13), the bounds on the variance are further reduced by $O(1 - (1/M))$. Experimental evidence in support of these theoretical results is given in Section 7.3.

7.2.4   Action Transition Policy Gradient

7.2.4.1   ATPG Approximation   The ATPG algorithm updates the policy gradient whenever the agent changes the action being executed (i.e. $a_t \neq a_{t+1}$), using the following approximation [2]:

$$P_t \equiv \left[\frac{\partial \pi(s_t, a_t; \theta)}{\partial \theta} q_t + \frac{\partial \pi(s_{t+1}, a_{t+1}; \theta)}{\partial \theta} q_{t+1} \right], \tag{7.15}$$

where, for M possible actions:

$$q_t = \left(\frac{1}{M}\right) \frac{\widehat{Q^\pi}(s_t, a_t) - \overline{Q_t}}{\pi(s_t, a_t;\ \theta)\ \pi(s_{t+1}, a_{t+1};\ \theta)}, \tag{7.16}$$

$$q_{t+1} = \left(\frac{1}{M}\right) \frac{\widehat{Q^\pi}(s_{t+1}, a_{t+1}) - \overline{Q_t}}{\pi(s_t, a_t;\ \theta)\ \pi(s_{t+1}, a_{t+1};\ \theta)}, \tag{7.17}$$

and $\widehat{Q^\pi}(s, a)$ is an unbiased estimate of $Q^\pi(s, a)$, and $\overline{Q_t}$ is the average of the state-action value functions:

$$\overline{Q_t} = \frac{1}{2}\left(\widehat{Q^\pi}(s_t, a_t) + \widehat{Q^\pi}(s_{t+1}, a_{t+1})\right). \tag{7.18}$$

The approximation defined by (7.15) through (7.18) has three main motivations. First, when estimating the gradient we only use value information obtained after action transition (i.e., (7.15)). Second, the actual value of the state action is not important, only the relative magnitude of $\widehat{Q^\pi}(s_t, a_t)$ and $\widehat{Q^\pi}(s_{t+1}, a_{t+1})$ matters. Therefore we subtract the average value from each such that $q_t = -q_{t+1}$; the result of this shifting of the Q values is that the gradient estimate will move toward increasing the probability of executing the more valuable action. Finally, normalizing by M and $\pi(s_t, a_t;\ \theta)\ \pi(s_{t+1}, a_{t+1};\ \theta)$ in (7.16) and (7.17) accounts for the averaging over M possible actions and the probability of executing a_t and then a_{t+1}.

An unbiased estimate of the state action value function Q^π at the end of each episode is given by:

$$\widehat{Q^\pi}(s_t, a_t) = \sum_{k=1}^{H-t} \gamma^{k-1} r_{t+k}, \tag{7.19}$$

where H is the number of time steps executed by the agent during the episode (i.e. $t = 1, ..., H$), and r_t are the rewards received by the agent.

Given (7.15), the estimate of the policy gradient after a single episode l is given by

$$\left[\frac{\widehat{\partial \rho}}{\partial \theta}\right]_l = \sum_{t=1}^{H} \varphi_t, \tag{7.20}$$

where

$$\varphi_t = \begin{cases} P_t & , \quad \text{if } a_t \neq a_{t+1}, \\ 0 & , \quad \text{otherwise.} \end{cases} \tag{7.21}$$

We can also estimate the policy gradient using L episodes, then, which gives the following ATPG Approximation:

$$\frac{\widehat{\partial \rho}}{\partial \theta} = \frac{1}{L} \sum_{l=1}^{L} \left[\frac{\widehat{\partial \rho}}{\partial \theta}\right]_l. \tag{7.22}$$

The following piecewise Lipschitz smoothness assumption on Q^π is used to prove the ATPG approximation:

$$
\begin{aligned}
&\forall s \in S, S \subseteq \Re^N, a \in A, \\
&\exists (k_1 > 0, \ k_1 \in \Re), \ \exists \left(\delta \in \Re^N\right), \exists (\epsilon \in \Re, \epsilon > 0), \ \text{s.t.,} \\
&\forall \|\delta\| \le \epsilon \to s + \delta \in S \wedge \\
&|Q^\pi (s, a) - Q^\pi (s + \delta, a)| \le k_1 \|\delta\|.
\end{aligned}
\tag{7.23}
$$

We further assume that $\pi(s, a; \ \theta)$ is piece-wise continuous with respect to s as follows:

$$
\begin{aligned}
&\forall s \in S, S \subseteq \Re^N, a \in A, \\
&\exists (k_2 > 0, \ k_2 \in \Re), \ \exists \left(\delta \in \Re^N\right), \exists (\epsilon \in \Re, \epsilon > 0), \ \text{s.t.,} \\
&\forall \|\delta\| \le \epsilon \to s + \delta \in S \wedge \\
&|\pi (s, a; \ \theta) - \pi (s + \delta, a; \ \theta)| \le k_2 \|\delta\|.
\end{aligned}
\tag{7.24}
$$

We now state the following theorem first presented in [1].

Theorem 7.2.3 ATPG Approximation *At each time t, let the step the agent takes be bounded by $(s_t - s_{t+1}) \le \delta$ for $s, \delta \in \Re^N$. Assume that Q^π satisfies the Lipschitz smoothness condition (7.23), that $\widehat{Q^\pi}$ is an unbiased estimate of Q^π, and that π satisfies (7.24) and is continuous w.r.t. θ. Assume also that the frequency of states visited under π is governed by $d^\pi(s)$. Then, as L becomes large in (7.22), and as $\|\delta\| \to 0$:*

$$
E\left[\widehat{\frac{\partial \rho}{\partial \theta}}\right] \to \frac{\partial \rho}{\partial \theta}.
\tag{7.25}
$$

Proof: See [1].

7.2.5   Boundary Localization: The η-Transform

In this section we demonstrate that any probabilistic policy

$$
(\pi (s, a; \ \theta) = Pr \{a_t = a \,|s_t = s; \ \theta \})
$$

can be transformed into approximately deterministic policies, while still preserving the policy gradient convergence results [1]. First consider a policy that consists of only two possible actions: $\pi(s, a_1; \ \theta)$ and $\pi(s, a_2; \ \theta)$. These policies can be mapped to boundary-localized stochastic policies, denoted by $\pi_d(s, a_1; \ \theta)$ and $\pi_d(s, a_2; \ \theta)$ respectively, using the following transformations:

$$
\pi_d (s, a_1; \ \theta) = \frac{1}{2} \left[1 + \tanh \left(\eta \left(\pi (s, a_1; \ \theta) - \pi (s, a_2; \ \theta)\right)\right)\right],
\tag{7.26}
$$

and

$$
\pi_d (s, a_2; \ \theta) = \frac{1}{2} \left[1 + \tanh \left(\eta \left(\pi (s, a_2; \ \theta) - \pi (s, a_1; \ \theta)\right)\right)\right],
\tag{7.27}
$$

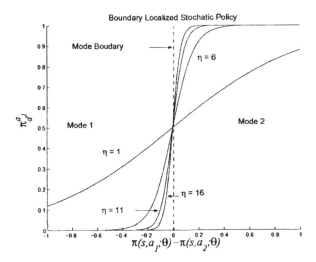

Fig. 7.2   The η-transformation.

where $\eta \rightarrow \infty$. These transformations are refered to as η-transformations. The effect of η on the probability distribution of the action a_1 (i.e. $\pi_c^{a_1} \equiv \pi_d(s, a_1; \theta)$) is demonstrated in Figure 7.2. Specifically, as $\eta \rightarrow \infty$ the probability of executing a_1 in regions of the state space where $(\pi(s, a_1; \theta) - \pi(s, a_2; \theta)) < 0$ becomes arbitrarily small. Similarly, as $\eta \rightarrow \infty$, the probability of executing action a_1 is arbitrarily close to 1 in regions of the state space where $(\pi(s, a_1; \theta) - \pi(s, a_2; \theta)) > 0$. As a result, the η-transformation transforms a policy $\pi(s, a_1; \theta)$ which is stochastic everywhere in the state space, to a policy $\pi_d(s, a_1; \theta)$ which is stochastic only near the boundaries defined by $(\pi(s, a_1; \theta) - \pi(s, a_2; \theta)) = 0$. These regions of the state space coincide with *mode boundary* regions (see Figure 7.1).

7.2.6   Boundary Localized Policy Gradient

The η-transformation is useful for improving the convergence properties of PGRL because it makes the policy gradient approaches zero everywhere except at mode boundaries. To see this, we differentiate the BL policy $\pi_d(s, a_1; \theta)$ with respect to the parameters θ as follows:

$$\frac{\partial \pi_d^{a_1}}{\partial \theta} = \frac{\eta}{2} \left(\text{sech}^2 \left(\eta \left(\pi^{a_1} - \pi^{a_2} \right) \right) \right) \left(\frac{\partial \pi^{a_1}}{\partial \theta} - \frac{\partial \pi^{a_2}}{\partial \theta} \right)$$
$$\stackrel{\Delta}{=} \Gamma \left(\eta, \left(\pi^{a_1} - \pi^{a_2} \right) \right) \left(\frac{\partial \pi^{a_1}}{\partial \theta} - \frac{\partial \pi^{a_2}}{\partial \theta} \right), \tag{7.28}$$

where, by definition, $\pi^{a_1} \equiv \pi(s, a_1; \theta)$, $\pi^{a_2} \equiv \pi(s, a_2; \theta)$, $\pi_d^{a_1} \equiv \pi_d(s, a_1; \theta)$ and $\pi_d^{a_2} \equiv \pi_d(s, a_2; \theta)$. From Eq. (7.28) we see that the performance gradient has

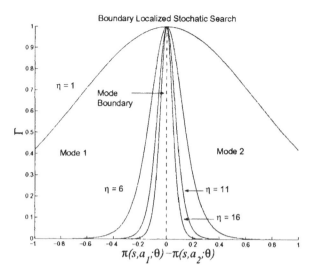

Fig. 7.3   The magnitude of the policy gradient goes to zero everywhere except mode boundaries as $\eta \to \infty$.

the following proportionality property:

$$\left| \frac{\partial \rho}{\partial \theta} \right| \propto \Gamma\left(\eta, \left(\pi^{a_1} - \pi^{a_2}\right)\right). \tag{7.29}$$

This equation is plotted in Figure 7.3, where we see that as $\eta \to \infty$, the policy gradient approaches zero everywhere except near mode boundaries. As a result, only regions in state space near mode boundaries need be stochastically searched when the agent uses BL policies (i.e., mode switching controllers). The direct result of this is that mode switching policies have a significantly reduced search space for learning, making them particularly suitable for high-dimensional RL problems.

It is easy to extend the argument presented above for a policy of two actions (or modes), to polices that have any finite number of actions (or modes). Therefore the η-transformation is valid for any finite set of policies, and one can transform any stochastic policy to a BL policy. Next we state the *Boundary Localized Policy Gradient Theorem* (first stated in [1]), which is a direct extension of the Policy Gradient theorem.

Theorem 7.2.4 Boundary Localized Policy Gradient *For any MDP, in either the average or discounted start-state formulations,*

$$\frac{\partial \rho}{\partial \theta} = \sum_s d^\pi(s) \sum_a \frac{\partial \pi_d(s, a; \theta)}{\partial \theta} Q^\pi(s, a). \tag{7.30}$$

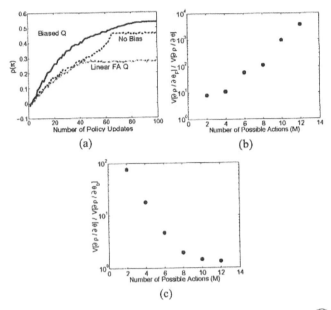

Fig. 7.4 Simulation Results: (a) Convergence of Algorithms; (b) $V[\widehat{\frac{\partial \rho}{\partial \theta}}_F]/V[\widehat{\frac{\partial \rho}{\partial \theta}}]$; (c) $V[\widehat{\frac{\partial \rho}{\partial \theta}}]/V[\widehat{\frac{\partial \rho}{\partial \theta}}_b]$.

Proof Sketch: If $\partial \pi/\partial \theta$ exists then because the η-transformation is continuously differentiable, so does $\partial \pi_d/\partial \theta$. The rest of the proof follows that of [5].

The significance of the BLPG theorem is that locally optimal mode switching polices can be learned using policy gradients. As a result, even though search is localized to a very small region of the state space, a policy gradient algorithm (7.30) will still converge to a locally optimum policy.

7.3   EXPERIMENTAL RESULTS

7.3.1   Verification of the Rate of Convergence Results

Simulated Environment: The experiments simulate an agent episodically interacting in a continuous two-dimensional environment. The agent starts each episode in the same state s_i, and executes a finite number of steps following a policy to a fixed goal state s_G. The stochastic policy is defined by a finite set of Gaussians, each associated with a specific action. The Gaussian associated with action a_m is defined as:

$$g_m(\mathbf{s}) = \exp\left[-\sum_{d=1}^{D} \frac{(s_d - c_{md})^2}{v_{md}}\right],$$

where $\mathbf{s} = (s_1, ..., s_D) \in \Re^D$, is the agents state, $c_{m1}, ..., c_{mD}$ is the Gaussian center, and $v_{m1}, ..., v_{mD}$ is the variance along each state space dimension. The probability of executing action a_m in state $\mathbf{s}$ is

$$\pi(\mathbf{s}, a_m; \theta) = \frac{g_m(\mathbf{s})}{\sum_{j=1}^{M} g_j(\mathbf{s})},$$

where $\theta = (c_{11}, ..., c_{1D}, v_{11}, ..., v_{1D}, ..., c_{M1}, ..., c_{MD}, v_{M1}, ..., v_{MD})$ defines the policy parameters that dictate the agent's actions. Action a_1 directs the agent toward the goals state s_G, while the remaining actions a_m (for $m = 2, ..., M$) direct the agent toward the corresponding Gaussian center $c_{m1}, ..., c_{mD}$.

Noise is modeled using a uniform random distribution between $(0, 1)$ denoted by $U(0, 1)$, such that the noise in dimension s_d is given by:

$$s_d^{obs} = s_d + \delta(U(0, 1) - 0.5),$$

where $\delta > 0$ is the magnitude of the noise, s_d^{obs} is the state the agent observes and uses to choose actions, and s_d is the actual state of the agent.

The agent receives a reward of $+1$ when it reaches the goal state, otherwise it receives a reward of:

$$r(\mathbf{s}) = -0.01 \exp\left[-\sum_{d=1}^{D} \frac{s_d^2}{4}\right].$$

Thus the agent gets negative rewards the closer it gets to the origin of the state space, and a positive reward whenever it reaches the goal state.

For further details we refer the reader to www.cis.upenn.edu/˜grudic/PGRLSim for source code and details.

Implementation Details: We sample $Q_{obs}^{\pi}(s, a_i)$ by executing action a_i in state s and thereafter following the policy. For the episodic formulation, where the agent executes a maximum of T steps during each episode, at the end of each episode, $Q_{obs}^{\pi}(s_t, a_t)$ for step t is evaluated using the following:

$$Q_{obs}^{\pi}(s_t, a_t) = \sum_{k=1}^{\infty} \gamma^{k-1} r_{t+k} | s_t = s, a_t = a, \pi.$$

Given that the agent executes a complete episode $((s_1, a_1), ..., (s_T, a_T))$ following the policy π, at the completion of the episode we can therefore calculate $(Q_{obs}^{\pi}(s_1, a_1), ..., Q_{obs}^{\pi}(s_T, a_T))$ using the above equation. This gives direct samples of T state action value pairs. From (7.5) we require a total of MT state action value function observations to estimate a performance gradient (for M possible actions). We obtain the remaining $(M-1)T$ observations of Q_{obs}^{π} by sending the agent out on $(M-1)T$ episodes, each time allowing it to follow the policy π for all T steps, with the exception that action $a_t = a_m$ is executed when $Q_{obs}^{\pi}(s_t, a_m)$ is being

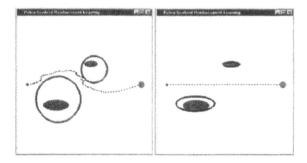

Fig. 7.5   Initial and typical final policy after learning in the smooth value function simulation.

observed. This sampling technique requires a total of $(M - 1)T - 1$ episodes and gives a complete set of Q^π_{obs} state action pairs for any path $((s_1, a_1), ..., (s_T, a_T))$. For the direct sampling algorithms in Section 7.2.3.2, these observations are directly used to estimate the performance gradient. For the linear basis function based PGRL algorithm in Section 7.2.3.1, these observations are first used to calculate the $w_{a_i, l}$ as defined in [5, 6], and then the performance gradient is calculated using (7.10).

Summary of Results: A plot of average $V[\widehat{\partial\rho/\partial\theta}_F]/V[\widehat{\partial\rho/\partial\theta}]$ values over 10,000 estimates of the performance gradient is shown in Figure 7.4b. As predicted by **Theorem 7.2.1** in Section 7.2.3.1 and **Theorem 7.2.2** in Section 7.2.3.2, as the number of actions M increases, this ratio also increases. One should note that Figure 7.4b plots average variance ratios, not the bounds in variance given in **Theorem 7.2.1** and **Theorem 7.2.2** (which have not been experimentally sampled). Thus the ML ratio predicted by the theorems is supported by the increase in the ratio as M increases. A plot of average $V[\widehat{\partial\rho/\partial\theta}]/V[\widehat{\partial\rho/\partial\theta}_b]$ values over 10,000 estimates of the performance gradient is shown in Figure 7.4c. This result also follows the predicted trends of **Theorem 7.2.1** and **Theorem 7.2.2**. Finally, Figure 7.4a shows the average reward over 100 runs as the three algorithms converge on a two action problem. Each algorithm is given the same number of Q^π_{obs} samples to estimate the gradient before each update. Because $\widehat{\partial\rho/\partial\theta}_b$ has the least variance, it allows the policy π to converge to the highest reward value $\rho(\pi)$. Similarly, because $\widehat{\partial\rho/\partial\theta}_F$ has the highest variance, its policy updates converge to the worst $\rho(\pi)$. Note that because all three algorithms will converge to the same locally optimal policy given enough samples of Q^π_{obs}, Figure 7.4a simply demonstrates that $\widehat{\partial\rho/\partial\theta}_F$ requires more samples than $\widehat{\partial\rho/\partial\theta}$, which in turn requires more samples than $\widehat{\partial\rho/\partial\theta}_b$.

7.3.2   Experimental Evaluation of the ATPG Algorithm

A simulated agent executes a policy in a continuous two-dimensional state space. The agent's environment has obstacles, starting positions and a goal position. The agent receives a reward of +1 if the goal is reached, and a negative reinforcement of -1 each time it hits an obstacle. The learning objective is to modify the policy (modelled by a set of Guassians as defined below) such that it avoids hitting obstacles while minimizing the time it takes to reach the goal. The agent uses a discount reward formulation.

The agent can execute one of two types of actions: move away from the center of a Gaussian or move toward the center of a Gaussian. Therefore the total number of actions is defined by the number of Gaussians in the agent's policy. There are four parameters per Gaussian: two defining its position (X, Y) in the two-dimensional state space, and two defining its width (S_x, S_y) (these make up the θ parameters in the policy $\pi(s, a;\ \theta)$). The agent chooses actions stochastically at each location in the state space, with the probability of executing an action being proportional to the relative magnitude of each Gaussian.

Typical paths followed by the agent are shown in Figure 7.5. The figure shows paths followed by the agent before and after learning as it moves along the dotted line from a starting point on the left-hand side to the goal (small circle) on the right, while avoiding the obstacles (shaded ellipsoid region). The agent's policy in Figure 7.5 is defined by three Gaussians, resulting in θ being a 12-dimensional vector that define $\pi(s, a;\ \theta)$. The location and width of two of the Gaussians is symbolized by regions enclosed by elliptical curves surrounding the shaded obstacles, which represent areas of the state space where the action associated with the Gaussian has the greatest probability of being executed. The ellipses symbolize the "move away from" the Gaussian center actions. The final Gaussian is centered at the goal position (the far right lightly shaded circle) and represents the "move toward" Gaussian center (i.e., goal position) action. The "move toward" goal action is most probably everywhere except within the black ellipse regions.

We compare the ATPG, PIFA and REINFORCE algorithms (see [2] for algorithm implementation details) on two types of simulated environments: one shown in Figure 7.5 which has a continuous value function with respect to changes in the policy (i.e., the function $Q^\pi(s, a)$ changes smoothly as the parameters θ of the policy change), and one shown in Figure 7.6 which has a discontinuous value function with respect to changes in the policy.

7.3.2.1 Smooth Value Function Simulation
The initial and learned policies of an agent in an environment where it never collides with an obstacle are shown in Figure 7.5. As learning progresses and the policy $\pi(s, a;\ \theta)$ is changed, the value function $Q^\pi(s, a)$ changes continuously with respect to θ. The left graphic in Figure 7.5 shows a typical path given an initial policy specification, and the right graphic in Figure 7.5 shows a typical path taken under the learned policy after either the ATPG or PIFA algorithms have converged. Both algorithms converge to a policy which takes the agent from the initial position to the final position in the shortest number of steps.

The ATPG and PIFA algorithms were compared for $L = 1, 3, 5, 7$, and 10. The average number of episodes over ten runs for the ATPG algorithm to converge ranged from 120 (standard deviation 10) for $L = 1$, to 1300 (standard deviation 150) for

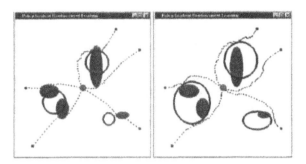

Fig. 7.6 Initial and typical final policy after learning in the discontinuous value function simulation.

$L = 10$. For the PIFA algorithm the number of episodes to converge ranged from 300 (standard deviation 20) for $L = 1$, to 5700 (standard deviation 980) for $L = 10$. Both algorithms obtained fastest convergence when the policy gradient was updated after each episode (i.e., $L = 1$). Therefore, for this smooth value function example, the PIFA algorithm required on average about twice as many episodes to converge as the ATPG algorithm. Convergence of the REINFORCE algorithm over 10 consecutive trails was first observed for $L = 94$, on average taking 710,000 episodes (s.d. 60,000).

7.3.2.2 Discontinuous Value Function Simulation   The initial policy of the discontinuous value function (with respect to changes in the policy parametrization θ) simulation is shown in the left graphic of Figure 7.6. The policy is defined by four Gaussians (i.e., a total of 16 parameters) and the agent has five possible starting positions (shown as dots on the outside of the workspace) and one goal (shown as a shaded circle in the center). Note that each of the paths initially collide with obstacles, which implies that the value function will become discontinuous with respect to the parameterization of the policy when the agent learns to avoid the obstacles. A typical policy after either the ATPG or PIFA algorithms has converged is shown in right graphic of Figure 7.6. Note that this learned policy typically has no paths which collide with obstacles.

The ATPG and PIFA algorithms were compared for $L = 1, 3, 5, 7$, and 10. The average number of episodes over ten runs for the ATPG algorithm to converge ranged from 230 (standard deviation 30) for $L = 1$, to 2000 (standard deviation 290) for $L = 10$. For the PIFA algorithm the number of episodes to converge ranged from 2400 (standard deviation 340) for $L = 1$, to 13000 (standard deviation 3000) for $L = 10$. Both algorithms obtained fastest convergence when the policy gradient was updated after each episode (i.e., $L = 1$). Therefore, for this discontinuous value function example, the PIFA algorithm required on average about ten times as many episodes to converge as the ATPG algorithm. We did not observe convergence of the REINFORCE algorithm for this simulation. Our simulations were stopped at $L = 200$ and 1,000,000 episodes.

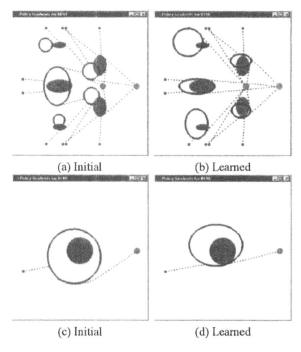

(a) Initial   (b) Learned

(c) Initial   (d) Learned

Fig. 7.7 Example of a simulated agent executing episodes in an environment. The agent begins at locations near the top, bottom, and/or left extremes of the environment and goes toward goal positions (small shaded circles) located at the right extreme or near the center. Dashed lines symbolize the agent's path and the obstacles are the larger gray areas. The agent can execute one of two possible actions: if it is executing a deterministic policy and if it is inside one of the regions delineated by a black ellipsoid, it moves away from the ellipsoid's center; otherwise it moves toward a goal position. If the agent is following a stochastic policy, then the ellipsoids indicate regions in state space where the "move away from" action is more probable.

7.3.3   Experimental Evaluation of the BLRL Algorithm

The simulations reported here use the same agents defined in the previous section. In the following, we present results on using the BLRL algorithm for both 2-D and N-D environments.

2-D Simulation: Figures 7.7a and b show a 2-D scenario which has ten possible starting positions, two goal positions, five obstacles, and six Gaussians for defining policies (five for "move away from" which are shown as ellipsoids, and one for "move toward goal", which is most probable everywhere except inside the ellipsoids). There are a total of 24 policy parameters θ.

Table 7.1   2-D Convergence Results with Standard Deviations

	Stochastic RL	Stochastic BLRL	Deterministic BLRL
Episodes to converge	6900 (sd 400)	600 (sd 90)	260 (sd 40)

Figure 7.7a shows the initial policy and the resulting paths through the environment. Note that four paths end before a goal is reached and eight paths have collisions with obstacles. Figure 7.7b shows the paths after the policy parameters have converged to stable values. The location and extent of the Gaussians has converged such that none of the paths now collide with obstacles, and the total distance travelled through state space is shorter than initial policy path.

Table 7.1 shows the average number of episodes (over ten runs) required for convergence for the three types of polices studied: stochastic, boundary localized stochastic ($\eta = 16$), and deterministic. Note that the purely stochastic polices take the greatest number of episodes to converge, while the deterministic policies take the fewest.

N-D Simulation: Our high-dimensional simulations include 4, 8, 16, 32, 64, and 128-dimensional environments, with the number of policy parameters θ ranging from 14 to 512 (i.e., 2 parameters per Gaussian per dimension). The projection of all of these environments into the two-dimensional XY plane is shown in Figure 7.7c and d. Figure 7.7c shows the starting policies, while Figure 7.7d shows policies after convergence. The convergence results are summarized in Figure 7.8 (over ten runs with standard deviation bars) for the three types of policies studied: stochastic, boundary localized stochastic ($\eta = 16$), and deterministic. For both the deterministic and boundary localized policies, convergence is essentially constant with dimension. However, for the stochastic policy, the convergence times explode with dimension. We only report convergence results up to 16 dimensions for stochastic policies - convergence on higher dimensions was still not achieved after 20,000 iterations, at which time the simulation was stopped.

7.4   CONCLUSION

The goal of this chapter is to propose a formulation for reinforcement learning that converges quickly in very-high-dimensional state spaces. We argue that the Policy Gradient Reinforcement Learning framework is an effective paradigm for large problem domains, and present theoretical and experimental results to support our position. Specifically, we show that gradients of increased reward in policy space can be efficiently estimated using the Action Transition Policy Gradient (ATPG) algorithm.

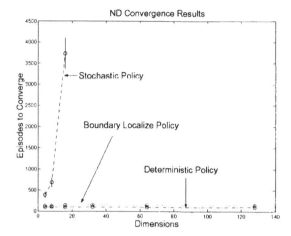

Fig. 7.8   N-D convergence results over ten runs with standard deviation bars.

In addition, we show that if the agent's control system (i.e. policy) belongs to the general class of widely used mode switching controllers, then Boundary Localized Reinforcement Learning (BLRL), can be used to give extremely fast convergence to locally optimal policies, in very-high-dimensional state spaces.

Acknowledgments

Thanks to Vijay Kumar and Jane Mulligan for discussing this work with us. This work was funded by the IRCS at the University of Pennsylvania, and by the DARPA ITO MARS grant no. DABT63-99-1-0017.

Bibliography

1. G. Z. Grudic and L. H. Ungar, Localizing search in reinforcement learning, *Proc. of the 17th National Conference on Artificial Intelligence*, vol. 17, pp. 590–595, 2000.

2. G. Z. Grudic and L. H. Ungar, Localizing policy gradient estimates to action transitions, *Proc. of the 17th International Conference on Machine Learning*, vol. 17, pp. 343–350, 2000.

3. G. Z. Grudic and L. H. Ungar, Rates of convergence of performance gradient estimates using function approximation and bias in reinforcement learning, *Advances in Neural Information Processing Systems*, vol. 14, pp. 1515–1522, 2002.

4. L. Baird and A. W. Moore, Gradient descent for general reinforcement learning, *Advances in Neural Information Processing Systems*, vol. 11, 1999.

5. R. S. Sutton, D. McAllester, S. Singh, and Y. Mansour, Policy gradient methods for reinforcement learning with function approximation, *Advances in Neural Information Processing Systems*, vol. 12, 2000.

6. V. R. Konda and J. N. Tsitsiklis, Actor-critic algorithms, *Advances in Neural Information Processing Systems*, vol. 12, 2000.

7. J. Baxter and P. L. Bartlett, Direct gradient-based reinforcement learning: I. Gradient estimation algorithms, *Technical Report of Computer Sciences Laboratory, Australian National University*, 1999.

8. L. C. Baird, Advantage updating, *Technical Report of Wright-Patterson Air Force Base*, 1993.

9. L. C. Baird, Residual algorithms: Reinforcement learning with function approximation, *Proc. of the 12th Int. Conference on Machine Learning*, pp. 30–37, 1995.

10. R. S. Sutton, Generalization in reinforcement learning: Successful examples using sparse coarse coding, *Advances in Neural Information Processing Systems*, vol. 8, pp. 1038–1044, 1996.

11. J. A. Boyan and A. W. Moore, Generalization in reinforcement learning: Safely approximating the value function, *Advances in Neural Information Processing Systems*, vol. 7, pp. 369–376, 1995.

12. R. J. Williams, Simple statistical gradient-following algorithms for connectionist reinforcement learning, *Machine Learning*, vol. 8, pp. 229–256, 1992.

13. R. J. Williams, A class of gradient-estimating algorithms for reinforcement learning in neural networks, *Proc. of the IEEE First International Conference on Neural Networks*, 1987.

14. L. P. Kaelbling, M. L. Littman, and A. W. Moore, Reinforcement learning: A survey, *Journal of Artificial Intelligence Research*, vol. 4, pp. 237–285, 1996.

15. G. J. Gordon, Stable function approximation in dynamic programming, *Proc. of the Twelfth Int. Conf. on Machine Learning*, vol. 4, pp. 261–268, 1995.

16. J. N. Tsitsiklis and B. Van Roy, Feature-based methods for large scale dynamic programming, *Machine Learning*, vol. 22, pp. 59–94, 1996.

17. D. P. Bertsekas and J. N. Tsitsiklis, *Neuro-dynamic programming*, Athena Scientific, Belmont, MA, 1996.

18. D. G. Lainiotis, A Unifying Framework for Adaptive Systems, I: Estimation, II, *Proc. of the IEEE*, vol. 64, pp. 1126–1134, 1182–1197, 1976.

19. B. Kuipers and K. J. Astrom, The composition and validation of heterogeneous control laws, *Automatica*, vol. 30, pp. 233–249, 1994.

20. K. S. Narendra, J. Balakrishnan and K. Ciliz, Adaptation and learning using multiple models, switching and tuning, *IEEE Control Systems Magazine*, vol. 15, pp. 37–51, 1995.

21. R. S. Sutton and A. G. Barto, *Reinforcement Learning: An Introduction*, MIT Press, Cambridge, MA, 1998.

22. R. S. Sutton, D. Precup and S. Singh, Between MDPs and semi-MDPs: A framework for temporal abstraction in reinforcement learning, *Artificial Intelligence*, vol. 112, pp. 181–211, 1999.

23. R. Brooks, A robust layered control system for a mobile robot, *IEEE Journal of Robotics and Automation*, vol. 2, pp. 14–23, 1986.

24. M. Kearns and S. Singh, Finite-sample convergence rates for Q-learning and indirect algorithms, *Advances in Neural Information Processing Systems*, vol. 11, 1999.

25. M. Minsky, Steps toward artificial intelligence, *Computers and Thought*, pp. 406–450, 1963.

26. J. Baxter and P. L. Bartlett, Reinforcement learning in POMDP's via direct gradient ascent, *Proc. of the 17th International Conference on Machine Learning (ICML'2000)*, pp. 41–48, 2000.

27. W. H. Press, B. P. Flannery, S. A. Teukolsky, and W. T. Vetterling, *Numerical Recipes in C*, Cambridge University Press, Cambridge, UK, 1998.

28. R. S. Sutton and A. G. Barto, Time-derivative models of pavlovian reinforcement, in M. Gabriel and J. Moore (eds.), *Learning and Computational Neuroscience: Foundations of Adaptive Networks*, pp. 497–537, 1990.

29. L. Peshkin, K.-E. Kim, N. Meuleau, and L. P. Kaelbling, Learning to cooperate via policy search, *Proc. of the Sixteenth International Conference on Uncertainty in Artificial Intelligence (UAI 2000)*, Stanford, CA, 2000.

30. R. Fierro, A. Das, J. Spletzer, Y. Hur, R. Alur, J. Esposito, G. Grudic, V. Kumar, I. Lee, J. P. Ostrowski, G. Pappas, J. Southall, and C. J. Taylor, A framework and architecture for multirobot coordination, *International Journal of Robotics Research*, vol. 21, no. 10-11, pp. 977–995, 2002.

Handbook of Learning and Approximate Dynamic Programming
Edited by Jennie Si, Andy Barto, Warren Powell and Donald Wunsch
Copyright © 2004 The Institute of Electrical and Electronics Engineers, Inc.

8 Hierarchical Decision Making

MALCOLM RYAN

University of New South Wales

Editor's Summary: As the field of reinforcement learning has advanced, interest in solving realistic control problems has increased. However, Markov Decision Process (MDP) models have not proven sufficient to the task. This has led to increased use of Semi-Markov Decision Process models and the development of Hierarchical Reinforcement Learning (HRL). This chapter is an overview of HRL beginning with a discussion of the problems with the standard MDP models, then presenting the theory behind HRL, and finishing with some actual HRL algorithms that have been proposed. To see some examples of how hierarchical methods perform, see Chapter 11.

8.1 INTRODUCTION

Reinforcement learning [24, 49] is a field of machine learning dedicated to building algorithms which learn to control the interaction of an agent with an environment, to achieve particular goals. For over a decade, since the advent of TD(λ) [46] and Q-Learning [54], most of the work in reinforcement learning has been based on the Markov Decision Process (MDP) model. While these algorithms have strong theoretical convergence properties, they have often performed poorly in practice. Optimal policies can be found for simple toy problems, but the algorithms are often difficult to scale up to realistic control problems.

Part of the problem is that MDPs model a system in fine detail. In recent years there has been a move from Markov Decision Processes to Semi-Markov Decision Processes, in an attempt to build models at higher levels of abstraction. This abstraction provides a number of advantages which allow us to apply reinforcement learning to a significantly larger collection of problems.

This field is known as Hierarchical Reinforcement Learning (HRL) and has been the subject of rapid growth in the past few years. Many techniques have arisen, all based on similar foundations but taking them in different directions. This profusion of different approaches can hide both the commonalities shared by these algorithms, and the different orthogonal improvements each approach contains. Many approaches combine several improvements into a single algorithm, which need not intrinsically be tied together.

In order to clarify this situation, we divide this review into three sections. In the first section we outline the problems with the standard MDP model for reinforcement learning in greater detail. In the second section we present the theory behind Hierarchical Reinforcement Learning and describe some of the different improvements it can offer. In the third section we describe some of the actual hierarchical reinforcement learning algorithms which have been published and try to show how they combine the different elements we have previously described.

This review tries to capture the motivations underlying the movement toward hierarchical reinforcement learning, and some of the prominent examples thereof. It is not exhaustive. For an alternative presentation, with several recent inventions that are omitted here, we refer the reader to Barto and Mahadevan's survey [3].

8.2 REINFORCEMENT LEARNING AND THE CURSE OF DIMENSIONALITY

Reinforcement learning models an agent interacting with an environment, trying to optimize its choice of action according to some reward criterion. The agent operates over a sequence of discrete time-steps $(t, t+1, t+2, \ldots)$. (One time-step indicates the duration of a single action. This may or may not correspond to a fixed unit of time in the real-world.)

At each step the agent observes the state of the environment s_t and selects an appropriate action a_t. Executing the action produces a change in the state of the environment to s_{t+1}. It is generally assumed that the sets of possible states S and available actions $\mathcal{A}$ are both finite. This is not always the case in practice, but it greatly simplifies the theory, so we shall follow this convention.

The mapping of states to actions is done by an internal policy π. The initial policy is arbitrarily chosen, generally random, and it is modified and improved based on the agent's experiences. Each experience $\langle s_t, a_t, s_{t+1} \rangle$ is evaluated according to some fixed reward function, yielding a reward $r_t \in \Re$. The agent's objective is to modify its policy to maximise its long-term reward. There are several possible definitions of "long-term reward" but the one most commonly employed is the *expected discounted return* given by:

$$\begin{aligned} R_t &= E\left\{r_t + \gamma r_{t+1} + \gamma^2 r_{t+2} + \ldots\right\} \\ &= E\left\{\sum_{i=0}^{\infty} \gamma^i r_{t+i}\right\}, \end{aligned} \tag{8.1}$$

where γ is the *discount rate* that specifies the relative weight of future rewards, with $0 \leq \gamma < 1$. Should the agent reach some terminal state s_T, then the infinite sum is cut short: all subsequent rewards $r_{T+1}, r_{T+2}, \ldots$ are considered to be zero. (Another, less commonly used, reward criterion is *average reward* [29, 41, 43].)

To ensure that this optimization problem is well-founded, most reinforcement learning algorithms place a strong constraint on the structure of the environment. They assume that it operates as a *Markov Decision Process* (MDP) [37]. An MDP describes a process that has no hidden state or dependence on history. The outcomes of every action, in terms of state transition and reward, obey fixed probability distributions that depend only on the current state and the action performed.

Formally an MDP can be described as a tuple $\langle S, A, T, R \rangle$ where S is a finite set of states, A is a finite set of actions, $T : S \times A \times S \to [0, 1]$ is a transition function and $R : S \times A \times \Re \to [0, 1]$ is a reward function with:

$$T\left(s'|s, a\right) = P\left(s_{t+1} = s' \mid s_t = s, a_t = a\right), \tag{8.2}$$

$$R\left(r|s, a\right) = P\left(r_t = r \mid s_t = s, a_t = a\right), \tag{8.3}$$

which respectively express the probability of ending up in state s' and receiving reward r after executing action a in state s. These probabilities must be independent of any criteria other than the values of s and a. This is called the *Markov Property*. An in-depth treatment of the theory of Markov Decision Processes can be found in any of [5–7, 19, 37, 49].

Given this simplifying assumption, the best action to choose in any state depends on that state alone. This means that the agent's policy can be expressed as a purely reactive mapping of states to actions, $\pi : S \to A$. Furthermore every state s can be assigned a value $V^\pi(s)$ that denotes the expected discounted return if the policy π is followed:

$$V^\pi(s) = E\left\{R_t \mid \varepsilon(\pi, s, t)\right\} \tag{8.4}$$

$$= E\left\{\sum_{i=0}^{\infty} \gamma^i r_{t+i} \mid \varepsilon(\pi, s, t)\right\} \tag{8.5}$$

$$= \int_{-\infty}^{+\infty} r\mathbf{R}_r s, a\mathrm{d}r + \gamma \sum_{s' \in S} \mathbf{T}_{s's}, aV^\pi(s'), \tag{8.6}$$

where $\varepsilon(\pi, s, t)$ denotes the event of policy π being initiated in state s at time t. V^π called the *state value function* for policy π.

An *optimal policy* can now be simply defined as a policy $\pi^\star$ that maximises $V^\pi(s)$ for all states s. The Markov property guarantees that there is such a globally optimal policy [49] although it may not be unique. We define the *optimal state-value function* $V^\star(s)$ as being the state value function of the policy $\pi^\star$:

$$\begin{aligned} V^\star(s) &= V^{\pi^\star}(s) \\ &= \max_\pi V^\pi(s). \end{aligned} \tag{8.7}$$

We can also define an *optimal state-action value function* $Q^\star(s, a)$ in terms of $V^\star(s)$ as:

$$Q^\star(s, a) = E\left\{r_t + \gamma V^\star(s_{t+1}) \mid s_t = s, a_t = a\right\}. \tag{8.8}$$

This function expresses the expected discounted return if action a is executed in state s and an optimal policy is followed thereafter. If such a function is known then an optimal policy can be extracted from it simply:

$$\pi^\star(s) = \arg\max_{a \in \mathcal{A}} Q^\star(s, a). \tag{8.9}$$

Thus the reinforcement learning problem can be transformed from learning an optimal policy $\pi^\star$ to learning the optimal state-action value function $Q^\star$. [1] This turns out to be a relatively straightforward dynamic programming problem, which could in principle be solved by conventional DP algorithms.

There is, however, one complication. Standard techniques such as value-iteration or policy-iteration [5] rely on us already having an accurate model of the underlying MDP. However, in reinforcement learning we assume that this model is not (initially) available. Our only source of information about the MDP is experimental interaction with the environment. Given this we can proceed in two ways: we can attempt to learn a model and then applied standard DP methods to construct a policy, or we can attempt to learn $Q^\star$ directly. These are called "model-based" and "model-free" approaches respectively. Much work has been done in both these areas, but model-free approaches tend to dominate. We perpetuate this bias in this chapter, focusing on model-free reinforcement learning, which is where most of the work in hierarchical learning has been taking place. We briefly summarize some of the model-based hierarchical learning toward the end of the chapter.

8.2.1   *Q*-Learning

Q-Learning [54] is widely regarded as the archetypal model-free reinforcement learning algorithm. It is an online incremental learning algorithm that learns an *approximate state-action value function* $Q(s, a)$ that converges to the optimal function $Q^\star$ in Eq. 8.8 above (under certain conditions outlined below). It is a simple algorithm which avoids the complexities of modelling the functions R and T of the MDP by learning Q directly from its experiences. It has significant practical limitations, but is theoretically sound and has provided a foundation for many more complex algorithms. Pseudocode for this algorithm is given in Algorithm 8.2.1.

The approximate Q-function is stored in a table. Its initial values may be arbitrarily chosen, typically they are all zero or else randomly assigned. At each time-step an

---

[1] In practice, it is only necessary for the agent to optimize its reward in the states it actually encounters. Complete optimal policies are not needed to do this as some states may never be reached. This is one way in which reinforcement learning differs from classical MDP solution methods.

Algorithm 8.2.1 Watkin's Q-Learning

---

function Q-LEARNING
    $t \leftarrow 0$
    Observe state s_t
    while s_t is not a terminal state **do**
        Choose action $a_t \leftarrow \pi(s_t)$ according to an exploration policy
        Execute a_t
        Observe resulting state s_{t+1} and reward r_t
        $Q(s_t, a_t) \overset{\alpha}{\longleftarrow} r_t + \gamma \max_{a \in \mathcal{A}} Q(s_{t+1}, a)$
        $t \leftarrow t + 1$
    end while
end Q-LEARNING

---

action is performed according to the policy dictated by the current Q-function:

$$a_t = \pi(s_t) = \arg\max_{a \in \mathcal{A}} Q(s_t, a). \tag{8.10}$$

The result of executing this action is used to update $Q(s_t, a_t)$, according to the temporal-difference rule:

$$Q(s_t, a_t) \leftarrow (1 - \alpha)Q(s_t, a_t) + \alpha(r_t + \gamma \max_{a \in \mathcal{A}} Q(s_{t+1}, a)), \tag{8.11}$$

where α is a learning rate, $0 \leq \alpha \leq 1$.[2]

---

[2]The expression in Eq. (8.11) is somewhat cumbersome. There are two operations being described simultaneously which are not clearly differentiated. The first operation is the temporal-difference step, which estimates the value of $Q(s_t, a_t)$ as:

$$Q_{new} = r_t + \gamma \max_{a \in \mathcal{A}} Q(s_{t+1}, a).$$

This value is the input to the second operation, which updates the existing value of $Q(s_t, a_t)$ toward this target value, using an exponentially weighted rolling average with learning rate α:

$$Q(s_t, a_t) \leftarrow (1 - \alpha)Q(s_t, a_t) + \alpha Q_{new}.$$

To simplify the equations we shall henceforth use the short-hand notation:

$$X \overset{\alpha}{\longleftarrow} Y$$

to indicate that the value of X is adjusted toward the target value Y via an exponentially weighted rolling average with decay factor α, that is:

$$X \leftarrow (1 - \alpha)X + \alpha Y.$$

Thus Eq. (8.11) shall be written as:

$$Q(s_t, a_t) \overset{\alpha}{\longleftarrow} r_t + \gamma \max_{a \in \mathcal{A}} Q(s_{t+1}, a). \tag{8.12}$$

This is non-standard notation, originally introduced by Baird in [2] but not widely adopted. I believe it captures the important elements of the formula more clearly and concisely.

The approximate state-action value function Q is proven to converge to the optimal function Q^* (and hence π to π^*) given certain technical restrictions on learning rates ($\sum_{t=1}^{\infty} \alpha_t = \infty$ and $\sum_{t=1}^{\infty} \alpha_t^2 < \infty$) and the requirement that all state-action pairs continue to be updated indefinitely [22, 55, 53]. This second requirement means that in executing the learned policy the agent must also do a certain proportion of non-policy actions for the purposes of exploration. Exploration is important in all the algorithms that follow. The simplest approach to exploration is the ε-greedy algorithm which simply takes an exploratory action with some small probability ε, and a policy action otherwise. A large number of more complex alternatives exist (see [52] for a summary).

8.2.2   Curse of Dimensionality

As stated above, Q-Learning is theoretically guaranteed to converge to an optimal policy, however the guarantee is only true in the limit, and in practical problems of any significant size it has often been found to be impracticably slow. Without doing a full analysis of the algorithm, we can observe certain factors which contribute to this failure.

To find an optimal policy, a Q-value must be learned for every state-action pair. This means, first of all, that every such pair needs to be explored at least once. So convergence time is at best $O(|S|.|A|)$. Real-world problems typically have large multi-dimensional state spaces. $|S|$ is exponential in the number of dimensions, so each extra dimension added to a problem multiplies the time it takes.

Furthermore, states are generally only accessible from a handful of close neighbors, so the distance between any pair of states in terms of action steps also increases with the size and dimensionality of the space. Yet a change in the value of one state may have consequences for the policy in a far distant state. As information can only propagate from one state to another through individual state transitions, the further apart two states are, the longer it will take for this information to be propagated. Thus the diameter of the state space is an additional factor in the time required to reach convergence.

A general-purpose solution to this problem has not yet been found. There have been many attempts to represent the table of Q-values more compactly by using one variety of function approximator or another. These have met with mixed success. Sometimes the resulting state-abstraction has enabled the learning algorithm to learn an effective (not necessarily optimal) policy in times faster than without abstraction by an order of magnitude or more for a particular domain (e.g., [4, 51, 56]), but no such approach has proven to be a general-purpose solution. What works well in one domain will fail spectacularly in another. Furthermore even the simplest and most conservative forms of function approximation have been shown to break the convergence proofs, causing Q-values to becoming wildly divergent in certain situations [2]. For a summary of such attempts, see [24].

As a result of these difficulties researchers have turned from seeking general-purpose to special-purpose solutions. It has been recognized that a number of the most successful applications of reinforcement learning have used significant task-specific background knowledge tacitly incorporated into the agent's representation of its states and actions. Focus is shifting toward creating an architecture by which this tacit information can become explicit and can be represented in a systematic way. The aim is to create systems that can benefit from the programmer's task-specific knowledge whilst maintaining desirable theoretical properties of convergence.

8.3   HIERARCHICAL REINFORCEMENT LEARNING IN THEORY

Significant attention has recently been given to hierarchical decomposition as a means to this end. "Hierarchical reinforcement learning" (HRL) is the name given to a class of learning algorithms that share a common approach to scaling up reinforcement learning.

Hierarchical decomposition has always been a natural approach to problem solving. "Divide-and-conquer" has long been a familiar motto in computer science. A complex problem can often be solved by decomposing it into a collection of smaller problems. The smaller problems can be solved more easily in isolation, and then recombined into a solution for the whole. Inspiration for hierarchical reinforcement learning came partly from behavior-based techniques for robot programming [10, 28, 31] and partly from the hierarchical methods used in symbolic planning [21, 25, 26, 40].

Hierarchical reinforcement learning accelerates learning by forcing a structure on the policies being learned. The reactive state-to-action mapping of Q-learning is replaced by a hierarchy of temporally abstract actions. These are actions that operate over several time-steps. Like a subroutine or procedure call, once a temporally abstract action is executed it continues to control the agent until it terminates, at which point control is restored to the main policy. These actions (variously called *subtasks*, *behaviors*, *macros*, *options*, *activities* or *abstract machines* depending on the particular algorithm in question) must themselves be further decomposed into one-step actions that the agent can execute. We shall henceforth refer to one-step actions as *primitive actions* and temporally-abstract actions as *behaviors*. Policies learned using primitive actions alone shall be called *monolithic* to distinguish them from *hierarchical* or *behavior-based* policies.

How does this decomposition aid us? There are two different ways. One, it allows us to limit the choices available to the agent, even to the point of hard-coding parts of the policy; and two, it allows us to specify local goals for certain parts of the policy. Different HRL algorithms implement these features in different ways. Some implement one and not the other. We shall postpone describing specific algorithms until Section 8.4, and for the moment present these features in more general terms, with the aid of an example.

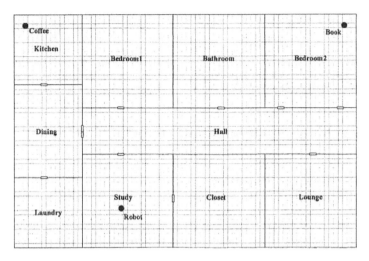

Fig. 8.1   An example environment.

8.3.1   Motivating Example

Figure 8.1 shows an example environment we shall use to illustrate the concepts in this chapter. Imagine that the learning agent is a house-hold robot in a house with the layout shown. Its purpose is to fetch objects from one room to another. It is able to know its location with a precision as shown by the cells of the grid, and its primitive actions enable it to navigate from a cell to any of its eight neighbors, with a small probability of error.

If the robot is in the same cell as an object, it can pick it up and carry it. There are two objects in the environment that we are interested in. In the kitchen in the northwest corner of the map is a machine which dispenses a cup of coffee. In the second bedroom there is a book, also indicated on the map. The robot starts at its docking location in the study. Its goal will vary from example to example as we consider different aspects of HRL (and later, of planning).

In this environment we have 15,000 states (75×50 cells, with two different states for each object, depending on whether the robot is holding it or not[3]) and 9 primitive actions (each of the 8 compass directions, plus the pickup action). This is not in itself a complex environment, and most goals will be relatively easy to complete, but it is certainly one that can be made simpler by providing an appropriate set of behaviors.

---

[3]The positions of the book and the coffee machine in their respective rooms are also part of the state but since these positions are fixed, and the robot cannot drop these objects elsewhere, this information can be omitted.

The obvious behaviors to specify are: Go(Room1, Room2) which moves the robot between two neighboring rooms, and Get(Object, Room) which moves toward and picks up the specified object when the robot is in the same room as it. We will discuss how these behaviors are implemented as we examine individual techniques.

8.3.2   Limiting the Agent's Choices

Since learning time is dominated by the number of state-action pairs that need to be explored, the obvious way to accelerate the process is to cut down the number of such pairs. Using background knowledge we can identify action choices which are plainly unhelpful and eliminate them from the set of possible policies. There is a variety of ways in which this can be done. Of course, such limitations must be applied with care, as overly broad limitations can prevent the agent from discovering policies that might otherwise be optimal.

Limiting Available Primitive Actions   The simplest solution is to hard-code portions of the policy. Some or all of the internal operation of a behavior can be written by hand by the system designer. This removes the need for the agent to do any kind of learning at all for significant portions of the state space, which will immediately improve performance. This assumes however that the system designer is able to do this. Part of the point of learning policies is to relieve the designer of the need to specify them, so this may be of limited use. Still, there are some situations in which simple behaviors might be wholly or partially specified, and algorithms have been designed to take advantage of this.

Less drastically, the internal policy of a behavior could be learned using only a limited subset of all available primitive actions. This is useful if the system designer knows that certain primitive actions are only suitable for particular behaviors and not for others. From the example, the Go() behaviors could reasonably be limited to only use the primitive actions which move the robot, and ignore the pickup action, which would be of no use to that behavior.

Note that it is not strictly necessary to use hierarchy to limit primitive action choice. A simple extension to the MDP formalism above would allow the admissible action set to be a function of the state. The Q-Learning algorithm can be successfully applied to this problem with only a slight modification. However adding hierarchy allows us to specify these restrictions on the wider context of the subtask we currently executing, as well as the immediate state.

Limiting Available Behaviors   Likewise, limits can be placed on which behaviors are available to the agent at different times. behaviors are generally limited in scope, so they often can only be executed from a subset of all possible states. For instance the Get() behavior can only be applied when the agent is it the same room as the target object. The set of states in which a behavior B can be applied is called its *applicability space*, which we shall denote B.pre. Learning algorithms should not allow the agent to choose a behavior in a state in which it is not applicable.

However this may not be limiting enough. As more ambitious problems are tackled, the repertoire of behaviors available to an agent is likely to become large, and many behaviors will have overlapping applicability spaces. It is of no use to limit the internal policy choices of behaviors if choosing between the behaviors becomes just as difficult.

To this end, most HRL algorithms implement some kind of *task hierarchy* to limit the choice of behaviors to those that are appropriate to the agent's situation. Consider the situation in the example environment when the robot is in hall with the goal of fetching both the book and the coffee. There are six applicable behaviors: Go(hall, study), Go(hall, dining), Go(hall, bedroom1), Go(hall, bathroom), Go(hall, bedroom2), and Go(hall, lounge). Of these, only two are appropriate: Go(hall, dining), if the agent decides to fetch the coffee first, and Go(hall, bedroom2) if the agent decides to fetch the book. Exploring the others is a waste of time. The system designer, who specified the behaviors, should realize this and incorporate it into the task hierarchy, limiting the agent's choices in this situation to one of these two behaviors. The larger an agent's repertoire of behaviors becomes, the more critical this kind of background knowledge.

Committing to behaviors   Finally, choices are limited by requiring long-term *commitment* to a behavior. It is conceivable that a learning algorithm could be written which implemented hard-coded behaviors but allowed the agent to choose a different behavior on every time step. Such an algorithm would hardly be any better than learning a primitive policy directly, and could easily be worse. Long-term commitment to behaviors has two benefits. First, a single behavior can traverse a long sequence of states in a single "jump", effectively reducing the diameter of the state-space and propagating rewards more quickly. In the grid-world, for example, fetching both the coffee and the book takes 126 primitive actions, but can be done with a sequence of just 10 behaviors.

Second, a behavior can "funnel" the agent into a particular set of terminating states. These states are then the launching points for new behaviors. If no behavior ever terminates in a given state, then no policy needs to be learned for that state. Again, referring to the grid-world, each Go() behavior terminates in one of the six cells surrounding a doorway, in one of four possible configuration of what the robot is holding. There are 10 doors, so this yields 240 states. Each Get() behavior terminates in the same location as the target object with 2 possible configurations of what the agent is holding, yielding a further 4 states. Plus 1 starting state gives a total of 245 states in which the agent needs to learn to choose a behavior, out of a possible 15,000. This is a significant reduction in the size of the policy-space and will result in much faster learning.

Flexible limitations   Limiting the policy space in this fashion will clearly have an effect on optimality. If the optimal policy does not fit the hierarchical structure, then any policy produced by a hierarchical reinforcement learner will be sub-optimal. This may well be satisfactory, but if not, it is possible to some degree to have the best of both worlds by imposing structure on the policy during the early phase of learning and relaxing it later. This allows the agent to learn a near-to-optimal policy quickly and then refine it to optimality in the long-term. Such techniques shall be described in more detail in Section 8.5.

8.3.3   Providing Local Goals

So far we have assumed that all choices the agent makes, at any point in the hierarchy, are made to optimize the one global reward function. Such a policy is said to be *hierarchically optimal* [15]. A hierarchically optimal policy is the best possible policy within the confines of the hierarchical structure imposed by the system designer.

Hierarchical optimality, however, contradicts part of the intuition of behavior-based decomposition of problems. The idea that a problem can be decomposed into several independent subparts which can be solved separately and recombined no longer holds true. The solution to each subpart must be made to optimize the whole policy, and thus depends on the solutions to every other subpart. The internal policy for a behavior depends on its role in the greater task.

Consider, for example, the behavior Go(hall, bedroom2) in the grid-world problem. Figure 8.2 shows two possible policies for this behavior. Assume, for the moment that diagonal movement is impossible. Which of these policies is hierarchically optimal? The answer depends on the context in which it is being used. If the agent's overall goal was to reach the room as soon as possible, then the policy in Figure 8.2(a) is preferable. If, on the other hand, the goal is to pick up the book, then the policy in Figure 8.2(b) is better, as it will result in a shorter overall path to the book (from some starting locations). (Note that, in both cases, the policy illustrated is only one of many available optimal policies.)

Furthermore, the same behavior may have different internal policies in different parts of the problem. For instance, if the agent's goal is to fetch the book, carry it to another room and then return to the bedroom, then the first instance of Go(hall, bedroom2) will use the policy in Figure 8.2(b) and the second instance will use the policy in Figure 8.2(a).

An alternative is to define local goals for each behavior in terms of a behavior-specific reward function. The behavior's internal policy is learned to optimize this local reward, rather than the global reward. This is called *recursive optimality* [15] and is a weaker form than hierarchical optimality. Recursively optimal policies make best use of the behaviors provided to them, but cannot control what the behaviors themselves do, and so cannot guarantee policies that are as efficient as hierarchically optimal policies.

The advantages of this approach, however, are several. First of all, learning an internal policy using a local reward function is likely to be much faster than learning with a global one. The behavior can be learned independently, without reference to the others. Local goals are generally simpler than global goals, and local rewards occur sooner than global ones. So each individual behavior will be learned more quickly.

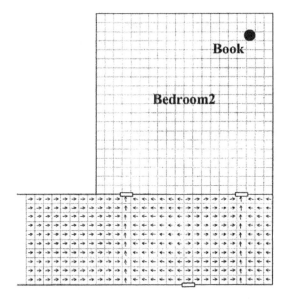

(a) A policy which optimizes the number of steps to enter the room

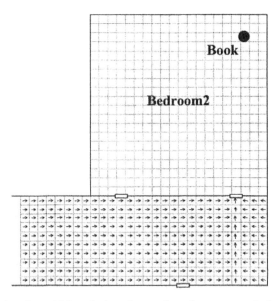

(b) A policy which optimizes the number of steps to reach the book

Fig. 8.2   Two different internal policies for the behavior Go(hall, bedroom2).

Furthermore, local goals often allow state abstraction. Elements of the state that are irrelevant to a local reward function can be ignored when learning the behavior. So, for example, if the Go(hall, bedroom2) behavior had a local reward function which rewarded the agent for arriving in the bedroom, then the internal policy for the behavior could ignore what the robot is carrying. This would reduce the size of the state space for this behavior by a factor of four.

Finally, local goals allow re-use. Once a behavior has been learned in one context, it can be used again in other contexts without having to re-learn its internal policy. This is useful when the same behavior is employed several different times within the one policy.

The decision whether or not to include local goals is a trade-off between optimality and learning speed. In the ideal case, when local rewards exactly match the projected global rewards, the policies learned will be identical. However this is unlikely to occur, and so we must decide which measure of performance is more important to us. In practice different researchers have chosen different approaches, as will become apparent in Section 8.4.

8.3.4   Semi-Markov Decision Processes: A Theoretical Framework

So far we have described hierarchical reinforcement learning in abstract terms. We have assumed that choosing between behaviors can be done in much the same way as choosing primitive actions in monolithic reinforcement learning, to optimize the expected discounted return. There is, however, a fundamental difference between monolithic and hierarchical reinforcement learning: behaviors are temporally extended where primitive actions are not. Executing a behavior will produce a sequence of state-transitions, yielding a sequence of rewards. The MDP model that was explained in Section 8.2 is limited insofar as it assumes each action will take a single time-step. A new theoretical model is needed to take this difference into account.

Semi-Markov Decision Processes [20] are an extension of the MDP model to include a concept of duration, allowing multiple-step actions. Formally an SMDP is a tuple $\langle S, B, T, R \rangle$, where S is a set of states, B is a set of behaviors (temporally abstract actions), $T : S \times B \times S \times \mathbb{N}^+ \rightarrow [0, 1]$ is a transition function (including duration of execution), and $R : S \times B \times \Re \rightarrow [0, 1]$ is a reward function:

$$T\left(s', k | s, B\right) = P\left(\begin{array}{c} B_t \text{ terminates in } s' \text{ at time } t + k \mid \\ s_t = s, B_t = B \end{array} \right), \quad (8.13)$$

$$R\left(r | s, B\right) = P\left(\sum_{i=0}^{k-1} \gamma^i r_{t+i} = r \mid s_t = s, B_t = B \right), \quad (8.14)$$

when B_t is the behavior executing at time t. T and R must both obey the Markov property, that is, they can only depend on the behavior and the state in which it was started. (This formulation of a SMDP is based on that described by Parr in [34].)

A policy is a mapping $\pi : \mathcal{S} \rightarrow \mathcal{B}$ from states to behaviors. A state-value function can be given as:

$$V^{\pi}(s) = \int_{-\infty}^{+\infty} r\mathbf{R}_r s, \pi(s) \mathrm{d}r + \sum_{s', k'} \mathbf{T}_{s', k'} s, \pi(s) \gamma^k V^{\pi}(s'). \qquad (8.15)$$

Semi-Markov Decision Processes are designed to model any continuous-time discrete-event system. Their purpose in hierarchical reinforcement learning is more constrained. Executing a behavior results in a sequence of primitive actions being performed. The value of the behavior is equal to the value of that sequence. Thus if behavior B is initiated in state s_t and terminates sometime later in state s_{t+k} then the SMDP reward value R_t is equal to the accumulation of the one-step rewards received while executing B:

$$R_t = r_t + \gamma r_{t+1} + \gamma^2 r_{t+2} + \cdots + \gamma^{k-1} r_{t+k-1}. \qquad (8.16)$$

Thus the state-value function in Eq. 8.15 above becomes:

$$V^{\pi}(s) = E\left\{ \sum_{i=0}^{\infty} \gamma^i r_{t+i} \mid \varepsilon(\pi, s, t) \right\}, \qquad (8.17)$$

which is identical to the state-value function for primitive policies shown previously in Eq. (8.4). We can define an optimal behavior-based policy π^* with the optimal state-value function V^{π^*} as:

$$V^{\pi^*}(s) = \max_{\pi} V^{\pi}(s). \qquad (8.18)$$

Since the value measure V^{π} for a behavior-based policy π is identical to the value measure V^{π} for a primitive policy we know that π^* yields the optimal primitive policy over the limited set of policies that our hierarchy allows.

8.3.5   Learning Behaviors

Learning internal policies of behaviors can be expressed along the same lines. Formally, let B.π be the policy of behavior B, and B.$\mathcal{A}$ be the set of sub-actions (either behaviors or primitives) available to B. Let Root indicate the root behavior, with reward function equal to that of the original (MDP) learning task. The recursively optimal policy has:

$$B.\pi^*(s) = \arg \max_{a \in B.\mathcal{A}} B.Q^*(s, a), \qquad (8.19)$$

where $B.Q^*(s, a)$ is the optimal state-action value function for behavior B according to its local reward function B.R (defined by the system designer in accordance to the behavior's goals) .

In contrast, the hierarchically optimal policy has

$$B.\pi^\star(stack, s) = \arg\max_{a \in B.\mathcal{A}} Root.Q^\star(stack, s, a), \qquad (8.20)$$

where $stack = \{Root, \ldots, B\}$ is the calling stack of behaviors and $Root.Q^\star$ is the state-action value function according to the root reward function. The stack is a necessary part of the input to an hierarchically optimal policy, as the behavior may operate differently in different calling contexts. (Hierarchically optimal policies do not allow local goals for behaviors, so $B.R$ and $B.Q^\star$ are not defined.)

8.4 HIERARCHICAL REINFORCEMENT LEARNING IN PRACTICE

We have discussed the expected benefits of hierarchical reinforcement learning in abstract terms without referring to any particular algorithm, to show what motivates its exploration. Historically a large number of different implementations have been proposed ([13, 23, 27]) but only recently have they been developed into a strong theoretical framework that has been commonly agreed upon. Even so, there are several current implementations that differ significantly in which elements they emphasise and how they approach the problem. We shall focus on four of the most recent offerings: SMDP Q-Learning, HSMQ-Learning, MAXQ-Q and HAMQ-Learning.

8.4.1 Semi-Markov Q-Learning

The simplest algorithm extends Watkins' Q-Learning to include temporally abstract behaviors. Bradtke and Duff [9] proposed such an algorithm, called *SMDP Q-Learning*. Assuming behaviors obey the Semi-Markov property, an optimal policy can be learned in a manner analogous to Watkins' Q-Learning, but discounting based on the time taken by the behavior, as shown in Algorithm 8.4.1.

Just as primitive Q-Learning learns a state-action value function, so SMDP Q-Learning learns a *state-behavior value function* $Q : \mathcal{S} \times \mathcal{B} \to \Re$, which is an approximation to the *optimal state-behavior value function* $Q^\star$:

$$Q^\star(s, B) = E\left\{ \sum_{i=0}^{k-1} \gamma^i r_{t+i} + \gamma^k V^\star(s_{t+k}) \mid \varepsilon(s, B, t) \right\}, \qquad (8.21)$$

where $\varepsilon(s, B, t)$ indicates the event of executing behavior B in state s at time t, and k is a random variable expressing the duration of the behavior B in this event (taken into account in the expectation).

The optimal policy is defined as before:

$$\pi^\star(s) = \arg\max_{B \in \mathcal{B}} Q^\star(s, B). \qquad (8.22)$$

Algorithm 8.4.1 SMDP Q-Learning

---

function SMDPQ

    $t \leftarrow 0$

    Observe state s_t

    while s_t is not a terminal state **do**

        Choose behavior $B_t \leftarrow \pi(s_t)$ according to an exploration policy

        $totalReward \leftarrow 0$

        $discount \leftarrow 1$

        $k \leftarrow 0$

        while B_t has not terminated **do**

            Execute B_t

            Observe reward r

            $totalReward \leftarrow totalReward + discount \times r$

            $discount \leftarrow discount \times \gamma$

            $k \leftarrow k + 1$

        end while

        Observe state s_{t+k}

        $Q(s_t, B_t) \xleftarrow{\alpha} totalReward + discount \times \max_{B \in \mathcal{B}} Q(s_{t+k}, B)$

        $t \leftarrow t + k$

    end while

end SMDPQ

The approximation $Q(s, B)$ can be learned via the update rule (analogous to the Q-Learning update rule in Eq. (8.12)):

$$Q(s_t, B_t) \xleftarrow{\alpha} R_t + \gamma^k \max_{B \in \mathcal{B}} Q(s_{t+k}, B), \qquad (8.23)$$

where k is the duration of B_t and R_t is a discounted accumulation of all single-step reward values received while executing the behavior:

$$R_t = \sum_{i=0}^{k-1} \gamma^i r_{t+i}. \qquad (8.24)$$

SMDP Q-Learning can be shown to converge to the optimal behavior-based policy under circumstances similar to those for 1-step Q-Learning [34].

Sutton, Precup and Singh [50] applied SMDP Q-Learning to behaviors they called "options". An option B has a fixed internal policy $B.\pi$ which is recursively constructed from other options and primitive actions. If this internal policy obeys the Markov property, then such behaviors are semi-Markov and can be used in SMDP Q-Learning. Other model-based dynamic programming techniques such as value-iteration and Monte Carlo methods can also be applied to options using the semi-Markov model [36]. Also, since different options are constructed from the same primitive building blocks, *intra-option* learning is possible, where experiences from

Algorithm 8.4.2 HSMQ-Learning

function HSMQ(state s_t, action a_t)
returns sequence of state transtions $\{\langle s_t, a_t, s_{t+1}\rangle, \ldots\}$
      if a_t is primitive **then**
            Execute action a_t
            Observe next state s_{t+1}
            return $\{\langle s_t, a_t, s_t + 1\rangle\}$
      else
            sequence $S \leftarrow \{\}$
            behavior B $\leftarrow a_t$
            $\mathcal{A}_t \leftarrow$ TASKHIERARCHY$(s_t, \mathbf{B})$
            while B is not terminated **do**
                  Choose action $a_t \leftarrow$ B.$\pi(s_t)$ from $\mathcal{A}_t$
                         according to an exploration policy
                  sequence $S' \leftarrow$ HSMQ(s_t, a_t)
                  $k \leftarrow 0$ $totalReward \leftarrow 0$
                  for each $\langle s, a, s'\rangle \in S'$ **do**
                        $totalReward \leftarrow totalReward + \gamma^k$B.$r(s, a, s')$
                        $k \leftarrow k + 1$
                  end for
                  Observe next state s_{t+k}
                  $\mathcal{A}_{t+k} \leftarrow$ TASKHIERARCHY$(s_{t+k}, \mathbf{B})$
                  B.$Q(s_t, a_t) \xleftarrow{\alpha} totalReward + \gamma^k \max_{a \in \mathcal{A}_{t+k}}$ B.$Q(s_{t+k}, a)$
                  $S \leftarrow S + S'$
                  $t \leftarrow t + k$
            end while
            return S
      end if
end HSMQ

executing one option can be applied to learning about another. This will be described in more detail in Section 8.6.

8.4.2 Hierarchical Semi-Markov Q-Learning

Hierarchical Semi-Markov Q-Learning (HSMQ) [16] is a recursively optimal learning algorithm that learns reactive behavior-based policies, with a designer-specified task hierarchy. As shown in Algorithm 8.4.2 it is a simple elaboration of the SMDP Q-Learning algorithm. The SMDPQ update rule given in Eq. (8.23) is applied recursively with local reward functions at each level of the hierarchy. TASKHIERARCHY is a function which returns a set of available actions (behaviors or primitives) that can be used by a particular behavior in a given state. This hierarchy is hand-coded by the system designer based on knowledge of which actions are appropriate on what occasions.

HSMQ-Learning converges to a recursively optimal policy with the same kinds of requirements as SMDP Q-Learning, provided also that the exploration policy for behaviors is greedy in the limit [44].

8.4.3 MAXQ-Q

A more sophisticated algorithm for learning recursively optimal policies is Dietterich's MAXQ-Q [15]. The policies it learns are equivalent to those of HSMQ, but it uses a special decomposition of the state-action value function in order to learn them more efficiently. MAXQ-Q relies on the observation that the value of a behavior B as part of its parent behavior P can be split into two parts: the reward expected while executing B, and the discounted reward of continuing to execute P after B has terminated. That is:

$$P.Q(s, B) = P.I(s, B) + P.C(P, s, B), \tag{8.25}$$

where $P.I(s, B)$ is the expected total discounted reward (according to the reward function of the parent behavior P) that is received while executing behavior B from initial state s, and $P.C(B_{parent}, s, B_{child})$ is the expected total reward of continuing to execute behavior B_{parent} after B_{child} has terminated, discounted appropriately to take into account the time spent in B_{child}. (Again with rewards calculated according to the behavior P.)

Furthermore the $I(s, B)$ function can be recursively decomposed into I and C via the rule:

$$P.I(s, B) = \max_{a \in B.\mathcal{A}} P.Q(s, a). \tag{8.26}$$

There are several advantages to this decomposition, primarily of value in learning recursively optimal Q-values. The I and C functions can each be represented with certain state abstractions that do not apply to both parts. The explanation is complex and beyond the scope of this review. For full details and pseudocode see [15].

8.4.4 Q-Learning with Hierarchies of Abstract Machines

Q-Learning with Hierarchies of Abstract Machines (HAMQ) [35] is an hierarchically optimal learning algorithm that uses a more elaborate model to structure the policy space. Behaviors are implemented as *hierarchies of abstract machines* (HAMs) which resemble finite-state machines, in that they include an internal *machine state*. The state of the machine dictates the action it may take. Actions include: (1) performing primitive actions, (2) calling other machines as subroutines, (3) making choices, (4) terminating and returning control to the calling behavior. Transitions between machine states may be deterministic, stochastic or may rely on the state of the environment. Learning takes place at choice states only, where the behavior must decide which of several internal state transitions to make. HAMs represent a compromise between hard-coded policies and fully-learned policies. Some transitions can be hard-coded into the machine while others can be learned. Thus they provide a means for background knowledge in the form of partial solutions to be specified.

Behaviors in HAMQ are merely a typographic convenience. In effect they are compiled into a single abstract machine, consisting of action nodes and choice nodes only. Algorithm 8.4.3 shows the Pseudocode for learning in such a machine.

Andre and Russell [1] have extended the expressive power of HAMs by introducing parameterisation, aborts and interrupts, and memory variables. These Programmable HAMs allow quite complex programmatic description of behaviors, while also providing room for exploration and optimisation of alternatives.

Algorithm 8.4.3 HAMQ-Learning

---

function HAMQ
> $t \leftarrow 0$
> $node \leftarrow$ starting node
> $totalReward \leftarrow 0$
> $k \leftarrow 0$
>   choice $a \leftarrow null$
>   choice state $s \leftarrow null$
>   choice node $n \leftarrow null$
>   **while** s is not a terminal state **do**
>>   **if** $node$ is an action node **then**
>>>   Execute action
>>>   Observe reward r
>>> $totalReward \leftarrow totalReward + \gamma^k r$
>>> $k \leftarrow k + 1$
>>> $node \leftarrow node.next$
>>   **else** $node$ is a choice node
>>>   Observe state s'
>>>   **if** $n \neq null$ **then**
>>>> $Q(n, s, a) \xleftarrow{\alpha} totalReward + \gamma^k \max_{a' \in \mathcal{A}} Q(node, s', a')$
>>>> $totalReward \leftarrow 0$
>>>> $k \leftarrow 0$
>>>   **end if**
>>> $n \leftarrow node$
>>> $s \leftarrow s'$
>>>   Choose transition $a \leftarrow \pi(n, s)$ according to an exploration policy
>>> $node \leftarrow a.destination$
>>   **end if**
>   **end while**
end HAMQ

8.5   TERMINATION IMPROVEMENT

In Section 8.3.2 we discussed the importance of long-term commitment to behaviors. Without this, much of the benefit of using temporally abstract actions is lost. However it can also be an obstacle in the way of producing optimal policies. Consider the situation illustrated in Figure 8.3. The task is to navigate to the indicated goal location. behaviors are represented by dotted circles and black dots indicating the applicability space and terminal states respectively. The heavy line shows a path from the starting location to the goal, using the behaviors provided. The path travels from one termination state to the next, indicating that each behavior is being executed all the way to completion.

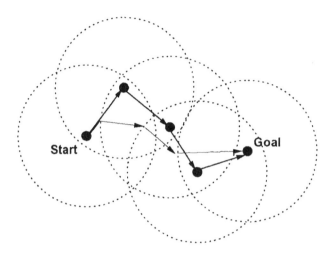

Fig. 8.3 A simple navigation task illustrating the advantage of termination improvement. The circles show the overlapping applicability spaces for a collection of hard-code navigation behaviors. Each behavior moves the agent toward the central landmark location (the black dots). The heavy line indicates the standard policy with commitment to behaviors. The lighter line indicates the path taken by a termination improved policy.

Compare this with the path shown by the lighter line. In this case each behavior is executed only until a more appropriate behavior becomes applicable. "Cutting corners" in this way results in a significantly shorter path, and a policy much closer to the optimal one.

This example is taken from the work of [45] who call this process *termination improvement*. They show how to produce such corner-cutting policies using hard-coded behaviors. Having already learned an optimal policy π using these behaviors, they transform it into an improved *interrupted policy* π' by prematurely interrupting an executing behavior B whenever $Q(s, B) < V(s)$, that is, when there is a better alternative behavior available. The resulting policy is guaranteed to be of equal or greater efficiency than the original.

A similar approach can be applied to policies learned using MAXQ-Q [15]. While MAXQ-Q is a recursively optimal learning algorithm, it nevertheless learns a value for each primitive action using the global reward function. In normal execution, actions are chosen on the basis of the local Q-value assigned to each by its calling behavior. However once such a recursively optimal policy has been learned, it can be improved by switching to selecting primitive actions based on their global Q-value instead. There is no longer any commitment to behaviors. Execution reverts to the reactive semantics of monolithic Q-learning, and the hierarchy serves only as a means to assign Q-values to primitives. This is called the *hierarchical greedy policy*, and is also guaranteed to be of equal or greater efficiency than the recursive policy. Furthermore, by continuing to update these Q-values, via *polling execution* [14, 23], this policy can be further improved.

In both these algorithms it is important that the transformation is made to the policy once an uninterrupted policy has already been learned. If behaviors can be arbitrarily interrupted from the outset then the advantages of using temporally abstract actions are lost. However in my own TRQ algorithm [39], I show that selective termination improvement can be done while learning is still ongoing, if we limit it to those occasions in which we can prove that the executing behavior is no longer appropriate. In TRQ this is done by giving behaviors abstract symbolic descriptions and using these descriptions to build high-level plans. A behavior is only appropriate as long as it is recommended by the plan. If a behavior is no longer appropriate then it may be interrupted before normal termination without adversing affecting learning performance.

8.6   INTRA-BEHAVIOR LEARNING

When behaviors are constructed from a common set of primitive actions, in a common Markov model, experiences gathered while executing one behavior may strongly overlap with those from another. If two behaviors B_1 and B_2 have overlapping applicability spaces and action sets, then when B_1 executes action $a \in B_1.\mathcal{A} \cap B_2.\mathcal{A}$ in state $s \in B_1.\text{pre} \cap B_2.\text{pre}$ then the resulting experience $\langle s, a, s' \rangle$ can often be used to learn about both B_1 and B_2.

There are two ways in which this information can be transferred from B_1 to B_2:

1. It can be used to update both the *internal* policy of B_2, through a method called *all-goals updating* [23], and

2. It can be used to update the *external* policy which is calling B_1 and B_2, through *intra-option learning* [36, 50].

8.6.1   All-Goals Updating

Q-Learning is what is termed an *off-policy* learning algorithm [49]. This means that the update rule for the state-action value function $Q(s,a)$ (Eq. (8.12) above) does not rely on the action a to be drawn from the current policy. In particular, if we are working in a recursively-optimal framework, in which there are behavior-specific reward functions specifying local goals, then any experience $\langle s, a, s' \rangle$ with

$s, s' \in$ B.pre and $a \in$ B.$\mathcal{A}$ can be used to update the state-action value function $B.Q(s, a)$ regardless of whether the experience was obtained while executing B itself or another behavior.

As a result, when multiple behaviors have overlapping applicability spaces and action sets, then experiences from executing one behavior can be used to improve the policy of others. This is what Kaelbling calls *all-goals updating* [23].

This is also the case for the hierarchical extensions of Q-learning which we have presented above, but is not true for all reinforcement learning algorithms. *On-policy* algorithms, such as SARSA [38, 48] are not amenable to this kind of experience sharing as they rely on the experiences used to update a policy being drawn from the execution of that policy.

8.6.2 Intra-Option Learning

Whereas all-goals updating concerns updating the internal policies of overlapping behaviors, *intra-option learning* [36, 50] updates the policy of the parent behavior which calls the subtasks B_1 and B_2. In this case B_1 and B_2 are assumed to be fixed-policy behaviors (options), although the technique can probably also be extended to hierarchically optimal or recursively optimal learned behaviors (no convergence proof is yet available).

Intra-option learning is based on the observation that if a behavior B dictates primitive action a_t in state s_t, and then continues executing then:

$$Q^*(s_t, B) = E(r_t) + \gamma \sum_{s_{t+1}} \mathbf{T}_{s_{t+1}s_t, a_t} Q^*(s_{t+1}, B), \qquad (8.27)$$

that is, the value of executing B in s_t is the immediate reward for executing a_t plus the discounted value of continuing to execute B thereafter. Alternatively, if B terminates in state s_{t+1} then:

$$Q^*(s_t, B) = E(r_t) + \gamma \sum_{s_{t+1}} \mathbf{T}_{s_{t+1}s_t, a_t} \max_{B' \in \mathcal{B}} Q^*(s_{t+1}, B'), \qquad (8.28)$$

that is, the value of executing B is the the immediate reward for executing a_t and then the discounted value of the subsequently chosen behavior.

Based on this fact, we can construct a *one-step intra-option Q-learning* update rule which updates the state-behavior value $Q(s, B)$ with $B.\pi(s) = a$ based on the experience $\langle s, a, r, s' \rangle$, as follows:

$$Q(s, B) \xleftarrow{\alpha} r + \gamma U(s_{t+1}, B), \qquad (8.29)$$

where

$$U(s, B) = (1 - \beta(B, s))Q(s, B) + \beta(B, s) \max_{B' \in \mathcal{B}} Q(s, B'),$$

where $\beta(B, s)$ denotes the probability B terminates in state s.

Once again, this update rule does not rely on the experience necessarily resulting from the execution of B as part of the parent policy. The experience may be drawn from the execution of behavior B_1 and used to update the value of B_2 provided that $B_1.\pi(s) = B_2.\pi(s)$. So in this way we can improve our estimations of the value of several "overlapping" behaviors based on the experience of executing only one.

Both this all-goals updating and intra-option learning increase the amount of computation done based on each experience gathered. This can potentially mean better performance with less experience, but it can also mean longer learning times due to excess computation. There is a tradeoff between the experience and computation. Generally it is assumed that experience is the bottleneck in learning, and that computation is relatively inexpensive, but as hierarchical methods grow more complex this is an issue that is going to need to be addressed.

8.7   CREATING BEHAVIORS AND BUILDING HIERARCHIES

As stated earlier, typically the hierarchy of behaviors is defined by a human designer. Many researchers have pointed to the desirability of automating this task (e.g., [8, 17]).

One approach to this problem is the HEXQ algorithm [18]. This algorithm is an extension of $MAXQ$-Q which attempts to automatically decompose a problem into a collection of subproblems. Sub-problems are created corresponding to particular variables in the state-vector. Variables that change infrequently inspire behaviors which aim to cause those variables to change.

A similar approach is used by acQuire [32]. It uses exploration to identify "bottlenecks" in the state-space – states which are part of many trajectories through the space. Bottleneck states are selected as subgoals for new behaviors.

Both these approaches implement a kind of uninformed behavior invention, based only on the dynamics of the environment without any background knowledge. RACHEL [39] adopts alternative approach. This system combines hierarchical reinforcement learning with symbolic planning. behaviors are specified by a training in abstract terms, in terms of their pre-conditions and goals. These specifications take the form of teleo-operators [33], which are used in conjunction with symbolic planning to build hierarchies of behaviors suited to a given task.

8.8   MODEL-BASED REINFORCEMENT LEARNING

As we said earlier, not all reinforcement learning algorithms are model-free like Q-Learning and it's derivatives. There are also many algorithms which attempt to learn models of the transition and reward functions $T_{s's}, a$ and $R_{s's}, a$, and then use these models to construct policies using dynamic programming methods such as value-iteration or policy-iteration (e.g., [47]). Such techniques have been somewhat less popular in practice as learning accurate models of T and R has been found to be more difficult than learning Q-values directly.

Nevertheless these technique have also been applied to the hierarchical reinforcement learning problem, and model-based hierarchical algorithms exist (e.g., H-Dyna [42], SMDP Planning [36], abstract MDPs [17], and discrete-event models [30]).

It is unfortunate that model-based methods have apparently acquired second-class status in this field as there are potentially many ways in which a complete model, at multiple levels of abstraction, could be exploited far beyond a particular task-specific set of Q-values. This is an area deserving further attention.

8.9   TOPICS FOR FUTURE RESEARCH

8.9.1   Recombination of Different Features

As we have described above, most hierarchical reinforcement learning algorithms are actually the combination of several independent features. These particular combinations are not necessarily forced, but are mainly accidental. Some of these features are:

1. Hierarchical vs. Recursive optimality,

2. The MAXQ decomposition of the state-action value function,

3. The HAMQ programmable behavior structure,

4. Termination improvement,

5. Intra-option learning.

It would certainly be worthwhile to investigate other recombinations of these features. There is no obvious reason, for example, why the MAXQ decomposition could not be applied to to a hierarchically optimal learning algorithm; or the HAMQ framework used to structure recursively optimal behaviors with termination improvement. Some combinations may not work, but we cannot be sure of this until we explore them.

8.9.2   Real-world Applications

As must be apparent from this chapter, the development of hierarchical reinforcement theory and algorithms has been proceeding strongly in recent years. However this development has far outstripped the use of these methods in non-trivial real-world situations. One piece of work stands out: Crites successful use of SMDP Q-learning to learn elevator control [11, 12]. Few others exist.

Hierarchical reinforcement learning is founded on the claim that hierarchy can help us overcome the curse of dimensionality. The field has matured, the algorithms are available. Now it is time to prove that these claims are not empty.

8.10   CONCLUSION

Hierarchy has proven to be a rich source of development in reinforcement learning, producing both theoretical and practical advances. It has found a solid theoretical model in Semi-Markov Decision Processes and has been growing strongly. Diverse algorithms have been built on this shared foundation, differing in terms of optimality criteria, action selection mechanisms, state-approximation, and, as always, terminology. In this chapter we have tried to unify these different approaches and show how their might be separated and recombined. It is our hope that this will give a coherent foundation for future development and the application of these new ideas to solving greater problems.

Acknowledgments

I would like to thank Andrew Barto for reviewing several drafts of this report and providing many valuable recommendations.

Bibliography

1. D. Andre and S. J. Russell, Programmable reinforcement learning agents, *Advances in Neural Information Processing Systems 12: Proc. of the 1999 Conference*, pp. 1019–1025, 2000.

2. L. C. Baird, Residual algorithms: Reinforcement learning with function approximation, *Proc. of the 12th International Conference on Machine Learning*, pp. 30–37, Morgan Kaufman Publishers, San Francisco, CA, 1995.

3. A. G. Barto and S. Mahadevan, Recent advances in hierarchical reinforcement learning, *Discrete-Event Dynamical Systems: Theory and Applications*, vol. 13, pp. 341–379, 2003.

4. J. Baxter, A. Tridgell, and L. Weaver, Knightcap: A chess program that learns by combining td(λ) with game-tree search, *Proc. of the 15th International Conference on Machine Learning*, pp. 28–36, Morgan Kaufman Publishers, San Francisco, CA, 1998.

5. R. Bellman, *Dynamic Programming*, Princeton University Press, Princeton, NJ, 1957.

6. D. P. Bertsekas, *Dynamic Programming: Deterministic and Stochastic Models*, Prentice-Hall, Englewood Cliffs, NJ, 1987.

7. D. P. Bertsekas and J. N. Tsitsiklis, *Neuro-Dynamic Programming*, Athena Scientific, Belmont, MA, 1996.

8. C. Boutilier, T. Dean, and S. Hanks, Decision-theoretic planning: Structural assumptions and computational leverage, *Journal of Artificial Intelligence Research*, vol. 11, pp. 1–94, 1999.

9. S. J. Bradtke and M. O. Duff, Reinforcement learning methods for continuous-time Markov decision problems, in G. Tesauro, D. Touretzky, and T. Leen, (eds.), *Advances in Neural Information Processing Systems*, vol. 7, pp. 393–400, MIT Press, Cambridge, MA, 1995.

10. R. A. Brooks, A robust layered control system for a mobile robot, *IEEE Journal of Robotics and Automation*, RA-2, no. 1, pp. 14–23, 1986.

11. R. H. Crites, *Large-Scale Dynamic Optimization Using Teams of Reinforcement Learning Agents,* University of Massachusetts, Amherst, 1996.

12. R. H. Crites and A. G. Barto, Elevator Group control using multiple reinforcement learning agents, *Machine Learning,* vol. 33, no. 2-3, pp. 235–262, 1998.

13. P. Dayan and G. E. Hinton, Feudal reinforcement learning, *Advances in Neural Information Processing Systems,* vol. 5, pp. 271–278, 1992.

14. T. G. Dietterich, The MAXQ method for hierarchical reinforcement learning, pp. 118–126, 1998.

15. T. G. Dietterich, Hierarchical reinforcement learning with the MAXQ value dunction decomposition, *Artitificial Intelligence,* vol. 13, pp. 227–303, 2000.

16. T. G. Dietterich, An overview of MAXQ hierarchical reinforcement learning, in B. Y. Choueiry and T. Walsh, (eds.), *Proc. of the Symposium on Abstraction, Reformulation and Approximation SARA 2000, Lecture Notes in Artificial Intelligence,* pp. 26–44, New York, 2000.

17. M. Hauskrecht, N. Meuleau, L. P. Kaelbling, T. Dean, and C. Boutilier, Hierarchical solution of Markov decision processes using macro-actions, *Uncertainty in Artificial Intelligence,* pp. 220–229, 1998.

18. B. Hengst, Discovering hierarchy in reinforcement learning with HEXQ, *Proc. of the Seventh International Conference on Machine Learning,* 1990.

19. R. A. Howard, *Dynamic Programming and Markov Processes,* MIT Press, Cambridge, MA, 1960.

20. R. A. Howard, *Dynamic Probabilistic Systems: Semi-Markov and Decision Processes,* Wiley, New York, 1971.

21. G. A. Iba, A heuristic approach to the discovery of macro-operators, *Machine Learning,* vol. 3, pp. 285–317, 1989.

22. T. Jaakkola, M. I. Jordan, and S. P. Singh, On the convergence of stochastic iterative dynamic programming algorithms, *Advances in Neural Information Processing Systems,* vol. 6, MIT Press, Cambridge, MA, 1994.

23. L. P. Kaelbling, Hierarchical learning in stochastic domains: Preliminary results, *Proc. of the Tenth International Conference on Machine Learning,* 1993.

24. L. P. Kaelbling, M. L. Littman, and A. P. Moore, Reinforcement learning: A survey, *Journal of Artificial Intelligence Research,* vol. 4, pp. 237–285, 1996.

25. C. A. Knoblock, Search reduction in hierarchical problem solving, *Proc. of the Ninth National Conference on Artificial Intelligence (AAAI-91),* vol. 2, pp.686–691, Anaheim, CA, 1991.

26. R. E. Korf, Planning as search: A quantitative approach, *Artificial Intelligence,* vol. 33, no. 1, pp. 65–88, 1987.

27. L.-J. Lin, *Reinforcement Learning for Robots Using Neural Networks,* Ph.D. Thesis, School of Computer Science, Carnegie Mellon University, Pittsburgh, PA, 1993.

28. P. Maes, How to do the right thing, *Connection Science Journal, Special Issue on Hybrid Systems,* vol. 1, 1990.

29. S. Mahadevan, Average reward reinforcement learning: Foundations, algorithms, and empirical results, *Machine Learning,* vol. 22, no. 1-3, pp. 159–195, 1996.

30. S. Mahadevan, N. Khaleeli, and N. Marchalleck, Designing agent controllers using discrete-event Markov models, *Working Notes of the AAAI Fall Symposium on Model-Directed Autonomous Systems,* Cambridge, MA, 1997.

31. M. J. Matarić, Behaviour based control: Examples from navigation, learning and group behaviour, *Journal of Experimental and Theoretical Artificial Intelligence,* vol. 9, no. 2–3, 1996.

32. A. McGovern and A. G. Barto, Automatic discovery of subgoals in reinforcement learning using diverse density, *Proc. 18th International Conference on Machine Learning,* pp. 361–368, San Francisco, CA, 2001.

33. N. J. Nilsson, Teleo-reactive programs for agent control, *Journal of Artificial Intelligence Research,* vol. 1, pp. 139–158, 1994.

34. R. Parr, *Hierarchical Control and Learning for Markov Decision Processes,* Ph.D. Thesis, University of California at Berkeley, 1998.

35. R. Parr and S. Russell, Reinforcement learning with hierarchies of machines, in M. I. Jordan, M. J. Kearns, and S. A. Solla, (eds.), *Advances in Neural Information Processing Systems,* vol. 10, MIT Press, Cambridge, MA, 1998.

36. D. Precup, *Temporal Abstraction in Reinforcement Learning,* University of Massachusetts, Amherst, 2000.

37. M. L. Puterman, *Markov Decision Processes: Discrete Stochastic Dynamic Programming,* Wiley, New York, 1994.

38. G. A. Rummery and M. Niranjan, Online Q-learning using connectionist systems, *Technical Report CUED/F-INFENG/TR 166,* Engineering Department, Cambridge University, 1994.

39. M. R. K. Ryan, *Hierarchical Reinforcement Learning: A Hybrid Approach,* Ph.D. Thesis, University of New South Wales, 2002.

40. E. D. Sacerdoti, Planning in a hierarchy of abstraction spaces, *Artificial Intelligence,* vol. 5, no. 2, pp. 115–135, 1974.

41. A. Schwartz, A reinforcement learning method for maximizing undiscounted rewards, *Proc. of the 10th International Conference on Machine Learning*, 1993.

42. S. P. Singh, Reinforcement learning with a hierarchy of abstract models, *Proc. of the 10th National Conference on Artifical Intelligence*, 1992.

43. S. P. Singh, Reinforcement learning algorithms for average-payoff Markovian decision processes, *National Conference on Artificial Intelligence*, pp. 700–705, 1994.

44. S. P. Singh, T. Jaakkola, M. L. Littman, and C. Szepesvari, Convergence results for single-step on-policy reinforcement-learning algorithms, *Machine Learning*, vol. 38, no. 3, pp. 287–308, 2000.

45. R. S. Sutton, S. Singh, D. Precup, and B. Ravindran, Improved switching among temporally abstract actions, *Advances in Neural Information Processing Systems*, vol. 11, MIT Press, Cambridge, MA, 1999.

46. R. S. Sutton, Implementation details for the TD(λ) procedure for the case of vector predictions and backpropagation, *Technical Report TN87-509.1*, GTE Laboratories, 1987.

47. R. S. Sutton, Integrated Architectures for Learning, Planning and Reacting Based on Approximating Dynamic Programming, *Proc. of the 7th International Conference on Machine Learning*, San Francisco, 1990.

48. R. S. Sutton, Generalisation in reinforcement learning successful examples using sparse coarse coding, *Advances in Neural Neural Information Processing Systems*, 1995.

49. R. S. Sutton and A. G. Barto, *Reinforcement Learning: An Introduction*, MIT Press, Cambridge, MA, 1998.

50. R. S. Sutton, D. Precup, and S. P. Singh, Between MDPs and semi-MDPs: A framework for temporal abstraction in reinforcement learning, *Artificial Intelligence*, vol. 112, no. 1–2, pp. 181–211, 1999.

51. G. Tesauro, TD-gammon, a self-teaching backgammon program achieves master-level play, *Neural Computation*, vol. 6, pp. 215–219, 1994.

52. S. B. Thrun, The role of exploration in learning control with neural networks, in D. A. White and D. A. Sofge, (eds.), *Handbook of Intelligent Control: Neural, Fuzzy and Adaptive Approaches*, Florence, KY, 1992.

53. J. N. Tsitsiklis, Asynchronous stochastic approximation and Q-learning, *Machine Learning*, vol. 16, no. 3, 1994.

54. C. J. C. H. Watkins, *Learning from Delayed Rewards*, Ph.D. Thesis, King's College, Cambridge University, UK, 1989.

55. C. J. C. H. Watkins and P. Dayan, *Q*-learning, *Machine Learning,* vol. 8, no. 3, pp. 279–292, 1992.

56. W. Zhang and T. G. Dietterich, A reinforcement learning approach to job-shop scheduling, *Proc. of the 14th International Joint Conference on Artificial Intelligence,* 1995.

Part II

Technical Advances

9 Improved Temporal Difference Methods with Linear Function Approximation

DIMITRI P. BERTSEKAS
Massachusetts Institute of Technology
VIVEK S. BORKAR
Tata Institute of Fundamental Research

ANGELIA NEDICH
Alphatech, Inc.

Editor's Summary: This chapter considers temporal difference algorithms within the context of infinite-horizon finite-state dynamic programming problems with discounted cost and linear cost function approximation. This problem arises as a subproblem in the policy iteration method of dynamic programming. Additional discussions of such problems can be found in Chapters 6 and 12. The method presented here is the first iterative temporal difference method that converges without requiring a diminishing step size. The chapter discusses the connections with Sutton's TD(λ) and with various versions of least-squares that are based on value iteration. It is shown using both analysis and experiments that the proposed method is substantially faster, simpler, and more reliable than TD(λ). Comparisons are also made with the LSTD method of Boyan, and Bradtke and Barto.

9.1 INTRODUCTION

In this chapter, we analyze methods for approximate evaluation of the cost-to-go function of a stationary Markov chain within the framework of infinite-horizon discounted dynamic programming. We denote the states by $1, \ldots, n$, the transition probabilities by p_{ij}, $i, j = 1, \ldots, n$, and the corresponding costs by $\alpha^t g(i, j)$, where α is a discount factor with $0 < \alpha < 1$. We want to evaluate the long-term expected cost corresponding to each initial state i, given by

$$J(i) = E\left[\sum_{t=0}^{\infty} \alpha^t g(i_t, i_{t+1}) \,\bigg|\, i_0 = i \right], \qquad \forall\, i = 1, \ldots, n,$$

where i_t denotes the state at time t. This problem arises as a subproblem in the policy iteration method of dynamic programming, and its variations, such as modified policy iteration, optimistic policy iteration, and λ-policy iteration (see Bertsekas and Tsitsiklis [4], Bertsekas [3], and Puterman [15] for extensive discussions of these methods).

The cost function $J(i)$ is approximated by a linear function of the form

$$\tilde{J}(i, r) = \phi(i)'r, \qquad \forall\, i = 1, \ldots, n,$$

where $\phi(i)$ is an s-dimensional feature vector, associated with the state i, with components $\phi_1(i), \ldots, \phi_s(i)$, while r is a weight vector with components $r(1), \ldots, r(s)$. (Throughout the chapter, vectors are viewed as column vectors, and a prime denotes transposition.)

Our standing assumptions are:

(a) The Markov chain has steady-state probabilities $\pi(1), \ldots, \pi(n)$ which are positive, that is,

$$\lim_{t \to \infty} P[i_t = j \mid i_0 = i] = \pi(j) > 0, \qquad \forall\, i, j.$$

(b) The matrix Φ given by

$$\Phi = \begin{bmatrix} -\ \phi(1)'\ - \\ \vdots \\ -\ \phi(n)'\ - \end{bmatrix}$$

has rank s.

The TD(λ) method with function approximation was originally proposed by Sutton [17], and its convergence has been analyzed by several authors, including Dayan [8], Gurvits, Lin, and Hanson [10], Pineda [14], Tsitsiklis and Van Roy [19], and Van Roy [18]. We follow the line of analysis of Tsitsiklis and Van Roy, who have also considered a discounted problem under the preceding assumptions on the existence of steady-state probabilities and rank of Φ.

The algorithm, described in several references, including the books by Bertsekas and Tsitsiklis [4], and Sutton and Barto [16], generates an infinitely long trajectory of the Markov chain $(i_0, i_1, \ldots)$ using a simulator, and at time t iteratively updates the current estimate r_t using an iteration that depends on a fixed scalar $\lambda \in [0, 1]$, and on the temporal differences

$$d_t(i_k, i_{k+1}) = g(i_k, i_{k+1}) + \alpha\phi(i_{k+1})'r_t - \phi(i_k)'r_t, \qquad \forall\, t = 0, 1, \ldots, \forall\, k \le t.$$

Tsitsiklis and Van Roy [19] have introduced the linear system of equations

$$Ar + b = 0,$$

where A and b are given by

$$A = \Phi'D(\alpha P - I) \sum_{m=0}^{\infty} (\alpha\lambda P)^m \Phi, \qquad b = \Phi'D \sum_{m=0}^{\infty} (\alpha\lambda P)^m \bar{g}, \qquad (9.1)$$

P is the transition probability matrix of the Markov chain, D is the diagonal matrix with diagonal entries $\pi(i)$, $i = 1, \ldots, n$,

$$D = \begin{pmatrix} \pi(1) & 0 & \cdots & 0 \\ 0 & \pi(2) & \cdots & 0 \\ & & \cdots & \\ 0 & 0 & \cdots & \pi(n) \end{pmatrix}, \qquad (9.2)$$

and $\bar{g}$ is the vector with components $\bar{g}(i) = \sum_{j=1}^{n} p_{ij} g(i,j)$. They have shown that TD(λ) converges to the unique solution $r^* = -A^{-1}b$ of the system $Ar + b = 0$, and that the error between the corresponding approximation Φr^* and the true cost-to-go vector J satisfies

$$\|\Phi r^* - J\|_D \leq \frac{1 - \alpha\lambda}{1 - \alpha} \|\Pi J - J\|_D,$$

where $\| \cdot \|_D$ is the weighted norm corresponding to the matrix D (i.e., $\|x\|_D = \sqrt{x'Dx}$), and Π is the matrix given by $\Pi = \Phi(\Phi'D\Phi)^{-1}\Phi'D$. (Note that $\Pi J - J$ is the difference between J and its projection, with respect to the weighted norm, on the range of the feature matrix Φ.)

The essence of the Tsitsiklis and Van Roy analysis is to write the TD(λ) algorithm as

$$r_{t+1} = r_t + \gamma_t(Ar_t + b) + \gamma_t(\Xi_t r_t + \xi_t), \qquad t = 0, 1, \ldots, \qquad (9.3)$$

where γ_t is a positive stepsize, and Ξ_t and ξ_t are some sequences of random matrices and vectors, respectively, that depend only on the simulated trajectory (so they are independent of r_t), and asymptotically have zero mean. A key to the convergence proof is that the matrix A is negative definite, so it has eigenvalues with negative real parts, which implies in turn that the matrix $I + \gamma_t A$ has eigenvalues within the unit circle for sufficiently small γ_t. However, in TD(λ) it is essential that the stepsize γ_t be diminishing to 0, both because a small γ_t is needed to keep the eigenvalues of $I + \gamma_t A$ within the unit circle, and also because Ξ_t and ξ_t do not converge to 0.

In this chapter, we focus on the λ-least squares policy evaluation method (λ-LSPE for short), proposed and analyzed by Nedić and Bertsekas [13]. This algorithm was motivated as a simulation-based implementation of the λ-policy iteration method, proposed by Bertsekas and Ioffe [2] (also described in Bertsekas and Tsitsiklis [4], Section 2.3.1). In fact the method of this chapter was also stated (without convergence analysis), and was used with considerable success by Bertsekas and Ioffe [2] [see also Bertsekas and Tsitsiklis [4], Eq. (8.6)] to train a tetris playing program — a challenging large-scale problem that TD(λ) failed to solve. In this chapter, rather than focusing on the connection with λ-policy iteration, we emphasize a connection with (multistep) value iteration (see Section 9.4).

The λ-LSPE method, similar to TD(λ), generates an infinitely long trajectory $(i_0, i_1, \ldots)$ using a simulator. At each time t, it finds the solution $\tilde{r}_t$ of a least squares problem,

$$\tilde{r}_t = \arg\min_r \sum_{m=0}^t \left(\phi(i_m)'r - \phi(i_m)'r_t - \sum_{k=m}^t (\alpha\lambda)^{k-m} d_t(i_k, i_{k+1}) \right)^2 , \quad (9.4)$$

and computes the new vector r_{t+1} according to

$$r_{t+1} = r_t + \gamma(\tilde{r}_t - r_t), \quad (9.5)$$

where γ is a positive stepsize. The initial weight vector r_0 is chosen independently of the trajectory $(i_0, i_1, \ldots)$.

It can be argued that λ-LSPE is a "scaled" version of TD(λ). In particular, from the analysis of Nedić and Bertsekas ([13], p. 101; see also Section 9.3), it follows that the method takes the form

$$r_{t+1} = r_t + \gamma(\Phi'D\Phi)^{-1}(Ar_t + b) + \gamma(Z_t r_t + \zeta_t), \qquad t = 0, 1, \ldots, \quad (9.6)$$

where γ is a positive stepsize, and Z_t and ζ_t are some sequences of random matrices and vectors, respectively, that converge to 0 with probability 1. It was shown in [13] that when the stepsize is diminishing rather than being constant, the method converges with probability 1 to the same limit as TD(λ), the unique solution r^* of the system $Ar + b = 0$ (convergence for a constant stepsize was conjectured but not proved).

One of the principal results of this chapter is that the scaling matrix $(\Phi'D\Phi)^{-1}$ is "close" enough to $-A^{-1}$ so that, based also on the negative definiteness of A, the stepsize $\gamma = 1$ leads to convergence for all $\lambda \in [0, 1]$, that is, the matrix $I + (\Phi'D\Phi)^{-1}A$ has eigenvalues that are within the unit circle of the complex plane. In fact, we can see that A may be written in the alternative form

$$A = \Phi'D(M - I)\Phi, \qquad M = (1 - \lambda) \sum_{m=0}^{\infty} \lambda^m (\alpha P)^{m+1},$$

so that for $\lambda = 1$, the eigenvalues of $I + (\Phi'D\Phi)^{-1}A$ are all equal to 0. We will also show that as λ decreases toward 0, the region where the eigenvalues of $I + (\Phi'D\Phi)^{-1}A$ lie expands, but stays within the interior of the unit circle.

By comparing the iterations (9.3) and (9.6), we see that TD(λ) and λ-LSPE have a common structure — a deterministic linear iteration plus noise that tends to 0 with probability 1. However, the convergence rate of the deterministic linear iteration is geometric in the case of λ-LSPE, while it is slower than geometric in the case of TD(λ), because the stepsize γ_t must be diminishing. This indicates that λ-LSPE has a significant rate of convergence advantage over TD(λ). At the same time, with a recursive Kalman filter-like implementation discussed in [13], λ-LSPE does not require much more overhead per iteration than TD(λ) [the associated matrix inversion at each iteration requires only $O(s^2)$ computation using the results of the inversion at the preceding iteration, where s is the dimension of r].

For some further insight on the relation of λ-LSPE with $\gamma = 1$ and TD(λ), let us focus on the case where $\lambda = 0$. TD(0) has the form

$$r_{t+1} = r_t + \gamma_t \phi(i_t) d_t(i_t, i_{t+1}), \tag{9.7}$$

while 0-LSPE has the form

$$r_{t+1} = \arg\min_r \sum_{m=0}^{t} \left(\phi(i_m)'r - \phi(i_m)'r_t - d_t(i_m, i_{m+1}) \right)^2 \tag{9.8}$$

[cf. Eq. (9.4)]. We note that the gradient of the least squares sum above is

$$-2 \sum_{m=0}^{t} \phi(i_m) d_t(i_m, i_{m+1}).$$

Asymptotically, in steady-state, the expected values of all the terms in this sum are equal, and each is proportional to the expected value of the term $\phi(i_t) d_t(i_t, i_{t+1})$ in the TD(0) iteration (9.7). Thus, TD(0) *updates r_t along the gradient of the least squares sum of 0-LSPE, plus stochastic noise that asymptotically has zero mean.* This interpretation also holds for other values of $\lambda \neq 0$, as will be discussed in Section 9.4.

Another class of temporal difference methods, parameterized by $\lambda \in [0, 1]$, has been introduced by Boyan [6], following the work by Bradtke and Barto [7] who considered the case $\lambda = 0$. These methods, known as Least Squares TD (LSTD), also employ least squares and have guaranteed convergence to the same limit as TD(λ) and λ-LSPE, as shown by Bradtke and Barto [7] for the case $\lambda = 0$, and by Nedić and Bertsekas [13] for the case $\lambda \in (0, 1]$. Konda [12] has derived the asymptotic mean squared error of a class of recursive and nonrecursive temporal difference methods [including TD(λ) and LSTD, but not including LSPE], and has found that LSTD has optimal asymptotic convergence rate within this class. The LSTD method is not iterative, but instead it evaluates the simulation-based estimates A_t and b_t of $(t+1)A$ and $(t + 1)b$, given by

$$A_t = \sum_{m=0}^{t} z_m \left(\alpha \phi(i_{m+1})' - \phi(i_m)' \right),$$

$$b_t = \sum_{m=0}^{t} z_m g(i_m, i_{m+1}), \qquad z_m = \sum_{k=0}^{m} (\alpha\lambda)^{m-k} \phi(i_k),$$

(see Section 9.3), and estimates the solution r^* of the system $Ar + b = 0$ by

$$\hat{r}_{t+1} = -A_t^{-1} b_t.$$

We argue in Section 9.5 that LSTD and λ-LSPE have comparable asymptotic perfor-mance, although there are significant differences in the early iterations. In fact, the iterates of LSTD and λ-LSPE converge to each other faster than they converge to r^*. Some insight into the comparability of the two methods can be obtained by verifying that the LSTD estimate $\hat{r}_{t+1}$ is also the unique vector $\hat{r}$ satisfying

$$\hat{r} = \arg\min_r \sum_{m=0}^{t} \left(\phi(i_m)'r - \phi(i_m)'\hat{r} - \sum_{k=m}^{t} (\alpha\lambda)^{k-m} \hat{d}(i_k, i_{k+1}; \hat{r}) \right)^2 , \quad (9.9)$$

where

$$\hat{d}(i_k, i_{k+1}; \hat{r}) = g(i_k, i_{k+1}) + \alpha\phi(i_{k+1})'\hat{r} - \phi(i_k)'\hat{r}.$$

While finding $\hat{r}$ that satisfies Eq. (9.9) is not a least squares problem, its similarity with the least squares problem solved by LSPE [cf. Eq. (9.4)] is evident.

We note, however, that LSTD and LSPE may differ substantially in the early iterations. Furthermore, LSTD is a pure simulation method that cannot take advantage of a good initial choice r_0. This is a significant factor in favor of λ-LSPE in a major context, namely optimistic policy iteration [4], where the policy used is changed (using a policy improvement mechanism) after a few simulated transitions. Then, the use of the latest estimate of r to start the iterations corresponding to a new policy, as well as a small stepsize (to damp oscillatory behavior following a change to a new policy) is essential for good overall performance.

The algorithms and analysis of the present chapter, in conjunction with existing research, support a fairly comprehensive view of temporal difference methods with linear function approximation. The highlights of this view are as follows:

(1) Temporal difference methods fundamentally emulate value iteration methods that aim to solve a Bellman equation that corresponds to a multiple-transition version of the given Markov chain, and depends on λ (see Section 9.4).

(2) The emulation of the kth value iteration is approximate through linear function approximation, and solution of the least squares approximation problem (9.4) that involves the simulation data $(i_0, i_1, \ldots, i_t)$ up to time t.

(3) The least squares problem (9.4) is fully solved at time t by λ-LSPE, but is solved only approximately, by a single gradient iteration (plus zero-mean noise), by TD(λ) (see Section 9.4).

(4) LSPE and LSTD have similar asymptotic performance, but may differ substan-tially in the early iterations. Furthermore, LSPE can take advantage of good initial estimates of r^*, while LSTD, as presently known, cannot.

The chapter is organized as follows. In Section 9.2, we derive a basic lemma regarding the location of the eigenvalues of the matrix $I + (\Phi'D\Phi)^{-1}A$. In Section 9.3, we use this lemma to show convergence of λ-LSPE with probability 1 for any stepsize γ in a range that includes $\gamma = 1$. In Section 9.4, we derive the connection of

λ-LSPE with various forms of approximate value iteration. Based on this connection, we discuss how our line of analysis extends to other types of dynamic programming problems. In Section 9.5, we discuss the relation between λ-LSPE and LSTD. Finally, in Section 9.6 we present computational results showing that λ-LSPE is dramatically faster than TD(λ), and also simpler because it does not require any parameter tuning for the stepsize selection method.

9.2 PRELIMINARY ANALYSIS

In this section we prove some lemmas relating to the transition probability matrix P, the feature matrix Φ, and the associated matrices D and A of Eqs. (9.2) and (9.1). We denote by R and C the set of real and complex numbers, respectively, and by R^n and C^n the spaces of n-dimensional vectors with real and with complex components, respectively. The complex conjugate of a complex number z is denoted $\hat{z}$. The complex conjugate of a vector $z \in C^n$, is the vector whose components are the complex conjugates of the components of z, and is denoted $\hat{z}$. The modulus $\sqrt{\hat{z}z}$ of a complex number z is denoted by $|z|$. We consider two norms on C^n, the standard norm, defined by

$$\|z\| = (\hat{z}'z)^{1/2} = \left(\sum_{i=1}^{n} |z_i|^2 \right)^{1/2}, \qquad \forall \, z = (z_1, \ldots, z_n) \in C^n,$$

and the weighted norm, defined by

$$\|z\|_D = (\hat{z}'Dz)^{1/2} = \left(\sum_{i=1}^{n} \pi(i)|z_i|^2 \right)^{1/2}, \qquad \forall \, z = (z_1, \ldots, z_n) \in C^n.$$

The following lemma extends, from $\Re^n$ to C^n, a basic result of Tsitsiklis and Van Roy [19].

Lemma 9.2.1 *For all $z \in C^n$, we have $\|Pz\|_D \leq \|z\|_D$.*

Proof: For any $z = (z_1, \ldots, z_n) \in C^n$, we have, using the defining property $\sum_{i=1}^{n} \pi(i)p_{ij} = \pi(j)$ of the steady-state probabilities,

$$
\begin{aligned}
\|Pz\|_D^2 &= \hat{z}'P'DPz \\
&= \sum_{i=1}^{n} \pi(i) \left(\sum_{j=1}^{n} p_{ij}\hat{z}_j \right) \left(\sum_{j=1}^{n} p_{ij}z_j \right)
\end{aligned}
$$

$$\leq \sum_{i=1}^{n} \pi(i) \left(\sum_{j=1}^{n} p_{ij} |z_j| \right)^2$$

$$\leq \sum_{i=1}^{n} \pi(i) \sum_{j=1}^{n} p_{ij} |z_j|^2$$

$$= \sum_{j=1}^{n} \sum_{i=1}^{n} \pi(i) p_{ij} |z_j|^2$$

$$= \sum_{j=1}^{n} \pi(j) |z_j|^2$$

$$= \|z\|_D^2,$$

where the first inequality follows since $\hat{x}y + x\hat{y} \leq 2|x| |y|$ for any two complex numbers x and y, and the second inequality follows by applying Jensen's inequality. ∎

The next lemma is the key to the convergence proof of the next section.

Lemma 9.2.2 *The eigenvalues of the matrix $I + (\Phi' D \Phi)^{-1} A$ lie within the circle of radius $\alpha(1 - \lambda)/(1 - \alpha\lambda)$.*

Proof: We have

$$A = \Phi' D (M - I) \Phi,$$

where

$$M = (1 - \lambda) \sum_{m=0}^{\infty} \lambda^m (\alpha P)^{m+1},$$

so that

$$(\Phi' D \Phi)^{-1} A = (\Phi' D \Phi)^{-1} \Phi' D M \Phi - I.$$

Hence

$$I + (\Phi' D \Phi)^{-1} A = (\Phi' D \Phi)^{-1} \Phi' D M \Phi.$$

Let β be an eigenvalue of $I + (\Phi' D \Phi)^{-1} A$ and let z be a corresponding eigenvector, so that

$$(\Phi' D \Phi)^{-1} \Phi' D M \Phi z = \beta z.$$

Letting

$$W = \sqrt{D} \Phi,$$

we have

$$(W'W)^{-1} W' \sqrt{D} M \Phi z = \beta z,$$

from which, by left-multiplying with W, we obtain

$$W(W'W)^{-1} W' \sqrt{D} M \Phi z = \beta W z. \tag{9.10}$$

The norm of the right-hand side of Eq. (9.10) is

$$\|\beta W z\| = |\beta| \, \|W z\| = |\beta| \, \sqrt{z \Phi' D \Phi z} = |\beta| \, \|\Phi z\|_D. \tag{9.11}$$

To estimate the norm of the left-hand side of Eq. (9.10), first note that

$$\|W(W'W)^{-1}W'\sqrt{D}M\Phi z\| \le \|W(W'W)^{-1}W'\| \, \|\sqrt{D}M\Phi z\|$$
$$\|W(W'W)^{-1}W'\| \, \|\sqrt{D}M\Phi z\| = \|W(W'W)^{-1}W'\| \, \|M\Phi z\|_D,$$

and then note also that $W(W'W)^{-1}W'$ is a projection matrix [i.e., for $x \in \Re^n$, $W(W'W)^{-1}W'x$ is the projection of x on the subspace spanned by the columns of W], so that $\|W(W'W)^{-1}W'x\| \le \|x\|$, from which

$$\|W(W'W)^{-1}W'\| \le 1.$$

Thus we have

$$\begin{aligned}
\|W(W'W)^{-1}W'\sqrt{D}M\Phi z\| &\le \|M\Phi z\|_D \\
&= \left\| (1-\lambda) \sum_{m=0}^{\infty} \lambda^m \alpha^{m+1} P^{m+1} \Phi z \right\|_D \\
&\le (1-\lambda) \sum_{m=0}^{\infty} \lambda^m \alpha^{m+1} \|P^{m+1} \Phi z\|_D \\
&\le (1-\lambda) \sum_{m=0}^{\infty} \lambda^m \alpha^{m+1} \|\Phi z\|_D \\
&= \frac{\alpha(1-\lambda)}{1-\alpha\lambda} \|\Phi z\|_D, \tag{9.12}
\end{aligned}$$

where the last inequality follows by repeated use of Lemma 9.2.1. By comparing Eqs. (9.12) and (9.11), and by taking into account that $\Phi z \ne 0$ (since Φ has full rank), we see that

$$|\beta| \le \frac{\alpha(1-\lambda)}{1-\alpha\lambda}.$$

$\blacksquare$

9.3   CONVERGENCE ANALYSIS

We will now use Lemma 9.2.2 to prove the convergence of λ-LSPE. It is shown in Nedić and Bertsekas [13] that the method is given by

$$r_{t+1} = r_t + \gamma B_t^{-1}(A_t r_t + b_t), \qquad \forall \, t, \tag{9.13}$$

where

$$B_t = \sum_{m=0}^{t} \phi(i_m)\phi(i_m)', \qquad A_t = \sum_{m=0}^{t} z_m\big(\alpha\phi(i_{m+1})' - \phi(i_m)'\big), \quad (9.14)$$

$$b_t = \sum_{m=0}^{t} z_m g(i_m, i_{m+1}), \qquad z_m = \sum_{k=0}^{m} (\alpha\lambda)^{m-k}\phi(i_k). \qquad (9.15)$$

[Note that if in the early iterations, $\sum_{m=0}^{t} \phi(i_m)\phi(i_m)'$ is not invertible, we may add to it a small positive multiple of the identity, or alternatively we may replace inverse by pseudoinverse. Such modifications are inconsequential and will be ignored in the subsequent analysis; see also [13].] We can rewrite Eq. (9.13) as

$$r_{t+1} = r_t + \gamma \overline{B}_t^{-1}(\overline{A}_t r_t + \overline{b}_t), \qquad \forall\, t,$$

where

$$\overline{B}_t = \frac{B_t}{t+1}, \qquad \overline{A}_t = \frac{A_t}{t+1}, \qquad \overline{b}_t = \frac{b_t}{t+1}.$$

Using the analysis of [13] (see the proof of Prop. 3.1, p. 108), it follows that with probability 1, we have

$$\overline{B}_t \to B, \qquad \overline{A}_t \to A, \qquad \overline{b}_t \to b,$$

where

$$B = \Phi' D\Phi,$$

and A and b are given by Eq. (9.1).

Thus, we may write iteration (9.13) as

$$r_{t+1} = r_t + \gamma(\Phi' D\Phi)^{-1}(Ar_t + b) + \gamma(Z_t r_t + \zeta_t), \qquad t = 0, 1, \ldots, \quad (9.16)$$

where

$$Z_t = \overline{B}_t^{-1}\overline{A}_t - B^{-1}A, \qquad \zeta_t = \overline{B}_t^{-1}\overline{b}_t - B^{-1}b.$$

Furthermore, with probability 1, we have

$$Z_t \to 0, \qquad \zeta_t \to 0.$$

We are now ready to prove our convergence result.

Proposition 9.3.1 *The sequence generated by the λ-LSPE method converges to $r^* = -A^{-1}b$ with probability 1, provided that the constant stepsize γ satisfies*

$$0 < \gamma < \frac{2 - 2\alpha\lambda}{1 + \alpha - 2\alpha\lambda}.$$

Proof: If we write the matrix $I + \gamma(\Phi'D\Phi)^{-1}A$ as

$$(1 - \gamma)I + \gamma\big(I + (\Phi'D\Phi)^{-1}A\big),$$

we see, using Lemma 9.2.2, that its eigenvalues lie within the circle that is centered at $1 - \gamma$ and has radius

$$\frac{\gamma\alpha(1 - \lambda)}{1 - \alpha\lambda}.$$

It follows by a simple geometrical argument that this circle is strictly contained within the unit circle if and only if γ lies in the range between 0 and $(2 - 2\alpha\lambda)/(1 + \alpha - 2\alpha\lambda)$. Thus for each γ within this range, the spectral radius of $I + \gamma(\Phi'D\Phi)^{-1}A$ is less than 1, and there exists a norm $\| \cdot \|_w$ over $\Re^n$ and an $\epsilon > 0$ (depending on γ) such that

$$\|I + \gamma(\Phi'D\Phi)^{-1}A\|_w < 1 - \epsilon.$$

Using the equation $b = -Ar^*$, we can write the iteration (9.16) as

$$r_{t+1} - r^* = \big(I + \gamma(\Phi'D\Phi)^{-1}A + \gamma Z_t\big)(r_t - r^*) + \gamma(Z_t r^* + \zeta_t), \qquad t = 0, 1, \ldots.$$

For any simulated trajectory such that $Z_t \to 0$ and $\zeta_t \to 0$, there exists an index $\bar{t}$ such that

$$\|I + \gamma(\Phi'D\Phi)^{-1}A + \gamma Z_t\|_w < 1 - \epsilon, \qquad \forall\, t \geq \bar{t}.$$

Thus, for sufficiently large t, we have

$$\|r_{t+1} - r^*\|_w \leq (1 - \epsilon)\|r_t - r^*\|_w + \gamma\|Z_t r^* + \zeta_t\|_w.$$

Since $Z_t r^* + \zeta_t \to 0$, it follows that $r_t - r^* \to 0$. Since the set of simulated trajectories such that $Z_t \to 0$ and $\zeta_t \to 0$ is a set of probability 1, it follows that $r_t \to r^*$ with probability 1.   ∎

Note that as λ decreases, the range of stepsizes γ that lead to convergence is reduced. However, this range always contains the stepsize $\gamma = 1$.

9.4 RELATIONS BETWEEN λ-LSPE AND VALUE ITERATION

In this section, we will discuss a number of value iteration ideas, which underlie the structure of λ-LSPE. These connections become most apparent when the stepsize is constant and equal to 1 ($\gamma \equiv 1$), which we will assume in our discussion.

9.4.1   The Case $\lambda = 0$

The classical value iteration method for solving the given policy evaluation problem is

$$J_{t+1}(i) = \sum_{j=1}^{n} p_{ij} \big(g(i,j) + \alpha J_t(j) \big), \qquad i = 1, \ldots, n, \qquad (9.17)$$

and by standard dynamic programming results, it converges to the cost-to-go function $J(i)$. We will show that approximate versions of this method are connected with three methods that are relevant to our discussion: TD(0), 0-LSPE, and the deterministic portion of the 0-LSPE iteration (9.16).

Indeed, a version of value iteration that uses linear function approximation of the form $J_t(i) \approx \phi(i)'r_t$ is to recursively select r_{t+1} so that $\phi(i)'r_{t+1}$ is uniformly (for all states i) "close" to $\sum_{j=1}^{n} p_{ij} \big(g(i,j) + \alpha J_t(j) \big)$; for example by solving a corresponding least squares problem

$$r_{t+1} = \arg\min_r \sum_{i=1}^{n} w(i) \left(\phi(i)'r - \sum_{j=1}^{n} p_{ij} \big(g(i,j) + \alpha\phi(j)'r_t \big) \right)^2,$$
$$t = 0, 1, \ldots, \qquad (9.18)$$

where $w(i)$, $i = 1, \ldots, n$, are some positive weights. This method is considered in Section 6.5.3 of Bertsekas and Tsitsiklis [4], where it is pointed out that divergence is possible if the weights $w(i)$ are not properly chosen; for example if $w(i) = 1$ for all i. It can be seen that the TD(0) iteration (9.7) may be viewed as a one-sample approximation of the special case of iteration (9.18) where the weights are chosen as $w(i) = \pi(i)$, for all i, as discussed in Section 6.5.4 of [4]. Furthermore, Tsitsiklis and Van Roy [19] show that for TD(0) convergence, it is essential that state samples are collected in accordance with the steady-state probabilities $\pi(i)$. By using the definition of temporal difference to write the 0-LSPE iteration (9.8) as

$$r_{t+1} = \arg\min_r \sum_{m=0}^{t} \big(\phi(i_m)'r - g(i_m, i_{m+1}) - \alpha\phi(i_{m+1})'r_t \big)^2, \qquad (9.19)$$

we can similarly interpret it as a multiple-sample approximation of iteration (9.18) with weights $w(i) = \pi(i)$. Of course, when $w(i) = \pi(i)$, the iteration (9.18) is not implementable since the $\pi(i)$ are unknown, and the only way to approximate it is through the on-line type of state sampling used in 0-LSPE and TD(0).

These interpretations suggest that the approximate value iteration method (9.18) should converge when the weights are chosen as $w(i) = \pi(i)$. Indeed for these weights, the method takes the form

$$r_{t+1} = \arg\min_r \| \Phi r - P(g + \alpha\Phi r_t) \|_D^2, \qquad (9.20)$$

which after some calculation, is written as

$$r_{t+1} = r_t + (\Phi'D\Phi)^{-1}(Ar_t + b), \qquad t = 0, 1, \dots, \qquad (9.21)$$

where A and b are given by Eq. (9.1), for the case where $\lambda = 0$. In other words *the deterministic linear iteration portion of the 0-LSPE method with $\gamma = 1$ is equivalent to the approximate value iteration (9.18) with weights $w(i) = \pi(i)$.* Thus, *we can view 0-LSPE as the approximate value iteration method (9.18), plus noise that asymptotically tends to 0.*

Note that the approximate value iteration method (9.20) can be interpreted as a mapping from the feature subspace

$$S = \{\Phi r \mid r \in \Re^s\}$$

to itself: it maps the vector Φr_t to its value iterate $P(g + \alpha \Phi r_t)$, and then projects [with respect to the norm $\|\cdot\|_D$ corresponding to the steady-state probabilities/weights $\pi(i)$] the result on S, as discussed by Tsitsiklis and Van Roy [19], who give an example of divergence when nonlinear function approximation is used. Related issues are discussed by de Farias and Van Roy [9], who consider approximate value iteration with linear function approximation, but multiple policies.

Figure 9.1 illustrates the approximate value iteration method (9.20) together with 0-LSPE, which is the same iteration plus asymptotically vanishing simulation error.

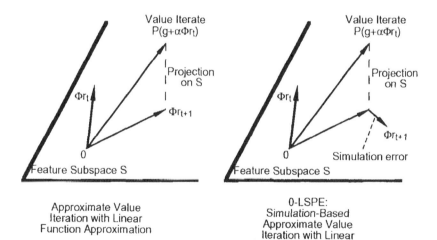

Fig. 9.1 Geometric interpretation of 0-LSPE as the sum of the approximate value iterate (9.20) plus asymptotically vanishing simulation error.

9.4.2   Connection with Multistep Value Iteration

In the case where $\lambda \in (0, 1)$, a similar connection with approximate value iteration can be derived, except that *each value iteration involves multiple state transitions* (see also the corresponding discussion by Bertsekas and Ioffe [2], and also Bertsekas and Tsitsiklis [4], Section 2.3). In particular, for $M \geq 1$, let us consider the M-transition Bellman's equation

$$J(i) = E\left[\alpha^M J(i_M) + \sum_{k=0}^{M-1} \alpha^k g(i_k, i_{k+1})) \,\Big|\, i_0 = i\right], \qquad i = 1, \dots, n. \quad (9.22)$$

This equation has the cost-to-go function J as its unique solution, and in fact may be viewed as Bellman's equation for a modified policy evaluation problem, involving a Markov chain where each transition corresponds to M transitions of the original, and the cost is calculated using a discount factor α^M and a cost per (M-transition) stage equal to $\sum_{k=0}^{M-1} \alpha^k g(i_k, i_{k+1})$. The value iteration method corresponding to this modified problem is

$$J_{t+1}(i) = E\left[\alpha^M J_t(i_M) + \sum_{k=0}^{M-1} \alpha^k g(i_k, i_{k+1}) \,\Big|\, i_0 = i\right], \qquad i = 1, \dots, n,$$

and can be seen to be equivalent to M iterations of the value iteration method (9.17) for the original problem. The corresponding simulation-based least-squares implementation is

$$r_{t+1} = \arg\min_r \sum_{m=0}^{t} \left(\phi(i_m)'r - \alpha^M \phi(i_{m+M})'r_t - \sum_{k=0}^{M-1} \alpha^k g(i_{m+k}, i_{m+k+1})\right)^2, \atop t = 0, 1, \dots,$$

or equivalently, using the definition of temporal difference,

$$r_{t+1} = \arg\min_r \sum_{m=0}^{t} \left(\phi(i_m)'r - \phi(i_m)'r_t - \sum_{k=m}^{m+M-1} \alpha^{k-m} d_t(i_k, i_{k+1})\right)^2,$$
$$t = 0, 1, \dots. \quad (9.23)$$

This method, which is identical to 0-LSPE for the modified policy evaluation problem described above, may be viewed as intermediate between 0-LSPE and 1-LSPE for the original policy evaluation problem; compare with the form (9.4) of λ-LSPE for $\lambda = 0$ and $\lambda = 1$.

Let us also mention the incremental gradient version of the iteration (9.23), given by

$$r_{t+1} = r_t + \gamma_t \,\phi(i_t) \sum_{k=t}^{t+M-1} \alpha^{k-t} d_t(i_k, i_{k+1}), \qquad t = 0, 1, \dots. \quad (9.24)$$

This method, which is identical to TD(0) for the modified (M-step) policy evaluation problem described above, may be viewed as intermediate between TD(0) and TD(1) [it is closest to TD(0) for small M, and to TD(1) for large M]. Note that temporal differences do not play a fundamental role in the above iterations; they just provide a convenient shorthand notation that simplifies the formulas.

9.4.3   The Case $0 < \lambda < 1$

The M-transition Bellman's equation (9.22) holds for a fixed M, but it is also possible to consider a version of Bellman's equation where M is random and geometrically distributed with parameter λ, that is,

$$\text{Prob}(M = m) = (1 - \lambda)\lambda^{m-1}, \qquad m = 1, 2, \ldots$$

This equation is obtained by multiplying both sides of Eq. (9.22) with $(1 - \lambda)\lambda^{m-1}$, for each m, and adding over m:

$$J(i) = \sum_{m=1}^{\infty} (1 - \lambda)\lambda^{m-1} E\left[\alpha^m J(i_m) + \sum_{k=0}^{m-1} \alpha^k g(i_k, i_{k+1})) \,\bigg|\, i_0 = i\right],$$
$$i = 1, \ldots, n. \quad (9.25)$$

Tsitsiklis and Van Roy [19] provide an interpretation of TD(λ) as a gradient-like method for minimizing a weighted quadratic function of the error in satisfying this equation.

We may view Eq. (9.25) as Bellman's equation for a modified policy evaluation problem. The value iteration method corresponding to this modified problem is

$$J_{t+1}(i) = \sum_{m=1}^{\infty} (1 - \lambda)\lambda^{m-1} E\left[\alpha^m J_t(i_m) + \sum_{k=0}^{m-1} \alpha^k g(i_k, i_{k+1})) \,\bigg|\, i_0 = i\right],$$
$$i = 1, \ldots, n,$$

which can be written as

$$
\begin{aligned}
J_{t+1}(i) &= J_t(i) + (1 - \lambda) \\
&\quad \sum_{m=1}^{\infty} \sum_{k=0}^{m-1} \lambda^{m-1}\alpha^k E\big[g(i_k, i_{k+1}) + \alpha J_t(i_{k+1}) - J_t(i_k) \mid i_0 = i\big] \\
&= J_t(i) + (1 - \lambda) \sum_{k=0}^{\infty} \\
&\quad \left(\sum_{m=k+1}^{\infty} \lambda^{m-1}\right) \alpha^k E\big[g(i_k, i_{k+1}) + \alpha J_t(i_{k+1}) - J_t(i_k) \mid i_0 = i\big]
\end{aligned}
$$

and finally,

$$J_{t+1}(i) = J_t(i) + \sum_{k=0}^{\infty} (\alpha\lambda)^k E\big[g(i_k, i_{k+1}) + \alpha J_t(i_{k+1}) - J_t(i_k) \mid i_0 = i\big],$$

$$i = 1, \ldots, n.$$

By using the linear function approximation $\phi(i)'r_t$ for the costs $J_t(i)$, and by replacing the terms $g(i_k, i_{k+1}) + \alpha J_t(i_{k+1}) - J_t(i_k)$ in the above iteration with temporal differences

$$d_t(i_k, i_{k+1}) = g(i_k, i_{k+1}) + \alpha\phi(i_{k+1})'r_t - \phi(i_k)'r_t,$$

we obtain the simulation-based least-squares implementation

$$r_{t+1} = \arg\min_r \sum_{m=0}^{t} \left(\phi(i_m)'r - \phi(i_m)'r_t - \sum_{k=m}^{t} (\alpha\lambda)^{k-m} d_t(i_k, i_{k+1}) \right)^2,$$

$$(9.26)$$

which is, in fact, λ-LSPE with stepsize $\gamma = 1$.

Let us now discuss the relation of λ-LSPE with $\gamma = 1$ and TD(λ). We note that the gradient of the least squares sum of λ-LSPE is

$$-2 \sum_{m=0}^{t} \phi(i_m) \sum_{k=m}^{t} (\alpha\lambda)^{k-m} d_t(i_k, i_{k+1}).$$

This gradient after some calculation, can be written as

$$-2\big(z_0 d_t(i_0, i_1) + \cdots + z_t d_t(i_t, i_{t+1})\big), \tag{9.27}$$

where

$$z_k = \sum_{m=0}^{k} (\alpha\lambda)^{k-m} \phi(i_m), \qquad k = 0, \ldots, t,$$

[cf. Eq. (9.15)]. On the other hand, TD(λ) has the form

$$r_{t+1} = r_t + \gamma_t z_t d_t(i_t, i_{t+1}).$$

Asymptotically, in steady-state, the expected values of all the terms $z_m d_t(i_m, i_{m+1})$ in the gradient sum (9.27) are equal, and each is proportional to the expected value of the term $z_t d_t(i_t, i_{t+1})$ in the TD(λ) iteration. Thus, TD(λ) *updates r_t along the gradient of the least squares sum of λ-LSPE, plus stochastic noise that asymptotically has zero mean.*

In conclusion, for all $\lambda < 1$, we can view λ-LSPE with $\gamma = 1$ as a least-squares based approximate value iteration with linear function approximation. However, each value iteration implicitly involves a random number of transitions with geometric distribution that depends on λ. The limit r^* depends on λ because the underlying Bellman's equation also depends on λ. Furthermore, TD(λ) and λ-LSPE may be viewed as stochastic gradient and Kalman filtering algorithms, respectively, for solving the least squares problem associated with approximate value iteration.

9.4.4   Generalizations Based on Other Types of Value Iteration

The connection with value iteration described above provides a guideline for developing other least-squares based approximation methods, relating to different types of dynamic programming problems, such as stochastic shortest path, average cost, and semi-Markov decision problems, or to variants of value iteration such as, for example, Gauss-Seidel methods. To this end, we generalize the key idea of the convergence analysis of Sections 9.2 and 9.3. A proof of the following proposition is embodied in the argument of the proof of Prop. 6.9 of Bertsekas and Tsitsiklis [4] (which actually deals with a more general nonlinear iteration), but for completeness, we give an independent argument that uses the proof of Lemma 9.2.2.

Proposition 9.4.1 *Consider a linear iteration of the form*

$$x_{t+1} = Gx_t + g, \qquad t = 0, 1, \ldots, \tag{9.28}$$

where $x_t \in \Re^n$, and G and g are given $n \times n$ matrix and n-dimensional vector, respectively. Assume that D is a positive definite symmetric matrix such that

$$\|G\|_D = \max_{\substack{\|z\|_D \leq 1 \\ z \in C^n}} \|Gz\|_D < 1,$$

where $\|z\|_D = \sqrt{\hat{z}'Dz}$, for all $z \in C^n$. Let Φ be an $n \times s$ matrix of rank s. Then the iteration

$$r_{t+1} = \arg\min_{r \in \Re^s} \|\Phi r - G\Phi r_t - g\|_D, \qquad t = 0, 1, \ldots \tag{9.29}$$

converges to the vector r^* satisfying

$$r^* = \arg\min_{r \in \Re^s} \|\Phi r - G\Phi r^* - g\|_D, \tag{9.30}$$

from every starting point $r_0 \in \Re^s$.

Proof: The iteration (9.29) can be written as

$$r_{t+1} = (\Phi'D\Phi)^{-1}(\Phi'DG\Phi r_t + \Phi'Dg), \tag{9.31}$$

so it is sufficient to show that the matrix $(\Phi'D\Phi)^{-1}\Phi'DG\Phi$ has eigenvalues that lie within the unit circle. The proof of this follows nearly verbatim the corresponding steps of the proof of Lemma 9.2.2. If r^* is the limit of r_t, we have by taking limit in Eq. (9.31),

$$r^* = \left(I - (\Phi'D\Phi)^{-1}\Phi'DG\right)^{-1}(\Phi'D\Phi)^{-1}\Phi'Dg.$$

It can be verified that r^* as given by the above equation, also satisfies Eq. (9.30). ∎

The above proposition can be used within various dynamic programming/function approximation contexts. In particular, starting with a value iteration of the form (9.28), we can consider a linear function approximation version of the form (9.29), as long as we can find a weighted Euclidean norm $\| \cdot \|_D$ such that $\|G\|_D < 1$. We may then try to devise a simulation-based method that emulates approximately iteration (9.27), similar to λ-LSPE. This method will be an iterative stochastic algorithm, and its convergence may be established along the lines of the proof of Prop. 9.3.1. Thus, Prop. 9.4.1 provides a general framework for deriving and analyzing least-squares simulation-based methods in approximate dynamic programming. An example of such a method, indeed the direct analog of λ-LSPE for stochastic shortest path problems, was stated and used by Bertsekas and Ioffe [2] to solve the tetris training problem [see also [4], Eq. (8.6)].

9.5   RELATION BETWEEN λ-LSPE AND LSTD

We now discuss the relation between λ-LSPE and the LSTD method that estimates $r^* = -A^{-1}b$ based on the portion $(i_0, \ldots, i_t)$ of the simulation trajectory by

$$\hat{r}_{t+1} = -A_t^{-1}b_t,$$

[cf. Eqs. (9.14) and (9.15)]. Konda [12] has shown that the error covariance $E\{(\hat{r}_t - r^*)(\hat{r}_t - r^*)'\}$ of LSTD goes to zero at the rate of $1/t$. Similarly, it was shown by Nedić and Bertsekas [13] that the covariance of the stochastic term $Z_t r_t + \zeta_t$ in Eq. (9.21) goes to zero at the rate of $1/t$. Thus, from Eq. (9.21), we see that the error covariance $E\{(r_t - r^*)(r_t - r^*)'\}$ of λ-LSPE also goes to zero at the rate of $1/t$.

We will now argue that a stronger result holds, namely that r_t "tracks" $\hat{r}_t$ in the sense that the difference $r_t - \hat{r}_t$ converges to 0 faster than $\hat{r}_t - r^*$. Indeed, from Eqs. (9.14) and (9.15), we see that the averages $\overline{B}_t$, $\overline{A}_t$, and $\overline{b}_t$ are generated by the slow stochastic approximation-type iterations

$$\overline{B}_{t+1} = \overline{B}_t + \frac{1}{t+2}\big(\phi(i_{t+1})\phi(i_{t+1})' - \overline{B}_t\big),$$

$$\overline{A}_{t+1} = \overline{A}_t + \frac{1}{t+2}\big(z_{t+1}(\alpha\phi(i_{t+2})' - \phi(i_{t+1})') - \overline{A}_t\big), \qquad (9.32)$$

$$\overline{b}_{t+1} = \overline{b}_t + \frac{1}{t+2}\big(z_{t+1}g(i_{t+1}, i_{t+2}) - \overline{b}_{t+1}\big). \qquad (9.33)$$

Thus they converge at a slower time scale than the λ-LSPE iteration

$$r_{t+1} = r_t + \overline{B}_t^{-1}(\overline{A}_t r_t + \overline{b}_t), \qquad (9.34)$$

where, for sufficiently large t, the matrix $I + \overline{B}_t^{-1}\overline{A}_t$ has eigenvalues within the unit circle, inducing much larger relative changes of $\overline{r}_t$. This means that the λ-LSPE iteration (9.34) "sees $\overline{B}_t$, $\overline{A}_t$, and $\overline{b}_t$ as essentially constant," so that, for large t, r_{t+1} is essentially equal to the corresponding limit of iteration (9.34) with $\overline{B}_t$, $\overline{A}_t$, and $\overline{b}_t$ held fixed. This limit is $-\overline{A}_t^{-1}\overline{b}_t$ or $\hat{r}_{t+1}$. It follows that the difference $r_t - \hat{r}_t$ converges to 0 faster than $\hat{r}_t - r^*$. The preceding argument can be made precise by appealing to the theory of two-time scale iterative methods (see e.g., Benveniste, Metivier, and Priouret [1]), but a detailed analysis is beyond the scope of this chapter.

Despite their similar asymptotic behavior, the methods may differ substantially in the early iterations, and it appears that the iterates of LSTD tend to fluctuate more than those of λ-LSPE. Some insight into this behavior may be obtained by noting that the λ-LSPE iteration consists of a deterministic component that converges fast, and a stochastic component that converges slowly, so in the early iterations, the deterministic component dominates the stochastic fluctuations. On the other hand, $\overline{A}_t$ and $\overline{b}_t$ are generated by the slow iterations (9.32) and (9.33), and the corresponding estimate $-\overline{A}_t^{-1}\overline{b}_t$ of LSTD fluctuates significantly in the early iterations.

Another significant factor in favor of LSPE is that LSTD cannot take advantage of a good initial choice r_0. This is important in contexts such as optimistic policy iteration, as discussed in the introduction. Figure 9.2 shows some typical computational results for two 100-state problems with four features, and the values $\lambda = 0$ and $\lambda = 1$. The four features are

$$\phi_1(i) = 1, \qquad \phi_2(i) = i, \qquad \phi_3(i) = I([81, 90]), \qquad \phi_4(i) = I([91, 100]),$$

where $I(S)$ denotes the indicator function of a set S [$I(i) = 1$ if $i \in S$, and $I(i) = 0$ if $i \notin S$].

The figure shows the sequence of the parameter values $r(1)$ over 1,000 iterations/simulated transitions, for three methods: LSTD, LSPE with a constant stepsize $\gamma = 1$, and LSPE with a time-varying stepsize given by

$$\gamma_t = \frac{t}{500 + t}.$$

While all three methods asymptotically give the same results, it appears that LSTD oscillates more that LSPE in the initial iterations. The use of the time-varying stepsize "damps" the noisy behavior in the early iterations.

9.6   COMPUTATIONAL COMPARISON OF λ-LSPE AND TD(λ)

We conducted some computational experimentation to compare the performance of λ-LSPE and TD(λ). Despite the fact that our test problems were small, the differences between the two methods emerged strikingly and unmistakably. The methods performed as expected from the existing theoretical analysis, and converged

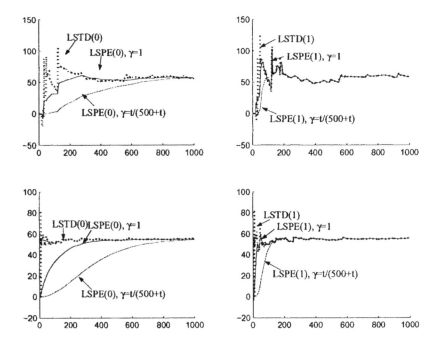

Fig. 9.2 The sequence of the parameter values $r(1)$ over 1,000 iterations/simulated transitions, for three methods: LSTD, LSPE with a constant stepsize $\gamma = 1$, and LSPE with a time-varying stepsize. The top figures correspond to a "slow-mixing" Markov chain (high self-transition probabilities) of the form

$$P = 0.9 * P_{random} + 0.1I,$$

where I is the identity and P_{random} is a matrix whose row elements were generated as uniformly distributed random numbers within $[0, 1]$, and were normalized so that they add to 1. The bottom figures correspond to a "fast-mixing" Markov chain (low self-transition probabilities):

$$P = 0.1 * P_{random} + 0.9I.$$

The cost of a transition was randomly chosen within $[0, 1]$ at every state i, plus $i/30$ for self-transitions for $i \in [90, 100]$.

to the same limit. In summary, the major observed differences between the two methods are:

(1) The number of iterations (length of simulation) to converge within the same small neighborhood of r^* was dramatically smaller for λ-LSPE than for TD(λ). Interestingly, not only was the deterministic portion of the λ-LSPE iteration much faster, but the noisy portion was faster as well, for all the stepsize rules that we tried for TD(λ).

(2) While in λ-LSPE there is no need to choose any parameters (we fixed the stepsize to $\gamma = 1$), in TD(λ) the choice of the stepsize γ_t was λ-dependent, and required a lot of trial and error to obtain reasonable performance.

(3) Because of the faster convergence and greater resilience to simulation noise of λ-LSPE, it is possible to use values of λ that are closer to 1 than with TD(λ), thereby obtaining vectors Φr^* that more accurately approximate the true cost vector J.

The observed superiority of λ-LSPE over TD(λ) is based on the much faster convergence rate of its deterministic portion. On the other hand, for many problems the noisy portion of the iteration may dominate the computation, such as for example when the Markov chain is "slow-mixing," and a large number of transitions are needed for the simulation to reach all the important parts of the state space. Then, both methods may need a very long simulation trajectory in order to converge. Our experiments suggest much better performance for λ-LSPE under these circumstances as well, but were too limited to establish any kind of solid conclusion. However, in such cases, the optimality result for LSTD of Konda (see Section 9.1), and the comparability of the behavior of LSTD and λ-LSPE, suggest a substantial superiority of λ-LSPE over TD(λ).

We will present representative results for a simple test problem with three states $i = 1, 2, 3$, and two features, corresponding to a linear approximation architecture of the form

$$\tilde{J}(i, r) = r(1) + ir(2), \qquad i = 1, 2, 3,$$

where $r(1)$ and $r(2)$ were the components of r. Because the problem is small, we can state it precisely here, so that our experiments can be replicated by others. We obtained qualitatively similar results with larger problems, involving 10 states and two features, and 100 states and four features. We also obtained similar results in limited tests involving the M-step methods (9.23) and (9.24).

We tested λ-LSPE and TD(λ) for a variety of problem data, experimental conditions, and values of λ. Figure 9.2 shows some results where the transition probability and cost matrices are given by

$$[p_{ij}] = \begin{pmatrix} 0.01 & 0.99 & 0 \\ 0.55 & 0.01 & 0.44 \\ 0 & 0.99 & 0.01 \end{pmatrix}, \qquad [g(i, j)] = \begin{pmatrix} 1 & 2 & 0 \\ 1 & 2 & -1 \\ 0 & 1 & 0 \end{pmatrix}.$$

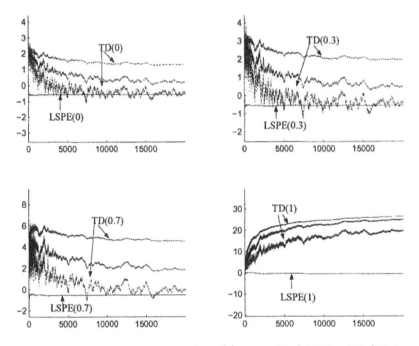

Fig. 9.3   The sequence of the parameter values $r(2)$ generated by λ-LSPE and TD(λ) [using the three stepsize rules (9.35)-(9.37)] over 20,000 iterations/simulated transitions, for the four values $\lambda = 0, 0.3, 0.7, 1$. All runs used the same simulation trajectory.

The discount factor was $\alpha = 0.99$. The initial condition was $r_0 = (0,0)$. The stepsize for λ-LSPE was chosen to be equal to 1 throughout. The stepsize choice for TD(λ) required quite a bit of trial and error, aiming to balance speed of convergence and stochastic oscillatory behavior. We obtained the best results with three different stepsize rules

$$\gamma_t = \frac{16(1 - \alpha\lambda)}{500(1 - \alpha\lambda) + t}, \tag{9.35}$$

$$\gamma_t = \frac{16(1 - \alpha\lambda)\sqrt{\log(t)}}{500(1 - \alpha\lambda) + t}, \tag{9.36}$$

$$\gamma_t = \frac{16(1 - \alpha\lambda)\log(t)}{500(1 - \alpha\lambda) + t}. \tag{9.37}$$

Rule (9.35) led to the slowest convergence with least stochastic oscillation, while rule (9.37) led to the fastest convergence with most stochastic oscillation.

It can be seen from Figure 9.3 that TD(λ) is not settled after 20,000 iterations/simulated transitions, and in the case where $\lambda = 1$, it does not even show signs of convergence. By contrast, λ-LSPE essentially converges within no more than 500 iterations, and with small subsequent stochastic oscillation. Generally, as λ becomes smaller, both TD(λ) and λ-LSPE converge faster at the expense of a worse bound on the error $\Phi r^* - J$. The qualitative behavior, illustrated in Figure 9.3, was replicated for a variety of transition probability and cost matrices, initial conditions, and other experimental conditions. This behavior is consistent with the computational results of Bertsekas and Ioffe [2] for the tetris training problem (see also Bertsekas and Tsitsiklis [4], Section 8.3). Furthermore, in view of the similarity of performance of λ-LSPE and LSTD, our computational experience is also consistent with that of Boyan [6].

Acknowledgments

Research supported by NSF Grant ECS-0218328 and Grant III.5(157)/99-ET from the Dept. of Science and Technology, Government of India. Thanks are due to Janey Yu for her assistance with the computational experimentation.

Bibliography

1. A. Benveniste, M. Metivier, and P. Priouret, *Adaptive Algorithms and Stochastic Approximations*, Springer-Verlag, New York, 1990.

2. D. P. Bertsekas and S. Ioffe, Temporal differences-based policy iteration and applications in neuro-dynamic programming, *Lab. for Info. and Decision Systems Report LIDS-P-2349*, MIT, Cambridge, MA, 1996.

3. D. P. Bertsekas, *Dynamic Programming and Optimal Control, 2nd Edition*, Athena Scientific, Belmont, MA, 2001.

4. D. P. Bertsekas and J. N. Tsitsiklis, *Neuro-Dynamic Programming*, Athena Scientific, Belmont, MA, 1996.

5. D. P. Bertsekas and J. N. Tsitsiklis, Gradient convergence in gradient methods with errors, *SIAM Journal Optimization*, vol. 10, pp. 627–642, 2000.

6. J. A. Boyan, Technical update: Least-squares temporal difference learning, *Machine Learning*, vol. 49, pp. 1–15, 2002.

7. S. J. Bradtke and A. G. Barto, Linear least-squares algorithms for temporal difference learning, *Machine Learning*, vol. 22, pp. 33–57, 1996.

8. P. D. Dayan, The convergence of TD(λ) for general λ, *Machine Learning*, vol. 8, pp. 341–362, 1992.

9. D. P. de Farias and B. Van Roy, On the existence of fixed points for approximate value iteration and temporal-difference learning, *Journal of Optimization Theory and Applications*, vol. 105, 2000.

10. L. Gurvits, L. J. Lin, and S. J. Hanson, Incremental learning of evaluation functions for absorbing Markov chains: New methods and theorems, Preprint, 1994.

11. V. R. Konda and J. N. Tsitsiklis, The asymptotic mean squared error of temporal difference learning, Unpublished Report, *Lab. for Information and Decision Systems*, MIT, Cambridge, MA, 2003.

12. V. R. Konda, *Actor-Critic Algorithms*, Ph.D. Thesis, Dept. of Electrical Engineering and Computer Science, MIT, Cambridge, MA, 2002.

13. A. Nedić and D. P. Bertsekas, Least squares policy evaluation algorithms with linear function approximation, *Discrete Event Dynamic Systems: Theory and Applications,* vol. 13, pp. 79–110, 2003.

14. F. Pineda, Mean-field analysis for batched TD(λ), *Neural Computation,* pp. 1403–1419, 1997.

15. M. L. Puterman, *Markov Decision Processes,* Wiley, New York, 1994.

16. R. S. Sutton and A. G. Barto, *Reinforcement Learning,* MIT Press, Cambridge, MA, 1998.

17. R. S. Sutton, Learning to predict by the methods of temporal differences, *Machine Learning,* vol. 3, pp. 9–44, 1988.

18. B. Van Roy, *Learning and Value Function Approximation in Complex Decision Processes,* Ph.D. Thesis, MIT, Cambridge, MA, 1998.

19. J. N. Tsitsiklis and B. Van Roy, An analysis of temporal-difference learning with function approximation, *IEEE Trans. on Automatic Control,* vol. 42, pp. 674–690, 1997.

Handbook of Learning and Approximate Dynamic Programming
Edited by Jennie Si, Andy Barto, Warren Powell and Donald Wunsch
Copyright © 2004 The Institute of Electrical and Electronics Engineers, Inc.

10 Approximate Dynamic Programming for High-Dimensional Resource Allocation Problems

WARREN B. POWELL   BENJAMIN VAN ROY
Princeton University   Stanford University

Editor's Summary: This chapter focuses on presenting a mathematical model for dynamic resource allocation problems and reviews computational methods including approximate dynamic programming that address them. Results from numerical examples are presented to illustrate the potential of ADP for addressing these complex problems.

10.1 INTRODUCTION

The allocation of human and physical resources over time is a fundamental problem that is central to management science. For example, a freight transportation company must manage personnel and equipment to move shipments in a timely manner in the presence of a variety of dynamic information processes: customer demands, equipment failures, weather delays, and failures of execution. This is a high-dimensional problem since it involves a large number of resources, each of which must be tracked as it is affected by decisions and uncertainties.

In principle, problems of dynamic resource allocation can be treated as Markov decision processes and solved using dynamic programming algorithms. Textbook dynamic programming algorithms — such as value iteration and policy iteration — typically require compute time and memory that grow exponentially in the number of state variables, the number of decision variables, and the number of random variables that affect the system in each time period. These three "curses of dimensionality" render such algorithms infeasible for problems of practical scale.

In this chapter, we focus on a formulation of dynamic resource allocation that was originally motivated by problems in transportation but also captures problems arising in a variety of other settings. Practical problems formulated in terms of our model typically involve thousands of state variables that describe current resources, thousands of decision variables that determine what is to be done with each resource, and thousands of random variables that influence the state. These random variables capture uncertainties from a variety of sources, such as customer demands, the physical network, and characteristics of the people and equipment used to provide services (e.g., equipment breakdowns and no-shows). Clearly, textbook dynamic programming algorithms are inapplicable, and solving such large-scale dynamic resource allocation problems has proven to be a terrific challenge. In this chapter, we present examples of approximate dynamic programming algorithms developed by the first author and collaborators to address such problems. We also discuss the relation between these algorithms and ideas studied in the broader approximate dynamic literature (including "neuro-dynamic programming" and "reinforcement learning" methods) and the mathematical programming literature.

This chapter is organized as follows: Section 10.2 presents a mathematical model for a broad class of resource allocation problems together with a simple illustrative example. Section 10.3 discusses how — in the context of our dynamic resource allocation model — the three curses of dimensionality prevent application of textbook dynamic programming algorithms. In Section 10.4, we present approximate dynamic programming algorithms designed for our model. We discuss in Section 10.5 models and algorithms that have emerged from the field of mathematical programming, and then review in Section 10.6 relations to ideas that have evolved from within the approximate dynamic programming literature. In Section 10.7, we present some experimental results.

10.2   DYNAMIC RESOURCE ALLOCATION

In this section, we present a formulation of dynamic resource allocation problems. After defining the formulation in Section 10.2.1, we discuss in Section 10.2.2 a simple example involving the management of a fleet of trucks. This example is a dramatically simplified version of a real trucking problem. However, it serves to illustrate how the formulation maps to a practical context while avoiding the intricacies of truly realistic models (see, e.g., [25, 28]).

10.2.1   Problem Formulation

We consider a system that evolves in discrete time over T periods. At each time $t = 0, \ldots, T$, the state of the system is described by a state variable R_t, taking the form:

$a =$ The vector of attributes that describe a single resource.

$\mathcal{A} =$ The set of possible attributes.

R_{ta} = The number of resources with attribute a at time t.

$R_t = (R_{ta})_{a \in \mathcal{A}}$.

For our purposes, it is enough to describe a single resource class such as trucks or people or product, but more complex problems can exhibit multiple resource classes (tractors, trailers and drivers; pilots, aircraft and passengers). For these problems, we have to define different resource classes $\mathcal{C}^R$. We then let $\mathcal{A}^c$ be the attribute space for a resource class $c \in \mathcal{C}^R$, and we let R_t^c be the vector of resources for class c. If we let $R_t = (R_t^c)_{c \in \mathcal{C}^R}$ then we can continue to let R_t be our resource state.

At each time $t = 0, \ldots, T - 1$, a decision is made on what to do with each resource. For any resource with attribute a, there is a set $\mathcal{D}_a$ of actions that can be applied to the resource. The collection of actions applied at time t is represented by a decision $x_t = (x_{tad})_{a \in \mathcal{A}, d \in \mathcal{D}_a}$, where x_{tad} is the number of resources with attribute a that action d is applied to. In a transportation example involving the management of equipment, a would capture the type and location of a piece of equipment and an action might be to move the equipment from one location to another. Needless to say, each decision variable x_{tad} must be a nonnegative integer and the decision variables must satisfy flow conservation constraints:

$$\sum_{d \in \mathcal{D}_a} x_{tad} = R_{ta} \quad \forall a \in \mathcal{A}, \tag{10.1}$$

which we write compactly as $A_t x_t = R_t$, where A_t is a linear operator. There may also be additional constraints on the decision variable. We assume that these constraints are linear, taking the form

$$U_t x_t \le u_t, \tag{10.2}$$

where U_t is a linear operator mapping decisions to R^{n_t} for some n_t, so that $U_t x_t$ and u_t are elements of $\Re^{n_t}$.

The change that a decision induces on the state R_t is a linear function of the decision. We define a linear operator Δ_t by:

$(\Delta_t x_t)_a$ = The number of resources that result in attribute a given the decision vector x_t.

The vector R_t of resources that we can act on in time period t is known as the *pre-decision state vector*. The resource vector after the resources have been acted on is given by $R_t^x = \Delta_t x_t$. We refer to R_t^x as the *post-decision* state vector, since it is the state of our system immediately after decisions have been made.

There are also random shocks — new resources appear and existing resources can disappear randomly. This randomness is driven by a sequence $\omega_1, \ldots, \omega_T$ of independent identically distributed random variables. The random change during the tth period in the number of resources with attribute a takes the form $\hat{R}_a(R_t^x, \omega_{t+1})$ for some function $\hat{R}_a$. Note that this quantity is a function of both the random variable ω_{t+1}

and the "post-decision" state $R_t^x = \Delta_t x_t$. Let $\hat{R}(R_t^x, \omega_{t+1}) = (\hat{R}_a(R_t^x, \omega_{t+1}))_{a \in \mathcal{A}}$. The dynamics of our system are described by the simple equation:

$$R_{t+1} \;=\; \Delta_t x_t + \hat{R}(R_t^x, \omega_{t+1}).$$

If our random changes are purely exogenous arrivals to the system (which do not depend on the state of the system), we will write $\hat{R}(\omega_t)$ as the exogenous arrivals to the system during time interval t.

At each time, we must select a feasible action x_t. For shorthand, we denote the feasible set by $\mathcal{X}_t(R_t)$ — this is the set of actions x_t with nonnegative integer components that satisfy (10.1) and (10.2). Each action x_t is selected based only on the current state R_t. Hence, we can think of decisions as being made by a *policy*, which is a function X_t^π that maps each state to a feasible action:

$$x_t = X_t^\pi(R_t) \in \mathcal{X}_t(R_t).$$

There are, of course, many different policies (functions) that we can use, so we let Π be our family of policies (decision functions) that we can choose from.

The contribution (or reward) generated in each tth period is a linear function $C_t x_t$ of the decision x_t. Here, C_t is a linear operator that gives the total contribution (if we are maximizing) of an action. Our objective is to select a policy that maximizes the expected contribution over the horizon:

$$\max_{\pi \in \Pi} E \left[\sum_{t=0}^{T-1} C_t X_t^\pi(R_t) \right].$$

10.2.2   Transportation Example

As an example to illustrate how our formulation might map to a real-world context, we discuss in this section a simplified model of a truckload motor carrier. For the sake of brevity, we will only discuss at a very high level how elements of the model might be used to capture features of the application.

Consider the management of N trucks moving among L cities over T time periods. Loads materialize at various cities, and each load is tagged with a destination. For the sake of simplicity, we assume that the only source of uncertainty is in demand (i.e., the loads that materialize in various cities).

We now discuss how one might formulate the problem in terms of our mathematical model. The attribute a specifies either that the resource is residing in a particular city, or that the resource is travelling, in which case the attribute a further specifies the next city where the resource will arrive and the time until arrival.

The random variable $\hat{R}(R_t^x, \omega_{t+1})$ captures new loads entering the system. The number of loads that materialize at each city in a given time period is independent of the numbers at other cities, and these arrivals are independent and identically distributed over time. Note that, in this special case, $\hat{R}$ only depends on ω_{t+1} and not R_t^x.

The set of possible actions $\mathcal{D}_a$ that can be applied to a resource depends on its current attribute. For example, a driver in one location can typically be assigned to move loads that are not too far away. Furthermore, the attribute vector might capture other characteristics of a driver that limit what a driver can do. For example, a Canadian-based driver in the United States can only accept a load that returns him to Canada.

Some additional constraints are placed on the choice of the aggregate collection of actions x_t. The flow conservation constraint (10.1) ensures that we only act on the trucks and loads that are available. Further requirements are that a load can only move if it is on a truck and that a truck can only carry a single load at a time. These requirements can be represented as a linear constraint $U_t x_t \leq u_t$ for appropriately defined U_t and u_t.

The operator Δ_t captures how resources migrate in the system and how loads leave the system as consequences of decisions. Each load generates a contribution upon delivery. The contribution is a decreasing function of the time taken to serve the load, as specified by its attribute a. The net contribution in a time period can be represented as a linear function $C_t x_t$ of x_t.

10.3   CURSES OF DIMENSIONALITY

The problem of dynamic resource allocation can — in principle — be addressed via dynamic programming. In particular, value functions $V_0, \ldots, V_T$ are computed by setting

$$V_T(R_T) = 0, \tag{10.3}$$

and applying the recursion

$$V_t(R_t) = \max_{x \in \mathcal{X}_t(R_t)} \left(C_t x + E_t \left[V_{t+1} \left(R_t^x + \hat{R}(R_t^x, \omega_{t+1}) \right) \right] \right), \tag{10.4}$$

where the expectation is taken over possible outcomes of $\hat{R}_{t+1}$. We use the subscript with the expectation operator to indicate the conditioning information. In particular, for any t, E_t denotes a conditional expectation, conditioned on $\omega_1, \ldots, \omega_t$ (if $t = 0$, this translates to an expectation with respect to the prior distribution). Optimal decisions x_t can then be generated according to

$$x_t \in \operatorname*{argmax}_{x \in \mathcal{X}_t(R_t)} \left(C_t x + E_t \left[V_{t+1} \left(R_t^x + \hat{R}(R_t^x, \omega_{t+1}) \right) \right] \right).$$

Three computational obstacles prevent use of this textbook approach when dealing with problems of practical scale. First, the number of possible state vectors R_t grows very quickly with the number $|\mathcal{A}|$ of possible attributes, making computation of $V_t(R_t)$ for every possible R_t unmanageable. Second, exact computation of the expectation is infeasible because the number of possible outcomes of ω_{t+1} typically becomes enormous as $|\mathcal{A}|$ grows. Finally, finding an optimal decision x_t from the discrete space $\mathcal{X}_t$ generally requires an exhaustive search over $\mathcal{X}_t$. This space once again grows extremely quickly as $|\mathcal{A}|$ grows, rendering an exhaustive search infeasible. We refer to these three obstacles as the three "curses of dimensionality."

We note that some authors have studied formulations similar to what we have presented and solved problem instances using exact dynamic programming techniques ([1, 16]). However, the problems instances treated involve very few variables — a far stretch from the scale of real transportation systems.

10.4   ALGORITHMS FOR DYNAMIC RESOURCE ALLOCATION

In this section, we present algorithms developed by the first author with collaborators to address large-scale dynamic resource allocation problems of the type presented in Section 10.2. These algorithms incorporate several ideas, one of which involves a reformulation of dynamic programming equations in terms of a "post-decision state." We discuss this reformulation in Section 10.4.1. The algorithms approximate a value function. This requires design of an approximation architecture and execution of algorithms that fit the approximation architecture to the value function. Two particular approximation architectures that have been successfully applied to transportation problems are discussed in Section 10.4.2. In Section 10.4.3 we present algorithms used to fit these approximation architectures.

10.4.1   Post-Decision State

Earlier we introduced the notion of a "post-decision state" $R_t^x = R_t + \Delta_t x_t$. This is the state of the system after being modified by the action x_t taken at time t but before being affected by the random variable $\hat{R}_{t+1}$. Let $V_t^x(R_t^x)$ be the maximal expected sum of contributions to be received from decisions $x_{t+1}, \dots, x_{T-1}$, given the post-decision state R_t^x, so that

$$V_t^x(R_t^x) = E_t \left[V_t \left(R_t^x + \hat{R}(R_t^x, \omega_{t+1}) \right) \right].$$

It is easy to show that V_t^x can be computed by setting

$$V_{T-1}^x(R_{T-1}^x) = 0, \tag{10.5}$$

and applying the recursion

$$V_t^x(R_t^x) = E_t \left[\max_{x \in \mathcal{X}_{t+1}(R_t^x + \hat{R}(R_t^x, \omega_{t+1}))} \left(C_{t+1}x + V_{t+1}^x \left(\Delta_{t+1}x \right) \right) \right]. \tag{10.6}$$

Further, optimal decisions can be generated according to

$$x_t \in \underset{x \in \mathcal{X}_t(R_t)}{\text{argmax}} \left(C_t x + V_t^x(\Delta_{t+1} x) \right).$$

The new recursion is similar to the standard dynamic programming recursion (10.4) used for computing V_t. An important difference, however, is that this new recursion (10.6) involves the expectation of a maximum whereas (10.4) requires computing the maximum of an expectation. This difference in the form of (10.6) can be critical for use of simulation-based methods, as we will discuss in the Section 10.4.3.

10.4.2   Approximation Architectures

Our approach to high-dimensional dynamic resource allocation problems involves approximating each value function V_t^x using an *approximation architecture*; that is, a family of functions $\bar{V}^x$ parameterized by a vector v of real values, so that each $\bar{V}^x(\cdot, v)$ is a function mapping states to real values. Approximation of the value function involves choosing an appropriate approximation architecture and then computing parameters v_t so that $\bar{V}^x(\cdot, v_t) \approx V_t^x$. Given such an approximation, suboptimal decisions x_t can be generated according to

$$x_t \in \underset{x \in \mathcal{X}_t(R_t)}{\text{argmax}} \left(C_t x + \bar{V}^x(\Delta_t x, v_t) \right). \tag{10.7}$$

When choosing an approximation architecture, one must strike a balance between computational efficiency and performance of the resulting policy. Some approximation architectures may amplify computational challenges. For example, a second or third order polynomial approximation to V_t^x may turn the maximization problem in (10.7) into a very difficult integer program. On the other hand, specially structured approximation architectures such as those we will discuss next may sometimes lead to computationally tractable integer programming problems; for example, in some cases the maximization problem in (10.7) may become an easily solved network optimization problem.

Two classes of approximation architectures have been successfully applied to a variety of problems in transportation operations. These are linear (in the state variable) architectures:

$$\bar{V}^x(R_t, v_t) \;\; = \;\; \sum_{a \in \mathcal{A}} v_{ta} R_{ta}, \tag{10.8}$$

and separable concave architectures:

$$\bar{V}^x(R_t, v_t) \;\; = \;\; \sum_{a \in \mathcal{A}} \left(\sum_{i=1}^{\lfloor R_{ta} \rfloor} v_{tai} + (R_{ta} - \lfloor R_{ta} \rfloor) v_{ta \lceil R_{ta} \rceil} \right), \tag{10.9}$$

where $v_{tai} \geq v_{taj}$ for all t, a, and $i \leq j$. In the linear case, the parameter vector takes the form $v_t = (v_{ta})_{a \in \mathcal{A}}$, whereas in the separable concave case, we have $v_t = (v_{tai})_{a \in \mathcal{A}, i \in \{1, R_{ta}^{\max}\}}$, where $R_{ta}^{\max}$ is an upper bound on the values that R_{ta}^{x} can take on. Though states are integer-valued, these approximation architectures assign values to all elements of the positive orthant. This is required to make the optimization problem (10.7) amenable to available software for solving integer programs. Note that the separable concave architecture can realize any separable concave function on the integers. Each non-integer point is assigned a value corresponding to a linear interpolation. Each of the two architectures offers specific advantages and disadvantages, as we now discuss.

Not surprisingly, linear approximations are the easiest to work with. They will work well when the value function is approximately linear over the range of interest, and also for discrete routing and scheduling problems where $R_{ta} \in \{0, 1\}$. If the one-period problem,

$$x_t \in \underset{x \in \mathcal{X}_t(R_t)}{\mathrm{argmax}} \ C_t x,$$

exhibits structure that facilitates efficient solution (e.g., network or near-network structure) a linear approximation to the value function will retain this structure in (10.7). Additionally, linear approximations offer computational advantages with regards to algorithms for computing appropriate parameters v_t. Linear approximations should generally be tried before other, more complex strategies, but they can sometimes work poorly. In such cases, one must move on to a richer approximation architecture.

For some problems, the value function is nonlinear but concave, and separable concave functions have proven to offer useful approximations. Although they are typically harder to work with than linear approximations, the decision optimization problem (10.7) often still retains network or near-network structure if the one-period problem exhibits such structure. For example, in certain problems involving management of homogeneous resources, the optimization problem (10.7) exhibits pure network structure so that solution of the linear programming relaxation naturally produces integer solutions (see, e.g., [10]). In more realistic problems, resources are heterogeneous. The resulting optimization problems do not generally exhibit pure network structure. Nevertheless, solution of the linear programming relaxation results in integer solutions the vast majority of the time, and non-integer solutions are often easy to resolve.

10.4.3 Algorithms

In this section, we describe algorithms for computing parameter values to fit linear and separable concave approximations to the value function. These algorithms are representative of ideas developed by the first author and collaborators to tackle large-scale problems in transportation and logistics. In order to deliver a brief and accessible exposition, we have chosen to present relatively simple versions of the algorithms. More complicated variations that improve on execution time and memory requirements can be found in ([10, 28, 33], and [26]). Such improvements are often critical for practical application to realistic large-scale problems.

The algorithms we will present are iterative and integrate integer programming, stochastic approximation, and function approximation. Each iteration involves an independent simulated sequence $\omega_1, \ldots, \omega_T$ and adjusts the parameters $v_0, \ldots, v_{T-2}$ by small amounts. The algorithms for linear and separable concave architectures are identical except for one step, so we describe them as one algorithm and only delineate the two cases when required. In stating the algorithm, we will use some new notation. For each $a, \bar{a} \in \mathcal{A}$, let $\delta_{\bar{a}}^{\bar{a}} = 1$ and $\delta_a^{\bar{a}} = 0$ for $a \neq \bar{a}$ (note that this can be thought of as a Dirac delta function). Also, let $\delta^{\bar{a}} = (\delta_a^{\bar{a}})_{a \in \mathcal{A}}$. The algorithm takes as a parameter a step size γ, which is generally chosen to be a small positive scalar. Let each element of v_t be initialized to 0. Each iteration of the algorithm executes the following steps:

1. **Generate random sequence.** Sample the random sequence $\omega_1, \ldots, \omega_T$.

2. **Simulate state trajectory.** For $t = 0, \ldots, T - 1$, let

$$
\begin{aligned}
x_t &\in \operatorname*{argmax}_{x \in \mathcal{X}_t(R_t)} \left(C_t x + \bar{V}^x(\Delta_t x, v_t) \right) \\
R_t^x &= \Delta_t x_t \\
R_{t+1} &= R_t^x + \hat{R}(R_t^x, \omega_{t+1}).
\end{aligned}
$$

3. **Estimate gradient.** For $t = 0, \ldots, T - 2$ and $a \in \mathcal{A}$ such that $R_{ta}^x < R_{ta}^{\max}$, let

$$
\begin{aligned}
q_t &= C_{t+1} x_{t+1} + \bar{V}^x(R_{t+1}^x, v_{t+1}) \\
R_{ta}^{x+} &= R_t^x + \delta_a \\
q_{ta}^+ &= \max_{x \in \mathcal{X}_{t+1}(R_t^x + \delta^a + \hat{R}(R_t^x + \delta^a, \omega_{t+1}))} \left(C_{t+1} x + \bar{V}^x(\Delta_{t+1} x, v_{t+1}) \right) \\
d_{ta} &= q_{ta}^+ - q_t.
\end{aligned}
$$

4. **Update approximation architecture parameters.** For $t = 0, \ldots, T - 2$ and $a \in \mathcal{A}$ such that $R_{ta}^x < R_{ta}^{\max}$, apply

   (a) **Linear case.**

   $$
   v_{ta} := (1 - \gamma) v_{ta} + \gamma d_{ta}.
   $$

   (b) **Separable concave case.** For the slope element $i = (R_{ta}^x + 1)$:

   $$
   v_{tai} := (1 - \gamma) v_{tai} + \gamma d_{ta},
   $$

which may produce a nonconcave function. Restore concavity according to:

$$
\begin{aligned}
&\text{if } v_{ta(R_{ta}^x+1)} > v_{ta R_{ta}^x} : \\
&\qquad v_{ta R_{ta}^x} := \frac{v_{ta R_{ta}^x} + v_{ta(R_{ta}^x+1)}}{2} \\
&\qquad v_{ta(R_{ta}^x+1)} := \frac{v_{ta R_{ta}^x} + v_{ta(R_{ta}^x+1)}}{2} \\
&\text{if } v_{ta(R_{ta}^x+1)} < v_{ta(R_{ta}^x+2)} : \\
&\qquad v_{ta(R_{ta}^x+1)} := \frac{v_{ta(R_{ta}^x+1)} + v_{ta(R_{ta}^x+2)}}{2} \\
&\qquad v_{ta(R_{ta}^x+1)} := \frac{v_{ta(R_{ta}^x+1)} + v_{ta(R_{ta}^x+2)}}{2} .
\end{aligned}
\tag{10.10}
$$

Let us now provide an interpretation of the algorithm. Each sampled sequence $\omega_1, \ldots, \omega_T$ is used in Step 2 to generate a sample trajectory of states R_0, R_0^x, ..., R_{T-1}, R_{T-1}^x, R_T. In addition to the random sequence, this state trajectory is influenced by decisions $x_0, \ldots, x_{T-1}$. As the state trajectory is generated, these decisions are selected to optimize an estimate of future contribution. This estimate is taken to be the sum $C_t x_t + \bar{V}^x(R_t + \Delta_t x_t, v_t)$ of immediate contribution and an approximation of subsequent contribution given by $\bar{V}^x(\cdot, v_t)$. Implicit in the statement of our algorithm is that we have access to a subroutine for optimizing this estimate over $x \in \mathcal{X}_t(R_t)$. The optimization problem is an integer program, and as discussed in the previous section, for applications in transportation and logistics, it often exhibits network or near-network structure. This facilitates use of available software for integer programming. Though there are no theoretical guarantees when the problem does not exhibit pure network structure, this approach appears to work well in practice.

For each attribute $a \in \mathcal{A}$ and post-decision state R_t^x, Step 3 estimates how availability of an additional unit of a resource with attribute a would have impacted the prediction of future contribution. The difference is denoted by d_{ta}. Step 4 is a stochastic approximation iteration that tunes the gradient of the approximation $\bar{V}^x(R_t^x, v_t)$ with respect to R_t^x to be closer to the estimated gradient d_t.

In the case of a separable concave approximation, an extra step after the tuning projects the approximation back into the space of concave functions, if the tuning step destroys concavity. Note that we cannot even allow intermediate functional approximations to be nonconcave because of the complexities this would introduce to the problem of optimizing over $x \in \mathcal{X}_t(R_t)$. For a discussion of methods for maintaining concavity, see [11, 27, 37].

We mentioned earlier that reformulation of the dynamic programming recursion in terms of a post-decision state could be critical to the use of simulation-based methods. Let us now discuss how this relates to the above algorithm. One could imagine developing an analogous algorithm to approximate the pre-decision value function $V_t(R_t)$ using an approximation architecture $\bar{V}(R_t, v_t)$. Again, we would generate a random sequence, simulate the corresponding state trajectory, estimate gradients of predicted contribution, and update the approximation architecture accordingly. However, a difficulty arises in the Step 3, as we now explain.

In Step 3, q_t can be viewed as a sample estimate of

$$E_t \left[\max_{x \in \mathcal{X}_{t+1}(R_t^x + \hat{R}(R_t^x, \omega_{t+1}))} \left(C_{t+1}x + \bar{V}^x(\Delta_{t+1}x, v_t) \right) \right]. \qquad (10.11)$$

It is easy to see that q_t is an unbiased estimate of this expectation, and this is important, since the stochastic approximation updates essentially average these estimates to better estimate the expectation.

Suppose now that we wish to adapt the algorithm so that it works with a pre-decision value function $\bar{V}(R_t, v_t)$. Then, in Step 3, we would want to generate an unbiased estimate of

$$\max_{x \in \mathcal{X}_t(R_t)} \left(C_t x + E_t \left[\bar{V}(R_t + \Delta_t x + \hat{R}(R_t + \Delta_t x, \omega_{t+1}), v_t) \right] \right), \qquad (10.12)$$

based on a sampled value of ω_{t+1}. In general, this is not possible. Unlike the expectation of the maximum (10.11), there is no simple way of generating an unbiased estimate of the maximum of an expectation (10.12) based on a single sample of ω_{t+1}. An exception to this, however, arises in the case of a linear approximation architecture if $\hat{R}$ depends on R_t instead of R_t^x. In this event, we have

$$\max_{x \in \mathcal{X}_t(R_t)} \left(C_t x + E_t \left[\bar{V} \left(\Delta_t x + \hat{R}(R_t, \omega_{t+1}), v_t \right) \right] \right)$$
$$= \max_{x \in \mathcal{X}_t(R_t)} \left(C_t x + \bar{V}(\Delta_t x, v_t) \right) + E_t \left[\bar{V} \left(\hat{R}(R_t, \omega_{t+1}), v_t \right) \right].$$

Clearly, an unbiased estimate can be generated by maximizing the first term on the right-hand side and adding $\bar{V}(\hat{R}(R_t, \omega_{t+1}), v_t)$ generated from a single sample ω_{t+1}. On the other hand, when the approximation architecture is nonlinear or $\hat{R}$ depends R_t^x, an unbiased estimate can no longer be generated based on a single sample.

10.5   MATHEMATICAL PROGRAMMING

The problem formulation and algorithms we have presented evolved from a line of work on solution methods for large-scale problems in transportation and logistics. As of this writing, this field is dominated by algorithms developed within the mathematical programming community for handling high-dimensional resource allocation problems. We briefly review a simple deterministic model (for which there is a vast array of algorithms) and then introduce two competing themes that have emerged from within this community for introducing uncertainty.

10.5.1   Deterministic Mathematical Programming

In the field of transportation and logistics, the most common modelling and algorithmic strategy is to formulate the problem deterministically and then use classical math programming algorithms to solve the resulting model. For example, consider a deterministic version of the problem we introduced in Section 10.2.1 resulting from setting $\hat{R}(R_t^x, \omega_{t+1}) = \hat{R}_{t+1}$ for a deterministic sequence $\hat{R}_1, \ldots, \hat{R}_T$. This would lead to a deterministic optimization problem:

$$\max_{x_0, \ldots, x_{T-1}} \sum_{t=0}^{T-1} C_t x_t \tag{10.13}$$

subject to, for $t = 0, \ldots, T - 1$:

$$A_t x_t = R_t \tag{10.14}$$

$$U_t x_t \leq u_t \tag{10.15}$$

$$R_{t+1} = \Delta_t x_t + \hat{R}_{t+1} \tag{10.16}$$

$$x_t \geq 0 \text{ and integer-valued} \tag{10.17}$$

As with our stochastic formulation from Section 10.2.1, Eq. (10.14) typically represents flow conservation. Equations (10.15) and (10.17) impose upper and lower bounds on the flow. Equation (10.16) captures the state dynamics.

Deterministic problems like the one we have formulated often result in large, time staged linear integer programs. These can sometimes be solved effectively. However, they completely ignore uncertainty and some of the more complex dynamics that can arise in real-world settings.

10.5.2   Stochastic Programming

The most common approach for modelling resource allocation problems in practice is to formulate the problem deterministically and apply mathematical programming algorithms developed for this setting. When uncertainty is introduced into mathematical programming, we find ourselves in a subfield called *stochastic programming*. In a sense, Markov decision processes might be viewed as an extension of stochastic systems models to incorporate optimization, while stochastic programming might be viewed as an extension of (deterministic) mathematical programming to incorporate uncertainty.

Two algorithmic strategies have emerged from the stochastic programming community. The first, covered in Section 10.5.2.1, uses an explicit representation of decisions in the future for each of a finite set of scenarios. This approach is fundamentally different approach from dynamic programming. The second strategy, called Benders decomposition (Section 10.5.2.2) uses a method for approximating the downstream impact of decisions. This is much closer in spirit to dynamic programming.

10.5.2.1   Scenario methods   The oldest strategy to solving linear programs under uncertainty is stochastic programming using scenario methods. The roots of this approach date to the work by Dantzig ([9]), but the earliest serious investigation of stochastic programs were primarily due to Wets ([43, 48–50]). For modern reviews of this field, see [5, 14, 15, 32]).

As an example, let us formulate a scenario-based stochastic program to solve a problem similar to that posed by our model of Section 10.2.1. Let uncertainty be represented in terms of a trajectory of random outcomes $\omega = (\omega_1, \ldots, \omega_T)$. Let Ω be the set of possible "scenarios" ω, and let $p(\omega)$ be the probability that each $\omega \in \Omega$ will occur. The random impact on resources at time t is assumed to take the form $\hat{R}_t(\omega)$. Using this representation of uncertainty, consider solving the following optimization problem:

$$\max_{x_0, \{x_t(\omega)|t=1,\ldots,T-1, \omega \in \Omega\}} C_0 x_0 + \sum_{\omega \in \hat{\Omega}} p(\omega) \sum_{t=1}^{T-1} C_t x_t(\omega), \tag{10.18}$$

subject to the following first stage constraints for $\omega \in \Omega$:

$$A_0 x_0 = R_0, \tag{10.19}$$
$$U_0 x_0 \leq u_0, \tag{10.20}$$
$$R_1(\omega) = \Delta_0 x_0 + \hat{R}_1(\omega), \tag{10.21}$$
$$x_0 \geq 0 \text{ and integer-valued}, \tag{10.22}$$

and the subsequent stage constraints, for $t = 1, \ldots, T - 1$ and $\omega \in \Omega$:

$$A_t x_t(\omega) = R_t(\omega), \tag{10.23}$$
$$U_t x_t(\omega) \leq u_t, \tag{10.24}$$
$$R_{t+1}(\omega) = \Delta_t x_t(\omega) + \hat{R}_{t+1}(\omega), \tag{10.25}$$
$$x_t(\omega) \geq 0 \text{ and integer-valued}. \tag{10.26}$$

This formulation allows us to make a different set of decisions for each outcome, which means we are allowing a decision to "see" into the future. To prevent this behavior, we incorporate an additional constraint for each $t = 1, \ldots, T - 1$ and pair of scenarios $\omega, \omega' \in \Omega$ for which $(\omega_1, \ldots, \omega_t) = (\omega'_1, \ldots, \omega'_t)$:

$$x_t(\omega) = x_t(\omega'). \tag{10.27}$$

The constraints (10.27) are called *nonanticipativity constraints*.

We note that there is no explicit use of a state variable. This approach has proven useful in the financial services sector where random processes (e.g., interest rates, currency fluctuations, stock prices) can be correlated in a very complex way. This technique is widely used in financial asset allocation problems for designing low-risk portfolios ([18, 21]), and a variety of specialized algorithms have been designed to help solve these problems ([17, 19, 20]).

The optimization problem represented by Eqs. (10.18)–(10.27) requires determining all decisions over all scenarios (over all time periods) at the same time. Not surprisingly, the resulting optimization problem is *much* larger than its deterministic counterpart. In addition, the formulation typically destroys integrality properties that might exist in the original problem. The solution of stochastic integer programs is an emerging area of research (see [31] for a thorough review of this field). This approach is generally not computationally tractable in the context of large scale resource allocation.

10.5.2.2 *Benders Decomposition*   An alternative strategy, which exploits the structure of linear programs, is to replace the second term on the right hand side of Eq. (10.18) with a series of cuts that captures the structure of this function. We discuss this approach in the context of a two-stage problem — that is, the case of $T = 2$. The idea is to iteratively solve a sequence of "master problems" of the form:

$$\max_{x_0, z} C_0 x_0 + z, \tag{10.28}$$

subject to the first stage constraints (10.19)–(10.22) and the constraints:

$$z - \beta_i x_0 \leq \alpha_i, \ \forall \ i = 1, ..., n, \tag{10.29}$$

where β_i and α_i are generated by solving the dual of a second stage problem:

$$\max_{x_1(\omega)} C_1 x_1(\omega), \tag{10.30}$$

subject to constraints (10.23)–(10.26). The second stage is solved for a single $\omega \in \Omega$, so the problem is no larger than a one-period deterministic problem. After solving the second stage problem for a single scenario, the dual information is used to update the cuts in Eq. (10.29). There are different strategies for creating and updating cuts. The "L-shaped decomposition algorithm" ([42]) solves two-stage problems with a finite set of scenarios. In one of the major breakthroughs in the field, [13] describes the "stochastic decomposition" algorithm that generalizes the L-shaped algorithm for problems with an infinite number of scenarios. [30] and [6] present variations of Benders for multistage problems. Figure 10.1 provides a sketch of the CUPPS algorithm in [6] which is basically a generalization of the L-shaped algorithm for multistage problems.

The appeal of Benders decomposition is seen by simply comparing the problem of solving Eq. (10.28) along with Eq. (10.30) to the challenge of solving Eq. (10.18). Benders decomposition exploits the structure of a linear program, however the rate of convergence remains an open question. [27], for example, found that Benders-based algorithms could require hundreds of iterations before it outperformed

Step 1. Solve the following *master problem*:

$$x_0^n \in \arg\max\{C_0 x + z : A_0 x_0 = R_0, \; z - \beta_k x \le \alpha_k, \; k = 1, \dots, n-1, \; x \ge 0\}$$

Step 2. Sample $\omega^n \in \Omega$ and solve the following dual *subproblem*:

$$v(x_0^n, \omega^n) \in \arg\min\{(\hat{R}_1(\omega^n) \mid \Delta_0 x_0^n)^T v : \Lambda_1^T v \ge C_1\}$$

Augment the set of dual vertices by:

$$\mathcal{V}^n = \mathcal{V}^{n-1} \bigcup \{v(x_0^n, \omega^n)\}$$

Step 3. Set:

$$v^n(\omega) \in \arg\min\{(\hat{R}_1(\omega) + \Delta_0 x_0^n) v : v \in \mathcal{V}^n\} \text{ for all } \omega \in \Omega$$

Step 4. Construct the coefficients of the nth cut to be added to the master problem by:

$$\alpha_n + \beta_n x_0 \equiv \sum_{\omega \in \Omega} p(\omega) \left(\hat{R}_1(\omega) + \Delta_0 x_0 \right)^T v^n(\omega)$$

Fig. 10.1   Sketch of the CUPPS algorithm

a simple deterministic approximation even for relatively small problems (the rate of convergence slows as the problem size increases).

Benders can be used for multistage problems by simply stepping forward through time, a technique that is sometimes referred to as nested Benders decomposition (see [23] for an application in an energy setting and [2] for an application to reservoir optimization). The basic strategy is identical to that used in approximate dynamic programming.

For an analysis of the convergence of these techniques see [6] and [30].

10.6   APPROXIMATE DYNAMIC PROGRAMMING

The algorithms we have presented emerged from a thread of research that grew to a large extent independently from the broader approximate dynamic programming literature — which includes, for example, work under the guise of "neuro-dynamic programming," "reinforcement learning," and "heuristic dynamic programming." We refer the reader to [4, 35, 40] for surveys of this literature. There are a number of ideas in common between the algorithm we have presented and those proposed in this approximate dynamic programming literature. In this section, we discuss these similarities and also some notable differences.

Like algorithms from the broader approximate dynamic literature, the algorithm presented in Section 10.4 approximates a dynamic programming value function for use in decision-making. Another striking commonality is the use of simulated state trajectories and stochastic approximation updates. In this respect, the algorithm of Section 10.4 is closely related to temporal-difference learning ([34]), Q-learning

([44] and [45]), and SARSA ([29]). Some work in the reinforcement learning literature highlights the importance of simulated state trajectories toward reducing approximation error ([36] and [39]). Perhaps this phenomenon has also contributed to the empirical success in transportation of algorithms like that of Section 10.4.

The formulation of dynamic programs around the pre-decision state is the most standard treatment in textbooks, though even Bellman's original treatment of dynamic programming ([3], p. 142) included an example of a dynamic programming recursion in terms of a post-decision state. In our context, such a reformulation of the dynamic programming recursion becomes necessary to enable computation of unbiased estimates of the right hand side via simulation. Such a reformulation is employed for the same reason by Q-learning and SARSA. One interesting difference, though, is that the post-decision state used by Q-learning and SARSA is taken to be the state-action pair — in our context, this would be (R_t, x_t). The algorithm we have presented, on the other hand, makes use of a more parsimonious representation $R_t^x = R_t + \Delta_t x_t$ of the post-decision state. This is possible because of special structure associated with our problem formulation, which makes R_t^x a sufficient statistic for predicting future contribution. Since Q-learning is designed for more general Markov decision processes that do not exhibit such structure, it can not make use of such a sufficient statistic. The idea of using a parsimonious sufficient statistic has also been proposed in an application of approximate dynamic programming to inventory management ([41]). The use of the post-decision state variable is implicit in the work of [24] and [7] for a stochastic fleet management problem, and is explicit in [10] for the same problem class, and [22] and [33] in other problem classes.

Another significant difference in the algorithm of Section 10.4 is that the updates adapt the derivative of the value function with respect to state, whereas most algorithms in the approximate dynamic programming literature adapt the values themselves. Clearly, it is the derivative of the value function that matters when it comes to decision-making, since the value function is only used to compare relative values among states. An interesting issue to explore may be whether there are fundamental advantages to adapting derivatives rather than values. Though the idea of adapting derivatives has not been a focus of the approximate dynamic programming literature, it has received some attention. For example, this idea has been promoted by Werbos, who proposed a number of algorithms with this objective in mind ([46] and [47]).

An important departure from the broader approximate dynamic programming literature is in the use of integer programming methods to solve the decision optimization problem

$$\max_{x \in \mathcal{X}_t(R_t)} \left(C_{t+1} x + \bar{V}^x(\Delta_{t+1} x, v_{t+1}) \right).$$

This approach has been successfully applied to problems involving thousands of decision variables. Though there is some work in the approximate dynamic programming literature on approaches for dealing with high-dimensional decision spaces (see, e.g., [4, 8, 12]), to our knowledge, no other applications of approximate dynamic programming have dealt with such large numbers of decision variables.

The use of separable linear and concave approximations facilitate the use of integer programming methods to optimize decisions. Though separable approximators have been used regularly in the approximate dynamic programming literature (see, e.g., [36]), there has not been a practice of restricting to linear or concave functions, and their use has not been motivated by a need to structure the decision optimization problem. The separable linear or concave structure is important — if more general nonlinear approximations were used this would likely destroy integrality properties of underlying linear programs. The result would be a decision optimization problem that could not be handled effectively by integer programming methods.

Finally, it is worth noting that the scale of problems tackled by the first author and collaborators far exceeds most other applications of approximate dynamic programming reported in the literature. In particular, there are typically thousands of state variables and thousands of decision variables. For problems of such enormous scale, it becomes important to use methods for which compute time and memory requirements grow slowly with the number of state variables. In this respect, separable approximation architectures are advantageous because the number of parameters involved grows linearly in the number of state variables.

10.7   EXPERIMENTAL COMPARISONS

The techniques described in this chapter have evolved in the context of solving an array of very large-scale resource allocation problems. One of the challenges that we always face in the field of approximate dynamic programming is evaluating how well our techniques work. Bounds can be useful, but tend to be quite loose in practice.

For our problem class, we can report on two types of experiments. The first focuses on two-stage problems (make a decision in the first stage, see information in the second stage, make a second decision, stop), for which computationally tractable, provably convergent algorithms already exist using the principle of Benders decomposition (which we reviewed in Section 10.5.2.2). The second set of experiments looks at multiperiod problems, and compare against rolling horizon approximations that use deterministic forecasts of the future.

Figure 10.2 shows comparisons of three variations of Benders against a variation of the concave, separable approximation method (described in Section 10.4.3) called the SPAR algorithm (see [27]). The algorithms were run on four sets of problems of increasing size (10, 25, 50 and 100 locations). The execution times are comparable (although the "L-shaped" algorithm is much slower than the others). But the results suggest that Benders exhibits higher errors after the same number of iterations, and that the error increases dramatically with problem size, whereas the SPAR algorithm actually improves. This is capturing the observation that in practice, large problems are reasonably approximated using separable approximations.

Benders decomposition can be run on multiperiod problems as well, but the rate of convergence appears to become prohibitively slow. The best competition comes from the standard engineering practice of solving sequences of deterministic approx-

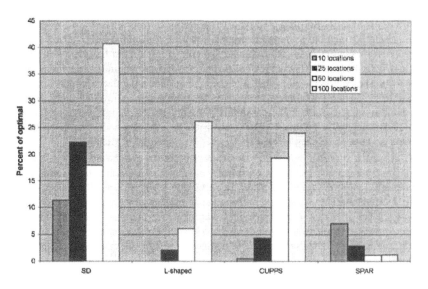

Fig. 10.2   The quality of the solution produced by variations of Benders and separable, concave approximations (SPAR) expressed as a percent of the optimal solution found using L-shaped decomposition. Based on results from [27].

imations using a rolling horizon procedure. [38] reports on comparisons using linear and piecewise linear, concave value function approximations against a deterministic, rolling horizon procedure. Table 10.1 summarizes one set of experiments that considered problems with 10, 20 and 40 locations. Each cell in this table gives the average objective function for each method divided by the average of all the posterior bounds obtained by calculating the (outcome-dependent) optimal solutions for each of the 100 sample realizations tested. This set of experiments focused on an application with different equipment types (such as box cars) which can be described purely based on the type of equipment and their location. It is possible to substitute different equipment types for the same order (at a cost), as well as substituting equipment of the same type from different locations. Separable concave value functions are estimated for each location and equipment type, which would suggest a significant potential for error. Just the same, separable concave approximations work quite well against the linear approximations (which works the worst) and the rolling horizon procedure.

As of this writing, we do not know exactly how well these methods work on general, multistage problems given the lack of effective, computationally tractable competing algorithms. There are many unanswered questions, but there has been enough progress to suggest that this is a promising line of investigation for this problem class.

Table 10.1 Comparisons of a deterministic rolling horizon procedure against value function approximations using linear and separable concave architectures, from [38]. Each cell gives the average objective function value normalized by the average of the posterior bounds for each sample realization.

Locations	Approx.	Mean	S.D.	Percentiles		
				5th	50th	95th
	Linear	84.55	2.339	80.94	84.45	88.43
10	Separable Concave	95.18	2.409	91.55	95.46	99.38
	Rolling Horizon	91.45	2.348	87.96	91.97	95.74
	Linear	80.52	2.463	76.91	80.45	85.25
20	Separable Concave	95.48	2.153	91.91	95.33	98.96
	Rolling Horizon	88.91	1.930	85.68	88.52	91.91
	Linear	74.13	1.816	72.02	74.14	77.56
40	Separable Concave	92.21	0.465	91.55	92.89	93.22
	Rolling Horizon	86.89	0.772	85.08	86.77	87.11

10.8 CONCLUSION

Dynamic resource allocation problems typically lead to very high-dimensional stochastic dynamic programs. We have presented approximation algorithms representative of some that have been successful in applications involving transportation and logistics. The algorithms incorporate several key ideas that address the three curses of dimensionality. Among these ideas are a reformulation of the dynamic programming recursion around a post-decision state variable, simulation-based stochastic approximation techniques, and continuous approximation architectures with functional forms that facilitate integer programming solutions to decision optimization sub-problems.

The approach has worked well for problems with thousands of discrete resources with tens of thousands of different attributes (in effect, a state vector with tens of thousands of dimensions). Furthermore, the rate of convergence of the learning of the value function approximations appears to be reasonably fast. In some applications, we found that the learning stabilizes within 20 to 40 iterations. This is important, since large problems can require as much as an hour per iteration.

Acknowledgments

This research of the first author was supported in part by grant AFOSR-F49620-93-1-0098 from the Air Force Office of Scientific Research. The second author was supported in part by the NSF under CAREER Grant ECS-9985229 and by the ONR under grant MURI N00014-00-1-0637.

Bibliography

1. J. Alden and R. Smith, Rolling horizon procedures in nonhomogeneous Markov decision processes, *Operations Research,* vol. 40, no. 2, pp. S183–S194, 1992.

2. T. W. Archibald, C. S. Buchanan, K. I. M. McKinnon, and L. C. Thomas, Nested benders decomposition and dynamic programming for reservoir optimisation, *Journal of the Operational Research Society,* vol. 50, pp. 468–479, 1999.

3. R. Bellman, *Dynamic Programming,* Princeton University Press, Princeton, NJ, 1957.

4. D. Bertsekas and J. Tsitsiklis, *Neuro-Dynamic Programming,* Athena Scientific, Belmont, MA, 1996.

5. J. Birge and F. Louveaux, *Introduction to Stochastic Programming,* Springer-Verlag, New York, 1997.

6. Z.-L. Chen and W. Powell, A convergent cutting-plane and partial-sampling algorithm for multistage linear programs with recourse, *Journal of Optimization Theory and Applications,* vol. 103, no. 3, pp. 497–524, 1999.

7. R. Cheung and W. B. Powell, An algorithm for multistage dynamic networks with random arc capacities, with an application to dynamic fleet management, *Operations Research,* vol. 44, no. 6, pp. 951–963, 1996.

8. R. Crites and A. Barto, Elevator group control using multiple reinforcement learning agents, Technical Report, 1994.

9. G. Dantzig and A. Ferguson, The allocation of aircrafts to routes: An example of linear programming under uncertain demand, *Management Science,* vol. 3, pp. 45–73, 1956.

10. G. Godfrey and W. B. Powell, An adaptive, dynamic programming algorithm for stochastic resource allocation problems I: Single period travel times, *Transportation Science,* vol. 36, no. 1, pp. 21–39, 2002.

11. G. A. Godfrey and W. B. Powell, An adaptive, distribution-free approximation for the newsvendor problem with censored demands, with applications to inventory

and distribution problems, *Management Science*, vol. 47, no. 8, pp. 1101–1112, 2001.

12. C. Guestrin, D. Koller, and R. Parr, Efficient solution algorithms for factored MDPs, Technical Report, 2003.

13. J. Higle and S. Sen, Stochastic decomposition: An algorithm for two stage linear programs with recourse, *Mathematics of Operations Research*, vol. 16, no. 3, pp. 650–669, 1991.

14. G. Infanger, *Planning under Uncertainty: Solving Large-scale Stochastic Linear Programs*, The Scientific Press Series, Boyd & Fraser, New York, 1994.

15. P. Kall and S. Wallace, *Stochastic Programming*, Wiley, New York, 1994.

16. A. Kleywegt and J. Papastavrou, Acceptance and dispatching policies for a distribution problem, *Transporation Science*, vol. 32, no. 2, pp. 127–141, 1998.

17. I. Lustig, J. Lustig, J. M. Mulvey, and T. Carpenter, Formulating stochastic programs for interior point methods, *Operations Research*, vol. 39, pp. 757–770, 1991.

18. J. M. Mulvey, Introduction to financial optimization: Mathematical Programming special issue, *Mathematical Programming: Series B*, vol. 89, pp. 205–216, 2001.

19. J. M. Mulvey and A. Ruszczynski, A diagonal quadratic approximation method for large scale linear programs, *Operations Research Letters*, vol. 12, pp. 205–215, 1991.

20. J. M. Mulvey and A. J. Ruszczynski, A new scenario decomposition method for large-scale stochastic optimization, *Operations Research*, vol. 43, no. 3, pp. 477–490, 1995.

21. J. M. Mulvey and H. Vladimirou, Stochastic network programming for financial planning problems, *Management Science*, vol. 38, no. 8, pp. 1642–1664, 1992.

22. K. Papadaki and W. B. Powell, A monotone adaptive dynamic programming algorithm for a stochastic batch service problem, *European Journal of Operational Research*, vol. 142, no. 1, pp. 108–127, 2002.

23. M. Pereira and L. Pinto, Multistage stochastic optimization applied to energy planning, *Mathematical Programming*, vol. 52, pp. 359–375, 1991.

24. W. B. Powell, A comparative review of alternative algorithms for the dynamic vehicle allocation problem, in B. Golden and A. Assad, (eds.), *Vehicle Routing: Methods and Studies*, North Holland, Amsterdam, pp. 249–292, 1988.

25. W. B. Powell, A stochastic formulation of the dynamic assignment problem, with an application to truckload motor carriers, *Transportation Science*, vol. 30, no. 3, pp. 195–219, 1996.

26. W. B. Powell and H. Topaloglu, Stochastic programming in transportation and logistics, in A. Ruszczynski and A. Shapiro, (eds.) *Handbook in Operations Research and Management Science*, Volume on *Stochastic Programming*, North Holland, Amsterdam, 2003.

27. W. B. Powell, A. Ruszczynski, and H. Topaloglu, Learning algorithms for separable approximations of stochastic optimization problems, Technical Report, Princeton University, Department of Operations Research and Financial Engineering, 2002a.

28. W. B. Powell, J. A. Shapiro, and H. P. Simão, An adaptive dynamic programming algorithm for the heterogeneous resource allocation problem, *Transportation Science*, vol. 36, no. 2, pp. 231–249, 2002b.

29. G. Rummery and M. Niranjan, On-line Q-learning using connectionist systems, Technical Report, Cambridge University Engineering Department, Technical Report CUED/F-INFENG/TR166, 1994.

30. A. Ruszczynski, Parallel decomposition of multistage stochastic programming problems, *Math. Programming*, vol. 58, no. 2, pp. 201–228, 1993.

31. S. Sen, Algorithms for stochastic, mixed-integer programs, in K. Aardal, G. L. Nemhauser, and R. Weismantel, (eds.), *Handbook in Operations Research and Management Science*, Volume on *Discrete Optimization*, North Holland, Amsterdam, 2004.

32. S. Sen and J. Higle, An introductory tutorial on stochastic linear programming models, *Interfaces*, vol. 29, no. 2, pp. 33–61, 1999.

33. M. Spivey and W. B. Powell, The dynamic assignment problem, *Transportation Science*, (to appear).

34. R. Sutton, Learning to predict by the methods of temporal differences, *Machine Learning*, vol. 3, pp. 9–44, 1988.

35. R. Sutton and A. Barto, *Reinforcement Learning*, MIT Press, Cambridge, MA, 1998.

36. R. S. Sutton, Generalization in reinforcement learning: Successful examples using sparse coarse coding, in M. E. H. D. S. Touretzky and M. C. Mozer, (eds.), *Advances in Neural Information Processing Systems*, vol. 19, pp. 1038–1044, 1996.

37. H. Topaloglu and W. B. Powell, An algorithm for approximating piecewise linear concave functions from sample gradients, *Operations Research Letters*, vol. 31, no. 1, pp. 66–76, 2003.

38. H. Topaloglu and W. B. Powell, Dynamic programming approximations for stochastic, time-staged integer multicommodity flow problems, Technical Report, Cornell University, Department of Operations Research and Industrial Engineering, 2003*b*.

39. J. Tsitsiklis and B. Van Roy, An analysis of temporal-difference learning with function approximation, *IEEE Trans. on Automatic Control,* vol. 42, pp. 674–690, 1997.

40. B. Van Roy, Neuro-dynamic programming: Overview and recent trends, in E. Feinberg and A. Shwartz, (eds.), *Handbook of Markov Decision Processes: Methods and Applications,* Kluwer, Boston, 2001.

41. B. Van Roy, D. P. Bertsekas, Y. Lee, and J. N. Tsitsiklis, A neuro-dynamic programming approach to retailer inventory management, *Proc. of the IEEE Conference on Decision and Control,* San Diego, CA, 1997.

42. R. Van Slyke and R. Wets, L-shaped linear programs with applications to optimal control and stochastic programming, *SIAM Journal of Applied Mathematics,* vol. 17, no. 4, pp. 638–663, 1969.

43. D. Walkup and R. Wets, Stochastic programs with recourse, *SIAM Journal of Applied Mathematics,* vol. 15, pp. 1299–1314, 1967.

44. C. Watkins, *Learning from Delayed Rewards,* Ph.D. Thesis, Cambridge University, Cambridge, UK, 1989.

45. C. Watkins and P. Dayan, *Q*-learning, *Machine Learning,* vol. 8, pp. 279–292, 1992.

46. P. J. Werbos, Approximate dynamic programming for real-time control and neural modelling, in D. A. White and D. A. Sofge, (eds.), *Handbook of Intelligent Control: Neural, Fuzzy, and Adaptive Approaches,* Van Nostrand Reinhold, New York, 1992*a*.

47. P. J. Werbos, Neurocontrol and supervised learning: an overview and valuation, in D. A. White and D. A. Sofge, (eds.), *Handbook of Intelligent Control: Neural, Fuzzy, and Adaptive Approaches,* Van Nostrand Reinhold, New York, 1992*b*.

48. R. Wets, Programming under uncertainty: The equivalent convex program, *SIAM Journal of Applied Mathematics,* vol. 14, pp. 89–105, 1966*a*.

49. R. Wets , Programming under uncertainty: The solution set, *SIAM Journal of Applied Mathematics,* vol. 14, pp. 1143–1151, 1966*b*.

50. R. Wets, Stochastic programs with fixed recourse: The equivalent deterministic problem, *SIAM Review,* vol. 16, pp. 309–339, 1974.

Handbook of Learning and Approximate Dynamic Programming
Edited by Jennie Si, Andy Barto, Warren Powell and Donald Wunsch
Copyright © 2004 The Institute of Electrical and Electronics Engineers, Inc.

11 Hierarchical Approaches to Concurrency, Multiagency, and Partial Observability

SRIDHAR MAHADEVAN
MOHAMMAD GHAVAMZADEH   GEORGIOS THEOCHAROUS
KHASHAYAR ROHANIMANESH
University of Massachusetts   MIT A. I. Laboratory

Editor's Summary: In this chapter the authors summarize their research in hierarchical probabilistic models for decision making involving concurrent action, multiagent coordination, and hidden state estimation in stochastic environments. A hierarchical model for learning concurrent plans is first described for observable single agent domains, which combines compact state representations with temporal process abstractions to determine how to parallelize multiple threads of activity. A hierarchical model for multiagent coordination is then presented, where primitive joint actions and joint states are hidden. Here, high-level coordination is learned by exploiting overall task structure, which greatly speeds up convergence by abstracting from low-level steps that do not need to be synchronized. Finally, a hierarchical framework for hidden state estimation and action is presented, based on multi-resolution statistical modeling of the past history of observations and actions.

11.1   INTRODUCTION

Despite five decades of research on models of decision-making, artificial systems remain significantly below human level performance in many natural problems, such as driving. Perception is not the only challenge in driving; at least as difficult are the challenges of executing and monitoring concurrent (or parallel) activities, dealing with limited observations, and finally modeling the behavior of other drivers on the road. We address these latter challenges in this chapter. To date, a general framework that jointly addresses concurrency, multiagent coordination, and hidden state estimation has yet to be fully developed. Yet, many everyday human activities, such as driving (see Figure 11.1), involve simultaneously grappling with these and other challenges.

Humans learn to carry out multiple concurrent activities at many abstraction levels, when acting alone or in concert with other humans. Figure 11.1 illustrates a familiar everyday example, where drivers learn to observe road signs and control steering, but also manage to engage in other activities such as operating a radio, or carrying on a cellphone conversation. Concurrent planning and coordination is also essential to many important engineering problems, such as flexible manufacturing with a team of machines to scheduling robots to transport parts around factories. All these tasks involve a hard computational problem: how to sequence multiple overlapping and interacting parallel activities to accomplish long-term goals. The problem is difficult to solve in general, since it requires learning a mapping from noisy incomplete perceptions to multiple temporally extended decisions with uncertain outcomes. It is indeed a miracle that humans are able to solve problems such as driving with relatively little effort.

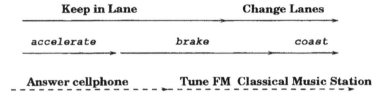

Fig. 11.1   Driving is one of many human activities illustrating the three principal challenges addressed in this chapter: concurrency, multiagency, and partial observability. To drive successfully, humans execute multiple parallel activities,while coordinating with actions taken by other drivers on the road, and use memory to deal with their limited perceptual abilities.

In this chapter, we summarize our current research on a hierarchical approach to concurrent planning and coordination in stochastic single agent and multiagent environments. The overarching theme is that efficient solutions to these challenges can be developed by exploiting multi-level temporal and spatial abstraction of actions and states. The framework will be elaborated in three parts. First, a hierarchical model for learning concurrent plans is presented, where, for simplicity, it is assumed that agents act alone, and can fully observe the state of the underlying process. The key idea here is that by combining compact state representations with temporal process abstractions, agents can learn to parallelize multiple threads of activity. Next, a hierarchical model for multiagent coordination is described, where primitive joint actions and joint states may be hidden. This partial observability of lower-level actions is indeed a blessing, since it allows agents to speedup convergence by abstracting from low-level steps that do not need to be synchronized. Finally, we present a hierarchical approach to state estimation, based on multi-resolution statistical modeling of the past history of observations and actions.

The proposed approaches all build on a common *Markov decision process* (MDP) modeling paradigm, which is summarized in the next section. Previous MDP-

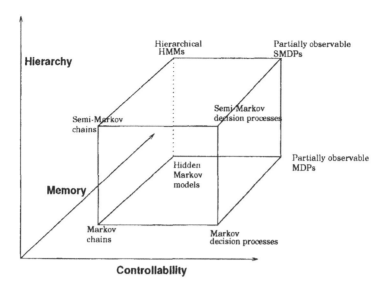

Fig. 11.2 A spectrum of Markov process models along several dimensions: whether agents have a choice of action, whether states are observable or hidden, and whether actions are unit-time (single-step) or time-varying (multi-step).

based algorithms have largely focused on sequential compositions of closed-loop programs. Also, earlier MDP-based approaches to learning multiagent coordination ignored hierarchical task structure, resulting in slow convergence. Previous finite memory and partially observable MDP-based methods for state estimation used flat representations, which scale poorly to long experience chains and large state spaces. The algorithms summarized in this chapter address these limitations in previous work, by using new spatiotemporal abstraction based approaches for learning concurrent closed-loop programs and abstract task-level coordination, in the presence of significant perceptual limitations.

11.2   BACKGROUND

Probabilistic finite state machines have become a popular paradigm for modeling sequential processes. In this representation, the interaction between an agent and its environment is represented as a finite automata, whose *states* partition the past history of the interaction into equivalence classes, and whose *actions* cause (probabilistic) transitions between states. Here, a state is a *sufficient statistic* for computing optimal (or best) actions, meaning past history leading to the state can be abstracted. This assumption is usually referred to as the *Markov* property.

Markov processes have become the mathematical foundation for much current work in reinforcement learning [36], decision-theoretic planning [2], information retrieval [8], speech recognition [11], active vision [22], and robot navigation [14]. In this chapter, we are interested in abstracting sequential Markov processes using two strategies: state aggregation/decomposition and temporal abstraction. State decomposition methods typically represent states as collections of *factored* variables [2], or simplify the automaton by eliminating "useless" states [4]. Temporal abstraction mechanisms, for example in hierarchical reinforcement learning [6, 25, 37], encapsulate lower-level observation or action sequences into a single unit at more abstract levels. For a unified algebraic treatment of abstraction of Markov decision processes that covers both spatial and temporal abstraction, the reader is referred to [29].

Figure 11.2 illustrates eight Markov process models, arranged in a cube whose axes represent significant dimensions along which the models differ from each other. While a detailed description of each model is beyond the scope of this chapter, we will provide brief descriptions of many of these models below, beginning in this section with the basic MDP model.

A *Markov decision process* (MDP) [28] is specified by a set of states S, a set of allowable actions $A(s)$ in each state s, and a transition function specifying the next-state distribution $P_{ss'}^a$ for each action $a \in A(s)$. A reward or cost function $r(s, a)$ specifies the *expected* reward for carrying out action a in state s. Solving a given MDP requires finding an optimal mapping or *policy* $\pi^* : S \rightarrow A$ that maximizes the long-term cumulative sum of rewards (usually discounted by some factor $\gamma < 1$) or the expected average-reward per step. A classic result is that for any MDP, there exists a stationary deterministic optimal policy, which can be found by solving a nonlinear set of equations, one for each state (such as by a successive approximation method called *value iteration*):

$$V^*(s) = \max_{a \in A(s)} \left(r(s, a) + \gamma \sum_{s'} P_{ss'}^a V^*(s') \right). \qquad (11.1)$$

MDPs have been applied to many real-world domains, ranging from robotics [14, 17] to engineering optimization [3, 18], and game playing [39]. In many such domains, the model parameters (rewards, transition probabilities) are unknown, and need to be estimated from samples generated by the agent exploring the environment. Q-learning was a major advance in direct policy learning, since it obviates the need for model estimation [45]. Here, the Bellman optimality equation is reformulated using *action values* $Q^*(x, a)$, which represent the value of the non-stationary policy of doing action a once, and thereafter acting optimally. Q-learning eventually finds the optimal policy asymptotically. However, much work is required in scaling Q-learning to large problems, and abstraction is one of the key components. Factored approaches to representing value functions may also be key to scaling to large problems [15].

11.3   SPATIOTEMPORAL ABSTRACTION OF MARKOV PROCESSES

We now discuss strategies for hierarchical abstraction of Markov processes, including temporal abstraction, and spatial abstraction techniques.

11.3.1   Semi-Markov Decision Processes

Hierarchical decision-making models require the ability to represent lower-level policies over primitive actions as primitive actions at the next level (e.g., in a robot navigation task, a "go forward" action might itself be composed of a lower-level actions for moving through a corridor to the end, while avoiding obstacles). Policies over primitive actions are "semi-Markov" at the next level up, and cannot be simply treated as single-step actions over a coarser time scale over the same states.

Semi-Markov decision processes (SMDPs) have become the preferred language for modeling temporally extended actions (for an extended review of SMDPs and hierarchical action models, see [1]). Unlike Markov decision processes (MDPs), the time between transitions may be several time units and can depend on the transition that is made. An SMDP is defined as a five tuple (S,A,P,R,F), where S is a finite set of states, A is the set of actions, P is a state transition matrix defining the single-step transition probability of the effect of each action, and R is the reward function. For continuous-time SMDPs, F is a function giving probability of transition times for each state-action pair until *natural termination*. The transitions are at decision epochs only. The SMDP represents snapshots of the system at decision points, whereas the so-called *natural process* [28] describes the evolution of the system over all times. For discrete-time SMDPs, the transition distribution is written as $F(s', N|s, a)$, which specifies the expected number of steps N that action a will take before terminating (naturally) in state s' starting in state s. For continuous-time SMDPs, $F(t|s, a)$ is the probability that the next decision epoch occurs within t time units after the agent chooses action a in state s at a decision epoch.

Q-learning generalizes nicely to discrete and continuous-time SMDPs. The Q-learning rule for discrete-time discounted SMDPs is

$$Q_{t+1}(s,a) \leftarrow Q_t(s,a)(1-\beta) + \beta \left(R + \gamma^k \max_{a' \in A(s')} Q_t(s',a') \right),$$

where $\beta \in (0,1)$, and action a was initiated in state s, lasted for k steps, and terminated in state s', while generating a total discounted sum of rewards of R.

Several frameworks for hierarchical reinforcement learning have been proposed, all of which are variants of SMDPs, including options [37], MAXQ [6], and HAMs [25]. We discuss some of these in more detail in the next section.

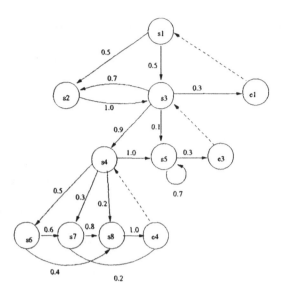

Fig. 11.3 An example hierarchical hidden Markov model. Only leaf nodes produce observations. Internal nodes can be viewed as generating sequences of observations.

11.3.2   Hierarchical Hidden Markov Models

Hidden Markov models (HMMs) are a widely-used probabilistic model for representing time-series data, such as speech [11]. Unlike an MDP, states are not perceivable, and instead the agent receives an observation o which can be viewed as being generated by a stochastic process $P(o|s)$ as a function of the underlying state s. HMMs have been widely applied to many time-series problems, ranging from speech recognition [11], information extraction [8], and bioinformatics [12]. However, like MDPs, HMMs do not provide any direct way of representing higher-level structure that is often present in many practical problems. For example, an HMM can be used as a spatial representation of indoor environments [34], but typically such environments have higher order structures such as corridors or floors which are not made explicit in the underlying HMM model. As in the case with MDPs, in most practical problems, the parameters of the underlying HMM have to be learned from samples. The most popular method for learning an HMM model is the Baum-Welch procedure, which is itself a special case of the more general Expectation-Maximization (EM) statistical inference algorithm.

Recently, an elegant hierarchical extension of HMMs was proposed [7]. The HHMM generalizes the standard hidden Markov model by allowing hidden states to represent stochastic processes themselves. An HHMM is visualized as a tree structure (see Figure 11.3) in which there are three types of states, production states (leaves of the tree) which emit observations, and internal states which are (unobservable) hidden states that represent entire stochastic processes. Each production state is associated with an observation vector which maintains distribution functions for each observation defined for the model. Each internal state is associated with a horizontal transition matrix, and a vertical transition vector. The horizontal transition matrix of an internal state defines the transition probabilities among its children. The vertical transition vectors define the probability of an internal state to activate any of its children. Each internal state is also associated with a child called an *end-state* which returns control to its parent. The end-states ($e1$ to $e4$ in Figure 11.3) do not produce observations and cannot be activated through a vertical transition from their parent.

Figure 11.3 shows a graphical representation of an example HHMM. The HHMM produces observations as follows:

1. If the current node is the root, then it chooses to activate one of its children according to the vertical transition vector from the root to its children.

2. If the child activated is a product state, it produces an observation according to an observation probability output vector. It then transitions to another state within the same level. If the state reached after the transition is the end-state, then control is returned to the parent of the end-state.

3. If the child is an abstract state then it chooses to activate one of its children. The abstract state waits until control is returned to it from its child end-state. Then it transitions to another state within the same level. If the resulting transition is to the end-state then control is returned to the parent of the abstract state.

The basic inference algorithm for hierarchical HMMs is a modification of the "inside-outside" algorithm for stochastic context-free grammars, and runs in $O(T^3)$, where T is the length of the observation sequence. Recently, Murphy developed a faster inference algorithm for hierarchical HMMs by mapping them onto a dynamic Bayes network [23].

11.3.3   Factored Markov Processes

In many domains, states are composed of collections of objects, each of which can be modeled as a multinomial or real-valued variable. For example, in driving, the state of the car might include the position of the accelerator and brake, the radio, the wheel angle, and so on. Here, we assume the agent-environment interaction can be modeled as a factored semi-Markov decision process, in which the state space is spanned by the Cartesian product of random variables $X = \{X_1, X_2, ..., X_n\}$, where each X_i takes on values in some finite domain $Dom(X_i)$. Each action is either a primitive (single-step) action or a closed-loop policy over primitive actions.

Dynamic Bayes networks (DBNs) [5] are a popular tool for modeling transitions across factored MDPs. Let X_i^t denote the state variable X_i at time t and X_i^{t+1} the variable at time $t+1$. Also, let A denote the set of underlying primitive actions. Then, for any action $a \in A$, the *Action Network* is specified as a two-layer directed acyclic graph whose nodes are $\{X_1^t, X_2^t, ..., X_n^t, X_1^{t+1}, X_2^{t+1}, ..., X_n^{t+1}\}$ and each node X_i^{t+1} is associated with a *conditional probability table (CPT)* $P(X_i^{t+1}|\phi(X_i^{t+1}), a)$ in which $\phi(X_i^{t+1})$ denotes the parents of X_i^{t+1} in the graph. The transition probability $P(X^{t+1}|X^t, a)$ is then defined by: $P(X^{t+1}|X^t, a) = \prod_i^n P(X_i^{t+1}|w_i, a)$ where w_i is a vector whose elements are the values of the $X_j^t \in \phi(X_i^{t+1})$.

Figure 11.4 shows a popular toy problem called the Taxi Problem [6] in which a taxi inhabits a 7-by-7 grid world. This is an episodic problem in which the taxi (with maximum fuel capacity of 18 units) is placed at the beginning of each episode in a randomly selected location with a randomly selected amount of fuel (ranging from 8 to 15 units). A passenger arrives randomly in one of the four locations marked as R(ed), G(reen), B(lue), and Y(ellow) and will select a random destination from these four states. The taxi must go to the location of the passenger (the "source"), pick up the passenger, move to the destination location (the "destination") and put down the passenger there. The episode ends when either the passenger is transported to the desired destination, or the taxi runs out of fuel. Treating each of taxi position, passenger location, destination and fuel level as state variables, we can represent this problem as a factored MDP with four state variables each taking on values as explained above. Figure 11.4 shows a factorial representation of taxi domain for *Pickup* and *Fillup* actions.

While it is relatively straightforward to represent factored MDPs, it is not easy to solve them because in general the solution (i.e., the optimal value function) is not factored. While a detailed discussion of this issue is beyond the scope of this article, a popular strategy is to construct an approximate factored value function as a linear summation of basis functions (see [15]). The use of factored representations is useful not only in finding (approximate) solutions more quickly, but also in learning a factored transition model in less time. For the taxi task illustrated in Figure 11.4, one idea that we have investigated is to express the factored transition probabilities as a mixed memory factorial Markov model [33]. Here, each transition probability (edge in the graph) is represented a weighted mixture of distributions, where the weights can be learned by an expectation maximization algorithm.

More precisely, the action model is represented as a weighted sum of *cross-transition* matrices:

$$P(x_{t+1}^i|X_t, a) = \sum_{j=1}^n \psi_a^i(j)\tau_a^{ij}(x_{t+1}^i|x_t^j), \tag{11.2}$$

where the parameters $\tau_a^{ij}(x'|x)$ are n^2 elementary $k \times k$ transition matrices and parameters $\psi_a^i(j)$ are positive numbers that satisfy $\sum_{j=1}^n \psi_a^i(j) = 1$ for every action $a \in A$ (here, $0 \leq i, j \leq n$, where n is the number of state variables). The number of free parameters in this representation is $O(|A|n^2k^2)$ as opposed to $O(|A|k^{2n})$ in the non-compact case. The parameters $\psi_a^i(j)$ measure the contribution of different state variables in the previous time step to each state variable in the current state. If the problem is completely factored, then $\psi^i(j)$ is the identity matrix whose i^{th} component is independent of the rest. Based on the amount of factorization that exists in an environment, different components of $\psi_a^i(j)$ at one time step will influence the ith component at the next. The cross-transition matrices $\tau_a^{ij}(x'|x)$ provide a compact way to parameterize these influences.

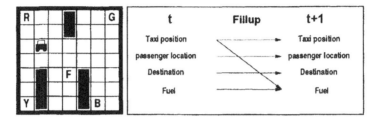

Fig. 11.4 The taxi domain is an instance of a factored Markov process, where actions such as fillup can be represented compactly using dynamic Bayes networks.

Figure 11.5 shows the learning of a factored MDP compared with a table-based MDP, averaged over 10 episodes of 50000 steps. Each point on the graph represents the RMS error between the learned model and the ground truth, averaged over all states and actions. The FMDP model error drops quickly in the early stages of the learning in both problems. Theoretically, the tabular maximum likelihood approach (which estimates each transition probability as the ratio of transitions between two states versus the number of transitions out of a state) will eventually learn the the exact model if every pair of states and action are visited infinitely often. However, the factored approach which uses a mixture weighted representation is able to generalize much more quickly to novel states and overall model learning happens much more quickly.

11.3.4 Structural Decomposition of Markov Processes

Other related techniques for decomposition of large MDPs have been explored, and some of these are illustrated in Figure 11.6. A simple decomposition strategy is to split a large MDP into sub-MDPs, which interact "weakly" [4, 25, 37]. An example of weak interaction is navigation, where the only interaction among sub-MDPs is the states that connect different rooms together. Another strategy is to decompose a large MDP using the set of available actions, such as in air campaign planning problem [21], or in conversational robotics [26]. An even more intriguing decomposition strategy is when sub-MDPs interact with each other through shared parameters. The transfer line optimization problem from manufacturing is a good example of such a parametric decomposition [44].

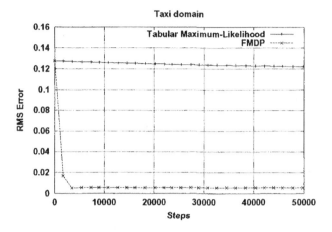

Fig. 11.5   Comparing factored versus tabular model learning performance in the taxi domain.

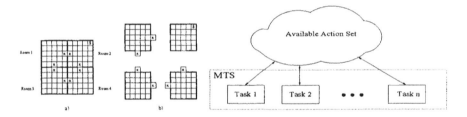

Fig. 11.6   State and action-based decomposition of Markov processes.

11.4   CONCURRENCY, MULTIAGENCY, AND PARTIAL OBSERVABILITY

This section summarizes our recent research on exploiting spatiotemporal abstraction to produce improved solutions to three difficult problems in sequential decision-making: (1) learning plans involving concurrent action, (2) multiagent coordination, and (3) using memory to estimate hidden state.

11.4.1   Hierarchical Framework for Concurrent Action

We now describe a probabilistic model for learning concurrent plans over temporally extended actions [30, 31]. The notion of concurrent action is formalized in a general way, to capture both situations where a single agent can execute multiple parallel processes, as well as the multi-agent case where many agents act in parallel.

The *Concurrent Action Model (CAM)* is defined as (S, A, T, R), where S is a set of states, A is a set of *primary* actions, T is a transition probability distribution $S \times \mathrm{wp}(A) \times S \times \mathbf{N} \to [0, 1]$, where $\mathrm{wp}(A)$ is the power-set of the primary actions and $\mathbf{N}$ is the set of natural numbers, and R is the reward function mapping $S \to \mathfrak{R}$. Here, a concurrent action is simply represented as a set of primary actions (hereafter called a *multi-action*), where each primary action is either a single step action, or a *temporally extended action* (e.g., modeled as a closed loop policy over single step actions [37]).

Figure 11.7 illustrates a toy example of concurrent planning. The general problem is as follows. The agent is given a set of primary actions, each of which can be viewed as a (fixed or previously learned) "subroutine" for choosing actions over a subspace of the overall state space. The goal of the agent is to learn to construct a closed-loop plan (or policy) that allows multiple concurrent subroutines to be executed in parallel (and in sequence) to achieve the task at hand. For multiple primary actions to be executed concurrently, their joint semantics must be well defined. Concurrency is facilitated by assuming states are not atomic, but structured as a collection of (discrete or continuous) variables, and the effect of actions on such sets of variables can be captured by a compact representation, such as a dynamic Bayes net (DBN) [5].

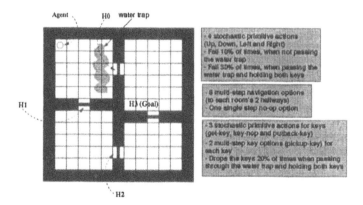

Fig. 11.7 A grid world problem to illustrate concurrent planning: the agent is given subroutines for getting to each door from any interior room state, and for opening a locked door. It has to learn the shortest path to the goal by concurrently combining these subroutines. The agent can reach the goal more quickly if it learns to parallelize the subroutine for retrieving the key before it reaches a locked door. However, retrieving the key too early is counterproductive since it can drop with some probability.

Since multiple concurrent primary actions may not terminate synchronously, the notion of a decision epoch needs to be generalized. For example, a decision epoch can occur when any one of the actions currently running terminates. We refer to this as the T_{any} termination condition (Figure 11.8, left). Alternatively, a decision epoch can be defined to occur when all actions currently running terminate, which we refer to as the T_{all} condition (Figure 11.8, middle). We can design other termination schemes by combining T_{any} and T_{all} : for example, another termination scheme called *continue* is one that always terminates based on the T_{any} termination scheme, but lets those primary actions that did not terminate naturally continue running, while initiating new primary actions if they are going to be useful (Figure 11.8, right).

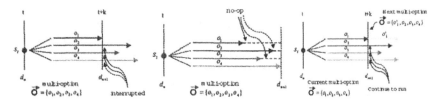

Fig. 11.8 Left: T_{any} termination scheme. Middle: T_{all} termination scheme. Right: $T_{continue}$ termination scheme.

For concreteness, we will describe the concurrent planning framework when the primary actions are represented as *options* [37]. The treatment here is restricted to options over discrete-time SMDPs and deterministic policies, but the main ideas extend readily to other hierarchical formalisms [6, 25] and to continuous-time SMDPs [9, 28]. More formally, an option o consists of three components: a policy $\pi : S \rightarrow A$, a termination condition $\beta : S \rightarrow [0, 1]$, and an initiation set $I \subseteq S$, where I denotes the set of states s in which the option can be initiated. For any state s, if option o is taken, then primitive actions are selected based on o until it terminates according to β. An option o is a *Markov option* if its policy, initiation set and termination condition depend stochastically only on the current state $s \in S$. An option o is *semi-Markov* if its policy, initiation set and termination condition are dependent on all prior history since the option was initiated. For example, the option *exit-room* in the grid world environment shown in Figure 11.7, in which states are the different locations in the room, is a Markov option, since for a given location, the direction to move to get to the door can be computed given the current state.

A hierarchical policy over primary actions or options can be defined as follows. The Markov policy over options $\mu : S \rightarrow O$ (where O is the set of all options) selects an option $o \in O$ at time t using the function $\mu(s_t)$. The option o is then initiated in s_t until it terminates at a random time $t + k$ in some state s_{t+k} according to a given termination condition, and the process repeats in s_{t+k}.

The multistep state transition dynamics over options is defined using the discount factor to weight the probability of transitioning. Let $p^o(s, s', k)$ denote the probability that the option o is initiated in state s and terminates in state s' after k steps. Then $p(s'|s, o) = \sum_{k=1}^{\infty} p^o(s, s', k)\gamma^k$ (note that when $\gamma < 1$, the transition model is not a stochastic matrix, since the distributions do not sum to 1). If multi-step models of options and rewards are known, optimal hierarchical plans can be found by solving a generalized Bellman equation over options similar to Eq. (11.1). Under either definition of the termination event (i.e., T_{any}, T_{all}, and $T_{continue}$), the following result holds.

Theorem 11.4.1 *Given a Markov decision process, and a set of concurrent Markov options defined on it, the decision process that selects only among multi-actions, and executes each one until its termination according to a given termination condition forms a semi-Markov decision process.*

The proof requires showing that the state transition dynamics $p(s', N | \vec{a}, s)$ and the rewards $r(s, \vec{a})$ over any concurrent action $\vec{a}$ defines a semi-Markov decision process [30]. The significance of this result is that SMDP Q-learning methods can be extended to learn concurrent plans under this model. The extended SMDP Q-learning algorithm for learning to plan with concurrent actions updates the multi-action-value function $Q(s, \vec{a})$ after each decision epoch where the multi-action $\vec{a}$ is taken in some state s and terminates in s' (under a specific termination condition):

$$
Q(s, \vec{a}) \leftarrow Q(s, \vec{a})(1 - \beta) + \beta \left[R + \gamma^k \max_{\vec{a}' \in O_{s'}} Q(s', \vec{a}') \right], \qquad (11.3)
$$

where k denotes the number of time steps between initiation of the multi-action $\vec{o}$ in state s and its termination in state s', and R denotes the cumulative discounted reward over this period. The result of using this algorithm on the simple grid world problem in shown in Figure 11.9. The figure illustrates the difference in performance under different termination conditions (T_{all}, T_{any}, and $T_{continue}$).

The performance of the concurrent action model also depends on the termination event defined for that model. Each termination event trades-off between the optimality of the learned plan and how fast it converges to its optimal policy. Let π^{*seq}, π^{*all} and π^{*any} denote the optimal policy when the primary actions are executed sequentially; when termination construct T_{all} is used; and when termination construct T_{any} is used, respectively. Also let $\pi_{continue}$ represent the policy learned based on the $T_{continue}$ termination construct. Intuitively, the models with a termination construct that imposes more frequent multi-action termination (such as T_{any} and $T_{continue}$), tend to *articulate* more frequently and should perform more optimally. However due to more interruption, they may converge more slowly to their optimal behavior. Based on the definition of each termination construct we can prove the following theorem:

Theorem 11.4.2 *In a concurrent action model and a set of termination schemes $\{T_{any}, T_{all}, continue\}$, the following partial ordering holds among the optimal policy based on T_{any}, the optimal policy based on T_{all}, the continue policy and the optimal sequential policy: $\pi^{*seq} \leq \pi^{*all} \leq \pi_{continue} \leq \pi^{*any}$, where $\leq$ denotes the partial ordering relation over policies.*

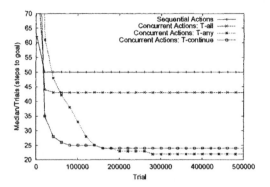

Fig. 11.9 This graph compares an SMDP technique for learning concurrent plans (under various termination conditions) with a slower "get-to-door-then-pickup-key" sequential plan learner. The concurrent learners outperform the sequential learner, but the choice of termination affects the speed and quality of the final plan.

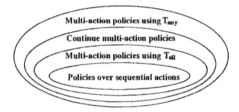

Fig. 11.10 Comparison of policies over multi-actions and sequential primary actions using different termination schemes.

Figure 11.10 illustrates the results defined by Theorem 11.4.2. According to this figure, the optimal multi-action policies based on T_{any} and T_{all}, and also *continue* multi-action policies dominate (with respect to the partial ordering relation defined over policies) the optimal policies over the sequential case. Furthermore, policies based on *continue* multi-actions dominate the optimal multi-action policies based on T_{all} termination scheme, while themselves being dominated by the optimal multi-action policies based on T_{any} termination scheme.

11.4.2   Learning Multiagent Task-Level Coordination Strategies

The second case study uses hierarchical abstraction to design efficient learning algorithms for *cooperative* multiagent systems [46]. Figure 11.11 illustrates a multiagent automated guided vehicle (AGV) scheduling task, where four AGV agents will maximize their performance at the task if they learn to coordinate with each other. The key idea here is that coordination skills are learned more efficiently if agents learn to synchronize using a hierarchical representation of the task structure [35]. In particular, rather than each AGV learning its response to low-level primitive actions of the other AGV agents (for instance, if AGV1 goes forward, what should AGV2 do), they learn high-level coordination knowledge (what is the utility of AGV1 delivering material to machine M3 if AGV2 is delivering assembly from machine M2, and so on). The proposed approach differs significantly from previous work in cooperative multiagent reinforcement learning [3, 38] in using hierarchical task structure to accelerate learning, and as well in its use of concurrent temporally extended actions.

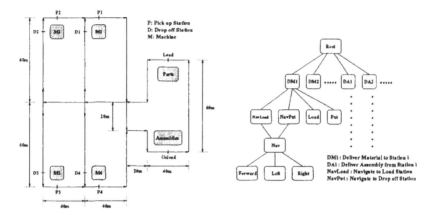

Fig. 11.11   A multiple automated guided vehicle (AGV) optimization task. There are four AGV agents (not shown) which carry raw materials and finished parts between the machines and the warehouse. The task graph of this problem is shown on the right hand side of this figure.

One general approach to learning task-level coordination is to extend the above concurrency model to the joint state action space, where base-level policies remain fixed. An extension of this approach is now presented, where agents learn coordination skills and the base-level policies simultaneously.

The hierarchical multiagent reinforcement learning algorithm described here can be implemented using other hierarchical reinforcement learning formalisms also, but for the sake of clarity, we use the MAXQ value function decomposition approach [6]. This decomposition is based on storing the value function in a distributed manner across all nodes in a task graph. The value function is computed on demand by querying lower-level (subtask) nodes whenever a high-level (task) node needs to be evaluated. The overall task is first decomposed into subtasks up to the desired level of details, and the task graph is constructed. We illustrate the idea using the above multiagent AGV scheduling problem. This task can be decomposed into subtasks and the resulting task graph is shown in Figure 11.11. All AGV agents are given the same task graph (homogeneous agents) and need to learn three skills. First, how to do each subtask, such as deliver parts to machine M1 or navigation to drop off station D3, and when to perform load or put action. Second, the agents also need to learn the order to do subtasks (for instance go to pick up station of a machine and pick up an assembly, before heading to the unload station). Finally, the agents also need to learn how to coordinate with other agents (i.e., AGV1 can deliver parts to machine M4 whereas AGV3 can deliver assemblies from machine M2). We can distinguish between two learning approaches. In the *selfish* case, the agents learn with the given task graph, but make no attempt to coordinate with each other. In the *cooperative* case, coordination skills among agents are learned by using joint actions at the level(s) immediately under the root task. Therefore, it is necessary to generalize the MAXQ decomposition from its original sequential single-agent setting to the concurrent multiagent coordination problem. We call this extension of MAXQ, *cooperative* MAXQ [19]. In this algorithm, each agent learns joint abstract action values by communicating with other agents only the high-level subtasks that they are doing. Since high-level tasks can take a long time to complete, communication is needed only fairly infrequently, which is a significant advantage over flat methods. A further advantage is that agents learn coordination skills at the level of abstract actions and it allows for increased cooperation skills as agents do not get confused by low-level details. In addition, each agent has only local state information and is ignorant about the other agent's state. Keeping track of just local information greatly simplifies the underlying reinforcement learning problem. This is based on the idea that in many cases, the state of the other agent might be roughly estimated just by knowing about the high-level action being performed by the other agent.

Let $\vec{s} = (s_1, \ldots, s_n)$ and $\vec{a} = (a_1, \ldots, a_n)$ denote a joint state and a concurrent action, where s_i is the local state and a_i is the action being performed by agent i. Let the joint action value function $Q(p, \vec{s}, \vec{a})$ represents the value of concurrent action $\vec{a}$ in joint state $\vec{s}$, in the context of executing parent task p.

The MAXQ decomposition of the Q-function relies on a key principle: the reward function for the parent task is essentially the value function of the child subtask. This principle can be extended to joint concurrent action values as shown below. The most salient feature of the *cooperative* MAXQ algorithm, is that the top level(s) (the level immediately below the root and perhaps lower levels) of the hierarchy is (are) configured to store the *completion function* values for joint abstract actions of all agents. The *completion function* $C(p, \vec{s}, \vec{a})$ is the expected cumulative discounted reward of completing parent task p after finishing concurrent action $\vec{a}$, which was invoked in state $\vec{s}$. The joint concurrent value function $V(p, \vec{s})$ is now approximated by each agent i (given only its local state s_i) as:

$$V^i(p, s_i) = \begin{cases} \max_{a_i} Q^i(p, s_i, \vec{a}) & \textit{if } \mathsf{p} \textit{ is a composite action,} \\ \sum_{s'_i} P(s'_i | s_i, p) R(s'_j | s_j, p) & \textit{if } \mathsf{p} \textit{ is a primitive action,} \end{cases}$$

where the action value function of agent i (given only its local state s_i) is defined as

$$Q^i(p, s_i, \vec{a}) \approx V^i(a_i, s_i) + C^i(p, s_i, \vec{a}). \tag{11.4}$$

The first term in Eq. (11.4), $V^i(a_i, s_i)$, refers to the discounted sum of rewards received by agent i for performing action a_i in local state s_i. The second term, $C^i(p, s_i, \vec{a})$, completes the sum by accounting for rewards earned for completing the parent task p after finishing subtask a_i. The completion function is updated in this algorithm from sample values using an SMDP learning rule. Note that the correct action value is approximated by only considering local state s_i and also by ignoring the effect of concurrent actions a_k, $k \neq i$ by other agents when agent i is performing action a_i. In practice, a human designer can configure the task graph to store joint concurrent action-values at the highest (or lower than the highest as needed) level(s) of the hierarchy.

To illustrate the use of this decomposition in learning multiagent coordination for the AGV scheduling task, if the joint action-values are restricted to only the highest level of the task graph under the root, we get the following value function decomposition for AGV1:

$$Q^1(Root, s_1, DM3, DA2, DA4, DM1) \approx V^1(DM3, s_1) + C^1(Root, s_1, DM3, DA2, DA4, DM1),$$

which represents the value of AGV1 performing task DM3 in the context of the overall root task, when AGV2, AGV3 and AGV4 are executing DA2, DA4 and DM1. Note that this value is decomposed into the value of AGV1 performing DM3 subtask itself and the completion sum of the remainder of the overall task done by all four agents.

Figure 11.12 compares the performance and speed of the *cooperative* MAXQ algorithm with other learning algorithms, including single-agent MAXQ and selfish multiagent MAXQ, as well as several well known AGV scheduling heuristics like "first come first serve", "highest queue first" and "nearest station first".

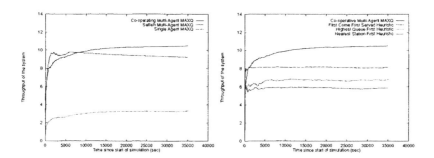

Fig. 11.12   This figure compares the performance of the *cooperative* MAXQ algorithm with other learning methods, including single-agent MAXQ and selfish multiagent MAXQ, as well as several well known AGV scheduling heuristics. The throughput is measured in terms of the number of finished assemblies deposited at the unload station per unit time.

11.4.3   Hierarchical Memory

When agents learn to act concurrently in real-world environments, the true state of the environment is usually hidden. To address this issue, we need to combine the above methods for learning concurrency and coordination with methods for estimating hidden state. We have explored two multiscale memory models [10, 42]. *Hierarchical Suffix Memory* (HSM) [10] generalizes the suffix tree model [20] to SMDP-based temporally extended actions. Suffix memory constructs state estimators from finite chains of observation-action-reward triples. In addition to extending suffix models to SMDP actions, HSM also uses multiple layers of temporal abstraction to form longer-term memories at more abstract levels. Figure 11.13 illustrates this idea for robot navigation for the simpler case of a linear chain, although the tree-based model has also been investigated. An important side-effect is that the agent can look back many steps back in time while ignoring the exact sequence of low-level observations and actions that transpired. Tests in a robot navigation domain showed that HSM outperformed "flat" suffix tree methods, as well as hierarchical methods that used no memory [10].

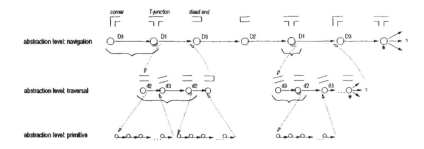

Fig. 11.13   A hierarchical suffix memory state estimator for a robot navigation task. At the abstract (navigation) level, observations and decisions occur at intersections. At the lower (corridor-traversal) level, observations and decisions occur within the corridor. At each level, each agent constructs states out of its past experience with similar history (shown with shadows).

Partially observable MDPs are theoretically more powerful than finite memory models, but past work on POMDPs has mostly studied "flat" models for which learning and planning algorithms scale poorly with model size. We have developed a new *hierarchical POMDP* framework termed H-POMDPs (see Figure 11.14) [42], by extending the hierarchical hidden Markov model (HHMM) [7] to include rewards, multiple entry/exit points into abstract states and (temporally extended) actions.

H-POMDPs can also be represented as Dynamic Bayesian networks [43], in a similar way that HHMMs can be represented as DBNs [23]. Figure 11.15 shows a Dynamic Bayesian net representation of H-POMDPs. This model differs from the

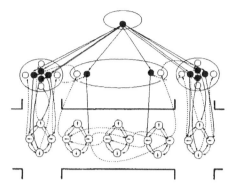

Fig. 11.14   State transition diagram of a hierarchical POMDP used to model corridor environments. Large ovals represent abstract states; the small solid circles within them represent entry states, and the small hollow circles represent exit states. The small circles with arrows represent production states. Arcs represent non-zero transition probabilities as follows: Dotted arrows from concrete states represent concrete horizontal transitions, dashed arrows from exit states represent abstract horizontal transitions, and solid arrows from entry states represent vertical transitions.

model described in [23] in two basic ways: the presence of action nodes A, and the fact that exit nodes X are no longer binary.

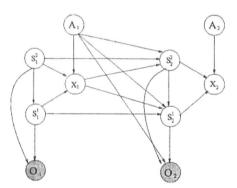

Fig. 11.15   A 2-level HPOMDP represented as a DBN.

In the particular navigation example shown in Figure 11.14, the exit node X_t can take on five possible values, representing (1) no-exit, (2) north-exit, (3) east-exit, (4) south-exit, and (5) west-exit. If X_t = no-exit, then we make a horizontal transition at the concrete level, but the abstract state is required to remain the same. If $X_t \neq$ no-exit, then we enter a new abstract state; this abstract state then makes a vertical transition into a new concrete state. The new concrete state, S_t^1, depends on the new abstract state, S_t^2, as well as the previous exit state, X_{t-1}. More precisely we can define the conditional probability distributions of each type of node in the DBN as follows: For the abstract nodes,

$$P(S_t^2 = s'|S_{t-1}^2 = s, X_{t-1} = x, A_{t-1} = a) = \begin{cases} \delta(s', s) & \text{if } x = \text{no-exit,} \\ T^{root}(s'_x|s_x, a) & \text{otherwise,} \end{cases}$$

where $T^{root}(s'_x|s_x, a)$ in the state representation of the HPOMDP model defines the transition probability from abstract state s and exit state x to abstract state s' and entry state x, where x defines the type of entry or exit state (north, east, west, south). S is the parent of s and s' in the state transition model.

For the concrete nodes,

$$P(S_t^1 = s'|S_{t-1}^1 = s, S_t^2 = S, X_{t-1} = x, A_{t-1} = a) = \begin{cases} T^S(s'|s, a) & \text{if } x = \text{no-exit,} \\ V(s'|S_x) & \text{otherwise,} \end{cases}$$

where $V(s'|S_x)$ defines the probability of a vertical transition from abstract state S and entry state of type x to concrete state s'.

For the exit nodes,

$$P(X_t = x|S_t^1 = s, S_t^2 = S, A_t = a) = T^S(S_x|s, a),$$

where $T^S(S_x|s, a)$ is the transition probability from production state s under abstract state S to exit from state S of type x.

For the sensor nodes,

$$P(O_t = z|S_t^1 = s, S_t^2 = S, A_{t-1} = a) = O^S(z|s, a),$$

where $O^S(z|s, a)$ is the probability of perceiving observation z at the $st h$ node under state S after action a.

One of the most important differences of Hierarchical HMMs/POMDPs and flat models are the results of inference. In a hierarchical model a transition to an abstract state at time t is zero, unless the abstract state is able to produce part of the remaining observations and actions in a given sequence. The inference algorithm for the state representation of HHMMs/H-POMDPs in [7, 42] achieves this by doing inference on all possible subsequences of observations under the different abstract states, which leads to $O(K^D T^3)$ time, where K is the number of states at each level of the hierarchy and D is the depth of the hierarchy. In a DBN representation we can achieve the same result as the cubic time algorithms by asserting that the sequence has finished. In our particular implementation we assert that at the last time slice the sequence has finished, and that there is a uniform probability of exit from any of the four orientations. Since we have a DBN representation, we can apply any standard Bayes net inference algorithm, such as junction tree, to perform filtering or smoothing which take in the worse case $O(K^{2D} T)$ time. Empirically it might be less, depending on the size of the cliques being formed, as was shown in [23].

Due to the cubic time complexity of the EM algorithm used in [40] we have developed various approximate training techniques such as "reuse-training", whereby submodels are trained separately and then combined into an overall hierarchy, and "selective-training" whereby only selected parts of the model are trained for every sequence. Even though these methods require knowledge as to which part of the model the data should be used for, they outperformed the flat EM algorithm in terms of fit to test data, robot localization accuracy, and capability of structure learning at higher levels of abstraction. However, a DBN-representation allows us to use longer training sequence. In [43] we show how the hierarchical model requires less data for training than the flat model, and also illustrate how combining the hierarchical and factorial representations outperforms both the hierarchical and flat models.

In addition to the advantages over flat methods for model learning, H-POMDPs have an inherent advantage in planning as well. This is because belief states can be computed at different levels of the tree, and there is often less uncertainty at higher levels (e.g., a robot is more sure of which corridor it is in, rather than exactly which low level state). A number of heuristics for mapping belief states to temporally extended actions (e.g., move down the corridor) provide good performance in robot navigation (e.g., the most-likely-state (MLS) heuristic assumes the agent is in the state corresponding to the "peak" of the belief state distribution) [14, 24, 34]. Such heuristics work much better in H-POMDPs because they can be applied at multiple levels, and probability distributions over abstract states usually have lower entropy (see Figure 11.16). For a detailed study of the H-POMDP model, as well as its application to robot navigation, see [40].

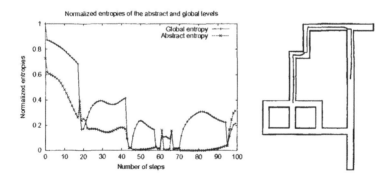

Fig. 11.16   This plot shows a sample robot navigation run whose trace is on the right, where positional uncertainty (measured by belief state entropy) at the abstract (corridor) level is less than at the product state level. Spatiotemporal abstraction reduces the uncertainty and requires less frequent decision-making, allowing the robot to get to goals without initial positional information.

11.5 SUMMARY AND CONCLUSIONS

In this chapter, we presented hierarchical models of decision-making involving concurrent actions, multiagent coordination, and hidden state estimation. The common thread which spanned solutions to these three challenges is that multi-level temporal and spatial abstraction of actions and states can be exploited to achieve effective solutions. The overall approach was presented in three phases, beginning with a hierarchical model for learning concurrent plans for observable single-agent domains. This concurrency model combined compact state representations with temporal process abstractions to formalize concurrent action. Multiagent coordination was addressed using a hierarchical model where primitive joint actions and joint states are abstracted by exploiting overall task structure, which greatly speeds up convergence since low-level steps are ignored that do not need to be synchronized. Finally, a hierarchical framework for hidden state estimation was presented, which used multi-resolution statistical models of the past history of observations and actions.

Acknowledgments

This research was supported in part by grants from the National Science Foundation (Knowledge and Distributed Intelligence program), the Defense Advanced Research Projects Agency (MARS, Distributed Robotics, and Robot-2020 programs), Michigan State University, and the University of Massachusetts at Amherst.

Bibliography

1. A. Barto and S. Mahadevan, Recent advances in hierarchical reinforcement learning, *Discrete-Event Systems: Theory and Applications,* vol. 13, pp. 41–77, 2003.

2. C. Boutilier, R. Dearden, and M. Goldszmidt, Stochastic dynamic programming with factored representations, *Artificial Intelligence,* vol. 121, no. 1–2, pp. 49–107, 2000.

3. R. H. Crites and A. G. Barto, Elevator group control using multiple reinforcement learning agents, *Machine Learning,* vol. 33, pp. 235–262, 1998.

4. T. Dean and R. Givan, Model minimization in Markov decision processes, *Proc. of AAAI,* 1997.

5. T. Dean and K. Kanazawa, A model for reasoning about persistence and causation, *Computational Intelligence,* vol. 5, no. 3, pp. 142–150, 1989.

6. T. G. Dietterich, Hierarchical reinforcement learning with the MAXQ value function decomposition, *International Journal of Artificial Intelligence Research,* vol. 13, pp. 227–303, 2000.

7. S. Fine, Y. Singer, and N. Tishby, The hierarchical hidden Markov model: Analysis and applications, *Machine Learning,* vol. 32, no. 1, 1998.

8. D. Freitag and A. K. McCallum, Information extraction with HMMs and shrinkage, *Proc. of the AAAI-99 Workshop on Machine Learning for Information Extraction,* Orlando, FL, 1999.

9. M. Ghavamzadeh and S. Mahadevan, Continuous-time hierarchical reinforcement learning, *Proc. of the 18th International Conference on Machine Learning,* Williamstown, MA, 2001.

10. N. Hernandez and S. Mahadevan, Hierarchical memory-based reinforcement learning, *Proc. of Neural Information Processing Systems,* 2001.

11. F. Jellinek, *Statistical Methods in Speech Recognition,* MIT Press, Cambridge, MA, 2000.

12. K. Karplus, C. Barrett, and R. Hughey, Hidden Markov models for detecting remote protein homologies, *Bioinformatics*, vol. 14, no. 10, pp. 846–856, 1998.

13. C. A. Knoblock, An analysis of ABSTRIPS, In J. Hendler (ed.), *Artificial Intelligence Planning Systems: Proc. of the First International Conference (AIPS 92)*, pp. 126–135, College Park, MD, 1992.

14. S. Koenig and R. Simmons, Xavier: A robot navigation architecture based on partially observable Markov decision process models, In D. Kortenkamp, P. Bonasso, and R. Murphy, (eds.), *AI-based Mobile Robots: Case-studies of Successful Robot Systems*, MIT Press, Cambridge, MA, 1997.

15. D. Koller and R. Parr, Computing factored value functions for policies in structured MDPs, *16th International Joint Conference on Artificial Intelligence (IJCAI)*, pp. 1332–1339, 1999.

16. M. Littman, Markov games as a framework for multi-agent reinforcement learning, *Proc. of the 11th International Conference on Machine Learning*, pp. 157–163, Morgan Kaufmann, San Francisco, CA, 1994.

17. S. Mahadevan and J. Connell, Automatic programming of behavior-based robots using reinforcement learning, *Artificial Intelligence*, vol. 55, pp. 311–365, 1992. Appeared originally as IBM TR RC16359, Dec 1990.

18. S. Mahadevan, N. Marchalleck, T. Das, and A. Gosavi, Self-improving factory simulation using continuous-time average-reward reinforcement learning, *Proc. 14th International Conference on Machine Learning*, pp. 202–210, Morgan Kaufmann, San Francisco, CA, 1997.

19. R. Makar, S. Mahadevan, and M. Ghavamzadeh, Hierarchical multiagent reinforcement learning, *Proc. 5th International Conference on Autonomous Agents*, pp. 246–253, ACM Press, New York, 2001.

20. A. K. McCallum, *Reinforcement Learning with Selective Perception and Hidden State*, Ph.D. Thesis, University of Rochester, 1995.

21. N. Meuleau, M. Hauskrecht, K. Kim, L. Peshkin, L. Kaelbling, T. Dean, and C. Boutilier, Solving very large weakly coupled Markov decision processes, *Proc. of the Conference on Uncertainty in Artificial Intelligence*, Madison, WI, 1998.

22. S. Minut and S. Mahadevan, A reinforcement learning model of selective visual attention, *Proc. of 5th International Conference on Autonomous Agents*, 2001.

23. K. Murphy and M. Paskin, Linear time inference in hierarchical hmms, in T. G. Dietterich, S. Becker, and Z. Ghahramani, (eds.), *Advances in Neural Information Processing Systems 14*, MIT Press, Cambridge, MA, 2002.

24. I. Nourbakhsh, R. Powers, and S. Birchfield, Dervish: An office-navigation robot, *AI Magazine*, vol. 16, no. 2, pp. 53–60, 1995.

25. R. E. Parr, *Hierarchical Control and Learning for Markov Decision Processes,* Ph.D. Thesis, University of California, Berkeley, 1998.

26. J. Pineau, N. Roy, and S. Thrun, A hierarchical approach to POMDP planning and execution, *Workshop on Hierarchy and Memory in Reinforcement Learning (ICML 2001),* Williamstown, MA, 2001.

27. A. Prieditis, Machine discovery of admissible heuristics, *Machine Learning* vol. 12, pp. 117–141, 1993.

28. M. L. Puterman, *Markov Decision Processes,* Wiley Interscience, New York, 1994.

29. B. Ravindran and A. Barto, Model minimization in hierarchical reinforcement learning, *Symposium on Abstraction and Reformulation (SARA 2002),* Springer-Verlag, Heidelberg, 2002.

30. K. Rohanimanesh and S. Mahadevan, Decision-theoretic planning with concurrent temporally extended actions, *Proc. of 17th Conference on Uncertainty in Artificial Intelligence,* Seattle, WA, 2001.

31. K. Rohanimanesh and S. Mahadevan, Incremental learning of factorial Markov decision processes, unpublished, 2002.

32. S. Russell and P. Norvig, *Artificial Intelligence: A Modern Approach,* Prentice-Hall, Englewood Cliffs, NJ, 1994.

33. L. K. Saul, and M. I. Jordan, Mixed memory Markov models: Decomposing complex stochastic processes as mixture of simpler ones, *Machine Learning,* vol. 37, pp. 75–87, 1999.

34. H. Shatkay and L. P. Kaelbling, Learning topological maps with weak local odometric information, *IJCAI,* vol. 2, pp. 920–929, 1997.

35. T. Sugawara and V. Lesser, Learning to improve coordinated actions in cooperative distributed problem-solving environments, *Machine Learning,* vol. 33, pp. 129–154, 1998.

36. R. Sutton and A. Barto, *An introduction to reinforcement learning,* MIT Press, Cambridge, MA, 1998.

37. R. Sutton, D. Precup, and S. Singh, Between MDPs and semi-MDPs: A framework for temporal abstraction in reinforcement learning, *Artificial Intelligence,* vol. 112, pp. 181–211, 1999.

38. M. Tan, Multi-agent reinforcement learning: Independent vs. cooperative agents, *Proc. of the 10th International Conference on Machine Learning,* pp. 330–337, Morgan Kaufmann, San Francisco, CA, 1993.

39. G. Tesauro, Practical issues in temporal difference learning, *Machine Learning,* vol. 8, pp. 257–278, 1992.

40. G. Theocharous, *Hierarchical Learning and Planning in Partially Observable Markov Decision Processes,* Ph.D. Thesis, Michigan State University, East Lansing, MI, 2002.

41. G. Theocharous and S. Mahadevan, Approximate planning with hierarchical partially observable Markov decision processs for robot navigation, *Proc. of the IEEE International Conference on Robotics and Automation (ICRA),* Washington, DC, 2002.

42. G. Theocharous, K. Rohanimanesh, and S. Mahadevan, Learning hierarchical partially observable Markov decision processs for robot navigation, *Proc. of the IEEE International Conference on Robotics and Automation (ICRA),* Seoul, Korea, 2001.

43. G. Theocharous, K. Murphy, and L. Kaelbling, Representing hierarchical POMDPs as DBNs for multi-scale robot localization, *IJCAI Workshop on Reasoning with Uncertainty in Robotics,* Acapulco, Mexico, 2003.

44. G. Wang and S. Mahadevan, Hierarchical optimization of policy-coupled semi-Markov decision processes, *Proc. of 16th International Conference on Machine Learning,* pp. 464–473, Morgan Kaufmann, San Francisco, CA, 1999.

45. C. Watkins, *Learning from Delayed Rewards,* Ph.D. Thesis, King's College, Cambridge University, UK, 1989.

46. G. Weiss, *Multiagent Systems: A Modern Approach to Distributed Artificial Intelligence,* MIT Press, Cambridge, MA, 1999.

12 Learning and Optimization — From a System Theoretic Perspective

XI-REN CAO

Hong Kong University of Science and Technology

Editor's Summary: Learning and optimization of stochastic systems is a multi-disciplinary area that attracts researchers in control systems, operations research, and computer science. Areas such as perturbation analysis (PA), Markov decision processes (MDP), and reinforcement learning (RL) share a common goal. This chapter offers an overview of the area of learning and optimization from a system theoretic perspective, and it is shown that these seemingly different fields are actually closely related. Furthermore, this perspective leads to new research directions, which are illustrated using a queueing example. The central piece of this area is the performance potentials, which can be equivalently represented as perturbation realization factors that measure the effects of a single change to a sample path on the system performance. Potentials or realization factors can be used as building blocks to construct performance sensitivities. These sensitivity formulas serve as the basis for learning and optimization.

12.1 INTRODUCTION

Learning and performance optimization of stochastic systems has been an active research topic for many years. It has attracted close attention from different research communities, including control systems (perturbation analysis, or PA [21, 28]), operations research (Markov decision processes, or MDP [1, 4, 7, 35, 37]), and artificial intelligence (reinforcement learning, or RL [30, 43]). These areas share a common goal, but have different perspectives, focuses, and perhaps different methodologies.

Perturbation analysis [28] was originally developed for estimating performance derivatives with respect to system parameters in stochastic systems with queueing structures (queueing networks, generalized semi-Markov processes, etc); the estimates can be obtained by analyzing a single sample path of such a system; it was shown that although the approach requires some conditions for the system structure [8], it is very efficient [27] since it utilizes the special dynamic properties of the system. The fundamental concept of PA, perturbation realization [10], has been extended to Markov processes. Recent research in this direction reveals a strong connection among PA, MDP, and RL [12].

In this chapter, we offer an overview of learning and optimization from a system theoretic perspective. We show how these seemingly different disciplines are closely related, how one topic leads to the others, and how this perspective may lead to new research topics and new results. Our discussion is based on the general model of discrete time Markov chains, which is used widely in these different disciplines to model stochastic systems. For simplicity, we discuss Markov chains with finite state space denoted as $\{1, 2, \cdots, M\}$. The central piece of learning and optimization is the performance potentials $g(i)$, $i = 1, \cdots, M$, or equivalently, perturbation realization factors $d(i, j) = g(j) - g(i)$ [15]. From a perturbation analysis point of view, a change in system parameters induces a series of perturbations on a sample path. The effect of a single perturbation on system performance can be measured by the realization factor of the perturbation, and the total effect of the parameter change on the performance is then the sum of the realization factors of all the perturbations induced by the parameter changes [10]. For Markov chains, parameters are the transition probabilities, a perturbation is a "jump" from one state i to another state j, and the realization factor equals the difference of the potentials at the two states. It has been shown that by the above principle, we can use potentials or realization factors as building blocks to construct performance sensitivities for many systems. When the changes are discrete, this leads to formulas for the performance difference of two Markov chains, and when the changes are infinitesimal, it leads to the formula for performance gradients [15, 17].

These two standard formulas are the basis for performance optimization [12]. Optimization can be achieved by combining the gradient estimate with stochastic approximation methods, or by policy iteration which can be easily derived from the performance difference formula (see Section 12.2). This leads to the following main research directions:

1. Develop efficient algorithms to estimate the potentials and/or the derivatives. Reinforcement learning, TD(λ), neuro-dynamic programming, etc, are efficient ways of estimating the performance potentials, realization factors, and related quantities such as Q-factors, etc., based on sample paths (Section 12.3.1). In addition, algorithms can be developed to estimate performance gradients directly from a single sample path (Sections 12.3.2, 12.3.3).

2. Develop efficient optimization algorithms with the potential or gradient estimates

   (a) Gradient-based optimization for parameterized systems; this approach combines the gradient estimates with stochastic gradient algorithms (Section 12.4).

(b) On-line policy iteration; this approach combines the potential estimates with stochastic approximation to implement policy iteration (Section 12.5).

(c) Gradient-based policy iteration; this is an open problem (Section 12.5.2).

3. Finally, in Section 12.6, we use an example to illustrate that potentials can be used to construct flexible performance sensitivities for systems with special structures. The sensitivity formula obtained is simpler than the standard one, and the quantities involved are aggregations of potentials and can be estimated on sample paths. Further research is going on in this direction.

Sample path-based estimation is also called "learning" in literature. It is worthwhile to mention the slight distinction between "online" learning and simulation-based learning. The former can be implemented on a single sample path obtained by observing the history of a real system without interfering the system operation; the latter requires a sample path which differs from the history of a system under normal operation and can be generated by simulation, for example, in Q-learning, it requires the sample path to visit every state-action pair (see [43]).

This chapter reviews the main principles of the above research topics and presents these results from a system point of view. In Section 12.2, we briefly review the concept of performance potentials and introduce the two formulas for performance differences and performance gradients. In Section 12.3, we discuss on-line estimation of potentials and performance gradients; in particular, we propose a general formula (Eq. (12.18)) for gradient estimation, which leads to a number of specific gradient estimates. In Section 12.4, we discuss gradient-based optimization. In Section 12.5, we deal with on-line policy iteration, which is in parallel to the gradient-based optimization, and a new research direction called gradient-based policy iteration. In Section 12.6, we present some recent works. We show that performance gradient-(or difference-) formulas can be constructed with potentials as building blocks; these formulas can be used as the basis for performance optimization. Potential aggregation are used to save computation. This approach can be applied flexibly to many practical systems. This is an on-going new research topic. For easy reference, the notations used in this paper are listed in Table 12.1.

12.2 GENERAL VIEW OF OPTIMIZATION

Consider an irreducible and aperiodic Markov chain $\mathbf{X} = \{X_n : n \geq 0\}$ on a finite state space $\mathcal{S} = \{1, 2, \cdots, M\}$ with transition probability matrix $P = [p(i,j)] \in [0,1]^{M \times M}$. Let $\pi = (\pi(1), \ldots, \pi(M))$ be the (row) vector representing its steady-state probabilities, and $f = (f(1), f(2), \cdots, f(M))^T$ be the (column) performance vector, where "T" represents transpose. We have $Pe = e$, where $e = (1, 1, \cdots, 1)^T$ is an M-dimensional vector whose components all equal 1, and $\pi e = 1$. The steady state probability flow balance equation is $\pi = \pi P$. The performance measure is the

Table 12.1   List of Notations

Notations	Definitions
$S = \{1, 2, \cdots, M\}$	State space of the Markov chain
$p(i, j), i, j \in S$	Transition probability from state i to state j
$P = [p(i, j)]$	Transition probability matrix
$Q = P' - P$	Difference of two transition probability matrix
$h = f' - f$	Difference of two performance vectors
$\pi = (\pi(1), \cdots, \pi(M))$	Steady state probability vector
$f = (f(1), \cdots, f(M))^T$	Performance vector
$e = (1, \cdots, 1)^T$	Column vector with all components being one
$\eta = \pi f$	Long-run average performance measure
$g = (g(1), \cdots, g(M))^T$	Potential vector
$g(i), i \in S$	Potential of state i
$g_L(i), i \in S$	Estimate of $g(i)$ based on L transitions
$d(i, j), i, j \in S$	Perturbation realization factor
$D = [d(i, j)]$	Realization matrix
A	Action space
$\mathcal{L} : S \rightarrow A$	Policy, superscript represents dependency on $\mathcal{L}$
α	discount factor

long-run average defined as

$$\eta = E_\pi(f) = \sum_{i=1}^M \pi(i)f(i) = \pi f = \lim_{L \to \infty} \frac{1}{L} \sum_{l=0}^{L-1} f(X_l), \quad w.p.1.$$

We start with the *Poisson equation*

$$(I - P)g + e\eta = f. \tag{12.1}$$

Its solution $g = (g(1), \cdots, g(M))^T$ is called a *performance potential* vector, and $g(i)$ is the potential at state i. g is also called the "value function" in dynamic programming, or the "differential" or "relative cost vector" [4], and "bias" [37]. The solution to (12.1) is only up to an additive constant; that is, if g is a solution to (12.1), then so is $g + ce$.

Let P' and π' be another irreducible transition probability matrix on the same state space and its steady state probability. Let f' be the performance function for the system with P', $Q = P' - P = [q(i, j)]$ and $h = f' - f$. Then $Qe = 0$. The steady state performance corresponding to P' is $\eta' = \pi' f'$. Multiplying both sides of (12.1) with π', we can verify that

$$\eta' - \eta = \pi'(Qg + h). \tag{12.2}$$

Now, suppose that P changes to $P(\delta) = P + \delta Q = \delta P' + (1 - \delta)P$, and f changes to $f(\delta) = f + \delta h$, with $\delta \in (0,1]$. Then the performance measure changes to $\eta(\delta) = \eta + \Delta\eta(\delta)$. The derivative of η in the direction of Q is defined as $\frac{d\eta}{d\delta} = \lim_{\delta \to 0} \frac{\Delta\eta(\delta)}{\delta}$. Taking $P(\delta)$ as the P' in (12.2), we have $\eta(\delta) - \eta = \pi(\delta)(\delta Q g + \delta h)$. Letting $\delta \to 0$, we get

$$\frac{d\eta}{d\delta} = \pi(Qg + h). \tag{12.3}$$

For references, see, for example, [12, 15]. Since $Qe = 0$, for any g satisfying (12.1) for any constant c, we have $Qg = Q(g+ce)$, thus both (12.3) and (12.2) still hold for $g' = g + ce$. This verifies again that potentials are determined only up to an additive constant; this is the same as the potential energy in physics.

In (12.3), a linear structure $P(\delta) = P + \delta Q$ is assumed. In general, the transition probability matrix may depend on an arbitrary parameter θ, which is normalized in $[0,1]$; that is, $P(\theta) = P + Q(\theta)$ with $Q(1) = P(1) - P = P' - P$. Similarly, we assume $f(\theta) = f + h(\theta)$. Thus, for $\theta << 1$, we have $P(\theta) = P + \{\frac{dQ}{d\theta}\}_{\theta=0}\theta$, and $f(\theta) = f + \{\frac{dh}{d\theta}\}_{\theta=0}\theta$; i.e., in the neighboring area of $\theta = 0$, $P(\theta)$ and $f(\theta)$ take a linear form. Replacing Q in (12.3) with $\{\frac{dQ}{d\theta}\}_{\theta=0}$ and h with $\{\frac{dh}{d\theta}\}_{\theta=0}$ and noting that $\frac{dP}{d\theta} = \frac{dQ}{d\theta}$ and $\frac{df}{d\theta} = \frac{dh}{d\theta}$ we get

$$\frac{d\eta}{d\theta}\Big|_{\theta=0} = \pi\left\{(\frac{dP}{d\theta})_{\theta=0}g + (\frac{df}{d\theta})_{\theta=0}\right\}. \tag{12.4}$$

Therefore, without loss of generality, we shall mainly discuss the linear case (12.3).

The two simple Eqs. (12.3) and (12.2) represent the performance sensitivity; (12.3) is the performance derivative (or gradient) with respect to continuous variables, and (12.2) is the performance difference for two discrete parameters (P and P'). Both of them depend mainly on the same quantity: the performance potential. Note that both depend on only the potential g (not g'), and π and g can be estimated based on a single sample path of the Markov chain with transition matrix P (see Section 12.3.1).

The two Eqs. (12.3) and (12.2) form the basis for performance optimization of Markov systems. Two basic approaches can be developed from them. The first one is gradient-based optimization, which combines gradient estimation based on (12.3) and stochastic approximation techniques. This approach applies to systems that can be parameterized by continuous variables. This is in the same spirit as the perturbation analysis (or PA) based optimization (see, e.g. [21, 22, 36], and [41]). The sensitivity formula (12.3) can indeed be derived by applying the PA principles. The second approach is the policy-iteration based optimization. It can be shown that policy iteration algorithms in Markov decision problems can be easily derived from (12.2) (see, e.g., [12]). The main issues here is to design fast policy iteration procedures that converge to the optimal policy (see [24]). Both approaches depend heavily on the estimation of potentials. Q-learning [48, 49], actor-critic type of algorithms [32, 44], and so on, are variants of this approach: they aim at to find directly the potentials (or the equivalent Q-factors) for the optimal policy. These are simulation based algorithms since they require the sample path to visit very state-action pair.

12.3   ESTIMATION OF POTENTIALS AND PERFORMANCE DERIVATIVES

As shown in Section 12.2, potentials and the two sensitivity formulas (12.2) and (12.3) play a crucial role in performance optimization. we first present in Section 12.3.1 some on-line algorithms for estimating potentials (or formulas based on which such algorithms can be easily derived). To overcome the well known difficulty of large state space, we observe that with (12.3) it is possible to estimate performance derivatives directly without estimating the potentials for all the states. This is discussed in Section 12.3.2. We first propose a basic equation for gradient estimation (12.18). Three different techniques developed by different authors for estimating gradients are presented as special cases of (12.18). Other algorithms can also be developed. Section 12.3.3 provides a historical view of the performance gradient estimation, early works on perturbation analysis (PA) and the likelihood ratio (LR) methods are discussed.

12.3.1   Estimation of Performance Potentials

We first show that the potentials of a Markov chain can be estimated with a single sample path of the Markov chain. Since g is only up to an additive constant, we may choose the one that satisfies $\pi g = \pi f = \eta$. Thus (12.1) becomes

$$(I - P + e\pi)g = f. \tag{12.5}$$

For ergodic chains, $I - P + e\pi$ is invertible; thus

$$
\begin{aligned}
g &= (I - P + e\pi)^{-1}f = \sum_{l=0}^{\infty}(P - e\pi)^l f \\
&= I + \sum_{k=1}^{\infty}(P^k f - \eta e). \tag{12.6}
\end{aligned}
$$

Consider a sample path of the Markov chain, denoted as $\mathbf{X} = \{X_0, X_1, \cdots, X_n, \cdots\}$. Ignoring the constant ηe in (12.6), we have

$$g(i) = E\left\{\sum_{k=0}^{\infty}[f(X_k) - \eta]|X_0 = i\right\}, \tag{12.7}$$

which can be approximated by (with a large integer L)

$$g_L(i) \approx E\left[\sum_{l=0}^{L-1}f(X_l)|X_0 = i\right] - L\eta. \tag{12.8}$$

Since g is determined only up to an additive constant, we may drop the constant $L\eta$ in (12.8) and simply use

$$g_L(i) \approx E\left[\sum_{l=0}^{L-1} f(X_l)|X_0 = i\right]. \tag{12.9}$$

Given a sample path $\mathbf{X} = \{X_0, X_1, \cdots\}$, π and η can be estimated easily. Let $\epsilon_i(x)$ be the indicator function for state i; that is, $\epsilon_i(x) = 1$ for $x = i$, and $= 0$ otherwise. Then

$$\pi(i) = \lim_{K\to\infty} \frac{1}{K}\sum_{k=0}^{K-1} \epsilon_i(X_k), \qquad w.p.1, \tag{12.10}$$

and

$$\eta = \lim_{K\to\infty} \frac{1}{K}\sum_{k=0}^{K-1} f(X_k), \qquad w.p.1. \tag{12.11}$$

With (12.9), the potential g can be estimated on a sample path in a similar way. By ergodicity, (12.8) and (12.9) lead to

$$g_L(i) = \lim_{K\to\infty}\left\{\frac{\sum_{k=0}^{K-L+1}\{\epsilon_i(X_k)[\sum_{j=0}^{L-1} f(X_{k+j})]\}}{\sum_{k=0}^{K-L+1}\epsilon_i(X_k)} - \frac{L}{K}\sum_{k=0}^{K-1} f(X_k)\right\},$$
$$w.p.1. \tag{12.12}$$

$$g_L(i) = \lim_{K\to\infty}\left\{\frac{\sum_{k=0}^{K-L+1}\{\epsilon_i(X_k)[\sum_{j=0}^{L-1} f(X_{k+j})]\}}{\sum_{k=0}^{K-L+1}\epsilon_i(X_k)}\right\}, \qquad w.p.1. \tag{12.13}$$

Although (12.13) is simpler, sometimes one would choose to use (12.12) because (12.13) may lead to a number that is too large to handle.

The convergence in (12.13) is not obvious, since the items $\sum_{j=0}^{n-1} f(X_{k+j})$ for different k may not be independent. The proof of (12.13) is based on a fundamental theorem on ergodicity [6]: Let $\mathbf{X} = \{X_k, k \geq 0\}$ be an ergodic process on state space S; $\phi(x_1, x_2, \cdots)$be a measurable function on S. Then the process $Z = \{Z_k, k \geq 0\}$ with $Z_k = \phi(X_k, X_{k+1}, \cdots)$ is also ergodic. In our case, we define $Z_k = \epsilon^i(X_k)[\sum_{j=0}^{n-1} f(X_{k+j})]$; then $\{Z_k, k \geq 0\}$ is ergodic. Thus (12.13) converges w.p.1. This theorem has been used to prove many similar result, see [20] and [2, 3].

It's sometimes easier to estimate the differences between the potentials at two states, called *perturbation realization factor* in PA, which is defined as

$$d(i,j) = g(j) - g(i), \qquad i,j \in S.$$

The matrix $D = [d(i, j)]$ is called a *realization matrix*. We have $D^T = -D$ and $D = eg^T - ge^T$. D satisfies the Lyapunov equation

$$D - PDP^T = F',$$

with $F = ef^T - fe^T$. (12.3) and (12.2) become

$$\frac{d\eta}{d\delta} = \pi(QD^T\pi^T + h).$$

$$\eta' - \eta = \pi'(QD^T\pi^T + h).$$

Now we consider a Markov chain $\mathbf{X} = \{X_k, k \geq 0\}$ with initial state $X_0 = i$, we define $L_i(j) = min\{n : n \geq 0, X_n = j\}$; that is, at $n = L_i(j)$, the Markov chain reaches state j for the first time. We have $E[L_i(j)|X_0 = i] < \infty$ [23], and from [15]

$$d(j, i) = E\left\{\sum_{k=0}^{L_i(j)-1} [f(X_k) - \eta]|X_0 = i\right\}. \tag{12.14}$$

Equation (12.14) relates $d(i, j)$ to a finite portion of the sample paths of $\mathbf{X}$. To develop an algorithm based on (12.14), we define $u_0 = 0$, and $u_{k+1} = min\{n : n > u_k, X_n = i\}$, $k \geq 0$, where i is a fixed state. $u_k, k \geq 0$ are regenerative points. For any $j \neq i$, define $v_k(j) = min\{n : u_{k+1} > n > u_k, X_n = j\}$ and $\chi_k(j) = 1$, if $\{u_{k+1} > n > u_k, X_n = j\} \neq \emptyset$; and $\chi_k(j) = 0$, otherwise. From (12.14), we have

$$d(i, j) = \lim_{K \to \infty} \frac{1}{\sum_{k=0}^{K-1} \chi_k(j)} \left\{ \left[\sum_{k=0}^{K-1} \{\chi_k(j) \sum_{n=v_k(j)}^{u_{k+1}-1} f(X_n)\} \right] \tag{12.15} \right.$$
$$\left. - \left[\sum_{k=0}^{K-1} \chi_k(j)[u_{k+1} - v_k(j)] \right] \eta \right\}, \qquad w.p.1,$$

where η can be simply estimated by (12.11).

Therefore, we have two ways to estimate the potentials; one is by (12.12) or (12.13) directly, the other is by (12.15) via the realization factor. After obtaining $d(i, j)$, we can choose any state i^* and set $g(i^*) = 0$ and $g(j) = d(i^*, j)$ for all $j \neq i^*$. Since only the differences among the potentials at different states are important, the potential estimates (12.8) and (12.9) contains more information than needed. In other words, using $d(i, j)$ in (12.15) to estimate potentials may be more efficient (with smaller variances). In addition, if we know the matrix D, we may use $g = D^T\pi^T$. Both (12.9) and (12.14) can be used to estimate the performance derivatives directly (see, Section 12.3.2).

Besides the "direct" estimates (12.12), (12.13) and (12.15), more efficient and/or convenient algorithms can be developed by applying stochastic approximation methods. For example, the *temporal difference (TD)* approach can be explained by

(12.7) and the standard stochastic approximation method. Given a sample path $\{X_0, \cdots, X_n, X_{n+1}, \cdots, \}$, at the nth transition, we have

$$g(X_n) = E\left\{\sum_{k=0}^{\infty}[f(X_{n+k}) - \eta]|X_n\right\} = [f(X_n) - \eta] + E[g(X_{n+1})|X_n].$$

(12.16)

Now suppose we observe a transition from state X_n to X_{n+1}. If $\hat{g}(X_n)$ is the estimate at time n, then $[f(X_n) - \eta] + \hat{g}(X_{n+1})$ may be a more accurate estimate than $\hat{g}(X_n)$ because it reflects the information at this transition. Define the *temporal difference (TD)* as

$$d_n = [f(X_n) - \eta + g(X_{n+1}) - g(X_n)],$$

which may represent the stochastic error observed at transition n. Based on (12.16) and with the stochastic approximation approach, some recursive on-line algorithms (called TD(λ) algorithms) for estimating $g(i)$ can be developed. [45, 47] present TD(λ) algorithms with linearly parameterized function approximation. For TD (λ) algorithms for discounted or total cost problems, see [5, 42, 43, 46].

In addition to the above approaches, *neuro-dynamic programming* is proposed to overcome the difficulty of the so-called "curse of dimensionality". Roughly speaking, in neuro-dynamic programming, we try to approximate the potential function $g(i)$ by $g(i, r)$, with a continuous parameter r. This generally involves two steps:

1. Develop an approximation architecture, e.g., a neuro-network, to represent $g(i, r)$,

2. Find a training algorithm for updating the parameter vector r, based on the information observed on a sample path.

After training, the parameter r reaches a proper value. The neuro-network will output an approximate value of $g(i)$ for an input integer i. For details and successful examples, see [5].

12.3.2 Gradient Estimation

There are two ways in estimating the performance measure η: we may estimate all $\pi(i)$ by (12.10) first and then use $\eta = \pi f$ to calculate the performance, or we may estimate η directly by (12.11). The situation is similar for the performance derivatives: we may first estimate g by the methods presented in Section 12.3.1, then obtain the derivatives by (12.3), or we may estimate $\pi Q g$ directly. The disadvantage of the former is that the state space is usually too large. In the following, we present a few "direct" algorithms for $\pi Q g$ (similar to (12.11) for η).

Consider a stationary Markov chain $\mathbf{X} = (X_0, X_1, \cdots,)$. (This implies the initial probability distribution is π.) Let E denote the expectation on the probability space generated by $\mathbf{X}$. Denote a generic time instant as k. Because it is impossible for a sample path with P to contain information about P', we need to use a standard technique in simulation called *importance sampling.* We have

$$
\begin{aligned}
\frac{d\eta}{d\delta} &= \pi(Qg + h) = \sum_{i \in S} \sum_{j \in S} \left\{ \pi(i) \left[p(i,j) \frac{q(i,j)}{p(i,j)} g(j) + h(i) \right] \right\} \\
&= E \left\{ \frac{q(X_k, X_{k+1})}{p(X_k, X_{k+1})} g(X_{k+1}) + h(X_k) \right\}.
\end{aligned}
\tag{12.17}
$$

Furthermore, if $\hat{g}$ is a random variable defined on $\mathbf{X}$ such that $E(\hat{g}) = g$ and $\hat{g}$ is independent of the transition from X_k to X_{k+1}, then we have (see [14])

$$
\frac{d\eta}{d\delta} = E \left\{ \frac{q(X_k, X_{k+1})}{p(X_k, X_{k+1})} \hat{g}(X_{k+1}) + h(X_k) \right\}.
\tag{12.18}
$$

Equation (12.18) is the fundamental equation for performance gradient estimation. Sample path based algorithms can be developed by using Eq. (12.18) and any estimate of g. In the following, we present three different techniques based on Eq. (12.18).

The first algorithm is based on Eq. (12.9). From Eq. (12.9), we can choose

$$
\hat{g}(X_0) := \sum_{l=0}^{L-1} f(X_l).
$$

Using this $\hat{g}$ in (12.18) (for simplicity, we set $h = 0$), we have

$$
\begin{aligned}
\frac{\partial \eta}{\partial \delta} &= E \left\{ \frac{q(X_k, X_{k+1})}{p(X_k, X_{k+1})} \left[\sum_{l=0}^{L-1} f(X_{k+l+1}) \right] + h(X_k) \right\} \\
&= \lim_{K \to \infty} \frac{1}{K - L + 1} \left\{ \sum_{k=0}^{K-L} \left\{ \frac{q(X_k, X_{k+1})}{p(X_k, X_{k+1})} \right\} \left[\sum_{l=0}^{L-1} f(X_{k+l+1}) \right] \right\}, w.p.1.
\end{aligned}
\tag{12.19}
$$

This equation is similar to (12.13) with one modification: a modifying factor $\frac{q(X_k, X_{k+1})}{p(X_k, X_{k+1})}$ is multiplied to the term $\sum_{l=0}^{L-1} f(X_{k+l+1})$. It can be shown that (12.19) is equivalent to [20]

$$
\frac{\partial \eta}{\partial \delta} \approx \lim_{K \to \infty} \frac{1}{K - L + 1} \sum_{k=0}^{K-L} \left\{ f(X_{k+L}) \sum_{l=0}^{L-1} \left[\frac{q(X_{k+l}, X_{k+l+1})}{p(X_{k+l}, X_{k+l+1})} \right] \right\}, \qquad w.p.1.
\tag{12.20}
$$

In Eqs. (12.19) and (12.20), g is approximated by truncation. In the second algorithm, we use an α-potential g_α, $0 < \alpha < 1$, to approximate g. g_α satisfies the following discounted Poisson equation [16]:

$$
(I - \alpha P + \alpha e\pi)g_\alpha = f.
$$

It is shown [16] that

$$\lim_{\alpha \to 1} g_\alpha = g.$$

Ignoring the constant term, we have (cf. (12.9))

$$g_{\alpha,L}(i) = E\left[\sum_{l=0}^{\infty} \alpha^l f(X_l)|X_0 = i\right].$$

Thus we can choose

$$\hat{g}(X_0) := \sum_{l=0}^{\infty} \alpha^l f(X_l)$$

as the $\hat{g}$ in (12.18), we get (cf. (12.19))

$$\frac{\partial \eta}{\partial \delta} \approx \lim_{K \to \infty} \frac{1}{K-L+1} \left\{ \sum_{k=0}^{K-L} \left\{ \frac{q(X_k, X_{k+1})}{p(X_k, X_{k+1})} \right\} \left[\sum_{l=0}^{\infty} \alpha^l f(X_{k+l+1}) \right] \right\}, \quad w.p.1. \tag{12.21}$$

This is equivalent to (c.f. (12.20))

$$\frac{\partial \eta}{\partial \delta} \approx \lim_{K \to \infty} \frac{1}{K} \sum_{k=0}^{K-1} \left\{ f(X_k) \sum_{l=0}^{k-1} \left[\alpha^{k-l-1} \frac{q(X_l, X_{l+1})}{p(X_l, X_{l+1})} \right] \right\}, \quad w.p.1. \tag{12.22}$$

An algorithm is developed in [2] to estimate $\frac{\partial \eta}{\partial \delta}$ using (12.22). It is easy to estimate $z_k := \sum_{l=0}^{k-1} \left[\alpha^{k-l-1} \frac{q(X_l, X_{l+1})}{p(X_l, X_{l+1})} \right]$ recursively:

$$z_{k+1} = \alpha z_k + \frac{q(X_k, X_{k+1})}{p(X_k, X_{k+1})}.$$

On the other hand, to estimate $\sum_{l=0}^{L-1} \left[\frac{q(X_{k+l}, X_{k+l+1})}{p(X_{k+l}, X_{k+l+1})} \right]$, one has to store L values.

In the third algorithm, we use (12.14) to obtain a $\hat{g}$. To this end, we first choose any regenerative state i^*. For convenience, we set $X_0 = i^*$ and define $u_0 = 0$, and $u_{m+1} = min\{n : n > u_m, X_m = i\}$ be the sequence of regenerative points. Set $g(i^*) = 0$. From (12.14), for any $X_n = i \neq i^*$ and $u_m \leq n < u_{m+1}$ we have

$$g(X_n) = d(i^*, i) = E\left\{ \sum_{l=n}^{t_{m+1}-1} [f(X_l) - \eta] \right\}.$$

With this and by (12.18), we have

$$\frac{d\eta}{d\delta} = E\left\{ \frac{q(X_k, X_{k+1})}{p(X_k, X_{k+1})} \left\{ \sum_{l=k+1}^{t_{m+1}-1} [f(X_l) - \eta] \right\} + h(X_k) \right\}. \tag{12.23}$$

Sample path based algorithms can then be developed, and we will not go into the details (the algorithms in [2] and [33] are in the same spirit as (12.23)).

12.3.3 Other Approaches for Gradient Estimation

Early works on single-sample-path-based performance gradient estimation include the perturbation analysis (PA) and the likelihood ratio (LR) (also called the score function (SF)) method.

PA has been successfully used for queueing networks. The basic principle of PA is as follows. A small change in a system parameter (such as the mean service time of a server) induces a series of changes on a sample path (such as changes in the service completion times); each change is called a *perturbation* of the sample path. The average effect of each perturbation on the system performance can be precisely measured by a quantity called perturbation *realization factor*. The total effect of the small change in the parameter on the system performance can then be calculated by adding together the average effects of all the perturbations induced by the parameter change. The sensitivity of the performance with respect to the parameter can then be determined. For more details, see [10, 21, 28].

The idea of perturbation realization was later extended to the sensitivity study of Markov processes. A perturbation on a sample path of a Markov process is a "jump" from a state i to another state j. The average effect of such a jump is measured by the realization factor $d(i, j)$, which equals $g(j) - g(i)$. The total effect of the change in a transition probability matrix can be measured by adding together the average effects of all the jumps induced by the change in the transition probability matrix. In fact, both sensitivity equations (12.2) and (12.3) can be derived using this approach. ((12.3) was first derived in [11] and [15] by the PA principles.) An advantage of this approach is that it can be used to derive sensitivity formulas for many problems flexibly; these formulas are otherwise difficult to conceive. (An example is given in Section 12.6.)

Another approach for gradient estimation is the LR or SF method. This approach is based on the importance sampling technique. The basic idea is: an event (a transition) that happens in a Markov chain with transition probability matrix P may happen in another Markov chain with transition matrix P' but with a different probability, which can be obtained by modifying the original probability with a weighting factor called the likelihood ratio, or the score function, which is similar to the $\frac{q(X_k, X_{k+1})}{p(X_k, X_{k+1})}$ term in our (12.17) to (12.19). For more details, see [25, 26, 38, 40].

PA is very efficient for many queueing systems; however it requires some "smoothness" properties from the sample performance function [8]. On the other hand, LR (SF) is widely applicable but suffers large variances. We will not elaborate more since both are wide topics. A comparison of PA and LR (SF) methods can be found in [9].

12.4   GRADIENT-BASED OPTIMIZATION

Any gradient estimate (PA, LR or SF, or the potential based estimates discussed in Section 12.3) can be used together with the standard stochastic gradient algorithms (Robbins-Monroe type, [39]) for optimizing the performance of Markov systems. For applications of PA and LR to the optimization problems, see, for example, [22, 27] and [26], respectively.

Reference [33] proposed a potential-based recursive algorithm for optimizing the average cost in finite state Markov reward processes that depend on a set of parameters denoted as θ. The approach is based on the regenerative structure of a Markov chain. The gradient estimate is similar to (12.23) except that the performance η is also estimated on the sample path and the gradient is not estimated explicitly in each step of the recursion, because its estimate is used in the stochastic gradient algorithm to determine the step size in a recursive procedure to reach the value θ at which the performance gradient is zero. The paper also provides a proof for the convergence of the algorithm. Variance reduction methods are proposed in [34].

The gradient based approach can be easily extended to the partially observable Markov decision processes (POMDP) (see, e.g., [2, 3]). The POMDP model in [2, 3] is described as follows. In addition to the state space $S = \{1, \cdots, M\}$, there is a control space denoted as $U = \{1, \cdots, N\}$ consisting of N controls and an observation space $Y = \{1, \cdots, L\}$ consisting of L observations. Each $u \in U$ determines a transition probability matrix P^u, which does not depend on the parameter θ. When the Markov chain is at state $i \in S$, an observation $y \in Y$ is obtained according to a probability distribution $\nu_i(y)$. For any observation y, we may choose a random policy $\mu_y(u)$, which is a probability distribution over the control space U. It is assumed that the distribution depends on the parameter θ and therefore is denoted as $\mu_y(\theta, u)$.

Given an observation distribution $\nu_i(y)$ and a random policy $\mu_y(\theta, u)$, the corresponding transition probabilities are

$$p_\theta(i, j) = \sum_{u,y} \left\{ \nu_i(y)\mu_y(\theta, u)p^u(i, j) \right\}.$$

Therefore,

$$\frac{d}{d\theta}p_\theta(i, j) = \sum_{u,y} \left\{ \nu_i(y)p^u(i, j)\frac{d}{d\theta}\mu_y(\theta, u) \right\}. \tag{12.24}$$

In POMDP, we assume that although the state $X_k, k = 0, 1, \cdots$, is not completely observable, the cost $f(X_k)$ is known. Thus, algorithms can be developed by replacing $q(i, j)$ with $\frac{d}{d\theta}p_\theta(i, j)$ of (12.24) in the algorithms developed for standard MDPs in Section 12.3.2. For example, if $h(i) = 0$ then (12.23) becomes

$$\frac{d\eta}{d\delta} = E\left\{ \frac{\frac{d}{d\theta}p_\theta(X_k, X_{k+1})}{p(X_k, X_{k+1})} \left\{ \sum_{l=k+1}^{t_{m+1}-1} [f(X_l) - \eta] \right\} \right\},$$

in which $f(X_k)$ is assumed to be observable.

A recursive algorithm called GPOMDP is presented in [2]. The algorithm uses a discount factor to approximate g (cf. (12.22)).

12.5   POLICY ITERATION

Policy iteration procedure in MDPs is a natural consequence of the performance difference equation (12.2). First, for two M-dimensional vectors a and b, we define $a = b$ if $a(i) = b(i)$ for all $i = 1, 2 \cdots, M$; $a \leq b$ if $a(i) < b(i)$ or $a(i) = b(i)$ for all $i = 1, 2 \cdots, M$; $a < b$ if $a(i) < b(i)$ for all $i = 1, 2 \cdots, M$; and $a \preceq b$ if $a(i) < b(i)$ for at least one i, and $a(j) = b(j)$ for other components. The relation $\leq$ includes $=$, $\preceq$, and $<$. Similar definitions are used for the relations $>$, $\succeq$, and $\geq$.

Next, we note that $\pi'(i) > 0$, $\forall i$, for ergodic chains. Thus, from (12.2), we know that if $Qg + h = (P' - P)g + (f' - f) \succeq 0$ then $\eta' - \eta > 0$. Thus, we have

Lemma 12.5.1 *If* $Pg + f \preceq P'g + f'$, *then* $\eta < \eta'$.

In an MDP, at any transition instant $n \geq 0$ of a Markov chain $\mathbf{X} = \{X_n, n \geq 0\}$, an action is chosen from an action space $\mathcal{A}$ and is applied to the Markov chain. We assume that the number of actions is finite, and we only consider stationary policies. A stationary policy is a mapping $\mathcal{L} : \mathcal{S} \rightarrow \mathcal{A}$, that is, for any state i, $\mathcal{L}$ specifies an action $\mathcal{L}(i) \in \mathcal{A}$. Let $\mathcal{E}$ be the policy space. If action α is taken at state i, then the state transition probabilities at state i are denoted as $p^\alpha(i, j)$, $j = 1, 2, \cdots, M$. With a policy $\mathcal{L}$, the Markov process evolves according to the transition matrix $P^\mathcal{L} = [p^{\mathcal{L}(i)}(i, j)]_{i=1}^M |_{j=1}^M$. We use the superscript $*^\mathcal{L}$ to denote the quantities associated with policy $\mathcal{L}$.

The cost depends on action and is denoted as $f(i, \alpha) = f(i, \mathcal{L}(i))$. The long-run average performance is $\eta^\mathcal{L}$. Our objective is to minimize this average cost over the policy space $\mathcal{S}$, i.e., to obtain $\min_{\mathcal{L} \in \mathcal{E}} \eta^\mathcal{L}$.

Define $f^\mathcal{L} = (f[1, \mathcal{L}(1)], \cdots, f[M, \mathcal{L}(M)])^T$. (12.5) becomes

$$(I - P^\mathcal{L} + e\pi^\mathcal{L})g^\mathcal{L} = f^\mathcal{L}, \qquad (12.25)$$

The following optimality theorem follows almost immediately from Lemma 12.5.1.

Theorem 12.5.1 *A policy $\mathcal{L}$ is optimal if and only if*

$$P^\mathcal{L} g^\mathcal{L} + f^\mathcal{L} \leq P^{\mathcal{L}'} g^\mathcal{L} + f^{\mathcal{L}'} \qquad (12.26)$$

for all $\mathcal{L}' \in \mathcal{E}$.

The optimality condition (12.26) is, of course, equivalent to the other conditions in the literature. To see this, we rewrite (12.25) in the following form:

$$\eta^{\mathcal{L}} e + g^{\mathcal{L}} = f^{\mathcal{L}} + P^{\mathcal{L}} g^{\mathcal{L}}. \tag{12.27}$$

Then Theorem 12.5.1 becomes: *A policy $\mathcal{L}$ is optimal if and only if*

$$\eta^{\mathcal{L}} e + g^{\mathcal{L}} = \min_{\mathcal{L}' \in \mathcal{E}} \{ P^{\mathcal{L}'} g^{\mathcal{L}} + f^{\mathcal{L}'} \}. \tag{12.28}$$

The minimum is taken for every component of the vector. (12.28) is the optimality equation, or the Bellman equation.

Policy iteration algorithms for determining the optimal policy can be easily developed by combining Lemma 12.5.1 and Theorem 12.5.1. Roughly speaking, at the kth step with policy $\mathcal{L}_k$, we set the policy for the next step (the $(k+1)$th step) as

$$\mathcal{L}_{k+1} = arg\{\min[P^{\mathcal{L}} g^{\mathcal{L}_k} + f^{\mathcal{L}}]\}, \quad (componentwise), \tag{12.29}$$

with $g^{\mathcal{L}_k}$ being the solution to the Poisson equation for $P^{\mathcal{L}_k}$. Lemma 12.5.1 implies that performance usually improves at each iteration. Theorem 12.5.1 shows that the minimum is reached when no performance improvement can be achieved. We shall not state the details here because they are standard. The extension of the above results to multi-chain Markov chains is presented in [19].

12.5.1 On-line Policy Iteration Algorithms

With the potentials estimated, we can implement policy iteration to obtain the optimal policy. The simplest way is that at each iteration k we run the system under policy $\mathcal{L}_k$ long enough to get an accurate estimate of $g^{\mathcal{L}_k}$, then determine the policy in the next iteration using (12.29) until no improvement can be made [13]. This approach requires a long sample path at each iteration. However, in order to make comparison of two policies to determine which one is better using (12.2), some errors in estimation are tolerable. This is especially true at the beginning of the iteration procedure since the performance is far from the optimum. (A precise study of this idea is formulated in the "ordinal optimization" approach, see [29].) Therefore, fast algorithms can be designed by combining the policy iteration method with stochastic approximation, with "rough" estimates of potentials at the beginning of the iterations.

Reference [24] proposes two such on-line optimization algorithms that implement policy iteration based on a sample path. In these algorithms, estimates of potentials are updated when the system visits a certain recurrent state. Since the policies are updated frequently, the potential estimates may reflect the values under the previous policies; thus, well-designed stochastic approximation schemes are used to guarantee the convergence of the policies to the optimal one. The difference between the two algorithms is that in one algorithm the entire policy is updated when the estimates of the potentials are made, and in the other the action for a particular state is updated only when the system visits the state. It is proved in the paper that starting from any initial value, the policy sequences obtained by these two algorithms will stop at the optimal one after a finite number of iterations. The on-line policy iteration approach is a counterpart of the online gradient based optimization approach presented in Section 12.4; the latter applies to parameterized systems and the former to systems within the MDP general structure.

12.5.2   Gradient-Based Policy Iteration

Comparing (12.2) with (12.3), we have the following observations [12, 18]:

1. The performance difference of any two policies P and P' (12.2) can be obtained by replacing π with π' in the formula for the performance derivative (12.3). For any P' (or any $Q = P' - P$) and h, the performance derivative can be calculated by solving for π and g for the system with P. But for performance difference, π' is needed for each P'.

2. At each iteration, one chooses the policy with the largest $|Qg + h|$, componentwise, as the policy in the next iteration. That is, policy iteration goes along the direction with the steepest performance gradient (with the largest $|\pi(Qg + h)|$).

3. Since π' is usually unknown, policy iteration must determine actions state by state; therefore, it does not apply to systems in which actions at different states are correlated. On the other hand, π is known at each iteration, we can calculate performance derivatives even when actions at some states are correlated.

With the above observations, [18] proposes a gradient-based policy iteration that may be applied to systems with correlated actions at different states. We treat the problem as performance optimization in a policy space with constraints for policies. The main idea is to iterate the policy along the steepest direction in the constrained region. However, there are some theoretical issues remaining. For example, the proposed approach may lead to a policy that is locally optimal, that is, at this point all the directional gradients to other policies are non-negative (in case of performance minimization). In addition, the convergence of the algorithms have to be proved.

The idea is best illustrated with an example. Consider an M/G/1/N queue, in which the service distribution is a Coxian distribution consisting of K stages, each is exponentially distributed with mean s_i, $i = 1, 2, \cdots, K$ ($K = 3$ in Figure 12.1). (For queueing theory, see [31].) After receiving service at stage k, a customer enters stage $k+1$ with probability a_k, and leaves the station with probability $b_k = 1 - a_k$, $a_K = 0$. Let n be the number of customers in the queue. N is the buffer size: when an arriving customer finds $n = N$, the customer is simply lost. Customers arrive with a Poisson process with rate λ. When there are n customers in the queue and the customer is served at stage k, an arriving customer has a probability of $\sigma(n, k)$ joining the queue, and $1 - \sigma(n, k)$ being rejected (the admission control problem); $\sigma(N, k) = 0$ for all k. This is equivalent to a load-dependent arrival rate $\lambda(n, k) = \lambda\sigma(n, k)$. The state space of the system is $\{0, (n, k), n = 1, 2, \cdots, N, k = 1, 2, \cdots, K\}$. The action

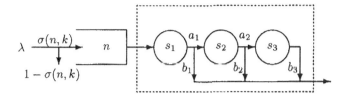

Fig. 12.1   An M/G/1/N Queue Admission Control

space is the set of values that $\sigma(n, k)$ may take; e.g., in the numerical example in [18], it is the set $\{0.1, 0.2, \cdots 0.9\}$.

Since the stage k is usually not observable, it is not feasible to implement a policy that depends on k. Denote the action as $\sigma(n, s)$ then we have the constraint

$$\sigma(n, 1) = \sigma(n, 2) = \sigma(n, 3) \equiv \sigma(n). \tag{12.30}$$

"n" is an "aggregated" state.

The problem can be converted to a discrete time MDP by uniformization. In the numerical example in [18], the cost function is set as $f(n, \lambda(n, k)) = n^2 + 9 \times (1/\lambda\sigma(n, k))^2$. Indeed, the standard policy iteration leads to an optimal policy which assigns different actions for different stages with the same n, which is not practically implementable.

To implement the gradient-based policy iteration, we need to calculate the performance gradient in the constraint policy space. In this example, let σ_0 be the current policy and σ be any other policy. Then the derivative along σ_0 to σ is

$$
\frac{d\eta}{d\delta} = \sum_{n=0}^{N} \sum_{i=1}^{K} \left\{ \pi^{\sigma_0}(n, i) [\sum_{m,j} \{ p^{\sigma}[(n, i), (m, j)] - p^{\sigma_0}[(n, i), (m, j)] \} g^{\sigma_0}(m, j) \right.
$$
$$
\left. + (f^{\sigma}(n, i) - f^{\sigma_0}(n, i))] \right\}. \tag{12.31}
$$

Defining a "W-factor" (cf. Q-factor)

$$
W(n, \sigma(n)) \equiv \sum_{i=1}^{K} \pi^{\sigma_0}(n, i) \left\{ \sum_{m,j} \{ p^{\sigma}[(n, i), (m, j)] g^{\sigma_0}(m, j) \} + f^{\sigma}(n, i) \right\}, \tag{12.32}
$$

we have

$$
\frac{d\eta}{d\delta} = \sum_{n=0}^{N} \{ W(n, \sigma(n)) - W(n, \sigma_0(n)) \}. \tag{12.33}
$$

For each n, $W(n, \sigma(n)) - W(n, \sigma_0(n))$ is called a *partial derivative*.

From the above discussion, with the current policy σ_0, we can calculate the W-factor for any action $\sigma(n)$ (for each aggregated state). Thus for any n we can choose an action with the smallest value of the W-factor (12.32), or equivalently the "steepest" partial derivative. From (12.33) we can get a policy having the steepest gradient and yet satisfying constraint (12.30). Using this policy as the policy in the next iteration, we can carry out policy iteration. Indeed, the numerical example in [18] shows that this procedure does stop at a feasible policy whose performance is very close to the one obtained by the standard policy iteration procedure (i.e., the optimal policy without the constraint (12.30)).

12.6   CONSTRUCTING PERFORMANCE GRADIENTS WITH POTENTIALS AS BUILDING BLOCKS

We have seen that potentials are the central piece in the performance sensitivity (difference and derivative) formulas. In fact, it can be shown that we can use potentials as building blocks to construct freely performance sensitivities for many special problems (see [17]). This "construction" approach is based on the following fundamental principle: changes in the values of system parameters induce "jumps" on a sample path; the effect of a jump from state i to j can be measured by the realization factors $d(i,j) = g(j) - g(i)$. In [17], it is shown that both (12.2) and (12.3) can be constructed in this way.

In this section, we use an example to show one application of this approach. Specifically, we show that in systems where only a part of the states is observable, we can aggregate the potentials on a set of states that share the same observable part of states and obtain the performance gradients. The aggregated potential can be estimated on a single sample path.

Consider the M/G/1/N queue shown in Figure 12.1. For illustrative purpose, we assume that the queue consists of K stages, all with the same mean $s_k \equiv s$, $k = 1, 2, \cdots, K$. Let $\mu = \frac{1}{s}$.

We are interested in the performance gradient with respect to $\sigma(n)$, for a given n, $n = 0, 1, \cdots, N - 1$. One way to get this gradient is to use (12.3); however, this requires us to model the system as a Markov chain and to estimate the potentials for all the states; many of the states may not be affected by the changes in $\sigma(n)$. In the following, we shall show that this gradient can be constructed directly by using potentials (or equivalently, realization factors) as building blocks without going through the standard Markov model. To illustrate the idea, we set $\lambda = \mu$ (assume $\mu > \lambda\sigma(n)$ for all n). With uniformization, we generate a Poisson process with rate $\lambda + K\mu = (K + 1)\lambda$, a point on this Poisson process is assigned to be the service completion time of one of the stages of the Coxian distribution or a customer arrival point with probability $\frac{\lambda}{\lambda + K\mu} = \frac{1}{K+1}$. When there are n customers in the system, with probability $\sigma(n)$, the arriving customer is accepted and the number of customers in the system increases to $n + 1$, and with probability $1 - \sigma(n)$, the arriving customer is rejected and the number of customers in the system remains to be n. Let the discretized Markov chain be denoted as $\mathbf{X} = \{X_0, X_1, \cdots, X_l, \cdots\}$.

We observe the system for a total of $L \gg 1$ transitions and obtain $\{X_0, X_1, \cdots, X_L\}$. Let $\tilde{p}(n)$ be the probability that a transition on the sample path is due to a customer arrival when there are n customers in the system before the arrival; we have

$$\tilde{p}(n) \approx \frac{L_n}{L},$$

where L_n is the number of such transitions in the observation period. Thus $L_n \approx L\tilde{p}(n)$, and the number of customers accepted by the system in the observation period when there are n customers in the system is $L_n^+ \approx L\tilde{p}(n)\sigma(n)$. Now suppose $\sigma(n)$ changes to $\sigma(n) + \Delta\sigma(n)$. Then $L\tilde{p}(n)\Delta\sigma(n)$ is the number of additional arrivals that are admitted to the system when there are n customers in it due to the change $\Delta\sigma(n)$. Next, we have

$$\tilde{p}(n) = \sum_{k=1}^{K} \tilde{p}(n, k),$$

where $\tilde{p}(n, k)$ is the probability that a transition in $\mathbf{X}$ is a customer arrival when the state is (n, k). Therefore, $L\tilde{p}(n, k)\Delta\sigma(n)$ is the number of additional arrivals that are admitted to the system when the state is (n, k) due to the change $\Delta\sigma(n)$. The average effect of such an additional arrival is $d[(n, k); (n + 1, k)]$. Thus the total effect of all these additional admitted customers on the performance is

$$\Delta F_L = \sum_{k=1}^{K} L\tilde{p}(n, k)d[(n, k); (n + 1, k)]\Delta\sigma(n).$$

Using $\Delta\eta = \frac{\Delta F_L}{L}$ and letting $\Delta\sigma(n) \to 0$, we get

$$
\begin{aligned}
\frac{d\eta}{d\sigma(n)} &= \sum_{k=1}^{K} \tilde{p}(n, k)d[(n, k); (n + 1, k)] \\
&= \tilde{p}(n) \sum_{k=1}^{K} \tilde{p}(k|n)d[(n, k); (n + 1, k)] \\
&= \tilde{p}(n) \left[\sum_{k=1}^{K} \tilde{p}(k|n)g(n + 1, k) - \sum_{k=1}^{K} \tilde{p}(k|n)g(n, k) \right] \\
&= \tilde{p}(n)[g^+(n) - g^-(n)], \quad\quad\quad\quad\quad (12.34)
\end{aligned}
$$

where $\tilde{p}(k|n) = \frac{\tilde{p}(n,k)}{\tilde{p}(n)}$ is the conditional probability, $g^+(n) = \sum_{k=1}^{K} \tilde{p}(k|n)g(n + 1, k)$ is the mean of the potentials in set $\mathcal{S}_{n+1} = \{(n + 1, k) : k = 1, \cdots, K\}$ given that n jumps to $n + 1$, and $g^-(n) = \sum_{k=1}^{K} \tilde{p}(k|n)g(n, k)$ is the mean of the potentials in set $\mathcal{S}_n$ given that n remains the same.

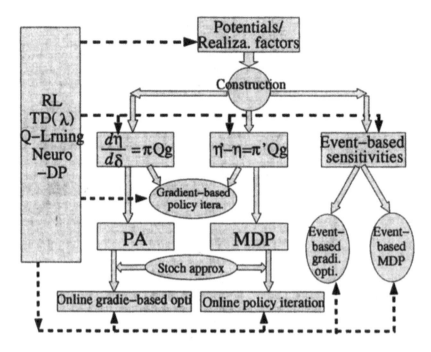

Fig. 12.2   A system point of view of learning and optimization.

Finally, $\tilde{p}(n)$, $g^+(n)$, and $g^-(n)$ can be estimated on a single sample path. This is simply because both arrivals and transitions from n to $n+1$ or from n to n are all observable events on the sample path. In Eq. (12.34), the potentials for states with the same state n are aggregated together; the number of potentials to be estimated reduces.

12.7   CONCLUSION

We have provided an overview of learning and optimization from a system point of view. It provides a unified framework for PA, MDP, and RL. This new perspective does lead to some new research directions, such as the gradient-based optimization and policy iteration and the event based sensitivity analysis by the construction method. Further research is needed for these topics. We summarize the results in Figure 12.2. The results can be easily extended to problems with discounted costs; see [16].

From a system theoretic perspective, Section 12.6 should be presented first: the two sensitivity formulas (12.3) and (12.2) can be constructed by using potentials as building blocks, this is a result from a system point of view. Other results can be derived from these two equations. However, to directly getting into the main subjects, we moved this section to the last.

Acknowledgments

This work was supported in part by a grant from Hong Kong UGC. Tel: (852) 2358-7048; Fax: (852) 2358-1485; Email: eecao@ee.ust.hk.

Bibliography

1. A. Arapostathis, V. S. Borkar, E. Fernandez-Gaucherand, M. K. Ghosh, and S. I. Marcus, Discrete-time controlled Markov processes with average cost criterion: A survey, *SIAM Journal Control and Optimization*, vol. 31, pp. 282–344, 1993.

2. J. Baxter and P. L. Bartlett, Infinite-horizon policy-gradient estimation, *Journal of Artificial Intelligence Research*, vol. 15, pp. 319–350, 2001.

3. J. Baxter, P. L. Bartlett, and L. Weaver, Experiments with infinite-horizon policy-gradient estimation, *Journal of Artificial Intelligence Research*, vol. 15, pp. 351–381, 2001.

4. D. P. Bertsekas, *Dynamic Programming and Optimal Control, Volume I and II*, Athena Scientific, Belmont, MA, 1995.

5. D. P. Bertsekas and J. N. Tsitsiklis, *Neuro-Dynamic Programming*, Athena Scientific, Belmont, MA, 1996.

6. L. Breiman, *Probability*, Addison-Wesley, Reading, MA, 1968. Springer-Verlag, New York, 1994.

7. P. Brémaud, *Markov Chains: Gibbs Fields, Monte Carlo Simulation, and Queues*, Springer-Verlag, New York, 1998.

8. X. R. Cao, Convergence of parameter sensitivity estimates in a stochastic experiment, *IEEE Trans. Automatic Control*, vol. AC-30, pp. 834–843, 1985.

9. X. R. Cao, Sensitivity estimates based on one realization of a stochastic system, *Journal of Statistical Computation and Simulation*, vol. 27, pp. 211–232, 1987.

10. X. R. Cao, *Realization Probabilities: The Dynamics of Queueing Systems*, Springer-Verlag, New York, 1994.

11. X. R. Cao, X. M. Yuan, and L. Qiu, A single sample path-based performance sensitivity formula for Markov chains, *IEEE Trans. Automatic Control*, vol. 41, pp. 1814–1817, 1996.

12. X. R. Cao, The relation among potentials, perturbation analysis, Markov decision processes, and other topics, *Journal of Discrete Event Dynamic Systems,* vol. 8, pp. 71–87, 1998.

13. X. R. Cao, Single sample path based optimization of Markov chains, *Journal of Optimization: Theory and Application,* vol. 100, no. 3, pp. 527–548, 1999.

14. X. R. Cao, A basic formula for on-line policy-gradient algorithms, submitted, 2004

15. X. R. Cao and H. F. Chen, Perturbation realization, potentials and sensitivity analysis of Markov processes, *IEEE Trans. Automatic Control,* vol. 42, pp. 1382–1393, 1997.

16. X. R. Cao, A unified approach to Markov decision problems and performance sensitivity analysis, *Automatica,* vol. 36, pp. 771–774, 2000.

17. X. R. Cao, Constructing performance sensitivities for Markov systems with potentials as building blocks, Proc. of the 42nd IEEE Conference on Decision and Control, Maui, Hawaii, 2003.

18. X. R. Cao and H. T. Fang, Gradient-based policy iteration: An example, *Proc. of 2002 IEEE Conference on Decision and Control,* pp. 3367–3371, 2002.

19. X. R. Cao and X. P. Guo, A unified approach to Markov decision problems and sensitivity analysis with discounted and average criteria: The multichain case, submitted.

20. X. R. Cao and Y. W. Wan, Algorithms for sensitivity analysis of Markov systems through potentials and perturbation realization, *IEEE Trans. Control System Tech,* vol. 6, pp. 482–494, 1998.

21. C. G. Cassandras and S. Lafortune, *Introduction to Discrete Event Systems,* Kluwer Academic Publishers, Boston, 1999.

22. E. K. P. Chong and P. J. Ramadge, Stochastic optimization of regenerative systems using infinitesimal perturbation analysis, *IEEE Trans. Automatic Control,* vol. 39, pp. 1400–1410, 1994.

23. E. Çinlar, *Introduction to Stochastic Processes,* Prentice-Hall, Englewood Cliffs, NJ, 1975.

24. H.-T. Fang and X. R. Cao, Potential-based on-line policy iteration algorithms for Markov decision processes, *IEEE Trans. Automatic Control,* vol. 49, no. 4, pp. 493–505, 2004.

25. P. W. Glynn, Likelihood ratio gradient estimation: An overview, in A. Thesen, H. Grant, and K. D. Kelton, (eds.), *Proc. of 1987 Winter Simulation Conference,* pp. 366–375, Society for Computer Simulation, San Diego, CA, 1988.

26. P. W. Glynn, Optimization of stochastic systems via simulation, in A. Thesen, H. Grant, and K. D. Kelton, (eds.), *Proc. of 1987 Winter Simulation Conference,* pp. 90–105, Society for Computer Simulation, San Diego, CA, 1988.

27. Y. C. Ho and X. R. Cao, Perturbation analysis and optimization of queueing networks, *Journal of Optimization Theory and Applications,* vol. 40, no. 4, pp. 559–582, 1983.

28. Y. C. Ho and X. R. Cao, *Perturbation Analysis of Discrete-Event Dynamic Systems,* Kluwer Academic Publisher, Boston, 1991.

29. L. H. Lee, E. T. K. Lau and Y. C. Ho, Explanation of goal softening in ordinal optimization, *IEEE Trans. Automatic Control,* vol. 44, pp. 94–99, 1999.

30. T. Jaakkola, S. P. Singh, and M. I. Jordan, Reinforcement learning algorithm for partially observable Markov decision problems, *Advances in Neural Information Processing Systems,* vol. 7, pp. 345–352, Morgan Kaufman, San Francisco, CA, 1995.

31. L. Kleinrock, *Queueing Systems, Vol. 1: Theory,* Wiley, New York, 1975.

32. V. R. Konda and V. S. Borkar, Actor-critic-type learning algorithms for Markov decision processes, *SIAM Journal of Control Optimization,* vol. 38, pp. 94–123, 1999.

33. P. Marbach and J. N. Tsitsiklis, Simulation-based optimization of Markov reward processes, *IEEE Trans. Automatic Control,* vol. 46, pp. 191–209, 2001.

34. P. Marbach and J. N. Tsitsiklis, Approximate gradient methods in policy-space optimization of Markov reward processes, *Journal of Discrete Event Dynamic Systems,l,* vol. 13, no. 1, pp. 111–148, 2003.

35. S. P. Meyn and R. L. Tweedie, *Markov Chains and Stochastic Stability,* Springer-Verlag, London, 1993.

36. E. L. Plambeck, B. R. Fu, S. M. Robinson, and R. Suri, Sample-path optimization of convex stochastic performance functions, *Math. Program. B,* vol. 75, pp. 137–176, 1996.

37. M. L. Puterman, *Markov Decision Processes: Discrete Stochastic Dynamic Programming,* Wiley, New York, 1994.

38. M. I. Reiman and A. Weiss, Sensitivity analysis via likelihood ratio, *Operations Research,* vol. 37, pp. 830–844, 1989.

39. H. Robbins and S. Monro, A stochastic approximation method, *Annals of Mathematical Statistics,* vol. 22, pp. 400–407, 1951.

40. R. V. Rubinstein, *Monte Carlo Optimization, Simulation, and Sensitivity Analysis of Queueing Networks,* Wiley, New York, 1986.

41. R. Suri and Y. T. Leung, Single run optimization of discrete event simulations — An empirical study using the M/M/1 queue, *IIE Trans.,* vol. 21, pp. 35–49, 1989.

42. R. S. Sutton, Learning to predict by the methods of temporal differences, *Machine Learning,* vol. 3, pp. 835–846, 1988.

43. R. S. Sutton and A. G. Barto, *Reinforcement Learning: An Introduction,* MIT Press, Cambridge, MA, 1998.

44. J. N. Tsitsiklis and V. R. Konda, Actor-critic algorithms, *Tech. Rep., Lab. Inform. Decision Systems,* MIT, Cambridge, MA, 2001, Preprint.

45. J. N. Tsitsiklis and B. Van Roy, Feature-based methods for large-scale dynamic programming, *Machine Learning,* vol. 22, pp. 59-94, 1994.

46. J. N. Tsitsiklis and B. Van Roy, An analysis of temporal-difference learning with function approximation, *IEEE Trans. Automatic Control,* vol. 42, pp. 674–690, 1997.

47. J. N. Tsitsiklis and B. Van Roy, Average cost temporal-difference learning, *Automatica,* vol. 35, pp. 1799–1808, 1999.

48. C. Watkins, *Learning from Delayed Rewards,* Ph.D. Thesis, Cambridge University, Cambridge, UK, 1989.

49. C. Watkins and P. Dayan, *Q*-learning, *Machine Learning,* vol. 8, pp. 279–292, 1992.

13 Robust Reinforcement Learning Using Integral-Quadratic Constraints

CHARLES W. ANDERSON   MATT KRETCHMAR
PETER YOUNG   DOUGLAS HITTLE

Colorado State University

Editor's Summary: This chapter presents a synthesis of robust control and reinforcement learning ideas. This chapter begins by discussing integral-quadratic constraints and stability analysis and then introduces reinforcement learning into the robust control framework. Several small demonstrations of this technique are provided in this chapter, and a substantial case study is done in Chapter 20 on a Heating, Ventilation, and Air Conditioning system.

13.1 INTRODUCTION

Modern control techniques rely upon mathematical models derived from physical systems as the basis for controller design. Problems arise when the mathematical models do not capture all aspects of the physical system. Optimal control design techniques can result in controllers that perform very well on the model, but may perform poorly on the physical plant. The field of robust control addresses this issue by incorporating *uncertainty* into the mathematical model. Robust optimization techniques are applied to the model to derive a controller that is guaranteed to maintain stability as long as the uncertainty bounds are not violated [2, 11, 14]. Robust controllers typically do not perform as well on the model because the uncertainty keeps them from exploiting all of the model's dynamics. However, a well-designed robust controller will often perform better than the optimal controller on the physical plant.

Thus robust control theory provides a framework of algorithms for designing feedback controllers for which stability may be guaranteed under broad conditions. Uncertainty in the knowledge of the parameters and relationships inherent in the system to be controlled result in the conservative nature of the controller's performance. Adaptive schemes for learning a better controller on-line—while interacting with the controlled system—hold promise for improvements in performance, but to maintain guarantees of stability, such schemes must be embedded within the robust control framework.

In this chapter, such a synthesis of robust control and learning is presented. A stability analysis technique based on integral-quadratic constraints (IQCs) on signals in the system and controller is combined with a reinforcement learning algorithm. The result is a robustly-designed feedback controller in parallel with a reinforcement learning agent that learns to improve performance on-line with guaranteed stability while learning.

Section 13.2 summarizes the use of IQCs for stability analysis. The embedding of the reinforcement learning agent within the IQC framework is described in Section 13.3. Following this, two demonstrations are presented in Section 13.4. The control task for the first demonstration is very simple and shows how the robust control constraints and reinforcement learning interact. The second demonstration is also simple, but complex enough to show the advantage provided by the robust control framework, without which the reinforcement learning agent temporarily learns an unstable controller before it converges on a stable controller. See Kretchmar et al. [4, 5] for further details on this approach.

13.2 INTEGRAL-QUADRATIC CONSTRAINTS AND STABILITY ANALYSIS

Integral Quadratic Constraint analysis (IQC) is a tool for verifying the stability of systems with uncertainty [6–8]. In this section, a brief overview of the main concepts is presented.

Consider the feedback interconnection shown in Figure 13.1. The upper block, M, is a known Linear-Time-Invariant (LTI) system, and the lower block, Δ, is a block-diagonal, structured uncertainty. An *Integral Quadratic Constraint* (IQC) is an inequality describing the relationship between the two signals, w and v, characterized by a Hermitian matrix function Π as:

$$\int_{-\infty}^{\infty} \left| \begin{array}{c} \hat{v}(j\omega) \\ \hat{w}(j\omega) \end{array} \right|^* \Pi(j\omega) \left| \begin{array}{c} \hat{v}(j\omega) \\ \hat{w}(j\omega) \end{array} \right| d\omega \geq 0, \tag{13.1}$$

where $\hat{v}$ and $\hat{w}$ are the Fourier Transforms of $v(t)$ and $w(t)$. The basic IQC stability theorem can be stated as follows.

Theorem 13.2.1 *Consider the interconnection system represented in Figure 13.1 and given by the equations*

$$v = Mw + f, \tag{13.2}$$

$$w = \Delta(v) + e. \tag{13.3}$$

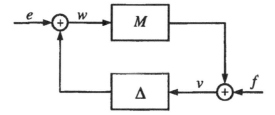

Fig. 13.1 Organization of a feedback system into a known, linear-time invariant system, M, and an unknown, block-diagonal, uncertainty, δ, defined by integral quadratic constraints defining the relationship between w and v. e and f are signals external to the feedback loop.

Assume that:

- *$M(s)$ is a stable, proper, real-rational transfer matrix, and Δ is a bounded, causal operator.*

- *The interconnection of M and $\tau\Delta$ is well-posed for all $\tau \in [0,1]$. (i.e., the map from $(v, w) \rightarrow (e, f)$ has a causal inverse)*

- *The IQC defined by Π is satisfied by $\tau\Delta$ for all $\tau \in [0,1]$.*

- *There exists an $\epsilon > 0$ such that for all ω:*

$$\left| \begin{array}{c} M(j\omega) \\ I \end{array} \right|^* \Pi(j\omega) \left| \begin{array}{c} M(j\omega) \\ I \end{array} \right| \leq -\epsilon I. \qquad (13.4)$$

Then the feedback interconnection of M and Δ is stable.

The power of this result lies in both its generality and its computability. Many system interconnections can be rearranged into the canonical form of Figure 13.1 [9]. Also, many types of uncertainty descriptions can be well captured as IQCs, including norm bounds, rate bounds, both linear and non-linear uncertainty, time-varying and time-invariant uncertainty, and both parametric and dynamic uncertainty. Hence this result can be applied in many situations, often without too much conservatism [7, 8]. Moreover, a library of IQCs for common uncertainties is available [6], and more complex IQCs can be built by combining the basic IQCs.

The computation involved to meet the requirements of the theorem is tractable, since the theorem requirements can be transformed into a Linear Matrix Inequality (LMI) as follows. Suppose that the IQCs that cover Δ are parameterized and hence are candidates to satisfy Theorem 13.2.1, as:

$$\Pi(j\omega) = \sum_{i=1}^{n} p_i \Pi_i(j\omega), \qquad (13.5)$$

where p_i are positive real parameters. Then the state space realizations of M and Π_i can be used to write the IQC components as:

$$
\left| \begin{array}{c} M(j\omega) \\ I \end{array} \right|^{*} \Pi_i(j\omega) \left| \begin{array}{c} M(j\omega) \\ I \end{array} \right| = \left| \begin{array}{c} (j\omega I - A)^{-1}B \\ I \end{array} \right|^{*} P_i \left| \begin{array}{c} (j\omega I - A)^{-1}B \\ I \end{array} \right|,
$$

(13.6)

where A is a Hurwitz matrix and P_i are real symmetric matrices. It follows from the Kalman-Yacubovich-Popov (KYP) lemma [10] that the existence of a solution to (13.4) is equivalent to the existence of a symmetric matrix Q such that

$$
\left[\begin{array}{cc} QA + A^T Q & QB \\ B^T Q & 0 \end{array} \right] + \sum_{i=1}^{n} p_i P_i < 0,
$$

(13.7)

which is a finite-dimensional LMI feasibility problem in the variables p_i and Q. As is well known, LMIs are convex optimization problems for which there exist fast, commercially available, polynomial time algorithms [3]. A Matlab IQC toolbox is available at http://web.mit.edu/~cykao/home.html that provides an implementation of an IQC library in Simulink[1], facilitating an easy-to-use graphical interface for setting up IQC problems. Moreover, the toolbox integrates an efficient LMI solver to provide a powerful comprehensive tool for IQC analysis. This toolbox was used for the demonstrations in this chapter.

13.3 REINFORCEMENT LEARNING IN THE ROBUST CONTROL FRAMEWORK

Reinforcement learning methods embody a general Monte Carlo approach to dynamic programming for solving optimal control problems. Sutton and Barto [12] and Bertsekas and Tsitsiklis [1] have written detailed introductions to reinforcement learning problems and algorithms. *Q-learning* procedures converge on value functions for state-action pairs that estimate the expected sum of future reinforcements defined to reflect behavior goals that might involve costs, errors, or profits [13].

To define the Q-learning algorithm, start by representing a system to be controlled as consisting of a discrete state space, S, and a finite set of actions, A, that can be taken in all states. A *policy* is defined by the probability, $\pi(s_t, a)$, that action a will be taken in state s_t. Let the reinforcement resulting from applying action a_t while the system is in state s_t be $R(s_t, a_t)$. $Q_\pi(s_t, a_t)$ is the value function given state s_t and action a_t, assuming policy π governs action selection from then on. Thus the

---

[1]Part of the Matlab commercial software package. See http://www.mathworks.com.

desired value of $Q_\pi(s_t, a_t)$ is

$$Q_\pi(s_t, a_t) = E_\pi \left\{ \sum_{k=0}^{T} \gamma^k R(s_{t+k}, a_{t+k}) \right\}, \tag{13.8}$$

where γ is a discount factor between 0 and 1 that weights reinforcement received sooner more heavily than reinforcement received later. R is later defined to be a function of the magnitude of the state tracking error, so actions are preferred with minimal Q values. From the above expression, a procedure for updating the Q values can be defined by deriving the following Monte Carlo version of value iteration:

$$\Delta Q_\pi(s_t, a_t) = \alpha_t \left[R(s_t, a_t) + \gamma \min_{a' \in A} Q_\pi(s_{t+1}, a') - Q_\pi(s_t, a_t) \right]. \tag{13.9}$$

This is what has become known as the Q-learning algorithm. Watkins [13] proves that it does converge to the optimal value function, meaning that selecting the action, a, that minimizes $Q(s_t, a)$ for any state s_t will result in the optimal sum of reinforcement over time. The proof of convergence assumes that the sequence of step sizes α_t satisfies the stochastic approximation conditions $\sum \alpha_t = \infty$ and $\sum \alpha_t^2 < \infty$. It also assumes that every state and action are visited infinitely often and every Q value is represented independently of all others.

The Q function implicitly defines the policy, π, defined as

$$\pi(s_t) = \operatorname*{argmin}_{a \in A} Q(s_t, a). \tag{13.10}$$

However, as Q is being learned, π will certainly not be an optimal policy. A variety of actions must be taken from every state in order to learn sufficiently accurate Q values for the state-action pairs that are encountered. In the following experiment, a random action was taken with probability p_t for step t, where $p_{t+1} = \lambda p_t$, $p_0 = 1$, and $0 < \lambda < 1$. Thus the value of p_t approaches 0 with time and the policy slowly shifts from a random policy to one determined by the learned Q function.

Here Q-learning is combined with a fixed, robust controller by adding the action chosen by the policy to the output of the fixed controller. The reinforcement learning objective is to find a policy that minimizes the sum of the magnitude of the difference between a reference input and the system's state variable that is meant to track that reference input. To embed Q-learning in the robust control framework, the policy must be defined explicitly in a form that can be included in stability analysis. To do so, an artificial neural network, called the actor network, is trained to duplicate the Q-learning policy. This architecture is shown in Figure 13.2. The Q function is learned by the critic component that, in the following demonstrations, is implemented as a table.

The Q function is not part of the feedback loop, so has no direct effect on the stability analysis. However, the actor network is part of the feedback loop and is part of the stability analysis. The activation functions in the units of the actor network are

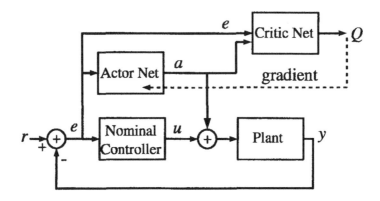

Fig. 13.2   Demonstration 1: Control System with Learning Agent

non-linear, and the modifiable weights in each unit are time varying. Both aspects must be represented with IQC's before IQC analysis can be applied to the whole system.

Consider a neural network with input vector $x = (x_1, ..., x_n)$ and output vector $a = (a_1, ..., a_m)$. The network has h hidden units, input weight matrix W_{hxn}, and output weight matrix V_{mxh}. The hidden unit activation function is the hyperbolic tangent function, which produces the hidden unit outputs as vector $\Phi = (\phi_1, \phi_2, \ldots, \phi_h)$. The neural network computes its output by

$$\Phi = Wx, \tag{13.11}$$

$$a = V \tanh(\Phi). \tag{13.12}$$

This can be rewritten as

$$\Phi = Wx, \tag{13.13}$$

$$g_j = \begin{cases} \frac{\tanh(\phi_j)}{\phi_j}, & \text{if } \phi_j \neq 0; \\ 1, & \text{if } \phi_j = 0, \end{cases} \tag{13.14}$$

$$\Gamma = \text{diag}\{g_j\}, \tag{13.15}$$

$$a = V\Gamma\Phi. \tag{13.16}$$

The function, g, computes the output of the hidden unit divided by the input of the hidden unit; this is the *gain* of the hyperbolic tangent hidden unit. The non-linearity is captured in the diagonal matrix, Γ. This matrix is composed of individual hidden unit gains, g, distributed along the diagonal, that act as non-linear gains via

$$w(t) = gv(t) = \left(\frac{\tanh(v(t))}{v(t)}\right) v(t) = \tanh(v(t)) \tag{13.17}$$

for input signal $v(t)$ and output signal $w(t)$. This non-linearity can be represented by the IQC, ψ_1, where $w(t) = \psi_1(v(t))$ and (see [6]):

$$\psi_1(-v) = -\psi_1(v), \tag{13.18}$$

$$\alpha(v_1 - v_2)^2 \leq (\psi_1(v_1) - \psi_1(v_2))(v_1 - v_2) \leq \beta(v_1 - v_2)^2 \tag{13.19}$$

Letting $\alpha = 0$ and $\beta = 1$, $\tanh(v)$ satisfies the odd condition

$$\tanh(-v) = -\tanh(v) \tag{13.20}$$

and the bounded slope condition

$$0 \leq (\tanh(v_1) - \tanh(v_2))(v_1 - v_2) \leq (v_1 - v_2)^2, \tag{13.21}$$

which is equivalent to (assuming without loss of generality that $v_1 > v_2$)

$$0 \leq (\tanh(v_1) - \tanh(v_2)) \leq (v_1 - v_2). \tag{13.22}$$

For this IQC,

$$\Pi(j\omega) = \begin{bmatrix} 0 & 1 + \frac{p}{j\omega+1} \\ 1 + \frac{p}{-j\omega+1} & -2\left(1 + \mathrm{Re}\left(\frac{p}{j\omega+1}\right)\right) \end{bmatrix} \tag{13.23}$$

with the additional constraint on the (otherwise free) parameter p that $|p| \leq 1$ (which is trivially reformulated as another IQC constraint on p). Note that this is the actual IQC used for analysis, and it is based on a scaling of $H(s) = \frac{1}{s+1}$, but one can attempt to get more accuracy at the expense of increased computation by using a more general scaling $H(s)$. In fact, it can be any transfer function whose $\mathcal{L}_1$ norm does not exceed one [7].

To represent the time-varying weights in the actor network, the *slowly time-varying, real scalar* IQC, ψ_2, is used, where $w(t) = \psi_2(t)v(t)$, and (see [7]):

$$|\psi_2(t)| \leq \beta, \tag{13.24}$$

$$|\dot{\psi}_2(t)| \leq \alpha. \tag{13.25}$$

The neural network learning rate determines the bounding constant, α, and β is assigned the largest allowable β for which stability can be proved. This determines a safe neighborhood in which the network is allowed to learn. Having determined α and β, the corresponding IQC's specialized to this problem can be stated as

$$\int_{-\infty}^{\infty} \begin{vmatrix} \hat{v}_{\mathrm{ext}}(j\omega) \\ \hat{w}_{\mathrm{ext}}(j\omega) \end{vmatrix}^* \begin{bmatrix} \beta^2 K_1 & M_1 \\ M_1^T & -K_1 \end{bmatrix} \begin{vmatrix} \hat{v}_{\mathrm{ext}}(j\omega) \\ \hat{w}_{\mathrm{ext}}(j\omega) \end{vmatrix} d\omega \geq 0, \tag{13.26}$$

and

$$\int_{-\infty}^{\infty} \begin{vmatrix} \hat{y}(j\omega) \\ \hat{u}(j\omega) \end{vmatrix}^* \begin{bmatrix} \alpha^2 K_2 & M_2 \\ M_2^T & -K_2 \end{bmatrix} \begin{vmatrix} \hat{y}(j\omega) \\ \hat{u}(j\omega) \end{vmatrix} d\omega \geq 0, \tag{13.27}$$

where the free parameters K_1, K_2, M_1, M_2 are subject to the additional (IQC) constraints that K_1, K_2 are symmetric positive definite matrices, and M_1, M_2 are skew-symmetric matrices. The signals $v_{\text{ext}}, w_{\text{ext}}$ are defined in terms of v, w and an additional (free) signal u as

$$\hat{v}_{\text{ext}}(s) = \left[\begin{array}{c} \frac{\hat{v}(s)}{s+1} \\ \hat{v}(s) \end{array} \right], \qquad \hat{w}_{\text{ext}}(s) = \left[\begin{array}{c} \frac{\hat{w}(s)}{s+1} + \frac{\hat{u}(s)}{s+1} \\ \hat{w}(s) \end{array} \right]. \qquad (13.28)$$

Note again that this is the actual IQC used for analysis, but in fact there are free scaling parameters in this IQC which are simply assigned as $\frac{1}{s+1}$. A more general statement of this IQC (with more general scalings) can be found in [6].

The remainder of this section describes the algorithm for robust, reinforcement learning control used for the demonstrations in the following section.

The input to the actor network is the error, e_t, between the reference input, r_t, and the actual state variable, y_t, that must track the reference:

$$e_t = r_t - y_t. \qquad (13.29)$$

Given e_t, the outputs of the hidden units, Φ_t, and the output unit, a_t, are

$$\Phi_t = \tanh(W_t e_t), \qquad (13.30)$$

$$a_t = \begin{cases} V_t \Phi_t, & \text{with probability } 1 - \epsilon_t; \\ V_t \Phi_t + a_{\text{rand}}, & \text{with probability } \epsilon_t, \text{ where } a_{\text{rand}} \text{ is a Gaussian} \\ & \text{random variable with mean 0 and variance 0.05.} \end{cases} \qquad (13.31)$$

Let the control law implemented by the robust, fixed controller be f. The outputs of the fixed controller and the actor network are summed and applied as the control signal u_t to the plant being controlled. Let h be the model of the plant that maps control input u to the next sampled value of y.

$$u_t = f(e_t) + a_t \qquad (13.32)$$

$$y_{t+1} = h(u_t) \qquad (13.33)$$

Again calculate the error, e_{t+1}, and the hidden and output values of the neural network, Φ_{t+1} and a_{t+1}.

$$e_{t+1} = r_{t+1} - y_{t+1} \qquad (13.34)$$

$$\Phi_{t+1} = \tanh(W_t e_{t+1}) \qquad (13.35)$$

$$a_{t+1} = \begin{cases} V_t \Phi_{t+1}, & \text{with probability } 1 - \epsilon_{t+1}; \\ V_t \Phi_{t+1} + a_{\text{rand}}, & \text{with probability } \epsilon_{t+1}, \text{ where } a_{\text{rand}} \text{ is a Gaussian} \\ & \text{random variable with mean 0 and variance 0.05} \end{cases} \qquad (13.36)$$

The reinforcement, R_{t+1}, is

$$R_{t+1} = |e_{t+1}|. \tag{13.37}$$

e_t and a_t are the inputs to the Q function table. Let Q_{index} be a function that maps e_t and a_t to the corresponding index into the Q table. To update the actor network, the optimal action, a_t^*, at step t is estimated by minimizing the value of Q for n different action inputs, a_i, in a local region of the action space centered on a_t, for which the estimate of the optimal action is given by

$$a_t^* = \operatorname*{argmin}_{a_i} Q_{Q_{\text{index}}(e_t, a_i)}. \tag{13.38}$$

Updates to the weights of the actor network are proportional to the difference between this estimated optimal action and the actual action:

$$V_{t+1} = V_t + \beta(a_t^* - a_t)\Phi_t^T, \tag{13.39}$$

$$W_{t+1} = W_t + \beta V^T(a_t^* - a_t) \cdot (1 - \Phi_t \cdot \Phi_t)e_t, \tag{13.40}$$

where $\cdot$ represents component-wise multiplication. The Q value for step t is updated by

$$Q_{Q_{\text{index}}(e_t, a_t)} = Q_{Q_{\text{index}}(e_t, a_t)} + \alpha(R_{t+1} + \gamma Q_{Q_{\text{index}}(e_{t+1}, a_{t+1})} - Q_{Q_{\text{index}}(e_t, a_t)}). \tag{13.41}$$

Now it must be determined whether or not the new weight values, W_{t+1} and V_{t+1}, remain within the stable region S. Initial values for W and V are random variables from a Gaussian distribution with mean zero and variance of 0.1. The stable region, S, is set to a rectangle in the multi-dimensional weight space and is initially centered at zero. The size of S is determined by an iterative expanding search described below involving small increases to the size and a stability analysis until a maximum size is reached or stability cannot be guaranteed. If the new weight values fall within S, S remains unchanged. Otherwise a new value for S is determined by the following procedure.

The actor network's weight values are collected into one vector, C. An initial guess at allowed weight perturbations, P, as factors of the current weights is defined to be proportional to the current weight values:

$$C = (W_t, V_t) = (c_1, c_2, \dots), \tag{13.42}$$

$$P = \frac{C}{\sum_i c_i}. \tag{13.43}$$

These perturbation factors are adjusted to estimate the largest factors for which the system remains stable. Let z_u and z_s be scalar multipliers of the perturbation factors for which the system is not-guaranteed and guaranteed stable, respectively. They are initialized to $z_u = z_s = 1$. The value of z_u is iteratively increased by $2z_u$ until the system with actor weights $C \pm z_u P \cdot C$ is not guaranteed stable. Similarly, the value of z_s is iteratively decreased by $\frac{1}{2} z_s$ until the system with actor weights $C \pm z_s P \cdot C$ is guaranteed stable. Now a finer search is performed to decrease the interval between z_s and z_u :

$$\text{While } \frac{z_u - z_s}{z_s} > 0.05 \text{ do} \tag{13.44}$$

$$z_m = \frac{z_u + z_s}{2} \tag{13.45}$$

$$\text{If not stable for } C \pm z_m P \cdot C, \text{ set } z_u = z_m, \tag{13.46}$$

$$\text{If stable for } C \pm z_m P \cdot C, \text{ set } z_s = z_m. \tag{13.47}$$

The resulting stable perturbations, $z_s P \cdot C$, now define the new set of stable weight regions, $S = C \pm z_s P \cdot C$.

13.4   DEMONSTRATIONS OF ROBUST REINFORCEMENT LEARNING

13.4.1   First Demonstration: A First-Order Positioning Task

The first demonstration involves a simple non-mechanical positioning task. A single input called the reference signal, r moves on the interval $[-1, 1]$ at random points in time. The plant output, y, must track r as closely as possible. The plant is a first order system and thus has one internal state variable x. A control signal u is provided by the controller to position y closer to r. The dynamics of the discrete-time system are given by:

$$x(k+1) \;=\; x(k) + u(k), \tag{13.48}$$

$$y(k) \;=\; x(k), \tag{13.49}$$

where k is the discrete time step representing 0.01 seconds of elapsed time. A simple proportional controller (the control output is proportional to the size of the current error) is implemented with $K_p = 0.1$:

$$e(k) \;=\; r(k) - y(k); \tag{13.50}$$

$$u(k) \;=\; K_p e(k). \tag{13.51}$$

For this demonstration, the critic is implemented using a lookup-table with inputs e and a and output, $Q(e, a)$. Each input is quantized into 25 intervals forming a 25×25 matrix. The actor network is a two-layer, feed-forward neural network with a single input, e. There are three $tanh$ hidden units, and one output unit providing action a. Each unit includes a bias weight. The output a is added to the output of the feedback controller. This arrangement is depicted in block diagram form in Figure 13.2.

For training, the reference input r is changed to a new value on the interval $[-1, 1]$ stochastically with an average period of 20 time steps (every half second of simulated

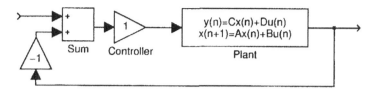

Fig. 13.3   Demonstration 1: Nominal System

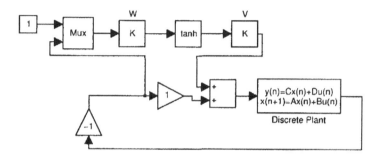

Fig. 13.4   Demonstration 1: With Neuro-Controller

time). Training proceeded for 2000 time steps at learning rates of $\alpha = 0.5$ and $\beta = 0.1$ for the critic and actor networks, respectively, then for an additional 2000 time steps with learning rates of $\alpha = 0.1$ and $\beta = 0.01$.

Figure 13.3 depicts a Simulink diagram for the nominal control system. This is referred to as the nominal system because there is no learning component added to the system. The plant is represented by a rectangular block that implements a discrete-time state space system. The simple proportional controller is implemented by a triangular gain block. Another gain block provides the negative feedback path. The reference input is drawn from the left and the system output exits to the right. Figure 13.4 shows the addition of the actor network in parallel to the nominal controller. The combination of a nominal controller and the actor network will be referred to as a neuro-controller. This diagram is suitable for conducting simulation studies in Simulink. However, this diagram cannot be used for stability analysis, because the neural network is not represented as an LTI system. For static stability analysis, assume constant gain matrices for the hidden layer weights, W, and the output layer weights, V. The static stability test will verify whether this particular neuro-controller implements a stable control system. Since the critic is not used in the stability analysis, it is not included in this Simulink diagram.

Figure 13.5 shows the LTI version of the same system. The non-linear *tanh* function is replaced by the single IQC block labeled *odd slope non-linearity* from

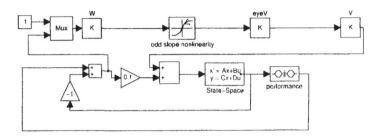

Fig. 13.5   Demonstration 1: With Neuro-Controller as LTI (IQC)

the IQC-β Matlab toolbox by Megretski et al. [6]. The *performance* block is another IQC block that must be included in all IQC Simulink diagrams.

There are actually *two* versions of the neuro-controller. In the first version, shown in Figure 13.4, the actor network includes its non-linearities. This is the actual actor network that will be used as a controller in the system. The second version of the system, shown in Figure 13.5, contains the actor network converted into the LTI framework; the non-linear *tanh* hidden layer has been replaced with an LTI uncertainty. This version of the neural network will never be implemented as a controller; the sole purpose of this version is to analyze stability. Because the LTI system *over-estimates* the gain of the non-linearity in the non-LTI system, a stability guarantee on the LTI version implies a stability guarantee on the non-LTI system.

The IQC analysis procedure performs a feasibility search for a matrix satisfying the IQC inequality. If the search is feasible, the system is guaranteed stable; if the search is infeasible, the system is not guaranteed to be stable. When applied to the Simulink diagram, if the feasibility constraints are satisfied, the neuro-controller is guaranteed to be stable.

Additional constraints are now imposed on the learning algorithm in order to ensure the network is dynamically stable while learning. The constraints are in the form of bounds on the perturbations, dW and dV, allowed on the actor network's weight values, as shown in Figure 13.6. The rate by which the weight values are allowed to change is also bounded. This constraint is implemented using the *slowly time-varying* IQC block as shown in Figure 13.6.

The matrices WA, WB, VA, and VB re-dimension the sizes of W and V; they have no effect on the uncertainty or norm calculations. In the diagram, dW and dV contain all the individual perturbations along the diagonal while W and V are not diagonal matrices. Thus, W_{hxn} and dW_{hnxhn} are not dimensionally compatible. By multiplying with WA and WB this dimensional incompatibility is fixed without affecting any of the numeric computations. Similarly, VA and VB are applied to V and dV.

The stability phase algorithm interacts with the Simulink diagram in Figure 13.6 to find the largest set of uncertainties (the largest perturbations) for which the system

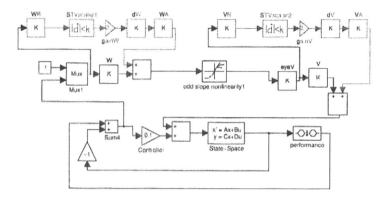

Fig. 13.6   Demonstration 1: Simulink Diagram for Dynamic IQC-analysis

is still stable. Furthermore, the system will remain stable if the actor network weight values are changed as long as the new weight values do not exceed the range specified by the perturbation matrices, dW and dV. In the learning phase, the reinforcement learning algorithm is applied until one of the network's weight values exceeds the range specified by the perturbations.

The results of applying the above robust, reinforcement learning controller to the first task are now described. Observations follow on the trajectory of the actor weights in the context of robust constraints.

The system is tested for a 10 second period (1000 discrete time steps with a sampling period of 0.01). The tracking performance of the trained system is shown in Figure 13.7. The top diagram shows the system performance with just the nominal controller, and the bottom graph shows performance with the trained neuro-controller.

The sum of the squared tracking error (SSE) is calculated over the 10 second interval. For the proportional only controller, the $SSE = 33.20$. Adding the neuro-controller reduced the SSE to 11.73. The reinforcement learning neuro-controller is able to improve the tracking performance. However, with this simple first-order system it is not difficult to construct a better performing proportional controller. In fact, setting the constant of proportionality to 1 ($K_p = 1$) achieves optimal control (minimal control error). A suboptimal controller is purposely chosen in this demonstration so that the neuro-controller has room to learn to improve control performance. The effect is similar to what would result if the value of a parameter of the plant was not accurately known, in which case the controller would not be as aggressive as it could be. K_p would have to be less than 1.

In order to understand the operation of the constrained reinforcement learning agent, a minimal actor net with a single hidden unit is used so that the results can be summarized in a two-dimensional plot. The actor network has two inputs (the bias = 1 and the tracking error e), one $tanh$ hidden unit, and one output (a). Thus, there are

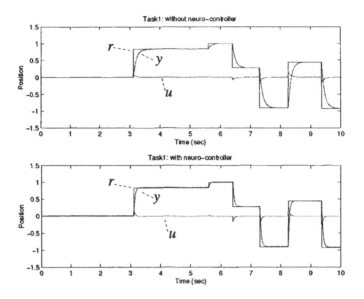

Fig. 13.7 Demonstration 1: Results of nominal controller (top graph) and of nominal controller plus the trained neuro-controller (bottom graph). Each graph includes the reference input r, the plant output y, and u, the sum of nominal controller and actor net outputs.

only two weights in this net with variable inputs. This network is still able to learn a relatively good control function, though three hidden units resulted in faster learning.

The trajectories of these weights in the actor network are tracked as they change during learning. Figure 13.8 depicts the two-dimensional weight space and the trajectory of these two weights during a typical training episode. The x-axis shows the second input weight W_2 while the y-axis represents the single output weight V. The trajectory starts in the lower center. Each point along the trajectory represents a weight pair (W_2, V) achieved at some point during the learning process.

An initial stability analysis is performed to determine the amount of uncertainty which can be added to the weights; the resulting perturbations, dW and dV, indicate how much learning can be performed and still remain stable. These perturbations are drawn as the rectangle centered on the initial weight values at lower center. Now the actor weights are iteratively modified by the Q-learning algorithm described earlier, forming the trajectory that rises to the upper boundary of the first perturbation rectangle. After the first learning phase, another stability phase is performed to compute new values for dW and dV. A second learning phase is then initiated that proceeds until a weight update results that exceeds the allowed range. This process of alternating stability and learning phases repeats. In the diagram of Figure 13.8 a total of five learning phases are shown, after which the weight changes become very small.

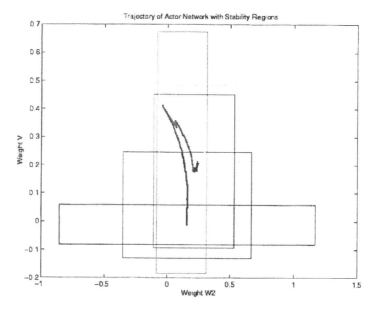

Fig. 13.8 Demonstraton 1: Trajectory of values of weights in the actor network while learning. Superimposed are the bounding boxes of regions for which weight values result in stable behavior.

It is important to note that if the trajectory reaches the edge of a bounding box, the weights may still be able to be adjusted in that direction. Recall that the edges of the bounding box are computed *with respect to the network weight values at the time of the stability phase*; these initial weight values are the point along the trajectory in the exact center of the bounding box. This central point in the weight space is the value of the actor network weights at the beginning of this particular stability/learning phase. Given that the current network weight values are that central point, the bounding box is the *limit* of weight changes that the network tolerates without forfeiting the stability guarantee. This is not to be confused with an *absolute* limit on the size of that network weight.

The third phase of the trajectory reveals some interesting dynamics. The trajectory stops near the edge of the box, and moves back toward the middle. Keep in mind that this trajectory represents the weight changes in the actor network. At the same time as the actor network is learning, the critic is also learning and adjusting its weights. During this phase in the training, the critic's predictions become more accurate. Because the gradient of the critic directs the weight changes for the actor network, the direction of weight changes in the actor network reverses. In the early part of the

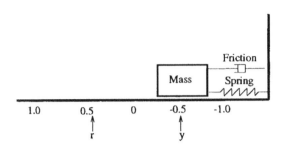

Fig. 13.9   Demonstration 2: Mass, Spring, Dampening System

learning the critic network gradient indicates that "upper left" is a desirable trajectory for weight changes in the actor network. By the third phase, the critic has changed to indicate that "upper-left" is now an undesirable direction for movement for the actor network.

Further dynamics are revealed in the last two phases. The actor network weights are not changing as rapidly as they did in the earlier learning phases. Optimal tracking performance according to the gradient in the critic is being achieved. The point of convergence of the actor network weights is a local optima in the value function of the critic. Training is halted at this point because the actor weights have ceased to change.

13.4.2   Second Demonstration: A Second-Order System

The second demonstration involves a second order mass/spring/dampener system, providing a more challenging and more realistic system in which to test the robust, reinforcement-learning algorithm. Once again, a single reference input r moves stochastically on the interval $[-1, 1]$; the single output of the control system y must track r as closely as possible. However, there are now friction, inertial, and spring forces acting on the system. Figure 13.9 depicts the different components of the system. The discrete-time update equations are given by

$$e(k) = r(k) - y(k), \tag{13.52}$$

$$u(k) = K_p e(k) + \int K_i e(k), \tag{13.53}$$

$$K_p = 0.01 \quad K_i = 0.001, \tag{13.54}$$

$$x(k+1) = \begin{bmatrix} 1 & 0.05 \\ -0.05 & 0.9 \end{bmatrix} x(k) + \begin{bmatrix} 0 \\ 1.0 \end{bmatrix} u(k), \tag{13.55}$$

$$y(k) = \begin{bmatrix} 1 & 0 \end{bmatrix} x(k). \tag{13.56}$$

Here, the nominal controller is a PI controller with both a *proportional* term and an *integral* term. This controller is implemented with its own internal state variable. The more advanced controller is required in order to provide reasonable nominal control for a system with second-order dynamics. The constant of proportionality, K_p, is 0.01, and the integral constant, K_i, is 0.001. Once again, a controller with suboptimal performance is chosen so that the RL agent has significant margin for improvement.

The architecture of the learning components is almost identical to that used in the first demonstration. The actor network consists of two inputs (the bias term and the current tracking error), three hidden units, and one output (the addition to the control signal).

The reference input r is changed to a new value on the interval $[-1, 1]$ stochastically with an average period of 20 time steps (every half second of simulated time). Due to the more difficult second-order dynamics, the training time is increased to 10,000 time steps at learning rates of $\alpha = 0.5$ and $\beta = 0.1$ for the critic and actor networks, respectively. Then an additional 10,000 steps of training are taken with learning rates of $\alpha = 0.1$ and $\beta = 0.01$.

Figure 13.10 shows the results of testing a trained system. The top portion of the diagram depicts the nominal control system (with only the PI controller) while the bottom half shows the same system with both the PI controller and the neuro-controller acting together. Recall that K_i and K_p parameters are *suboptimal* so that the neural network has opportunity to improve the control system. The addition of the neuro-controller clearly does improve system tracking performance. The total squared tracking error for the nominal system is $SSE = 246.6$ while the total squared tracking error for the neuro-controller is $SSE = 76.3$.

To demonstrate the advantage provided by stability analysis, two training scenarios are analyzed. The first training scenario involved the stable reinforcement learning algorithm with IQC-analysis. In the second training scenario, training proceeds without the constraints imposed by the stability analysis. Both scenarios result in similar control performance; both produce similar weights for the actor network. While both scenarios result in a stable controller as an end product (the final neural network weight values), only the scenario trained with IQC constraints enforced retains stability *throughout* the training. The stand-alone reinforcement learning scenario actually produces unstable intermediate neuro-controllers during the learning process.

An example of this behavior is shown in Figure 13.11. Clearly, the actor net is not implementing a good control solution; the system has been placed into an unstable limit cycle, because of the actor network. Notice the scale of the y-axis compared to the stable control diagram of Figure 13.10. This is exactly the type of scenario that must be avoided if neuro-controllers are to be useful in industrial control applications. To verify the instability of this system, these temporary actor network weights are used for a static stability test. IQC-analysis is unable to find a feasible solution, indicating that the system is indeed unstable.

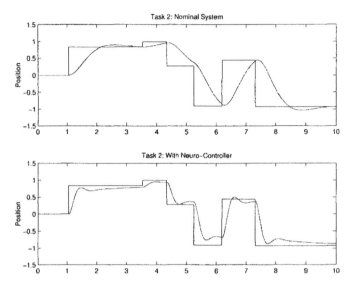

Fig. 13.10   Demonstration 2: Tracking performance of nominal controller and nominal plus trained neuro-controller. Reference signal, r, and plant output, y, are shown.

In summary, the purpose of this demonstration is to construct a control system with dynamics adequately simple to be amenable to introspection, but also adequately complex to introduce the possibility of learning/implementing unstable controllers. In this task the restrictions imposed on weights from the dynamic stability analysis are necessary to keep the neuro-control system stable during reinforcement learning.

13.5   CONCLUSIONS

The primary objective of this work is an approach to robust control design and adaptation, in which reinforcement learning and robust control theory are combined to implement a learning neuro-controller guaranteed to provide stable control. Robust control can be overly conservative, and thus sacrifice some performance. Some of this performance loss may be reclaimed by adding a reinforcement learning agent to the robust controller to optimize the performance of the control system.

This chapter summarizes a *static stability* test to determine whether a neural network controller, with a specific fixed set of weights, implements a stable control system. While a few previous research efforts have achieved similar results to the static stability test, here a *dynamic stability* test is also developed in which the neuro-controller provides stable control even while the neural network weights are changing during the learning process.

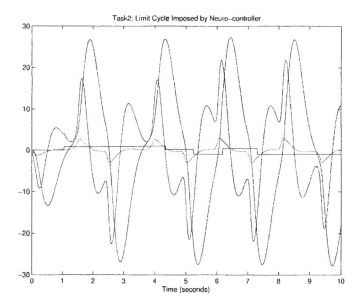

Fig. 13.11 Demonstration 2: System becomes unstable at an intermediate stage while learning. Two state variables, the control signal, u, and the reference signal, r, are shown.

A secondary objective is to demonstrate that the robust reinforcement learning approach is practical to implement in real control situations. Dynamic stability analysis leads directly to the stable reinforcement learning algorithm. The algorithm is essentially a repetition of two phases. In the stability phase, IQC-analysis is used to compute the largest amount of weight uncertainty the neuro-controller can tolerate without being unstable. Then the weight uncertainty is used in the reinforcement learning phase as a restricted region in which to change the neural network weights.

This approach to robust reinforcement learning remains to be tested on realistic control tasks more difficult than the simple demonstrations described in this chapter. In real applications, one problem may be the training time on the physical system which could be prohibitively expensive as the system must be driven through all of its dynamics multiple times. The advantage of this approach is limited in situations where the physical plant and plant model closely match each other or cases in which differences between the model and plant do not greatly affect the dynamics.

In current work the robust reinforcement learning procedure is being extended to more difficult control problems (Chapter 20). Additional ways of capitalizing on the theory of static and dynamic stability using IQC's are being developed for other categories of action function approximators and for including robust stability objectives in the reinforcement function.

Acknowledgments

This work was supported by the National Science Foundation through grants CMS-9401249 and CISE-9422007.

Bibliography

1. D. P. Bertsekas and J. N. Tsitsiklis, *Neuro-Dynamic Programming,* Athena Scientific, Belmont, MA, 1995.

2. J. C. Doyle, B. A. Francis, and A. R. Tannenbaum, *Feedback Control Theory,* Macmillan Publishing Company, New York, 1992.

3. P. Gahihet, A. Nemirovski, A. J. Laub, and M. Chilali, *LMI Control Toolbox,* MathWorks Inc., 1995.

4. R. M. Kretchmar, *A Synthesis of Reinforcement Learning and Robust Control Theory,* Ph.D. Thesis, Colorado State University, Department of Computer Science, Fort Collins, CO, 2000.

5. R. M. Kretchmar, P. M. Young, C. W. Anderson, D. Hittle, M. Anderson, C. Delnero, and J. Tu, Robust reinforcement learning control with static and dynamic stability, *International Journal of Robust and Nonlinear Control*, vol. 11, pp. 1469–1500, 2001.

6. A. Megretski, C.-Y. Kao, U. Jonsson, and A. Rantzer, *A Guide to IQCβ: Software for Robustness Analysis,* MIT / Lund Institute of Technology, http://www.mit.edu/people/ameg/home.html, 1999.

7. A. Megretski and A. Rantzer, System analysis via integral quadratic constraints, *IEEE Trans. Automatic Control*, vol. 42, no. 6, pp. 819–830, 1997.

8. A. Megretski and A. Rantzer, System analysis via integral quadratic constraints: Part II, Technical Report ISRN LUTFD2/TFRT–7559–SE, Lund Institute of Technology, September 1997.

9. A. Packard and J. Doyle, The complex structured singular value, *Automatica*, vol. 29, no. 1, pp. 71–109, 1993.

10. A. Rantzer, On the Kalman-Yacubovich-Popov lemma, *Systems & Control Letters*, vol. 28, pp. 7–10, 1996.

11. S. Skogestad and I. Postlethwaite, *Multivariable Feedback Control,* Wiley, New York, 1996.

12. R. S. Sutton and A. G. Barto, *Reinforcement Learning: An Introduction,* MIT Press, Cambridge, MA, 1998.

13. C. J. C. H. Watkins, *Learning with Delayed Rewards,* Ph.D. Thesis, Cambridge University Psychology Department, Cambridge, UK, 1989.

14. K. Zhou and J. C. Doyle, *Essentials of Robust Control,* Prentice-Hall, Englewood Cliffs, NJ, 1998.

14 Supervised Actor-Critic Reinforcement Learning

MICHAEL T. ROSENSTEIN   ANDREW G. BARTO

University of Massachusetts

Editor's Summary: Chapter 7 introduced policy gradients as a way to improve on stochastic search of the policy space when learning. This chapter presents supervised actor-critic reinforcement learning as another method for improving the effectiveness of learning. With this approach, a supervisor adds structure to a learning problem and supervised learning makes that structure part of an actor-critic framework for reinforcement learning. Theoretical background and a detailed algorithm description are provided, along with several examples that contain enough detail to make them easy to understand and possible to duplicate. These examples also illustrate the use of two kinds of supervisors: a feedback controller that is easily designed yet sub-optimal, and a human operator providing intermittent control of a simulated robotic arm.

14.1 INTRODUCTION

Reinforcement learning (RL) and supervised learning are usually portrayed as distinct methods of learning from experience. RL methods are often applied to problems involving sequential dynamics and optimization of a scalar performance objective, with online exploration of the effects of actions. Supervised learning methods, on the other hand, are frequently used for problems involving static input-output mappings and minimization of a vector error signal, with no explicit dependence on how training examples are gathered. As discussed in Chapter 2 by Barto and Dietterich, the key feature distinguishing RL and supervised learning is whether training information from the environment serves as an evaluation signal or as an error signal, and in this chapter, we are interested in problems where both kinds of feedback are available to a learning system *at the same time*.

As an example, consider a young child learning to throw a ball. For this motor task, as well as many others, there is no substitute for ongoing practice. The child repeatedly throws a ball under varying conditions and with variation in the executed motor commands. Bernstein [5] called this kind of trial-and-error learning "repetition without repetition." The outcome of each movement, such as the visual confirmation of whether the ball reached a nearby parent, acts as an evaluation signal that provides the child with feedback about the quality of performance—but with no specific information about what corrections should be made. In addition, the parent may interject error information in the form of verbal instruction or explicit demonstration of what went wrong with each preceding throw. In reality, the feedback may be much more subtle than this. For instance, the final position of the ball reveals some directional information, such as too far to the left or the right; a learned forward model [13] can then be used to make this corrective information more specific to the child's sensorimotor apparatus. Similarly, the tone of the parent's voice may provide evaluative praise simultaneously with the verbal error information. In any case, the two kinds of feedback play interrelated, though complementary roles [2]: The evaluation signal drives skill optimization, whereas the error signal provides a standard of correctness that helps ensure a certain level of performance, either on a trial-by-trial basis or for the learning process as a whole.

The richness of the training information in this example, which we believe is the rule rather than the exception in realistic learning problems, is not effectively used by systems that rely on RL alone. This contributes to some of the difficulties that have been observed when attempting to apply RL to practical problems. Learning can require a very large number trials, and system behavior during learning can lead to unacceptable risks. For this reason, in most large-scale applications RL is applied to simulated rather than real experience. To overcome some of these difficulties, a number of researchers have proposed the use of supervisory information that effectively transforms a learning problem into one which is easier to solve. Common examples involve shaping [10, 18, 20], learning from demonstration [14, 16, 25, 29], or the use of carefully designed controllers [12, 15, 21]. Approaches that explicitly model the role of a supervisor include ASK FOR HELP [7], RATLE [17], and the mentor framework [23]. In each case, the goal of learning is an optimal policy, that is, a mapping from states to actions that optimizes some performance criterion. Despite the many successful implementations, none of these approaches combines both kinds of feedback as described shortly. Either supervised learning precedes RL during a separate training phase, or else the supervisory information is used to modify a *value function* rather than a policy. Those methods based on Q-learning [32], for instance, build a value function that ranks the actions available in a given state. The corresponding policy is then represented implicitly, usually as the action with the best ranking for each state.

The approach taken in this chapter involves an actor-critic architecture for RL [3]. Actor-critic architectures differ from other value-based methods in that separate data structures are used for the control policy (the "actor") and the value function (the "critic"). One advantage of the actor-critic framework is that action selection requires

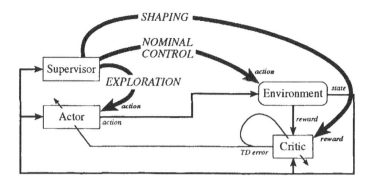

Fig. 14.1   Actor-critic architecture and several pathways for supervisor information.

minimal computation [31]. Methods that lack a separate data structure for the policy typically require a repeated search for the action with the best value, and this search can become computationally prohibitive, especially for real-valued actions as in the examples of Section 14.3.

Another important advantage of the actor-critic framework is that the policy can be modified directly by standard supervised learning methods. In other words, the actor can change its behavior based on state-action training pairs provided by a supervisor, without the need to calculate the values of those training data. The critic (or some other comparable mechanism) is still required for optimization, whereas the supervisor helps the actor achieve a level of proficiency whenever the critic has a poor estimate of the value function. In the next section we describe a *supervised* actor-critic architecture where the supervisor supplies not only error information for the actor, but also actions for the environment.

14.2   SUPERVISED ACTOR-CRITIC ARCHITECTURE

Figure 14.1 shows a schematic of the usual actor-critic architecture [31] augmented by three major pathways for incorporating supervisor information. Along the "shaping" pathway, the supervisor supplies an additional source of evaluative feedback, or *reward*, that essentially simplifies the task faced by the learning system. For instance, the critic may receive favorable evaluations for behavior which is only approximately correct given the original task. As the actor gains proficiency, the supervisor then gradually withdraws the additional feedback to shape the learned policy toward optimality for the true task. With "nominal control" the supervisor sends control signals (*actions*) directly to the controlled system (the *environment*). For example, the supervisor may override bad commands from the actor as a way to ensure safety and to guarantee a minimum standard of performance. And along the "exploration" pathway, the supervisor provides the actor with hints about which actions may or

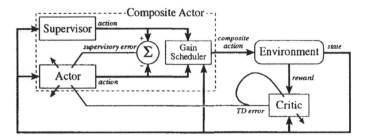

Fig. 14.2   The supervised actor-critic architecture.

may not be promising for the current situation, thereby altering the exploratory nature of the actor's trial-and-error learning. In this section, we focus on the latter two pathways, and we examine the use of supervised learning which offers a powerful counterpart to RL methods.

The combination of supervised learning with actor-critic RL was first suggested by Clouse and Utgoff [8] and independently by Benbrahim and Franklin [4]. Their approach has received almost no attention, yet the more general problem of how to combine a teacher with RL methods has become quite popular. In their work, Benbrahim and Franklin [4] used pre-trained controllers, called "guardians," to provide safety and performance constraints for a biped robot. Joint position commands were sent to the robot by a central pattern generator, but those commands were modified by the guardians whenever a constraint was violated. Superimposed with the joint position commands were exploratory actions generated according to Gullapalli's SRV algorithm [11]. In effect, the central pattern generator learned not only from exploration, but from the guardians as well.

Figure 14.2 shows our version of the supervised actor-critic architecture, which differs from previous work in a key way described in Section 14.2.2. Taken together, the actor, the supervisor and the gain scheduler[1] form a "composite" actor that sends a composite action to the environment. The environment responds to this action with a transition from the current state, s, to the next state, s'. (Appendix 1 to this chapter contains a list of symbols used throughout the remainder of this chapter.) The environment also provides an evaluation called the immediate reward, r. The job of the critic is to observe states and rewards and to build a value function, $V^{\pi}(s)$, that accounts for both immediate and future rewards received under the composite policy, π. This value function is defined recursively as

$$V^{\pi}(s) = \sum_{s' \in \mathcal{S}} \Pr(s'|s,a)\{R(s') + \gamma V^{\pi}(s')\},$$

---

[1]"Gain scheduling" refers to the construction of a global nonlinear controller by interpolation, or scheduling, of local linear controllers [27]. We use the term in a broader sense to mean the blending of two or more sources of control actions.

where $R(s')$ is the expected value of r, $\gamma \in [0, 1]$ is a factor that discounts the value of the next state, and $\Pr(s'|s, a)$ is the probability of transitioning to state s' after executing action $a = \pi(s)$. Here we focus on deterministic policies, although this work also generalizes to the stochastic case where π represents a distribution for choosing actions probabilistically. For RL problems, the expected rewards and the state-transition probabilities are typically unknown. Learning, therefore, must proceed from samples, that is, from observed rewards and state transitions. For RL algorithms, temporal-difference (TD) methods [30] are commonly used to update the state-value *estimates*, $V(s)$, by an amount proportional to the TD error, defined as

$$\delta = r + \gamma V(s') - V(s).$$

14.2.1 Gain Scheduler

For deterministic policies and real-valued actions, the gain scheduler computes the composite action, a, as simply a weighted sum of the actions given by the component policies. In particular,

$$a \leftarrow ka^E + (1 - k)a^S,$$

where a^E is the actor's exploratory action and a^S is the supervisor's action, as given by policies π^E and π^S, respectively. (The supervisor's actions are observable but its policy is unknown.) We also denote by a^A the actor's greedy action determined by the corresponding policy, π^A. Typically, π^E is a copy of π^A modified to include an additive random variable with zero mean. Thus each exploratory action is simply a noisy copy of the corresponding greedy action, although we allow for the possibility of more sophisticated exploration strategies.

The parameter $k \in [0, 1]$ interpolates between π^E and π^S, and therefore k determines the level of control, or autonomy, on the part of the actor.[2] In general, the value of k varies with state, although we drop the explicit dependence on s to simplify notation. The parameter k also plays an important role in modifying the actor's policy, as described in more detail below. We assume that π^A is given by a function approximator with the parameter vector w, and after each state transition, those parameters are updated according to a rule of the form

$$w \leftarrow w + k\Delta w^{RL} + (1 - k)\Delta w^{SL}, \tag{14.1}$$

where Δw^{RL} and Δw^{SL} are the individual updates based on RL and supervised learning, respectively. Thus k also interpolates between two styles of learning.

The use of k—a single state-dependent parameter that trades off between two sources of action and learning—allows for a wide range of interactions between actor and supervisor. If the actor has control of the gain scheduler, for instance, then the actor can set the value of k near 0 whenever it needs help from its supervisor, cf. ASK

---

[2] For the stochastic case, k gives the probability that the gain scheduler chooses the actor's exploratory action rather than the supervisor's action.

FOR HELP [7]. Similarly, if the supervisor has control of the gain scheduler, then the supervisor can set $k = 0$ whenever it loses confidence in the autonomous behavior of the actor, cf. RATLE [17]. The gain scheduler may even be under control of a third party. For example, a linear feedback controller can play the role of supervisor, and then a human operator can adjust the value of k as a way to switch between actor and supervisor, perhaps at a longer time scale than that of the primitive actions.

14.2.2   Actor Update Equation

To make the reinforcement-based adjustment to the parameters of π^A we compute

$$\Delta w^{RL} \leftarrow \alpha \delta (a^E - a^A) \nabla_w \pi^A(s), \qquad (14.2)$$

where α is a step-size parameter. Equation (14.2) is similar to the update used by the *REINFORCE* class of algorithms [33], although we utilize the gradient of the deterministic policy π^A rather than that of the stochastic exploration policy π^E. When the TD error is positive, this update will push the greedy policy evaluated at s closer to a^E, that is, closer to the exploratory action which led to a state with estimated value better than expected. Similarly, when $\delta < 0$, the update will push $\pi^A(s)$ away from a^E and in subsequent visits to state s the corresponding exploratory policy will select this unfavorable action with reduced probability.

To compute the supervised learning update, Δw^{SL}, we seek to minimize in each observed state the supervisory error

$$E = \frac{1}{2} [\pi^S(s) - \pi^A(s)]^2.$$

Locally, this is accomplished by following a steepest descent heuristic, that is, by making an adjustment proportional to the negative gradient of the error with respect to w:

$$\Delta w^{SL} \leftarrow -\alpha \nabla_w E(s).$$

Expanding the previous equation with the chain rule and substituting the observed actions gives the usual kind of gradient descent learning rule:

$$\Delta w^{SL} \leftarrow \alpha (a^S - a^A) \nabla_w \pi^A(s). \qquad (14.3)$$

Finally, by substituting Eqs. (14.2) and (14.3) into Eq. (14.1) we obtain the desired actor update equation:

$$w \leftarrow w + \alpha [k\delta(a^E - a^A) + (1 - k)(a^S - a^A)] \nabla_w \pi^A(s). \qquad (14.4)$$

---

input

    critic value function, $V(s)$, parameterized by θ

    actor policy, $\pi^A(s)$, parameterized by w

    exploration size, σ

    actor step size, α, and critic step size, β

    discount factor, $\gamma \in [0,1]$

    eligibility trace decay factor, λ

initialize θ, w arbitrarily

repeat for each trial

    $e \leftarrow 0$ (clear the eligibility traces)

    $s \leftarrow$ initial state of trial

    repeat for each step of trial

        $a^A \leftarrow$ action given by $\pi^A(s)$

        $a^E \leftarrow a^A + N(0,\sigma)$

        $a^S \leftarrow$ action given by supervisor's unknown policy, $\pi^S(s)$

        $k \leftarrow$ interpolation parameter from gain scheduler

        $a \leftarrow ka^E + (1-k)a^S$

        $e \leftarrow \gamma\lambda e + \nabla_\theta V(s)$

        take action a, **observe** reward, r, and next state, s'

        $\delta \leftarrow r + \gamma V(s') - V(s)$

        $\theta \leftarrow \theta + \beta\delta e$

        $w \leftarrow w + \alpha\left[k\delta(a^E - a^A) + (1-k)(a^S - a^A)\right]\nabla_w\pi^A(s)$

        $s \leftarrow s'$

    until s is terminal

Fig. 14.3 The supervised actor-critic learning algorithm for deterministic policies and real-valued actions.

Equation (14.4) summarizes a steepest descent algorithm where k trades off between two sources of gradient information:[3] one from a performance surface based on the evaluation signal and one from a quadratic error surface based on the supervisory error. Figure 14.3 gives a complete algorithm.

As mentioned above, the architecture shown in Figure 14.2 is similar to one suggested previously by Benbrahim and Franklin [4] and by Clouse and Utgoff [8]. However, our approach is novel in the following way: In the figure, we show a direct connection from the supervisor to the actor, whereas the supervisor in both [4] and [8] influences the actor indirectly through its effects on the environment as well as

---

[3]In practice, an additional parameter may be needed to scale the TD error. This is equivalent to using two step-size parameters, one for each source of gradient information.

the TD error. Using our notation the corresponding update equation for these other approaches, for example [4, Eq. (1)], essentially becomes

$$w \leftarrow w + \alpha[k\delta(a^E - a^A) + (1 - k)\delta(a^S - a^A)]\nabla_w\pi^A(s) \qquad (14.5)$$

$$= w + \alpha\delta[ka^E + (1 - k)a^S - a^A]\nabla_w\pi^A(s). \qquad (14.6)$$

The key attribute of Eq. (14.5) is that the TD error modulates the supervisory error, $a^S - a^A$. This may be a desirable feature if one "trusts" the critic more than the supervisor, in which case one should view the supervisor as an additional source of exploration. However, Eq. (14.5) may cause the steepest descent algorithm to ascend the associated error surface, especially early in the learning process when the critic has a poor estimate of the true value function. Moreover, when δ is small, the actor loses the ability to learn from its supervisor, whereas in Eq. (14.4) this ability depends primarily on the interpolation parameter, k.

14.3   EXAMPLES

In this section we present several examples that illustrate a gradual shift from full control of the environment by the supervisor to autonomous control by the actor. In each case, the supervisor enables the composite actor in Figure 14.2 to solve the task *on the very first trial*, and on every trial while it improves, whereas the task is virtually impossible to solve with RL alone. The first three examples are targeting tasks—each with a stable controller that brings the system to target, although in a sub-optimal fashion. The final example involves a human supervisor that controls a simulated robot during a peg insertion task.

For each example we used the learning algorithm in Figure 14.3 with step-size parameters of $\alpha = 0.1$ for the actor and $\beta = 0.3$ for the critic. To update the critic's value function, we used the TD(λ) algorithm [30] with $\lambda = 0.7$. We implemented both actor and critic by a tile coding scheme, that is, CMAC [1], with a total of 25 tilings, or layers, per CMAC. (Appendix 2 to this chapter provides a brief description of CMAC function approximators.)

14.3.1   Ship Steering Task

For our first experiment we adapted a standard problem from the optimal control literature where the task is to steer a ship to a goal in minimum time [6]. The ship moves at a constant speed of $C = 0.01$ km·s^{-1}, and the real-valued state and action are given, respectively, by the ship's two-dimensional position and scalar heading. For this problem, the supervisor is a hand-crafted controller that always steers directly toward the center of the goal region, that is, toward the origin $(0, 0)$. Under full supervision, this strategy ensures that the ship will reach the goal eventually—but not in minimum time due to a water current that complicates the task. More specifically,

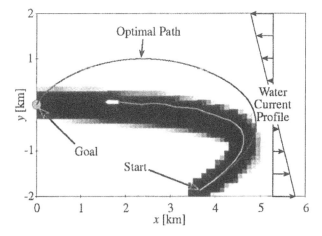

Fig. 14.4 Ship steering task simulator after 1000 learning trials. The grayscale region indicates the level of autonomy, from $k = 0$ (white) to $k = 1$ (black).

the equations of motion are

$$\dot{x} = C\,(\cos\phi - y), \quad \dot{y} = C\sin\phi, \tag{14.7}$$

where ϕ is the ship's heading. Notice that the water current acts along the horizontal direction, x, yet depends solely on the vertical position, y. The start location is $x_0 = 3.66$, $y_0 = -1.86$, and the goal region has a of radius 0.1 km. A convenient feature of this test problem is that one can solve for the optimal policy analytically [6], and the darker curve in Figure 14.4 shows the corresponding optimal path. Under the optimal policy the minimum time to goal is 536.6 s while the supervisor's time to goal is 1111 s.

We integrated Eq. (14.7) numerically using Euler's method with a step size of 1 s. Control decisions by the gain scheduler were made every 25 s, at which time the ship changed heading instantaneously. Exploratory actions, a^E, were Gaussian distributed with a standard deviation of 10 degrees and a mean equal to the greedy action, a^A. The CMAC tiles were uniform with a width of 0.5 km along each input dimension. The actor CMAC was initialized to steer the ship leftward while the critic CMAC was initialized to $V(s) = -1000$, for all s. Rewards were -1 per time step, and the discount factor was $\gamma = 1$ (no discounting).

To make the interaction between supervisor and actor dependent on state, the interpolation parameter, k, was set according to a state-visitation histogram, also implemented as a CMAC with 25 uniform tilings. At the end of each trial, the weights from the "visited" histogram tiles were incremented by a value of 0.0008, for a total increment of 0.02 over the 25 layers. During each step of the simulation, the value of k was set to the CMAC output for the current state, with values cut off at a maximum of $k = 1$ (at full autonomy). Thus the gain scheduler made a gradual shift from full supervision to full autonomy as the actor and critic acquired enough control knowledge to reach the goal reliably. A decay factor of 0.999 was also used to downgrade the weight of each CMAC tile; in effect, autonomy "leaked away" from infrequently visited regions of state space.

Figure 14.5 shows the effects of learning for each of two cases. One case corresponds to the supervised actor-critic algorithm in Figure 14.3, with the parameter values described above. The other case is from a two-phase learning process where 1000 trials were used to seed the actor's policy as well as the critic's value function, followed by RL alone, cf. [14, 16, 25, 29]. That is, $k = 0$ for the first 1000 trials, and $k = 1$ thereafter. Both cases show rapid improvement during the first 100 trials of RL, followed by slower convergence toward optimality. In panel (a) the two-phase process appears to give much improved performance—if one is willing to pay the price associated with an initial learning phase that gives no immediate improvement. Indeed, if we examine cumulative reward instead, as in panel (b), the roles become reversed with the two-phase process performing worse. Performance improves somewhat if we reduce the number of seed trials, although with fewer than 500 seed trials of supervised learning, the actor is able to reach the goal either unreliably (300 and 400 trials) or else not at all (100 and 200 trials).

14.3.2 Manipulator Control

Our second example demonstrates that the style of control and learning used for the ship steering task is also suitable for learning to exploit the dynamics of a simulated robotic arm. The arm was modeled as a two-link pendulum with each link having length 0.5 m and mass 2.5 kg, and the equations of motion [9] were integrated numerically using Euler's method with a step size of 0.001 s. Actions from both actor and supervisor were generated every 0.75 s and were represented as two-dimensional velocity vectors with joint speed limits of ±0.5 rad/sec. The task was to move with minimum effort from the initial configuration with joint angles of -90 and 0 degrees to the goal configuration with joint angles of 135 and 90 degrees. For this demonstration, effort was quantified as the total integrated torque magnitude.

Similar to the ship steering task, the supervisor in this example is a hand-crafted controller that moves the arm at maximum speed directly toward the goal in configuration space. Therefore, actions from the supervisor always lie on a unit square centered at the origin, whereas the actor is free to choose from the entire set of admissible actions. In effect, the supervisor's policy is to follow a straight-line path to the goal—which is time-optimal given the velocity constraints. Due to the dynamics of the robot, however, straight-line paths are not necessarily optimal with respect to other performance objectives, such as minimum energy.

A lower-level control system was responsible for transforming commanded velocities into motor torques for each joint. This occurred with a control interval of 0.001 s and in several stages: First, the commanded velocity was adjusted to account for acceleration constraints that eliminate abrupt changes in velocity, especially at

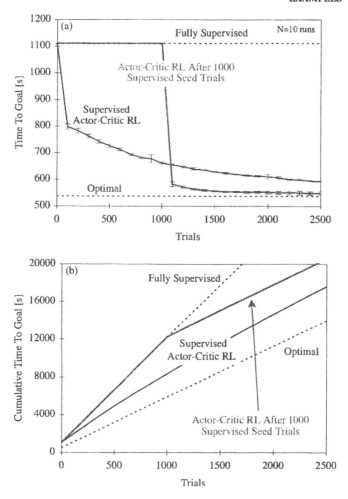

Fig. 14.5 Ship steering task effects of learning averaged over 10 runs of 2500 trials each: (a) time to goal, and (b) cumulative time to goal. For the supervised learning seed trials the initial position of the ship was chosen randomly from the region $0 \leq x \leq 6$, $-2 \leq y \leq 2$.

the beginning and end of movement. The adjusted velocity, along with the current position, was then used to compute the desired position at the end of the next control interval. Third, a proportional-derivative controller converted this target position into joint torques, but with a target velocity of zero rather than the commanded velocity. And finally, a simplified model of the arm was used to adjust the feedback-based torque to include a feed-forward term that compensates for gravity. This scheme is intended to match the way some industrial manipulators are controlled once given a higher-level movement command, for example velocity as used here. Gravity compensation guarantees stability of the lower-level controller [9], and the target velocity of zero helps ensure that the arm will stop safely given a communications failure with the higher level.

The above control scheme also holds an advantage for learning. Essentially, the manipulator behaves in accordance with a tracking controller—only the desired trajectory is revealed gradually with each control decision from the higher level. At this level, the manipulator behaves like an overdamped, approximately first-order system, and so policies need not account for the full state of the robot. That is, for both actor and supervisor it suffices to use reduced policies that map from positions to velocity commands, rather than policies that map from positions *and* velocities to acceleration commands. As is common with tracking controllers, this abstraction appears to cancel the dynamics we intend to exploit. However, by designing an optimal control problem, we allow the dynamics to influence the learning system by way of the performance objective, that is, through the reward function.

For the RL version of this optimal control problem, rewards were the negative effort accumulated over each 0.75 s decision interval. As with the ship steering task, the discount parameter was set to $\gamma = 1$ and the exploratory actions, a^E, were Gaussian distributed with a mean equal to the greedy action (although a^E was clipped at the joint speed limits). The standard deviation of the exploratory actions was initially 1.0 rad/sec, but this value decayed exponentially toward zero by a factor of 0.999 after each trial. CMAC tiles were uniform with a width of 25 degrees along each input dimension; the actor CMAC was initialized to zero whereas the critic CMAC was initialized to -300. Like the previous example, a third CMAC was used to implement a state-visitation histogram that stored the value of the interpolation parameter, k. As above, the histogram increment was 0.02 over the 25 layers and the decay factor was 0.999.

Figure 14.6(a) shows the configuration of the robot every 0.75 s along a straight-line path to the goal. The proximal joint has more distance to cover and therefore moves at maximum speed, while the distal joint moves at a proportionately slower speed. The total effort for this fully supervised policy is 258 Nm·s. Figures 14.6(b) and 14.6(c) show examples of improved performance after 5000 trials of learning, with a final cost of 229 and 228 Nm·s, respectively. In each of the left-hand diagrams, the corresponding "spokes" from the proximal link fall in roughly the same position, and so the observed improvements are due to the way the distal joint modulates its movement around the straight-line path, as shown in the right-hand diagrams.

Figure 14.7 shows the effects of learning averaged over 25 runs. The value of the optimal policy for this task is unknown, although the best observed solution has a cost of 216 Nm·s. Most improvement happens within 400 trials and the remainder of learning shows a drop in variability as the exploration policy "decays" toward the greedy policy. One difficulty with this example is the existence of many locally optimal solutions to the task. This causes the learning system to wander among solutions, with convergence to one of them only when forced to do so by the reduced exploration.

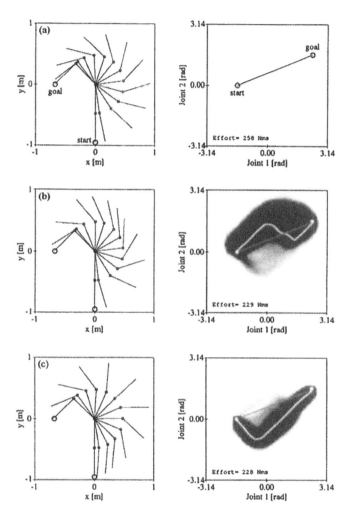

Fig. 14.6 Simulated two-link arm after (a) no learning and (b,c) 5000 learning trials. Configuration-space paths after learning are shown in white, and the grayscale region indicates the level of autonomy, from $k = 0$ (white) to $k = 1$ (black).

14.3.3 Case Study With a Real Robot

To demonstrate that the methods in this chapter are suitable for real robots, we replicated the previous example with a seven degree-of-freedom whole arm manipulator (WAM; Barrett Technology Inc., Cambridge, MA). Figure 14.8 shows a sequence of several postures as the WAM moves from the start configuration (far left frame) to

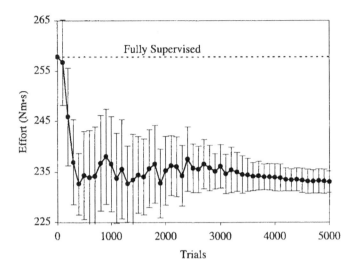

Fig. 14.7 Effects of learning for the simulated two-link arm averaged over 25 runs of 5000 trials each.

the goal configuration (far right frame). As with the previous example the task was formulated as a minimum-effort optimal control problem—utilizing a stable tracking controller and a supervisor that generates straight-line trajectories to the goal in configuration space. The joint speed limits for this example were increased to ±0.75 rad/sec rather than ±0.5 rad/sec as used above. The learning algorithm was virtually identical to the one in the previous example, although several parameter values were modified to encourage reasonable improvement with very few learning trials. For instance, the histogram increment was increased from 0.02 to 0.10, thereby facilitating a faster transition to autonomous behavior. Also, the level of exploration did not decay, but rather remained constant, and a^E was Gaussian distributed with a standard deviation of 0.25 rad/sec.

Fig. 14.8 Representative configurations of the WAM after learning.

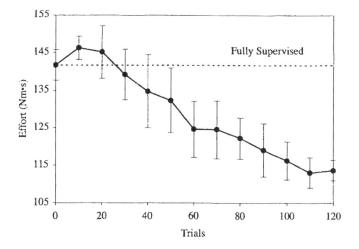

Fig. 14.9   Effects of learning for the WAM averaged over 5 runs of 120 trials each.

Figure 14.9 shows the effects of learning averaged over 5 runs. Performance worsens during the first 10 to 20 trials due to the initialization of the actor's policy. More specifically, at the start of learning the actor's policy maps all inputs to the zero velocity vector, and so the actor cannot move the robot until it has learned how to do so from its supervisor. The drawback of this initialization scheme—along with a fast transition to autonomous behavior—is that early in the learning process the supervisor's commands become diminished when blended with the actor's near-zero commands. The effect is slower movement of the manipulator and prolonged effort while raising the arm against gravity. However, after 60 trials of learning the supervised actor-critic architecture shows statistically significant improvement ($p < 0.01$) over the supervisor alone. After 120 trials, the overall effect of learning is approximately 20% reduced effort despite an increased average movement time from 4.16 s to 4.34 s (statistically significant with $p < 0.05$).

14.3.4   Peg Insertion Task

One goal of our ongoing work is to make the techniques described in this chapter applicable for telerobotics, that is, for remote operation of a robot by a human supervisor, possibly over great distances. Figure 14.10 shows the setup for our current tool-use experiment. For such applications, the human operator is often responsible for setting immediate goals, for managing sub-tasks, for making coarse-grained motor decisions (e.g., grasp a tool with the palm up rather than down) and also for executing fine-grained control of the robot. Moreover, these many responsibilities on the part of the human are just one source of the operator fatigue which hampers the effectiveness of virtually all telerobotic systems. The operator deals not only with a hierarchy of

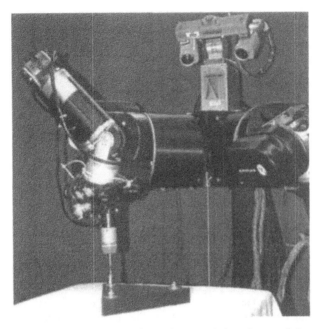

Fig. 14.10   Humanoid robot using a tool to perform a variation of a canonical parts insertion task.

control objectives but also with limited, and sometimes confusing feedback across the interface between human and machine. The potential contribution from machine learning is a way to push the human's involvement further up the hierarchy as the machine gains competence at each level of control. In particular, a supervised actor-critic architecture allows the human supervisor to remain "in the loop" as the actor learns about the details of the robot's task, especially those details which are difficult to convey across the user interface, for example, tactile feedback.

As a preliminary example, Figure 14.11 shows several snapshots during a simu-lated peg insertion task. With no initial control knowledge about the task, the actor is completely dependent upon input from its human supervisor (via a mouse). After just 10–20 trials, the actor has gathered sufficient information with which to pro-pose actions. Short bars in the figure depict the effects of such actions, as projected forward in time by prediction through a kinematic model. In Figure 14.11(a), for instance, the bars indicate to the operator that the actor will move the gripper toward the (incorrect) middle slot. In this scenario, the target slot is hidden state for the actor, and so brief input from the operator (panel b) is needed to push the actor into the basin of attraction for the upper target. Full autonomy by the actor is undesirable for this task. The human supervisor remains in control of the robot, while short intervals of autonomous behavior alleviate much of the operator's fatigue.

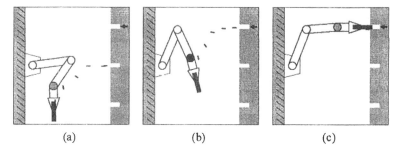

(a)   (b)   (c)

Fig. 14.11   Human-robot interface for a simulated peg insertion task after 20 learning trials. The arrow marks the target slot, and small bars indicate predicted gripper positions under autonomous control by the actor. (a) After successful re-grasp of the peg, the actor begins movement toward the middle slot. (b) A momentary correction by the human supervisor places the robot on track for the upper target, after which (c) the actor completes the sub-task autonomously.

14.4   CONCLUSIONS

The examples in Section 14.3 demonstrate a gradual shift from full supervision to full autonomy—blending two sources of actions and learning feedback. Much like the examples by Clouse [7] and by Maclin and Shavlik [17], this shift happens in a state-dependent way with the actor seeking help from the supervisor in unfamiliar territory. Unlike these other approaches, the actor also clones the supervisor's policy very quickly over the visited states. This style of learning is similar to methods that seed an RL system with training data, for example, [25, 29], although with the supervised actor-critic architecture, the interpolation parameter allows the seeding to happen in an incremental fashion at the same time as trial-and-error learning. Informally, the effect is that the actor knows what the supervisor knows, but only on a need-to-know basis.

One drawback of these methods for control of real robots is the time needed for training. By most standards in the RL literature, the supervised actor-critic architecture requires relatively few trials, at least for the examples presented above. However, some robot control problems may permit extremely few learning trials, say 10 or 20. Clearly, in such cases we should not expect optimality; instead we should strive for methods that provide gains commensurate with the training time. In any case, we might tolerate slow optimization if we can deploy a learning robot with provable guarantees on the worst-case performance. Recent work by Kretchmar et al. [15] and by Perkins and Barto [22] demonstrates initial progress in this regard.

With regard to telerobotic applications, our results thus far are promising, although several key challenges remain. First, our simulated peg insertion task is somewhat simplified—in terms of the noise-free sensors and actuators, the user interface, the surface contact model, and so forth—and so our present efforts are focused on a more convincing demonstration with the humanoid robot shown in Figure 14.10. Another difficulty is that input from the supervisor can quickly undo any progress made by the RL component. Consequently, we are also exploring principled ways to weaken the effects of the supervised learning aspect without necessarily weakening the human operator's control over the robot. A third challenge is related to the way a human-robot interface introduces constraints during the learning process. For example, the interface may restrict the supervisor's control of the robot to various subsets of its degrees of freedom. In turn, this biases the way training data are gathered, such that the actor has difficulty learning to coordinate all degrees of freedom simultaneously.

Despite the challenges when we combine supervised learning with an actor-critic architecture, we still reap benefits from both paradigms. From actor-critic architectures we gain the ability to discover behavior that optimizes performance. From supervised learning we gain a flexible way to incorporate domain knowledge. In particular, the internal representations used by the actor can be very different from those used by the supervisor. The actor, for example, can be an artificial neural network, while the supervisor can be a conventional feedback controller, expert knowledge encoded as logical propositions, or a human supplying actions that depend on an entirely different perception of the environment's state. Moreover, the supervisor can convey intentions and solution strategies to the actor, and so this work is similar in spirit to work on imitation learning, for example, [19, 26]. And presumably the supervisor has a certain proficiency at a given task, which the actor exploits for improved performance throughout learning.

Appendix 1: Nomenclature

a	composite action	π^E	exploratory policy
a^A	actor action	π^S	supervisor policy
a^E	exploratory action	r	immediate reward
a^S	supervisor action	R	expected immediate reward
α	learning step size (actor)	s	current state
β	learning step size (critic)	s'	next state
δ	TD error	σ	exploration size
E	supervisory error	θ	CMAC weight vector (critic)
e	eligibility trace	V^π	value function under policy π
γ	discount factor	V	estimate of V^π
k	interpolation parameter	w	CMAC weight vector (actor)
λ	eligibility trace decay factor	Δw^{RL}	reinforcement learning update
π	composite policy	Δw^{SL}	supervised learning update
π^A	actor policy		

Appendix 2: Cerebellar Model Arithmetic Computer

A CMAC, or *cerebellar model arithmetic computer* [1], is a kind of artificial neural network inspired by the anatomy and physiology of the cerebellum. Much like radial basis function (RBF) networks, CMACs have processing units that are localized in different regions of an input space. In contrast with RBF networks—where each unit computes a scalar value based on how close an input is to the center of the unit—CMAC units instead compute a binary value for each input. In either case, the computed values affect how much the associated weight parameters contribute to the network's output. With CMAC units, their binary nature therefore determines whether a particular weight "participates" entirely or not at all in the output calculation.

Also in contrast to RBF networks—with radially symmetric processing units—CMAC units are hyper-rectangles arranged as a *tiling*, that is, as a grid-like tessellation of the input space. This leads to computationally efficient implementations similar to lookup tables. Another advantage of CMACs is that multiple tilings, with each one offset from the rest, can be used to improve resolution, while relatively large hyper-rectangles can be used to improve generalization. And for modern RL algorithms, a function approximator with binary units allows one to take advantage of replacing eligibility traces [28], as we do with the algorithm in Figure 14.3. See [24] for more information about CMACs, including an empirical evaluation of their use for RL problems.

Acknowledgments

We thank Andrew Fagg, Mohammad Ghavamzadeh, and Balaraman Ravindran for helpful discussions. This work was supported by NASA under award No. NAG-9-1379. Any opinions, findings, and conclusions or recommendations expressed in this material are those of the authors and do not necessarily reflect the views of the National Aeronautics and Space Administration.

Bibliography

1. J. Albus, *Brains, Behavior, and Robotics,* Byte Books, Peterborough, NH, 1981.

2. A. G. Barto, Reinforcement learning in motor control, in M. A. Arbib, (ed.), *The Handbook of Brain Theory and Neural Networks, Second Edition,* pp. 968–972, MIT Press, Cambridge, MA, 2003.

3. A. G. Barto, R. S. Sutton, and C. W. Anderson, Neuronlike elements that can solve difficult learning control problems, *IEEE Trans. Systems, Man, and Cybernetics,* vol. 13, pp. 835–846, 1983.

4. H. Benbrahim and J. A. Franklin, Biped dynamic walking using reinforcement learning, *Robotics and Autonomous Systems,* vol. 22, pp. 283–302, 1997.

5. N. A. Bernstein, *The Co-ordination and Regulation of Movements,* Pergamon Press, Oxford, 1967.

6. A. E. Bryson and Y.-C. Ho, *Applied Optimal Control,* Hemisphere Publishing Corp., New York, 1975.

7. J. A. Clouse, *On Integrating Apprentice Learning and Reinforcement Learning,* Ph.D. Thesis, University of Massachusetts, Amherst, 1996.

8. J. A. Clouse and P. E. Utgoff, A teaching method for reinforcement learning, *Proc. Ninth International Conference on Machine Learning,* pp. 92–101, Morgan Kaufmann, San Francisco, CA, 1992.

9. J. J. Craig, *Introduction To Robotics: Mechanics and Control,* Addison-Wesley, Reading, MA, 1989.

10. M. Dorigo and M. Colombetti, Robot shaping: developing autonomous agents through learning, *Artificial Intelligence,* vol. 71, no. 2, pp. 321–370, 1994.

11. V. Gullapalli, A stochastic reinforcement learning algorithm for learning real-valued functions, *Neural Networks,* vol. 3, no. 6, pp. 671–692, 1990.

12. M. Huber and R. A. Grupen, A feedback control structure for on-line learning tasks, *Robotics and Autonomous Systems,* vol. 22, no. 3-4, pp. 303–315, 1997.

13. M. I. Jordan and D. E. Rumelhart, Forward models: Supervised learning with a distal teacher, *Cognitive Science,* vol. 16, no. 3, pp. 307–354, 1992.

14. M. Kaiser and R. Dillmann, Building elementary robot skills from human demonstration, *Proc. IEEE International Conference on Robotics and Automation,* pp. 2700–2705, IEEE, Piscataway, NJ, 1996.

15. R. M. Kretchmar, P. M. Young, C. W. Anderson, D. C. Hittle, M. L. Anderson, C. C. Delnero, and J. Tu, Robust reinforcement learning control with static and dynamic stability, *International Journal of Robust and Nonlinear Control,* vol. 11, pp. 1469–1500, 2001.

16. L.-J. Lin, Self-improving reactive agents based on reinforcement learning, planning and teaching, *Machine Learning,* vol. 8, no. 3-4, pp. 293–321, 1992.

17. R. Maclin and J. W. Shavlik, Creating advice-taking reinforcement learners, *Machine Learning,* vol. 22, no. 1-3, pp. 251–281, 1996.

18. M. J. Mataric, Reward functions for accelerated learning, *Proc. Eleventh International Conference on Machine Learning,* pp. 181–189, San Francisco, CA, 1994.

19. M. J. Mataric, Sensory-motor primitives as a basis for imitation: linking perception to action and biology to robotics, in C. Nehaniv and K. Dautenhahn, (eds.), *Imitation in Animals and Artifacts,* MIT Press, Cambridge, MA, 2000.

20. A. Y. Ng, D. Harada, and S. Russell, Policy invariance under reward transformations: Theory and applications to reward shaping, *Proc. Sixteenth International Conference on Machine Learning,* pp. 278–287, Morgan Kaufmann, San Francisco, CA, 1999.

21. T. J. Perkins and A. G. Barto, Lyapunov-constrained action sets for reinforcement learning, in C. Brodley and A. Danyluk, (eds.), *Proc. 18th International Conference on Machine Learning,* pp. 409–416, Morgan Kaufmann, San Francisco, CA, 2001.

22. T. J. Perkins and A. G. Barto, Lyapunov design for safe reinforcement learning, *Journal of Machine Learning Research,* vol. 3, pp. 803–832, 2002.

23. B. Price and C. Boutilier, Implicit imitation in multiagent reinforcement learning, in I. Bratko and S. Dzeroski, (eds.), *Proc. 16th International Conference on Machine Learning,* pp. 325–334, Morgan Kaufmann, San Francisco, CA, 1999.

24. J. C. Santamaria, R. S. Sutton, and A. Ram, Experiments with reinforcement learning in problems with continuous state and action spaces, *Adaptive Behavior,* vol. 6, pp. 163–217, 1997.

25. S. Schaal, Learning from demonstration, in M. C. Mozer, M. I. Jordan, and T. Petsche, (eds.), *Advances In Neural Information Processing Systems 9,* pp. 1040–1046, MIT Press, Cambridge, MA, 1997.

26. S. Schaal, Is imitation learning the route to humanoid robots? *Trends in Cognitive Science,* vol. 3, pp. 233–242, 1999.

27. J. S. Shamma, Linearization and gain-scheduling, in W. S. Levine, (ed.), *The Control Handbook,* pp. 388–396, CRC Press, Boca Raton, FL, 1996.

28. S. P. Singh and R. S. Sutton, Reinforcement learning with replacing eligibility traces, *Machine Learning,* vol. 22, no. 1-3, pp. 123–158, 1996.

29. W. D. Smart and L. P. Kaelbling, Effective reinforcement learning for mobile robots, *Proc. IEEE International Conference on Robotics and Automation,* pp. 3404–3410, IEEE, Piscataway, NJ, 2002.

30. R. S. Sutton, Learning to predict by the method of temporal differences, *Machine Learning,* vol. 3, pp. 9–44, 1988.

31. R. S. Sutton and A. G. Barto, *Reinforcement Learning: An Introduction,* MIT Press, Cambridge, MA, 1998.

32. C. J. C. H. Watkins and P. Dayan, Q-learning, *Machine Learning,* vol. 8, no. 3-4, pp 279–292, 1992.

33. R. J. Williams, Simple statistical gradient-following algorithms for connectionist reinforcement learning, *Machine Learning,* vol. 8, pp. 229–256, 1992.

15 Backpropagation Through Time and Derivative Adaptive Critics — A Common Framework for Comparison †

DANIL V. PROKHOROV

Ford Motor Company

Editor's Summary: This chapter compares and contrasts derivative adaptive critics (DAC) such as dual heuristic programming (DHP), which was first introduced in Chapter 1 and also discussed in Chapter 3 with back-propagation through time (BPTT). A common framework is built and it is shown that both are techniques for determining the derivatives for training parameters in recurrent neural networks. This chapter goes into sufficient mathematical detail that the reader can understand the theoretical relationship between the two techniques. The author presents a hybrid technique that combines elements of both BPTT and DAC and provides detailed pseudocode. Computational issues and classes of challenging problems are discussed.

15.1 INTRODUCTION

Heterogeneous ordered networks or, more specifically, recurrent neural networks (RNN) are convenient and flexible computational structures applicable to a broad spectrum of problems in system modeling and control. They require efficient algorithms to achieve successful learning for the given task. Backpropagation through time (BPTT), henceforth assumed to be its truncated version with a sufficient depth of truncation, and derivative adaptive critics (DAC) are two seemingly quite different approaches to solve temporal differentiable optimization problems with continuous

---

† Portions of this chapter were previously published in [4, 7, 9, 12–14, 23].

variables. In fact, both BPTT and DAC are means to obtain derivatives for RNN parameters.

We show that both approaches are related. BPTT is used in DAC to obtain targets for derivative critic adaptation in RNN training. DAC can be interpreted as a method to reduce the need for introducing a potentially large truncation depth in BPTT by providing estimates of derivatives from the future time steps. This realization allows us to establish a common framework for comparison of derivatives of BPTT and those of DAC and summarize their differences. The main difference stems from the fact that derivatives provided by DAC are learned via a representation (critic network), and such derivatives can be averages of derivatives provided by BPTT. It should be kept in mind that some derivative averaging naturally occurs in the training process during which RNN parameters are being adjusted (usually incrementally). Depending on the problem setup and critic training mechanics, DAC derivatives may or may not be sufficiently accurate for successful RNN training.

Both BPTT and DAC must also be equipped with a parameter adjustment rule or algorithm. BPTT equipped with various forms of the Kalman filter algorithm has shown its power in tackling difficult RNN training problems. Training a single RNN to model or control multiple systems with its weights fixed upon completion of the training is particularly remarkable because it defies successful applications of feed-forward or time-delay NN. In spite of several successful demonstrations of DAC and in contrast to BPTT, DAC has mostly been restricted to training feed-forward NN (neurocontrollers) using a gradient descent rule. It is crucial for DAC to reinforce itself with more powerful architectures and training algorithms to be capable of solving truly difficult optimization problems.

Sufficiently detailed comparisons between BPTT and DAC as training approaches to RNN are essentially lacking. Careful comparisons should be based not only on the results of comprehensive testing of the solutions but also on assessments of the computational requirements of the approaches. It is noteworthy that the critic network is discarded as soon as RNN parameter training is finished, which is wasteful. Furthermore, comparisons for a clearly formulated and easily accessible modeling problem may be preferable over comparisons for control problems because modeling problems usually have a relatively simple setup. We suggest a nonstationary system modeling problem as a possible benchmark for comparing BPTT and DAC.

This chapter is structured as follows. In Section 15.2 we show a relationship between BPTT and DAC that gives rise to a common framework for comparison of the two methods. In Section 15.3 we discuss critic representation. In Section 15.4 we propose a hybrid between BPTT and DAC which can be useful for comparative studies. In Section 15.5 we emphasize the need to base comparisons of the two methods not only on the final result but also on computational requirements for each method. We discuss two classes of challenging problems which could form a core of future comparative studies of BPTT and DAC in Section 15.6.

15.2   RELATIONSHIP BETWEEN BPTT AND DAC

We would like to show how BPTT is used within the DAC approach. We consider differentiable optimization with criterion

$$J(k) \;=\; \frac{1}{2}\sum_{t=k}^{k+h}\gamma^{t-k}\sum_{j=N_1}^{N_2} U_j^2(t), \tag{15.1}$$

where $0 < \gamma \le 1$ is a discount, depth (horizon) h is as large as required, $U_j(t)$ is an instantaneous cost (or utility) function. Without loss of generality, each U_j is assumed to be a function of state variables of the following *ordered* heterogeneous network

$$x_i(t) \;=\; x_i^{ext}(t), \quad 1 \le i \le m, \tag{15.2}$$

$$net_i(t) \;=\; \sum_{j=1}^{i-1} W_{i,j}(t)x_j(t) + \sum_{j=m+1}^{N} W_{i,j}^1(t)x_j(t-1), \tag{15.3}$$

$$x_i(t) \;=\; f_i(net_i(t)), \quad m+1 \le i \le N, \tag{15.4}$$

where $x_i(\cdot) \in R$, $f_i(\cdot) \in C^1$, m is a number of external inputs x_i^{ext} to the network, and $m + 1 \le (N_1, N_2) \le N$ is a set of indexes for which U_j are defined. The execution order is assumed to be from node 1 to node N. We want to determine parameters $W_{i,j}, W_{i,j}^1$ delivering a minimum to (15.1) in the mean square sense in the domain of interest $X : x_i \in X$.

Ordered derivatives [1] of the criterion J with respect to x_i are determined by taking into account (15.2)–(15.4):

$$F\_x_i(t) \;=\; E + F + R, \tag{15.5}$$

where

$$E \;=\; U_i(t)\frac{\partial U_i(t)}{\partial x_i(t)}, \tag{15.6}$$

$$F \;=\; \sum_{j=i+1}^{N} W_{j,i}(t)\frac{\partial f_j(net_j(t))}{\partial net_j(t)}F\_x_j(t), \tag{15.7}$$

$$R \;=\; \gamma\sum_{j=m+1}^{N} W_{j,i}^1(t+1)\frac{\partial f_j(net_j(t+1))}{\partial net_j(t+1)}F\_x_j(t+1). \tag{15.8}$$

Equation (15.5) is run backwards in *both* space ($i = N, N-1, ..., 1$) and time ($t = k+h, k+h-1, ..., k$) initializing $F\_x_i(k+h+1) = 0$ and $W_{j,i}^1(k+h+1) = W_{j,i}^1(k+h)$.

The expression for $F\_x_i(t)$ consists of three components. The term E of (15.6) is an explicit derivative of $1/2U_i^2(t)$ with respect to $x_i(t)$ (if exists). If the node x_i

feeds other nodes through feed-forward connections, then F of (15.7) should reflect all such connections. Likewise, R of (15.8) reflects all time-delayed connections through which the node x_i feeds others.

Ordered derivatives with respect to parameters $\mathbf{W}$ and $\mathbf{W}^1$ are determined using $F\_x_i(t)$ and (15.2)–(15.4):

$$F\_W_{i,j}(t) = F\_x_i(t)\frac{\partial f_i(net_i(t))}{\partial net_i(t)}x_j(t), \qquad (15.9)$$

$$F\_W_{i,j}^1(t) = F\_x_i(t)\frac{\partial f_i(net_i(t))}{\partial net_i(t)}x_j(t-1). \qquad (15.10)$$

Equations (15.5)–(15.10) are called BPTT [1]. Here they express a truncated form of BPTT, henceforth denoted as BPTT(h) [2].

Updates of parameters $\mathbf{W}$ and $\mathbf{W}^1$ can be made by using $\sum_{t=t_0}^{t_f} F\_W_{i,j}(t)$ and $\sum_{t=t_0}^{t_f} F\_W_{i,j}^1(t)$, respectively, where t_0, t_f are suitably chosen time steps of the trajectory (e.g., $t_0 = k$, $t_f = k + h$, another possibility is $t_0 = t_f = k$).

We now consider another optimization criterion (cf. (15.1)):

$$J'(t) = \frac{1}{2}\sum_{j=N_1}^{N_2} U_j^2(t) + \gamma J'(t+1). \qquad (15.11)$$

The criterion J of (15.1) is an approximation of J' which becomes increasingly more accurate for sufficiently large h and $\gamma < 1$. On the other hand, the criterion (15.11) is approximated by J critic in approximate (heuristic) dynamic programming [3].

Let us examine derivatives of $J'(t)$ with respect to $x_i(t)$ taking into account (15.2)–(15.4):

$$\lambda_i'(t) = E' + F' + R', \qquad (15.12)$$

where

$$E' = U_i(t)\frac{\partial U_i(t)}{\partial x_i(t)}, \qquad (15.13)$$

$$F' = \sum_{j=i+1}^{N} W_{j,i}(t)\frac{\partial f_j(net_j(t))}{\partial net_j(t)}\lambda_j'(t), \qquad (15.14)$$

$$R' = \gamma \sum_{j=m+1}^{N} W_{j,i}^1(t+1)\frac{\partial f_j(net_j(t+1))}{\partial net_j(t+1)}\lambda_j'(t+1), \qquad (15.15)$$

where $\lambda_j' \equiv \partial J'/\partial x_i$. Note that Eq. (15.12) is *nearly identical to (15.5) expressing backpropagation between two consecutive moments in time (BPTT(1)) [4].* (Replacing $\lambda_i'(t)$ and $\lambda_j'(t+1)$ in (15.12),(15.14),(15.15) with $F\_x_i(t)$ and $F\_x_j(t+1)$, respectively, makes them identical.) Eqs. (15.9)–(15.10) may also be used for minimization of (15.11) except that $F\_x_i(t)$ should be replaced by $\lambda_i'(t)$.

In a popular DAC approach called *dual heuristic programming* (DHP), each λ_i' is to be approximated by output λ_i of λ critic [3]. One can demonstrate that traditional DHP equations (see, e.g., Eqs. (7) and (8) in [5]) are a special case of (15.12). The λ critic is expressed as a suitable representation $\lambda(\mathbf{x}(t), \mathbf{W}_C)$ with outputs $\lambda_i(\cdot)$ and adjustable weights $\mathbf{W}_C$. The critic is supposed to be trained with the error between $\lambda_i'(t)$ from (15.12) and its corresponding output $\lambda_i(t)$. (Each $\lambda_j'(t+1)$ in (15.15) is replaced by the appropriate output $\lambda_j(t+1)$ of the critic.)

We just revealed the similarity between a particular form of BPTT and a popular DAC formulation for the network (15.2)–(15.4), but this is also valid for much more general networks (see [4] and Section 15.4 of this chapter) and systems including those with distributed parameters [6]. Recognizing this similarity enables us to create a hybrid of BPTT and DAC in which DAC may act as a means to reduce depth h in BPTT(h) by providing estimates of derivatives from the future time steps (see Section 15.4).

Many researchers have pointed out similarities between BPTT and the Euler-Lagrange/Hamiltonian formulation (see, e.g., [17, 19]). The expression (15.12) or (15.5) may be recognized as the Lagrange multipliers (partial derivatives of a Hamiltonian with respect to state variables) for the network (15.2)–(15.4) with the criterion (15.11) [17, 18]. The derivatives (15.9), (15.10), or partial derivatives of a Hamiltonian with respect to controls $\mathbf{W}$ and $\mathbf{W}^1$, can be used to update incrementally the network parameters in order to minimize the criterion (15.11).

We describe a typical application of BPTT to functional minimization. First, we initialize the state variables of (15.2)–(15.4). We run (15.2)–(15.4) forward for one or more time steps and compute the appropriate values of $U(t) : U(t), U(t+1), ..., U(t+k)$, where U is a known function of state variables and their targets (e.g., U is a tracking error). We then compute the derivatives according to (15.5) by backpropagating from $t+k$ to t (BPTT(k)) and perform the incremental updates based on (15.9) and (15.10).

Compared with this description of BPTT application, the DAC approach requires the utilization of only two adjacent time steps, t and $t+1$, according to (15.12). We run (15.2)–(15.4) forward for one time step, obtain $U(t)$, and invoke (15.12) to prepare for critic training. The right-hand side of (15.12) serves as both the set of instantaneous targets for critic training and the input to the parameter updates (15.9) and (15.10) (in place of $F\_x_i(t)$). If gradient descent is employed, critic training is also incremental, and it may be based on the expression $(\lambda_i'(t) - \lambda_i(t))\frac{\partial \lambda_i(t)}{\mathbf{W}_C}$, where $\mathbf{W}_C$ is a vector of critic weights.

For either the BPTT or the DAC approach, we continue the training process for the next point $t+1$ along the trajectory. We can train for a segment of the trajectory, then reinitialize the state variables (15.2)–(15.4) and move on to training on another segment of the trajectory. Meanwhile, our weights $\mathbf{W}$, $\mathbf{W}^1$ (and $\mathbf{W}_C$ for DAC) serve as the long-term memory, incorporating the effects of adaptation averaged over many instances. For an adequately chosen training strategy, we can reasonably expect that application of BPTT or DAC will result in the triplet $(\mathbf{W}, \mathbf{W}^1, \mathbf{W}_C)^*$ such that J is approximately minimized over X. As with any numerical and (generally) nonconvex optimization, all we can guarantee is that the proper training process should result in attaining a local minimum of J (most of the time such a minimum is satisfactory). What is proper remains problem dependent, but determining the training strategy includes choosing training parameters (e.g., learning rates) for updates based on (15.9) and (15.10), the length and the assignment of trajectory segments and the initialization of state variables. For DAC we need to add the choice of critic training parameters and the coordination scheme between critic and network training processes [4, 15].

The training process based on BPTT also resembles a form of model predictive control (MPC) with receding horizon (see, e.g., [20]). As in the MPC, we run the system forward for several time steps collecting values of U. Our horizon $(t + k)$ recedes once the weight updates are carried out except that our updates are incremental, not "greedy" as in the receding-horizon MPC.

We summarize the differences between derivatives obtained by BPTT and those of DAC [7]:

1) BPTT derivatives are computed directly, while critic derivatives are computed from a representation, e.g., a neural network, with its parameters to be learned.

2) BPTT(h) derivatives generally involve a finite time horizon (equal to the chosen *truncation depth* h), while critic derivatives are estimates for an infinite horizon; it is not uncommon, however, to employ a discount factor in (15.11), which may be interpreted as a gentle truncation ($\gamma < 1$). Large h are permissible because BPTT computations scale linearly with h, but frequently small h ($h < 10$) suffice.

3) BPTT derivatives necessarily compute the effect of changing a variable in the *past*, while a derivative critic may be used to estimate the effect of a change at the present time. If critics are used only to adjust network parameters, this distinction is *irrelevant*. On the other hand, recognizing it poses interesting possibilities for alternative or supplementary methods of control, as discussed in Section 6 of [7].

4) A BPTT derivative is essentially exact for the specific trajectory for which it is computed, while a critic derivative necessarily represents an average of some kind, e.g., an average over trajectories that begin with a given system state. Such an estimate may be quite accurate (if the critic has been well adapted or trained and if exogenous inputs to the system are either small or statistically well behaved) or may be essentially worthless (if future operation is completely unpredictable due to arbitrary inputs or targets).

Item 4 is discussed in more details in the next section.

15.3 CRITIC REPRESENTATION

Critic predicts the effect of a change in a variable of the system or network (15.2)–(15.4) on its future operation. Critic is thus a function of the system state, and it is important to include in critic representation as much information as available about the system. What is encompassed by the system depends on the context. In the context of indirect model reference adaptive control [8], the system is interpreted quite generally as consisting of: 1) controller network, 2) reference model, 3) object to be controlled (plant), and 4) model of the plant. State variables or their estimates of *all* of these components should be provided as inputs to the critic (the plant model often serves as the plant state estimator). However, the only adjustable quantities are the weights of controller and (sometimes) parameters of the model.

It is convenient to consider all main components of the control system as parts of a *single* heterogeneous recurrent network, perhaps similar to (15.2)–(15.4). This viewpoint equates control problems with modeling problems (employing RNN) since both types of problems feature feedback.

In the modeling context, we suggest connecting all recurrent nodes or time-delayed elements to the critic because they reflect the *state* of the system. In the example below we illustrate that adding a state variable is indeed beneficial [4].

Consider a system

$$x(k+1) = x^{ext}(k+1) + wx(k), \qquad (15.16)$$
$$x^d(k+1) = x^{ext}(k+1) + w^d x^d(k), \qquad (15.17)$$

where $0 < |w| < 1$, $0 < |w^d| < 1$, and $x^{ext}(k), \forall k$ are i.i.d. random variables with the same mean x^{ext}. The system (15.16)-(15.17) is a simple illustration of the network (15.2)–(15.4), with the controller part expressed as (15.16). The Eq. (15.17) is an example of the reference model, and it describes the desired behavior of (15.16).

The minimization criterion is

$$J(k) = \frac{1}{2}\sum_{t=0}^{\infty}\gamma^t U^2(k+t) = \frac{1}{2}\sum_{t=0}^{\infty}\gamma^t (x^d(k+t) - x(k+t))^2. \qquad (15.18)$$

Differentiating $J(k)$ with respect to $x(k)$ yields

$$F\_x(k) = \sum_{t=0}^{\infty}(\gamma w)^t (x(k+t) - x^d(k+t)). \qquad (15.19)$$

Substituting (15.16) and (15.17) into the equation for $F\_x(k)$ above, averaging for all $x^{ext}(\cdot)$ and recognizing appropriate power series results in

$$<F\_x(k)> = \frac{x(k)}{1-\gamma w^2} - \frac{x^d(k)}{1-\gamma ww^d}$$
$$+\frac{x^{ext}\gamma w}{1-\gamma w}\left(\frac{1}{1-\gamma w^2} - \frac{1}{1-\gamma ww^d}\right), \qquad (15.20)$$

where $<F\_x(k)>$ is an (ensemble) average ordered derivative of $J(k)$ with respect to $x(k)$.

We wish to show that a linear λ critic is sufficient to recover $< F\_x(k) >$. The critic representation is

$$\lambda(t) = Ax(t) + Bx^d(t) + C, \tag{15.21}$$

where A, B, and C are the critic weights. Similar to (15.12), we can write

$$\lambda(t) = E' + R', \tag{15.22}$$

where

$$E' = x(t) - x^d(t), \tag{15.23}$$

$$R' = \gamma\lambda(t+1)\frac{\partial x(t+1)}{x(t)}. \tag{15.24}$$

Here (15.23) corresponds to (15.13), and (15.24) corresponds to (15.15) (there is no feed-forward part like (15.14)). Substituting the representation (15.21) into the equation for $\lambda(t)$ above yields

$$Ax(t) + Bx^d(t) + C = x(t) - x^d(t) \\ + \gamma w(Ax(t+1) + Bx^d(t+1) + C). \tag{15.25}$$

When the weights converge they become

$$A = \frac{1}{1 - \gamma w^2}, \tag{15.26}$$

$$B = -\frac{1}{1 - \gamma w w^d}, \tag{15.27}$$

$$C = \frac{x^{ext}\gamma w}{1 - \gamma w}\left(\frac{1}{1 - \gamma w^2} - \frac{1}{1 - \gamma w w^d}\right), \tag{15.28}$$

which, incidentally, indicates that $< F\_x(k) >$ is restored exactly.

A reasonable question to ask is what happens if we do not have access to state(s) of the data generator/reference model x^d. Such situation may happen when training a recurrent network from a file of input-output pairs. We suggest approximating the states of the missing reference model by its time-delayed outputs and provide these estimates as inputs to the critic. (The order of the system which produced the training data might be known and used to determine how many time-delayed inputs to employ.)

One can certainly argue that even a simpler critic representation lacking some difficult-to-find inputs might suffice for a problem at hand. However, we should keep in mind that excluding a state or its estimate from the critic input set effectively turns such a state into a disturbance which tends to decrease the likelihood of getting an accurate critic and handicaps the critic as compared to BPTT.

Many representations for critics are possible. For example, each output can be a separate function or neural network $\lambda_i(\mathbf{x}(t), \mathbf{W}_C)$. In the example above we used the linear critic. Extremely simple (bias weight only) representations are also viable for some problems [9], as illustrated below.

Consider the following system

$$x_1(t) \quad = \quad bx^{ext}(t), \tag{15.29}$$
$$x_2(t+1) \quad = \quad x_1(t), \tag{15.30}$$
$$x_3(t+1) \quad = \quad 0.5x_3(t) + x_1(t+1) + x^{ext}(t+1) - 2x_2(t), \tag{15.31}$$

or, in a compact form,

$$x_3(t+1) \quad = \quad 0.5x_3(t) + (1+b)x^{ext}(t+1) - 2bx^{ext}(t-1). \tag{15.32}$$

The reference input x^{ext} has a piecewise constant pattern with a long enough dwell time (e.g., 50 time steps). The goal is to adjust the parameter b so as to minimize $(x_3^d(t) - x_3(t))^2$, where $x_3^d(t) \equiv 0, \forall t$, in the mean square sense in which case $b = 1$. BPTT(2) must be used to accomplish this (the training process diverges if BPTT(h) with $h < 2$ is used).

We show how the use of two DAC (λ) critics eliminates the need for more than the (computational) equivalent of BPTT(1). The BPTT equations for this system are similar to (15.5):

$$F\_x_3(t) \quad = \quad -e(t) + 0.5F\_x_3(t+1), \tag{15.33}$$
$$F\_x_2(t) \quad = \quad -2F\_x_3(t+1), \tag{15.34}$$
$$F\_x_1(t) \quad = \quad F\_x_3(t) + F\_x_2(t+1), \tag{15.35}$$

where $e(t) = 0 - x_3(t) = -x_3(t)$. In the first equation $-e(t)$ is equivalent to (15.6), and $0.5F\_x_3(t+1)$ is equivalent to (15.8) (it is obtained from the equation for $x_3(t+1)$). The equation for $x_3(t+1)$ is also used to obtain $F\_x_2(t)$ and $F\_x_3(t)$ in the expression for $F\_x_1(t)$. The last equation (for $F\_x_1(t)$) is obtained by combining $F\_x_1(t) = F\_x_2(t+1)$ with $F\_x_1(t+1) = F\_x_3(t+1)$ at time t.

The BPTT equations above are to be repeated two times (BPTT(2)) to produce F_x_1 suitable for updating b correctly:

$$F\_b(t) \quad = F\_x_1(t)x^{ext}(t). \tag{15.36}$$

According to (15.12), we replace $F\_x_3(t+1)$ and $F\_x_2(t+1)$ with $\lambda_3(t+1)$ and $\lambda_2(t+1)$, respectively, and obtain

$$\lambda_3'(t) \quad = \quad -e(t) + 0.5\lambda_3(t+1), \tag{15.37}$$
$$\lambda_2'(t) \quad = \quad -2\lambda_3(t+1), \tag{15.38}$$
$$F\_x_1(t) \quad = \quad \lambda_3'(t) + \lambda_2(t+1), \tag{15.39}$$

where $F\_x_1(t)$ is used in (15.36) to update b. It turns out that the bias-weight-only critics suffice for this problem, and they may be updated as follows

$$\lambda_3 \ +=\ \eta(\lambda_3'(t) - \lambda_3), \tag{15.40}$$

$$\lambda_2 \ +=\ \eta(\lambda_2'(t) - \lambda_2), \tag{15.41}$$

where learning rate $\eta > 0$ is reasonably chosen, and the C-language notation "$+=$" indicates that the quantity on the right hand side is to be added to the previously computed value of the left hand side. It can be shown that performing updates of the critics and the weight b results in convergence of b to its desired value of unity if the system is sufficiently excited by x^{ext}.

The system (15.32) could easily be changed to represent D delays which would either require BPTT(D) or D critics trained in a fashion similar to this example.

As illustrated with the example (15.29)–(15.41), even a trivial critic can be effective and competitive with BPTT when dealing with a predictable system. (A good example of a predictable system is a system with a constant (or fixed-statistics random) disturbance.) The system above is predictable because it is driven by a slowly-varying excitation. However, any critic-based training approach will have difficulties if the system is subject to x^{ext} changing significantly at every time step. For this and other systems with recurrence and block delays affected by fast-changing disturbances or excitations it is better to use a hybrid of temporal backpropagation [10], BPTT and, possibly, DAC proposed in [11].

15.4   HYBRID OF BPTT AND DAC

Here we provide the reader with C-language-style pseudocode describing a hybrid of BPTT and DAC which is less efficient in handling block delays than that in [11] but easier to implement.

The forward equations for an ordered network with n_in inputs and n_out outputs may be expressed very compactly in a pseudocode format [13]. Let the network consist of n_nodes nodes, of which n_in serve as receptors for the external inputs. The bias input, which we denote formally as node 0, is not included in the node count. The bias input is set to the constant 1.0. The array **I** contains a list of the input nodes; e.g., I_j is the number of the node that corresponds to the jth input, in_j. Similarly, a list of the nodes that correspond to network outputs out_p is contained in the array **O**. We allow network outputs and targets to be advanced or delayed with respect to node outputs by assigning a phase τ_p to each output. For example, if we wish to associate the network output p with the output of some system five steps in the future, we would have $\tau_p = 5$. Node i receives input from n_con(i) other nodes and has activation function $f_i(\cdot)$; n_con(i) is zero if node i is among the nodes listed in the input array **I**. The array **c** specifies connections between nodes; $c_{i,j}$ is the node number for the jth input of node i. Inputs to a given node may originate at the current or past time steps, as specified by delays contained in the array d, and through weights for time step t contained in the array $\mathbf{W}(t)$.

Prior to beginning network operation, all appropriate memory is initialized. Normally, such memory (except weights) will be set to zero. (In some cases memory that corresponds to the network initial state may be set to specified values.)

At the beginning of each time step, we execute the following buffer operations on weights and node outputs (in practical implementation, a circular buffer and pointer arithmetic may be employed). Here dmax is the largest value of delay represented in the array d, and h is the truncation depth of the BPTT gradient calculation described in a pseudocode form later.

```
for i = 1 to n_nodes {
 for i_t = t-h-dmax to t-1 {
```

$$\mathbf{W}(i_t) = \mathbf{W}(i_t + 1) \tag{15.42}$$
$$y_i(i_t) = y_i(i_t + 1) \tag{15.43}$$

```
 } /* end i_t loop */
} /* end i loop */
```

Then, the actual network execution is expressed as

```
for i = 1 to n_in {
```

$$y_{I_i}(t) = \text{in}_i(t) \tag{15.44}$$

```
}
for i = 1 to n_nodes {
 if n_con(i) > 0 {
```

$$a_i(t) = \left[\begin{array}{c} \sum_{j=1}^{\text{n\_con}(i)} W_{i,j}(t) y_{c_{i,j}}(t - d_{i,j}) \\ \prod_{j=1}^{\text{n\_con}(i)} (W_{i,j}(t) + y_{c_{i,j}}(t - d_{i,j})) \end{array} \right. \tag{15.45}$$

$$y_i(t) = f_i(a_i(t)) \tag{15.46}$$

```
 }
}
for p = 1 to n_out {
```

$$\text{out}_p(t + \tau_p) = y_{O_p}(t) \tag{15.47}$$

```
}
```

Most commonly, we take the (differentiable) activation function $f_i(\cdot)$ to be either linear or a bipolar sigmoid, though we also can make use of other functions, e.g., sinusoids, for special purposes. In the pseudocode above, the top portion of the right-hand side of (15.45) is invoked whenever the node i performs a summation of its inputs weighted by the appropriate elements of $\mathbf{W}$. The bottom portion of the right-hand side of (15.45) is invoked if the node i is a product (multiplicative) node.

The pseudocode above is very general, and it can be used to describe a great deal of neural and non-neural computational structures including (15.2)–(15.4).

In the pseudocode for a hybrid of *aggregate* BPTT [12] and DAC below, $F^p_{-}y$ denotes an ordered derivative of $\frac{1}{2} \sum_{i_h=0}^{h} \gamma^{i_h} (U_p(t - h + i_h))^2$ (cf. (15.1)), where U_p is a component of the utility vector $\mathbf{U}$ usually expressed as a deviation between appropriate target and output of the network. This pseudocode can be invoked at each time step only after the completion of forward propagation at time step t.

```
for p = 1 to n_out {
 for i = 1 to n_nodes {
  for k = 1 to n_con(i) {
```

$$F^p_{-}W_{i,k} \quad = \quad 0 \tag{15.48}$$

```
 } /* end k loop */
 for i_t = t to t-h-dmax {
```

$$F^p_{-}y_i(i_t) \quad = \quad 0 \tag{15.49}$$

```
  if i_t = t {
```

$$F^p_{-}y_i(i_t) \quad = \quad \kappa^p_i(i_t) \tag{15.50}$$

```
  }
 } /* end i_t loop */
 } /* end i loop */
 for i_h = 0 to h {
```

$$
\begin{aligned}
i_1 &= \max(t - i_h, 0) & (15.51) \\
U_p(i_1) &= \mathrm{tgt}_p(i_1 + \tau_p) - \mathrm{out}_p(i_1 + \tau_p) & (15.52) \\
F^p_{-}y_{0_p}(i_1) &+= -U_p(i_1) & (15.53)
\end{aligned}
$$

```
 for i = n_nodes to 1 {
  if n_con(i) > 0 {
```

```
for k = n_con(i) to 1 {
```

$$j \quad = \quad c_{i,k} \tag{15.54}$$

$$i_2 \quad = \quad \max(i_1 - d_{i,k}, 0) \tag{15.55}$$

$$\mathrm{F\_}y_j(i_2) \quad += \quad \gamma^{d_{i,k}} \mathrm{F\_}y_i(i_1) f_i'(a_i(i_1))$$

$$\times \left[\frac{W_{i,k}(i_1)}{\prod_{m=1,m\neq k}^{n\_con(i)}(W_{i,m}(i_1) + y_{c_{i,m}}(i_1 - d_{i,m}))} \right. \tag{15.56}$$

$$\mathrm{F\_}W_{i,k} \quad += \quad \mathrm{F\_}y_i(i_1) f_i'(a_i(i_1))$$

$$\times \left[\frac{y_j(i_2)}{\prod_{m=1,m\neq k}^{n\_con(i)}(W_{i,m}(i_1) + y_{c_{i,m}}(i_1 - d_{i,m}))} \right. \tag{15.57}$$

```
  } /* end k loop */
 }
 } /* end i loop */
 } /* end i_h loop */
} /* end p loop */
```

Here $f_i'(a_i(i_1)) = \partial f_i(a_i(i_1))/\partial a_i(i_1)$. The loops for (15.48) and (15.49) serve as initializations. We use κ_i^p to denote the output i of derivative critic for component p of the utility vector $\mathbf{U}$. This is done to avoid confusion with λ_i discussed above.

By virtue of "+ =" notation, the appropriate derivatives are distributed from a given node to all nodes and weights that feed it in the forward direction, with due allowance for any delays that might be present in each connection. The simplicity of the formulation reduces the need for visualizations such as unfolding in time or signal-flow graphs.

If the node i is one of the summation nodes (as in the top portion of the right-hand side of (15.45)), then the derivatives with respect to y_j and $W_{i,k}$ are computed according to the top portions of the left-bracketed terms in (15.56) and (15.57), respectively. For a multiplicative node (as in the bottom portion of the right-hand side of (15.45)), these derivatives are computed according to the bottom portions of the left-bracketed terms in (15.56) and (15.57).

DAC errors and targets are computed after completion of the pseudocode above:

```
for p = 1 to n_out {
 for i = n_nodes to 1 {
  if n_con(i) > 0 {
```

```
for k = n_con(i) to 1 {
```

$$j = c_{i,k} \qquad (15.58)$$

$$i_2 = t - h - d_{i,k} \qquad (15.59)$$

$$\kappa_j^{*p}(i_2) = F\_y_j^p(i_2) \qquad (15.60)$$

$$\epsilon_j^p(i_2) = \kappa_j^{*p}(i_2) - \kappa_j^p(i_2) \qquad (15.61)$$

```
} /* end k loop */
 }
} /* end i loop */
} /* end p loop */
```

A special case with $h = 0$ and dmax $= 1$ is what may be called the pure DAC algorithm (cf. Eq. (15.12)). However, there is one crucial difference. In the usual formulation (15.12), derivative critics designated by λ are trained to estimate derivatives of J', including the contribution from the current step, k. The pseudocode above separates the estimate of the future from that of the present, and it is a generalization of the algorithm proposed in [14] (cf. C critics in [14]).

The difference between the values of the two critic forms is precisely the quantity that results when the derivative of error at each output node $\frac{\partial J'}{\partial y_{out}(k)}$ is backpropagated to $y(k)$. (In the simple case of a single-node network, the new critic $\kappa(k)$ is related to the usual critic as follows: $\lambda(k) = \frac{\partial J'}{\partial y(k)} + \kappa(k)$.) The κ critic is thus not required to estimate quantities which can be computed exactly. Limited experimentation suggests that the use of κ critics may lead to faster training than that of λ critics.

Critic can be trained using the error (15.61). For example, a gradient descent update of critic weights may look like this:

$$\mathbf{W}_C^p \ += \ \eta^p \epsilon_j^p(t-1) \frac{\partial \kappa_j^p(t-1)}{\partial \mathbf{W}_C^p}, \qquad (15.62)$$

where $\mathbf{W}_C^p$ is a vector of weights of the critic p, and $\eta^p > 0$ ($h = 0$ and dmax $=$ 1). (We assume in (15.62) that an individual critic is used to approximate ordered derivatives of the discounted sum of U_p^2 with respect to node output y_j. One can also combine all such critics into one network.) The critic outputs may correspond to different time steps when $d_{i,k}$ is different for different RNN nodes, as follows from (15.61).

No critic training happens if all critic outputs κ_j^p and their learning rates are fixed at zero. In such a case, our pseudocode amounts to carrying out the aggregate BPTT(h). If the gradient descent training of weights $\mathbf{W}$ is desired, then we use $F\_W_{i,k}^p$ of (15.57) to adjust $W_{i,k}(t)$. For extended Kalman filter (EKF) training [13], the error injection (15.53) should be modified for consistency with mechanics of the EKF recursion (see [12]).

If $h > 0$ and critic training is enabled, then (15.62) may be invoked to train the critic (alternatively, the EKF algorithm can be used). While limited experiments with utilizing derivatives (15.57) in conjunction with EKF updates have been carried out successfully, further experimentation is needed since such derivatives are different from those usually employed by the EKF, especially when backpropagation to the RNN weights, as in (15.57), is performed only for $i_1 = t - h$ [14].

Concluding this section, it should be mentioned that our hybrid can be used even when components of U are not defined for all time steps. Furthermore, a differentiable approximation of U may be used if the true U is not well defined. Such an approximation capturing an essential relationship between the network variables and the desired instantaneous utility or the final outcome (e.g., in a game setting) could be learned in a separate training task prior to invoking the hybrid equations of this section [7].

15.5   COMPUTATIONAL COMPLEXITY AND OTHER ISSUES

We wish to compare the overhead associated with computations of $F_{-}W_{i,j}$ and κ_i for DAC with the cost of computing $F_{-}W_{i,j}$ for BPTT(h).

We assume that the cost of forward and back-propagations through a network is dominated by a linear term proportional to the number of its weights. For a critic with N_{W_C} weights, the cost of carrying out static BP (BPTT(0)) is $O(N_{W_C})$. If we use total N_C data points to train a critic, the critic training cost is proportional to both N_{W_C} and N_W because backpropagation through a RNN with N_W weights to obtain critic targets (15.60) has $O(N_W)$ complexity for each of N_C data points. Training a network with DAC on N_A data points incurs a cost proportional to both N_W and N_{W_C} because the critic is to be executed with $O(N_{W_C})$ complexity for each of N_A data points. Thus the total computational cost of DAC algorithm is proportional to $(N_C + N_A)N_{W_C}$ and $(N_C + N_A)N_W$.

The cost of BPTT(h) is $O(N_W h)$. If we use N_B data points to train the network, then the total cost is proportional to $N_B N_W h$.

Our simple analysis does not take into account the cost of updating **W** in both the network and the critic which can be significant, especially for second order methods. Let us assume that the network weight updating incurs the same cost for both methods. If

$$\alpha(N_C + N_A)N_{W_C} + \beta(N_C + N_A)N_W + W_C \text{ update cost} \quad < \quad \omega N_B N_W h,$$
(15.63)

where α, β, ω are some problem-dependent constants, then the DAC approach is more efficient computationally than the BPTT(h) approach *provided* that the updates of **W** result in the same network performance upon completion of its training.

It appears possible to simplify the DAC approach for some problems, for example, when the system is affine in controls a:

$$\mathbf{x}(t+1) \;\; = \;\; \mathbf{F}(\mathbf{x}(t)) + \mathbf{G}(\mathbf{x}(t))\mathbf{a}(t), \tag{15.64}$$

where $\mathbf{F}$ is a vector function, $\mathbf{G}$ is a control matrix. If $U(t)$ is quadratic in a, that is, includes the term $\mathbf{a}^T(t)\mathbf{R}\mathbf{a}(t)$, $\mathbf{R} > 0$, then implementable parameterization of the optimal controller may be expressed using the critic $\boldsymbol{\lambda}$ (vector) as the following product

$$\mathbf{a}(t) \;\; = \;\; -\mathbf{R}^{-1}\mathbf{G}^T(\mathbf{x}(t))\boldsymbol{\lambda}(\mathbf{x}(t), \mathbf{W}_C). \tag{15.65}$$

Thus unlike the usual DAC approach featuring training of two entities (critic and controller), no controller training is necessary here. In such a case the expression for computational cost comparison above is changed to

$$\zeta N_C N_{\mathbf{W}_C} + \mathbf{W}_C \text{ update cost} \;\; < \;\; \omega N_B N_{\mathbf{W}} h, \tag{15.66}$$

where ζ is another constant.

Derivatives from DAC (values of λ_i or κ_i) represent averages of derivatives $F_{-}x_i$, as shown in the example (15.16)–(15.28). That example also touches upon the following important issue. To obtain the accurate average $< F_{-}x(k) >$, we have to wait until convergence of critic weights A, B and C. Even more importantly, the parameters w and w^d must be kept fixed. If training of w is in progress (or w^d is changing), critic accuracy will also depend on how these parameters are being changed, because every change in w or w^d results in a change to the system or network to which the critic is being adapted. In a general case of many weights changing in a network, if their updates are to be closely interleaved with critic updates, a relatively simple critic representational structure might be warranted, so that the critic can be quickly adapted to changes in the system.

The proper coordination of critic training with network (controller) training has been a research topic. We have found that, in some cases, it is possible to update both network weights and critic weights at every training step, although such a strategy may not work well in the presence of network weight updates of larger size, as frequently result from second order training procedures. Alternating the training processes in blocks is a reasonable option, since holding the network fixed while adapting the critic generally leads to greater critic accuracy. The drawback is that once the critic is held fixed and the network is changed, the critic may become very inaccurate and compromise training with poor derivatives. Hence a better approach might be to carry out the training processes concurrently but to monitor the critic error (the error used to update the critic) and to suspend network training for some number of steps if a specified critic error is exceeded. Of special interest are case studies in [15, 16] where various alternatives for coordinated updates of critic and network have been analyzed. One alternative includes training of more than one critic (termed "shadow critic/controller" method), with periodic alternations between the critics to improve convergence. Yet, no comparison from the standpoint of computational cost akin to the analysis of this section have been made.

Efficiency of critic adaptation is paramount because the critic is discarded as soon as the controller training is finished. Interestingly, very few attempts have been made to analyze critic training accuracy or critic usability after obtaining the required controller. As for the latter, we discuss the use of a critic to analyze Lyapunov stability of the closed-loop system in [21].

15.6   TWO CLASSES OF CHALLENGING PROBLEMS

BPTT(h) equipped with EKF algorithm [13] has shown its power in dealing with difficult training problems requiring the use of RNN. Recently progress has been made in nonlinear Kalman filters in a joint estimation framework with promise to eliminate not only BPTT, but also the necessity to calculate derivatives in the system. Currently the new method's only drawback is extra complexity as compared to that of the standard EKF method [22].

More challenging of RNN training problems solved with Kalman filter methods can be categorized in two broad classes [23]. Class I encompasses neural approximation of multiple input-output mappings of the following form

$$\mathbf{y}^d(t) = \mathbf{f}_\theta(\mathbf{z}_\theta(t-1), \mathbf{x}_\theta(t)), \tag{15.67}$$

where $\mathbf{f}_\theta$ is a discrete or continuous set of mappings with the output vector $\mathbf{y}^d(t)$ at time t, $\mathbf{x}_\theta$ is a vector of inputs, and $\mathbf{z}_\theta$ is the mapping's state vector (evolution of $\mathbf{z}_\theta$ may be represented by a separate equation which is avoided in our notation as it is assumed to be a part of $\mathbf{f}_\theta$). The RNN approximating $\mathbf{y}^d$ for all t in the mean square sense has the form

$$\hat{\mathbf{y}}(t) = \mathbf{f}(\mathbf{z}(t-1), \mathbf{x}_\theta(t)), \tag{15.68}$$

where $\mathbf{z}$ is its state vector. Sometimes none of the mappings have states $\mathbf{z}_\theta$, as in [24] and [25]. Furthermore, the input $\mathbf{x}_\theta(t)$ may include the previous value of the target output $\mathbf{y}^d(t-1)$ to provide the network with appropriate context.

Class II includes problems in which accurate control of multiple distinct systems $\mathbf{g}_\theta$ (or plants) is required:

$$\hat{\mathbf{y}}(t) = \mathbf{g}_\theta(\mathbf{z}_\theta(t-1), \mathbf{f}(\hat{\mathbf{y}}(t-1), \mathbf{z}(t-1), \mathbf{x}_\theta(t))). \tag{15.69}$$

Here the system's output $\hat{\mathbf{y}}(t)$ should closely track the target output $\mathbf{y}^d(t)$ produced by a reference model (e.g., $\mathbf{y}^d(t)$ can be zero at all times, as in [26]). The input $\hat{\mathbf{y}}(t-1)$ of the controller RNN $\mathbf{f}$ may or may not include $\mathbf{z}_\theta(t-1)$ (or part thereof). Another input $\mathbf{x}_\theta(t)$ includes $\mathbf{y}^d(t)$ and, possibly, other external signals.

We briefly describe examples of class-I problems. In [27], a single recurrent multilayer perceptron (RMLP) with three fully recurrent hidden layers (21 states) is trained to make good one-time-step predictions of 13 different time series (periodic and chaotic). The fixed-weight RMLP is demonstrated to be capable of good generalization to time series with somewhat different sets of generating parameters as well as to those corrupted by noise. In [28], achieving good one-time-step predictions of five different time series from a two-hidden layer RMLP (14 states) via training is combined with two conditioning tasks. The trained network must remember which of the two tasks it dealt with in the past (Henon maps, type 1 or 2) in order to activate one of the two appropriate output responses for the random input. This problem is impossible (or, at least, very inefficient) to solve with feed-forward network equipped with a tapped-delay line because of the need to correctly maintain a potentially arbitrarily long response to the random input.

Two problems below are examples of class-II problems. In [26], a two-hidden-layer RMLP (14 states) is trained to act as a stabilizing controller for three distinct and unrelated systems, without explicit knowledge of system identity. This problem, too, has a feature which makes it very difficult (if not impossible) to apply successfully a controller based on a feed-forward network equipped with a tapped-delay line. Specifically, the steady state values of controls for all three plants are quite different, yet the stabilization is required around the same equilibrium point (the origin).

In [29], training an RMLP with 10 states is accomplished to achieve robust control of more than 10,000 systems derived from a single nominal system by parametric perturbations. The robustness results of RMLP-based controller are shown to be much better than those of a controller based on a feed-forward network.

These results obtained with BPTT(h) and EKF for clearly formulated and easily accessible control problems may serve as benchmarks for future comparison studies with DAC. Indeed, in spite of several successful demonstrations of DAC, they have mostly been limited to training controllers based on feed-forward/time-delay neural networks using the gradient descent algorithm (see, e.g., [16, 30–32]), with the notable exception of [33].

Both class I and class II represent problems that are important and frequently observed in practice. For example, a physical system to be modeled or controlled is usually known only to within parametric or structural (possibly time-varying) uncertainties. (Such uncertainties amount to different, discrete or continuous sets of mappings f_θ and g_θ.) One approach is to employ an adaptive system whose parameters would adapt in response to differences between the model and the reality. Another approach discussed in this section is to employ a RNN with *fixed* weights whose recurrent nodes would act as counterparts of parameters of the conventional adaptive system. This approach has an advantage of bypassing the thorny issue of adapting weights on-line.

In general, setup of class-II problems is more complex than that of class-I problems, therefore we suggest to try comparative studies initially on problems of class I. They share the same feature (feedback) with class-II problems due to the presence of recurrent connections in the network. We discuss one of class-I problems below.

15.6.1   Learning All Quadratic Functions

The problem of learning *all* quadratic functions of two variables is proposed in [24]. The quadratic functions are $y^d(t) = a(k)x_1{}^2(t) + b(k)x_2{}^2(t) + c(k)x_1(t)x_2(t) + d(k)x_1(t) + e(k)x_2(t) + f(k)$ where ranges for $a, b, c, d, e, f, x_1, x_2$ are the same: $[-1.0, 1.0]$. The index k (function index) changes discretely and much less frequently than the index t (example index). A special form of RNN called *long short-term memory* (LSTM) is explored in [24]. The LSTM has three inputs $(x_1(t), x_2(t)$ and $y^d(t - 1))$, one output $(\hat{y}(t))$, and it consists of 5,373 weights. Its training set is a time series of 128,000 points (128 different quadratic functions of 1000 examples each). The root mean square (RMS) error reaches 0.025 by the end of training. The final LSTM demonstrates the test RMS error of 0.026. It is claimed that other recurrent networks can not match performance of LSTM on this and other *metalearning* problems.

   We can interpret the quadratic function problem as a modeling problem of a nonstationary time series. Indeed, only x_1, x_2 and y^d are observed, and $(a, b, c, d, e, f)_k$ forms a hidden state changing every so often. We want to train a RMLP in the same setting as that of the LSTM training experiment [25]. Our RMLP has the same three inputs, $x_1(t)$, $x_2(t)$ and $y^d(t - 1)$, and architecture 3-30R-10R-1L with output $\hat{y}(t)$. (The notation 3-30R-10R-1L stands for RMLP with three inputs, 30 nodes in the first

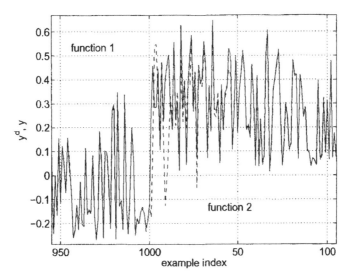

Fig. 15.1 Typical behavior of the trained network during testing. A fragment of the test time series with two different quadratic functions is shown. The target is solid, and the network output is dashed. The function change is clearly visible. The transient subsides within 50 example presentations.

hidden fully recurrent layer, 10 nodes in the second hidden fully recurrent layer, and one linear output node.) It has 1441 weights. Values of y^d and $\hat{y}$ are scaled to be approximately within the range ± 1.0. One epoch of training consists of the following steps. First, we randomly choose 20 segments of 1040 consecutive points each within the time series of 128,000 points. The initial 40 points of each segment are used to let the network develop its states (*priming* operation) from their initial zero values, rather than for training weights. Next, we apply the 20-stream global EKF to update weights, with derivatives being computed by BPTT(40). We use 20×1000 points for training in each epoch. Our training session lasts for 1620 epochs, during which each data point was presented to the network approximately 250 times. The first 600 epochs are carried out with the parameter $R_0 = 100$ (measurement noise or inverse learning rate) and the parameter $Q = 0.01/R_0$. The process noise Q is decreased to $0.003/R_0$ and $0.001/R_0$ at epoch numbers 601 and 1401, respectively. The RMS error attained after 600 epochs of training is equal to 0.0273, and it is equal to 0.020 by the end of training [23]. The final network is tested on many new time series 128,000 points long (examples of totally new quadratic functions) resulting in RMS errors of less that 0.025.

Figure 15.1 illustrates typical behavior of the trained network on a test time series. Just after the function change occurs, the network makes relatively large errors. It requires presenting 50 examples of the new function to the network to reduce the error to an acceptable level.

We observe that, for this particular problem, node-decoupled EKF training seems to result in much worse performance than that of the global EKF-trained RMLP (the gradient descent seems utterly hopeless). Likewise, a significantly shorter truncation depth h of BPTT(h) appears to be insufficient to deliver acceptable performance.

In Figure 15.2 we show the absolute instantaneous error averaged over 128 functions for the 1000-example segment. The average error is about 0.1 after presentation of 10 examples. It decreases significantly to about 0.01 after presentation of 100 examples. It is clear that the network spends a fairly significant number of examples to figure out what function it deals with, but eventually results in a good steady state solution.

It is interesting to contemplate application of DAC to this problem. First, we note that it may be necessary to have a critic with as many as 40 outputs (one per each recurrent node of the RMLP). Second, we may need to use more powerful procedures for both the critic and the network training because the gradient descent does not suffice. And we also need to decide how to treat the coefficients a, b, c, d, e, f: whether (1) to use them as inputs to the critic, or (2) interpret them as disturbances and ignore them, or (3) use a separate critic for each type of quadratic functions. The second option might result in an insufficiently accurate critic, whereas the third option could be impractical. The first option offers a tradeoff between critic accuracy and complexity of training.

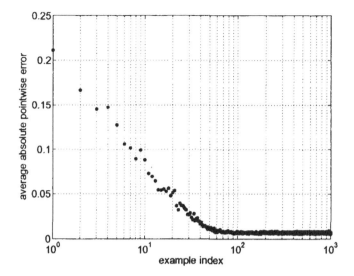

Fig. 15.2 The absolute pointwise error averaged over 128 functions for the 1000-example segment for a typical test time series. For each point in 1000-example segment the absolute instantaneous error was computed and averaged over all 128 functions.

15.7 CONCLUSION

We demonstrate that BPTT and DAC are closely related. This enables us to establish a common framework for comparison of the two methods featuring (1) analysis of critic representation, (2) a hybrid for smooth integration of BPTT and DAC, and (3) computational cost comparison. In our framework, both methods are considered in application to a heterogeneous recurrent network subsuming essential modules of the control system, that is, plant, its model, reference model and feedback controller. This viewpoint is immediately applicable to modeling problems for which the use of RNN is advantageous. We also propose avenues for future comparative studies.

Acknowledgments

The author is very grateful to Dr. Lee Feldkamp, for many helpful discussions and his past contributions to research on this topic.

Bibliography

1. P. J. Werbos, Backpropagation through time: What it does and how to do it, *Proc. of the IEEE*, vol. 78, no. 10, pp. 1550–1560, 1990.

2. R. J. Williams and D. Zipser, Gradient-based learning algorithms for recurrent networks and their computational complexity, in Chauvin and Rumelhart, (eds.), *Backpropagation: Theory, Architectures and Applications*, New York: Erlbaum, pp. 433–486, 1995.

3. P. J. Werbos, Approximate dynamic programming for real-time control and neural modeling, in D. A. White and D. A. Sofge, (eds.), *Handbook of Intelligent Control: Neural, Fuzzy and Adaptive Approaches*, pp. 493–525, 1992.

4. D. V. Prokhorov, *Adaptive Critic Designs and their Applications*, Ph.D. Thesis, Department of Electrical Engineering, Texas Tech University, Lubbock, 1997.

5. D. V. Prokhorov and D. C. Wunsch, Adaptive Critic Designs, *IEEE Trans. Neural Networks*, vol. 8, no. 5, pp. 997–1007, 1997.

6. D. V. Prokhorov, Optimal Neurocontrollers for Discretized Distributed Parameter Systems, *Proc. American Control Conference*, pp. 549–554, Denver, CO, 2003.

7. L. A. Feldkamp and D. V. Prokhorov, Observations on the practical use of adaptive critics, *Proc. 1997 IEEE SMC Conference*, pp. 3061–3066, Orlando, FL, 1997.

8. K. S. Narendra and A. M. Annaswamy, *Stable Adaptive Systems*, Prentice-Hall, Englewood Cliffs, NJ, 1989.

9. D. V. Prokhorov and L. A. Feldkamp, Primitive adaptive critics, *Proc. 1997 IEEE International Conference on Neural Networks*, pp. IV-2263–2267, Houston, TX, 1997.

10. E. A. Wan, Temporal backpropagation for FIR neural networks, *Proc. 1990 International Joint Conference on Neural Networks*, pp. I-575–580, San Diego, CA, 1990.

11. L. A. Feldkamp and D. V. Prokhorov, Phased backpropagation: A hybrid of TB and BPTT, *Proc. International Joint Conference on Neural Networks*, pp. 2262–2267, Anchorage, AK, 1998.

12. L. A. Feldkamp, D. V. Prokhorov, C. F. Eagen, and F. Yuan, Enhanced multi-stream Kalman filter training for recurrent networks, in J. Suykens and J. Vande-walle, (eds.), *Nonlinear Modeling: Advanced Black-Box Techniques*, pp. 29–53, Kluwer, Boston, 1998.

13. L. A. Feldkamp and G. V. Puskorius, A signal processing framework based on dynamic neural networks with application to problems in adaptation, filtering and classification, *Proc. of the IEEE*, vol. 86, no. 11, pp. 2259–2277, 1998.

14. L. A. Feldkamp, G. V. Puskorius, and D. V. Prokhorov, Unified formulation for training recurrent networks with derivative adaptive critics, *Proc. 1997 International Conference on Neural Networks*, pp. IV-2268–2272, Houston, 1997.

15. G. Lendaris and C. Paintz, Training strategies for critic and action neural nets in dual heuristic programming method, *Proc. International Conference on Neural Networks*, pp. 712–717, Houston, 1997.

16. G. Lendaris, T. T. Shannon, and A. Rustan, A comparison of training algorithms for DHP adaptive critic neuro-control, *Proc. International Conference on Neural Networks (IJCNN'99)*, Washington, DC, 1999.

17. E. S. Plumer, Optimal control of terminal processes using neural networks, *IEEE Trans. Neural Networks*, vol. 7, no. 2, pp. 408–418, 1996.

18. R. Stengel, *Optimal Control and Estimation*, Dover, New York, 1994.

19. S. W. Piché, Steepest descent algorithms for neural network controllers and filters, *IEEE Trans. Neural Networks*, vol. 5, no. 2, pp. 198–212, 1994.

20. *Nonlinear Model Predictive Control*, F. Allgöwer and A. Zheng, (eds.), Progress in Systems and Control Theory Series, vol. 26, Birkhauser Verlag, Basel, 2000.

21. D. V. Prokhorov and L. A. Feldkamp, Analyzing for Lyapunov stability with adaptive critics, *Proc. 1998 IEEE Conference on Systems, Man, and Cybernetics*, pp. 1658–1661, San Diego, CA, 1998.

22. L. A. Feldkamp, T. M. Feldkamp, and D. V. Prokhorov, Recurrent neural network training by nprKF joint estimation, *Proc. International Joint Conference on Neural Networks '02*, Honolulu, HI, 2002.

23. D. Prokhorov, L. Feldkamp, and I. Tyukin, Adaptive behavior with fixed weights in RNN: Overview, *Proc. International Joint Conference on Neural Networks '02*, Honolulu, HI, 2002.

24. S. Younger, S. Hochreiter, and P. Conwell, Meta-Learning with backpropagation, *Proc. International Joint Conference on Neural Networks*, pp. 2001–2006, 2001.

25. S. Hochreiter, S. Younger, and P. Conwell, Learning to learn using gradient descent, *Proc. ICANN*, pp. 87–94, 2001.

26. L. A. Feldkamp and G. V. Puskorius, Fixed weight controller for multiple systems, *Proc. 1997 International Joint Conference on Neural Networks,* vol. II, pp. 773–778, Houston, 1997.

27. L. A. Feldkamp, G. V. Puskorius, and P. C. Moore, Adaptive behavior from fixed weight networks, *Information Sciences,* vol. 98, pp. 217–235, 1997.

28. L. A. Feldkamp, D. Prokhorov, and T. Feldkamp, Simple and conditioned adaptive behavior from Kalman filter trained recurrent neural network, *Neural Networks,* vol. 16, pp. 683–689, 2003.

29. D. V. Prokhorov, G. V. Puskorius, and L. A. Feldkamp, Dynamical neural networks for control, in J. Kolen and S. Kremer, (eds.), *A Field Guide to Dynamical Recurrent Networks,* pp. 257–289, IEEE Press, 2001.

30. D. V. Prokhorov, R. A. Santiago, and D. C. Wunsch, Adaptive critic designs: A case study for neurocontrol, *Neural Networks,* vol. 8, no. 9, pp. 1367–1372, 1995.

31. S. N. Balakrishnan and V. Biega, Adaptive critic based neural networks for aircraft optimal control, *Journal of Guidance, Control and Dynamics,* vol. 19, no. 4, pp. 893–898, 1996.

32. G. K. Venayagamoorthy, R. G. Harley, and D. C. Wunsch, Comparison of heuristic dynamic programming and dual heuristic programming adaptive critics for neurocontrol of a turbogenerator, *IEEE Trans. Neural Networks,* vol. 13, no. 3, pp. 764–773, 2002.

33. P. H. Eaton, D. V. Prokhorov, and D. C. Wunsch, Neurocontroller alternatives for "fuzzy" ball-and-beam systems with nonuniform nonlinear friction, *IEEE Trans. Neural Networks,* vol. 11, no. 2, pp. 423–435, 2000.

Part III

Applications

16 Near-Optimal Control Through Reinforcement Learning and Hybridization

AUGUSTINE O. ESOGBUE   WARREN E. HEARNES II

Georgia Institute of Technology

Editor's Summary: This chapter focuses on learning to act in a near-optimal manner through reinforcement learning for problems that either have no model or whose model is very complex. The emphasis here is on continuous action space (CAS) methods. Monte-Carlo approaches are employed to estimate function values in an iterative, incremental procedure. Derivative-free line search methods are used to find a near-optimal action in the continuous action space for a discrete subset of the state space. This near-optimal policy is then extended to the entire continuous state space using a fuzzy additive model. To compensate for approximation errors, a modified procedure for perturbing the generated control policy is developed. Convergence results, under moderate assumptions and stopping criteria, are established. References to successful applications of the controller are provided.

16.1 INTRODUCTION

As real-world control problems become more complex, the application of traditional analytical and statistical control techniques requiring mathematical models of the plant becomes less appealing and appropriate. These methods assume that the system to be controlled can be modeled correctly and precisely. Clearly, the accuracy of a model affects the validity of the derived control policy. In many cases, the assumption of certainty in the models is made not necessarily for validity, but for the need to obtain simpler and more readily solvable formulations.

When a model of the system is known, traditional well-developed theories such as optimal control or adaptive control may be appropriate. Without a model however, these methods may not be adequate. In such situations, a model-free approach is required. Several approaches to model-free control exist. These include neural

407

networks, fuzzy logic, and reinforcement learning. Model-free techniques learn the control policy through either supervised or unsupervised means [29, 37, 38]. Supervised learning, on the other hand, requires some sort of teacher to provide the desired response at each time period. However, a teacher or expert may not always be available. Consequently, model-free control methods that can learn a control policy for a complex system through online experience have been proposed [3, 4, 17, 25, 31]. These are generally classified as reinforcement learning algorithms.

Model-free reinforcement learning methods are increasingly being used as potent and capable learning algorithms in intelligent autonomous systems for various applications in fields such as control theory, optimization, and computer science. A number of successful applications of model-free reinforcement learning controllers have been demonstrated. Illustrative examples include [3, 4, 17, 21, 31, 34, 35].

Complex control processes are often characterized by inherent uncertainties in behavior or measurement. An intelligent control algorithm for such processes must deal explicitly with these uncertainties. Practical applications that may introduce problems better suited for model-free intelligent control include those with severe nonlinearities, time-varying or unknown dynamics, partially observable states, or other complexities. These prevent the direct use of standard control techniques of variational calculus, classical dynamic programming, or global optimization.

The intelligent controller discussed in this chapter, an extension of the Continuous Action Space (CAS) controller first proposed in [16] and developed further in [23], involves a hybrid approach that combines some of the advantages of reinforcement learning, fuzzy sets, dynamic programming methods, neural networks, and nonlinear optimization. The result is a hybrid algorithm that intelligently searches continuous action spaces, learns a near-optimal control policy for continuous state spaces, and uses heuristic methods to dynamically allocate nodes during the learning phase.

16.2   TERMINAL CONTROL PROCESSES

The focus of this chapter is on closed-loop terminal control processes. In certain terminal control processes, the performance or objective function is not measured for a fixed time interval $[0, T]$. Rather, the termination of the control process may depend on the state. More specifically, a set of terminal states, S_T, is defined. This set is the set-theoretic union of all successful termination states, S_+, and all failure termination states, S_-:

$$S_T = S_+ \cup S_-. \tag{16.1}$$

The control process then terminates when either a success or a failure state is reached. In general, terminal control problems are generally easier to solve using dynamic programming rather than classical control methods [6]. Throughout this chapter, it is instructive to assume that the system in question is *set-point controllable*. In other words, in a deterministic system there exists an action sequence $\{a(1), \ldots, a(k-1)\}$ of admissible control actions for each initial state $s(1) \in S$ that drives the system from $s(1)$ to a successful termination state $s(k) \in S_+$ in finite time. Analogously, for systems with random disturbances, there exists an action sequence that drives any initial state to a successful termination state with some positive probability. This ensures that at least one control policy exists that will drive the system from any given initial state to a successful termination state or set-point.

In set-point regulation problems, a goal state or set-point, s^*, is usually defined. The objective then, is to drive the system from any initial state $s \in S$ to the set-point s^* in an optimal manner with respect to some defined scalar return function. Some classical measures, for such systems, include long-run expected reward with respect to external reinforcement signals, settling time or rise time with regard to transient response, or integral of absolute error (IAE) with regard to performance indices. A number of practical control systems, such as a manipulator for robot arms on the space shuttle, an inverted pendulum on a movable cart, or a gain scheduler to reduce oscillations in power generators, can be formulated as set-point regulator problems.

Consider a possibly stochastic dynamical system with scalar returns. The set of all possible states is represented by the set S while the set of all possible actions is represented by A. The m-dimensional state of the system at any time step k is the m-vector $s(k) \in S$ and the n-dimensional action taken at time step k is the n-vector $a(k) \in A$. The new state $s(k+1)$ is defined by the transition function τ:

$$s(k+1) = \tau\left(s(k), a(k), \omega(k)\right), \qquad (16.2)$$

where $k = 0, 1, \ldots$ and $\omega(k)$ is some random disturbance. For ease of exposition, represent the probability that $s(k+1) = j$ when $s(k) = i$ and $a(k) = a$ as $p_{ij}(a)$.

It is further assumed that the transition function τ for the process is not known a priori and that $n = 1$. This investigation is restricted to policies that are time-invariant. For any given control policy $\pi : S \mapsto A$, the objective function for an infinite horizon control problem can be defined [5, 6, 7] as

$$V_\pi(s) = E_\pi\left[\sum_{k=0}^{\infty} \gamma^k R(s(k), a(k)) | s(0) = s\right] \qquad \forall s \in S, \qquad (16.3)$$

and let

$$V(s) = \inf_\pi V_\pi(s) \quad \forall s \in S, \qquad (16.4)$$

where $\gamma \in [0, 1)$ is a discount factor and E_π is the conditional expectation using policy π. $V_\pi(s)$ represents the expected discounted total return using policy π and starting in state s. A policy π^* is γ-optimal if

$$V_{\pi^*}(s) = V(s) \quad \forall s \in S. \qquad (16.5)$$

Formulating the optimal control problem, as a dynamic program, results in the following functional equation:

$$V(s) = \inf_{a \in A} \left[R(s, a) + \gamma \sum_{s' \in S} p_{ss'}(a) V(s') \right] \quad \forall s \in S, \qquad (16.6)$$

which represents the minimum expected discounted return when starting in state s and always following an optimal policy.

The following assumptions are made:

Assumption 16.2.1 *The return,* $R(s, a)$, *for action* a *taken in state* s *is determined immediately or in some fixed time interval. Further,* $R(s, a)$ *is bounded from below by 0 and above by some number* B.

Assumption 16.2.2 *The system is set-point controllable. Furthermore, there exists a policy* π *such that* $V_\pi(s) < \infty$ *for all* $s \in S$.

Assumption 16.2.3 *Multiple trials can be run on the system.*

Approximate dynamic programming methods rely on repeated experimental runs. Systems must be allowed to experience both success—reaching a state in S_+—and failure—reaching a state in S_-—without damaging the plant.

Assumption 16.2.4 *The transition function* $\tau(s, a, \omega)$ *is not known. Consequently, the transition probability,* $p_{ss'}(a)$, *from state* s *to state* s' *when action* a *is taken is not known.*

16.3 A HYBRIDIZATION: THE GCS-Δ CONTROLLER

Several issues make the current traditional and intelligent methods for controlling set-point regulation processes inadequate for some complex systems. Many processes and operations involve systems that are highly nonlinear, have a large number of interacting variables, or much of what is known about their modeling and control is either imprecise or expressible mostly in linguistic terms. The model of the plant is either not known with certainty or cannot be easily solved. For example, the tethered satellite system retrieval problem has some very complex models, all of which are coarse approximations since including all parameters governing the physical movements of a satellite on a tether is far too complex [13]. Yet, even with an approximate model the solution method is very cumbersome. Therefore, a control method that is not based on knowledge of the plant dynamics can theoretically provide solutions in cases where traditional methods fail. In practical applications, such as the tethered satellite retrieval problem, failures are costly and prohibitive. Simulations of the plant can help jump-start the learning process in these cases.

16.3.1   Reinforcement Learning via Approximate Dynamic Programming

Reinforcement learning and approximate dynamic programming are the foundation of a number of intelligent control methods. In reinforcement learning, the objective is for the *agent* to learn a mapping or *policy*, $\pi : S \mapsto A$, from the state space S to the action space A that maximizes some scalar reinforcement signal $r : S \times A \mapsto R$ over a specified period of time. The reinforcement learning problem can be formulated as a dynamic programming problem since, given the initial state of the system, a sequence of actions that maximizes r is desired.

Among the two main approaches to reinforcement learning reported in the literature are temporal differences and Q-learning. Sutton's method of temporal differences (TD) is a form of the policy evaluation method in dynamic programming. Specifically, a control policy π_0 is chosen. The prediction problem becomes that of learning the expected discounted rewards, $V^\pi(i)$, for each state $i \in S$ using π_0. With the learned expected discounted rewards, a new policy π_1 can be determined that improves upon π_0. Under this alternating improvement procedure, the algorithm may eventually converge to some optimal policy as in Howard's algorithm [40].

Watkins' Q-learning algorithm is a form of the successive approximations technique of dynamic programming [35]. While policy iteration fixes a policy and determines the corresponding value functions, successive approximations determines the optimal value functions directly. Utilizing the Q-learning notation, assign a value $Q_k(i, a)$ to each state-action pair. This value represents the estimate of the optimal value function after k updates of starting in state i, taking action a, and proceeding in an optimal manner as estimated by the current Q-values. The Q-values are updated through a model-based update equation:

$$Q_{k+1}(i, a) = \alpha_k Q_k(i, a) + (1 - \alpha_k) \sum_j p_{ij}(a) \left(R(i, a) + \gamma \left[\min_{a'} Q_k(j, a') \right] \right),$$

(16.7)

where

$$\sum_{k=1}^{\infty} \alpha_{k(s,a)} = \infty \quad \forall s, a,$$

(16.8)

and

$$\sum_{k=1}^{\infty} \alpha_{k(s,a)}^2 < \infty \quad \forall s, a.$$

(16.9)

Q-learning converges with probability 1 to the optimal Q-values, $Q^\star$,

$$Q^\star(i, a) = \sum_{j=1}^{N} p_{ij}(a) \left(R(i, a) + \gamma V_{\pi^\star}(j) \right).$$

(16.10)

if each state-action pair is visited infinitely often and the update parameter α satisfies Eqs. (16.8) and (16.9) [3, 12, 35, 36].

Watkins' Q-learning has been modified by a number of researchers. Berenji [10] for example, proposed an extension of Q-learning into fuzzy environments and termed it Fuzzy Q-learning. This has the advantage of improvements in the learning rate but has the disadvantage of requiring some initial rulebase. Similarly, Oh et al. [32] developed an approach that initializes the conventional Q-values with Q-values derived from a fuzzy rulebase. With a better initial starting point, the Q-learning algorithm reinforced with a beneficial knowledge base and intelligence, may take less time to converge to the optimal Q-values. Wiering and Schmidhuber [39] modify Peng and Williams $Q(\lambda)$ procedure [33] by allowing Q-value updates to be postponed until they are needed. By doing this "lazy learning" process, the complexity of the lookup table is bounded by the number of actions instead of the number of state-action pairs.

Several methods utilize a hybrid approach to control in order to take advantage of the separate strengths of each method such as neural networks, fuzzy control, and reinforcement learning to compensate for the weaknesses of the others. Berenji extended his own Generalized Approximate Reasoning-based Intelligent Control (GARIC) model by using a set of intelligent agents controlled, at the top level, by Fuzzy Q-learning and, at the local level, by the GARIC architecture [10]. Taking the modifications one step further, Berenji and Sarar [11], allow collaboration among these intelligent agents for a specified period of time in their MULTI-GARIC-Q method. A critic, or action evaluator, then chooses the agent that has learned the "best" control policy to determine which agent ultimately controls the system. On a different tangent, Kandadai and Tein modify Berenji's original GARIC architecture [26]. Realizing that a disadvantage to the GARIC model is the assumption of an existing linguistic rulebase, Kandadai and Tien use a supervised learning scheme using both reinforcement learning and backpropagation to generate this linguistic rulebase. The result is a fuzzy-neural system that can generate its own control policy through reinforcements.

The foundation for the theory of the Statistical Fuzzy Associative Learning Controller (SFAL-C), introduced in 1993 by Esogbue and Murrell [17], was developed extensively in [31]. The SFAL-C is a hybrid fuzzy controller and neural network with a dynamic programming like optimizer. A variation of temporal differences is used to estimate the long-run reward for choosing a specific reference action in each state. After the learning process has ended, the long-run reward estimates for each action are transformed into a fuzzy set membership function for the proper action at that particular state. Fuzzy inference is used to generalize these learned controls to the entire state space [15, 17–19, 21]. Additional properties of this controller are provided in [14]. The successful operation of the fuzzy adaptive controller was demonstrated on several environments.

16.3.2   Review of CAS Controller

The controller which forms the focal attention of this chapter is an extension of the Continuous Action Space (CAS) algorithm proposed in [16, 23]. To motivate the current work, it is instructive to first present a brief outline of the structure and properties of this algorithm. The CAS algorithm uses Q-learning as the evaluator for a discrete subset of actions for continuous action space control. It begins with a representative finite discrete subset $a_j, j = 1, \ldots, A$, for each state i, spanning some interval of uncertainty (IoU) regarding the location of the optimal control action in the continuous action space $\mathbf{A}$. The Q-learning algorithm determines the optimal policy using only this action subset. Based on the optimal policy using only actions in the representative discrete subset a_j, the interval of uncertainty regarding the location of the optimal action can be reduced for selected states. For the new interval of uncertainty, the locations of the reference actions are adjusted and the Q-values are re-evaluated. As the CAS algorithm continues, the intervals of uncertainty for each state are reduced toward 0, centering on the optimal action in the continuous action space if certain assumptions are maintained. The choice of the reduction parameter β, learning rate α, threshold N, and initial Q-value M all affect the rate of convergence.

The key to the efficiency of the CAS algorithm is that the optimal policy can, in many cases, be determined before the Q-learning algorithm converges to the optimal functional values. For example, consider a discrete approximation to the inverted pendulum balancing problem [2]. With the objective to minimize the discounted sum of the squared error from the set-point, one discretization of the state space may consist of $51 \times 5 \times 51 \times 5$ discrete states for θ, $\Delta\theta$, x, and Δx, respectively. Each $Q_n(i, a)$ value is arbitrarily initialized at some large integer. Using a full backup for each iteration, 3401 iterations are necessary before the Q-values converge to within $\epsilon = 0.0001$ of the optimal functional values. Yet, the algorithm determines the optimal policy after only 36 iterations, as shown in Figure 16.1, which plots these policy changes during the 3401 iterations, a 98.94% reduction in the computational effort. Detailed investigation of possible computational savings is reported in [16, 23]. Basing the policy improvement procedure on the learned optimal policy at the earlier stages, before the optimal value is learned, is equivalent to waiting for the Q-learning algorithm to converge.

In the problems of interest in this chapter, the action space $\mathbf{A}$ is not discrete but continuous over a compact interval of $\mathbb{R}$. Each discrete state i (defined by a discretization of the continuous state space into a subset of regions that serve as an approximation to the continuous state space) has an interval of uncertainty (IoU) which contains the true optimal action $\pi^\star(i)$. In the CAS algorithm, the estimates of the $Q(i, a_j)$ values for each state i serve as a guide for reducing the interval of uncertainty. Initially, this interval is the entire action space $\mathbf{A}$. Each reduction is by a factor of $0 < \beta < 1$. Therefore, the interval can be made arbitrarily small using successive reductions. Let the reference action subset for state i be defined

$$\mathbf{A}_A(i) = \{a_1, a_2, \ldots, a_A\} \subset \mathbf{A}. \qquad (16.11)$$

Further, define the transformation

$$B(\beta, \mathbf{A}_A(i), k) = \{\beta[a_1 - a_{j^\star}] + a_{j^\star}, \ldots, \beta[a_A - a_{j^\star}] + a_{j^\star}\}, \qquad (16.12)$$

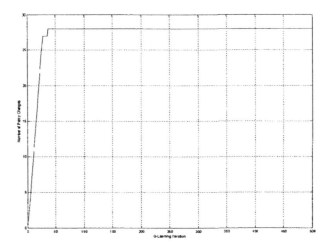

Fig. 16.1 Optimal policy is found in considerably fewer iterations than the optimal functional values in an inverted pendulum balancing example.

where

$$j^{\star} = \text{argmin}_j Q_k(i, \mathbf{a}_j),$$ (16.13)

as the Successive Reduction procedure. If the initial interval of uncertainty, IoU_0, is denoted as

$$IoU_0 = \sup_a \{a \in \mathbf{A}\} - \inf_a \{a \in \mathbf{A}\},$$ (16.14)

then it requires at least

$$\frac{\ln \epsilon - \ln IoU_0}{\ln \beta},$$ (16.15)

successive reductions of the interval of uncertainty to reach any specified accuracy $\epsilon > 0$.

There is a positive probability, however small, that the optimal action on the reference subset $\mathbf{a}_j$ will not be the basis for the successive reduction. This implies that there is a positive probability that the optimal action on the reference subset will not be contained in the interval of uncertainty for the next iteration of the CAS algorithm.

Consider the sequence of estimates $\{Q_0(i, a), Q_1(i, a), \ldots, Q'_k(i, a)\}$ for some state-action pair i, a. Since the true optimal Q-value, $Q(i, a)$ is not known, one possible measure of the quality of the estimation is the mean and sample variance of the last n estimates. Define

$$\bar{Q}'_k(i,a) \equiv \frac{\sum_{k=k'-n+1}^{k'} Q_k(i,a)}{n} \quad \forall i,a, \tag{16.16}$$

and

$$s^2_{Q'_k(i,a)} \equiv \frac{\sum_{k=k'-n+1}^{k'} (Q_k(i,a) - \bar{Q}'_k(i,a))^2}{n-1} \quad \forall i,a. \tag{16.17}$$

By appropriately choosing statistical hypotheses and a level of significance, the Successive Reduction procedure could be modified such that the interval of uncertainty regarding the optimal action would not eliminate any reference action in the subset a_j that could not be shown to be statistically significantly non-optimal.

In general, the above modification results have one of the following two effects. Either the Successive Reduction Procedure would continue to reduce the interval of uncertainty to an arbitrarily small interval, or, the interval of uncertainty would become fixed around some subset of actions which had estimated Q-values such that their long-run performance was not statistically different. In the first case, the optimal action is found by the reduction of the interval. In the second case, the optimal action is not necessarily found, but the action determined is near-optimal. The theoretical convergence is guaranteed in deterministic systems by the appropriate choice of which states to first apply the successive reduction procedure. The constructed sequence, $S^1, \ldots, S^N$, accomplishes this but, under Assumption 16.2.4, this classification of states cannot be made explicitly. Consequently, the suggested approach is to employ heuristics that attempt to approximate these subsets of S.

16.3.3   GCS-Δ Controller

The CAS algorithm demonstrates the ability to efficiently learn an ϵ-optimal control law for the inverted pendulum and power system stabilization problems, using heuristics to focus its search and eliminate some unnecessary computations [16]. In this section, we extend the controller, with some additional enhancements, to problems involving approximating continuous state spaces using fuzzy sets.

The CAS algorithm is an efficient method for applying approximate dynamic programming methods to problems with continuous action spaces. To achieve a very fine approximation to a continuous state space, many discrete states are needed and the efficiency of the CAS algorithm deteriorates. A generalized continuous space (GCS) algorithm is proposed that uses fuzzy sets to generalize the learned information throughout the continuous state space. The GCS algorithm learns the proper control for each of S reference states $\{s_1, \ldots, s_S\}$ with the CAS algorithm. These learned values of $Q_n(s_i, a_j)$ for the discrete state space serve as the required expert knowledge to generate a rulebase. The state space S is partitioned into a set of S fuzzy sets corresponding to the center of the discrete states in the CAS algorithm. A Gaussian membership function $\mu_{\tilde{s}_i}$ is defined for each fuzzy subset $\tilde{s}_i$ of the state space:

$$\mu_{\tilde{s}_i}(x) = \exp(-\|x - s_i\|^2_\infty / 2\sigma_i^2) \quad i = 1, \dots, S, \qquad (16.18)$$

where s_i is the center or prototype member of the fuzzy set $\tilde{s}_i$, $\| \cdot \|_\infty$ is the sup norm, and σ_i defines its point of inflection. Denote the set of S prototype members, $\{s_1, s_2, \dots, s_S\}$, as the reference state space S_S.

For each of the S discrete states in the CAS algorithm, corresponding values for $\min_j Q(i, a_j)$ and $\pi^*(i)$ are estimated. The rulebase, therefore, is

$$
\begin{array}{lll}
& \text{IF s is } \tilde{s}_1 & \text{THEN a is } a_{j_1^*}, \\
\text{OR} & \text{IF s is } \tilde{s}_2 & \text{THEN a is } a_{j_2^*}, \\
& \quad \vdots & \\
\text{OR} & \text{IF s is } \tilde{s}_S & \text{THEN a is } a_{j_S^*},
\end{array}
$$

where $j_i^* = \operatorname{argmin}_j Q(i, a_j)$. Therefore, a fuzzy system, based on the center points of each discrete region of the state space and the learned action obtained from the CAS algorithm, provides an interpolation mechanism that generalizes the learned actions to the continuous state space. After the convergence of the CAS algorithm, the generalized continuous space (GCS) procedure tests various spread factors σ for the fuzzy partition of the state space.

An interpolation scheme for generalizing the reference states to a continuous state space should possess certain properties. Let the estimated policy $\hat{\pi}_\sigma$ be the output of the fuzzy additive model:

$$\hat{\pi}_\sigma(s) = \frac{\sum_{i=1}^S \mu_{\tilde{s}_i}(s) a_{j_i^*}}{\sum_{i=1}^S \mu_{\tilde{s}_i}(s)}, \qquad (16.19)$$

where

$$\mu_{\tilde{s}_i}(x) = \exp(-\|x - s_i\|^2 / 2\sigma^2) \quad i = 1, \dots, S. \qquad (16.20)$$

Using this function, the σ which minimizes the expected discounted return is determined.

The GCS algorithm is as follows:

Algorithm 16.3.1

   1 Determine optimal policy π^* via CAS algorithm.
   2 Select some scalar performance measure $P : \pi \to \mathbb{R}$.
   3 Determine $\min_{\sigma \in (0,1]} P(\hat{\pi}_\sigma)$.

Determining $\min_{\sigma \in (0,1]} P(\hat{\pi}_\sigma)$ is done via a set of N test runs at various levels of σ that estimate $P(\hat{\pi}_\sigma)$. The best value of σ is then used in the fuzzy additive model to generalize the discrete state space of the CAS algorithm to the continuous state space of the set-point regulation problem.

An important property of the GCS algorithm is that it performs at least as well as the output of the CAS algorithm.

Proposition 16.3.1 *If $\pi^\star$ is the output of the CAS algorithm and $\hat{\pi}_\sigma$ is the output of the GCS algorithm based on some performance measure $P : \pi \to \mathbb{R}$, then*

$$P(\hat{\pi}_\sigma) \geq P(\pi^\star). \tag{16.21}$$

Proof: The membership function value of s in each fuzzy partition is

$$\mu_{\tilde{s}_i}(s) = \exp(-\|s - s_i\|^2 / 2\sigma^2) \quad i = 1, \ldots, S, \tag{16.22}$$

where s_i is the center of the i-th fuzzy partition. Consider any two fuzzy partitions, i and j. The distances from s to s_i and s_j are $\|s - s_i\|$ and $\|s - s_j\|$, respectively. Without loss of generality, let

$$\|s - s_i\| \geq \|s - s_j\|. \tag{16.23}$$

Therefore,

$$\|s - s_i\| = \|s - s_j\| + \epsilon, \tag{16.24}$$

for some $\epsilon \geq 0$. The membership of s in fuzzy partition $\tilde{s}_i$ is

$$
\begin{aligned}
\mu_{\tilde{s}_i}(s) &= \exp(-\|s - s_i\|^2 / 2\sigma^2) \\
&= \exp(-(\|s - s_j\| + \epsilon)^2 / 2\sigma^2) \\
&= \exp(-(\|s - s_j\|^2 + 2\|s - s_j\|\epsilon + \epsilon^2) / 2\sigma^2) \\
&= \mu_{\tilde{s}_j}(s) \exp(-(2\|s - s_j\|\epsilon + \epsilon^2) / 2\sigma^2).
\end{aligned}
\tag{16.25}
$$

As $\sigma \to 0^+$, $\exp(-(2\|s - s_j\|\epsilon + \epsilon^2) / 2\sigma^2) \to 0$ for any value of $\epsilon \geq 0$. Therefore,

$$\lim_{\sigma \to 0^+} \left(\frac{\mu_{\tilde{s}_i}(s)a^\star(i) + \mu_{\tilde{s}_j}(s)a^\star(j)}{\mu_{\tilde{s}_i}(s) + \mu_{\tilde{s}_j}(s)} \right) \to a^\star(i). \tag{16.26}$$

The argument can be extended to any finite number of fuzzy partitions with the same result. Therefore, for any performance measure P,

$$\lim_{\sigma \to 0^+} P(\hat{\pi}_\sigma) \to P(\pi^\star), \tag{16.27}$$

and $P(\hat{\pi}_\sigma) \not< P(\pi^\star)$. $\triangle$

In this section, we wish to discuss an aspect of the CAS procedure relative to the maintenance of the integrity of its output. We note that the CAS algorithm learns an optimal action for a particular discrete state. When fuzzy sets are used to generalize this learned control law, an improper choice of fuzzy membership functions can result in a defuzzified output that is different from the learned optimal action. Consider the state s such that $\mu_{\tilde{s}_1}(s) = 1$. If the CAS algorithm determines that $\pi^\star(1) = a^\star$, then it is a desirable property that $\hat{\pi}_\sigma(s) = a^\star$, also. With standard center-of-area defuzzification procedure, if the value of σ is chosen to be too large, it is likely that this is not true. A preliminary analysis into this phenomenon generates a range of σ where this property generally holds. For $\hat{\pi}_\sigma(s_i) \approx \pi^\star(i)$ for state i, the following must hold:

$$\mu_{\bar{s}_j}(s_i) \ll 1 \quad \forall j \neq i. \tag{16.28}$$

With a one-dimensional state space normalized to [0,1], the distance between reference states is

$$s_{i+1} - s_i = \frac{1}{S-1} \quad i = 1, \dots, S-1. \tag{16.29}$$

Therefore, for any $\epsilon > 0$,

$$\begin{aligned} \mu_{\bar{s}_j}(s_i) &= \epsilon, \\ \exp\left(\frac{-(S-1)^2\sigma^{*2}}{2}\right) &= \epsilon, \\ \sigma^* &= \sqrt{\frac{-1}{2(S-1)^2 \ln \epsilon}}. \end{aligned} \tag{16.30}$$

Generally, a value of σ^* provides a balance between interpolation between reference states and consistency with the CAS output, π^*, near the reference states. This value is chosen for initial use in the GCS algorithm.

We next discuss the concept of distance measure in a gaussian membership function which plays an important role in our algorithm. The distance measure chosen determines the shape of the resulting membership function. This shape has a significant impact on the level of support. In some controllers in the literature, such as the SFAL-C [17, 31], membership functions are defined using the L_2-norm

$$\|s\| = \sqrt{s_1^2 + \dots + s_m^2}, \tag{16.31}$$

where s_i is the i-th element of the state vector s. For $m = 2$, this generates a fuzzy partition with the contour plot as shown in Figure 16.2. The problem caused by the

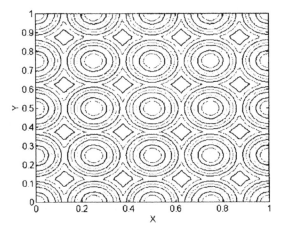

Fig. 16.2   Contour plot of a $[0, 1] \times [0, 1]$ state space into 25 fuzzy sets using Gaussian membership functions defined with L_2-norm.

L_2-norm in higher dimensions is the lack of *support* for a large portion of the state space. Due to the negative exponential nature, the theoretical support of each fuzzy state $\tilde{s}_i$ is S. However, in practice, the support as the set of all states $s \in S$ such that $\mu_{\tilde{s}_i}(s) \geq \epsilon$ for some $\epsilon > 0$ should be considered. In practical applications of fuzzy control, it has been empirically determined that an odd number of fuzzy sets is desired and that every state s should be a member of one fuzzy partition to degree 0.5:

$$\max_i \mu_{\tilde{s}_i}(s) \geq 0.5 \quad \forall s \in S. \tag{16.32}$$

With membership functions defined by the L_2-norm and a value of σ defined as in (16.30), the radius of the support circle around each s_i that meets (16.32) is

$$r_{0.5} = \frac{1}{(S)^{\frac{1}{m}} - 1} \sqrt{\frac{\ln 0.5}{\ln \epsilon}}, \tag{16.33}$$

where m is the number of dimensions in the state vector and ϵ is the one used to determine σ in (16.30). With a uniform distribution of s_i, $i = 1, \ldots, 25$ for $m = 2$, the total area that meets Eq. (16.32) is

$$\left[4 \cdot \frac{1}{4} + 4 \cdot 3 \cdot \frac{1}{2} + 3 \cdot 3 \right] \pi r_{0.5}^2 = 16\pi r_{0.5}^2 = 0.4729. \tag{16.34}$$

Therefore, 52.71% of the state space is underrepresented by the L_2-norm Gaussian membership functions.

Defining the membership functions with the L_∞-norm gives the fuzzy partitions illustrated in Figure 16.3. The L_∞-norm for state vector s is defined as:

$$\|s\|_\infty = \sup\{|s_1|, \ldots, |s_m|\}, \tag{16.35}$$

where s_i is the i-th element of the state vector s. Under this representation, the contours for various levels of membership function values are rectangles instead of circles. Therefore, with a uniform distribution of s_i, $i = 1, \ldots, 25$ for $m = 2$, the total area that meets (16.32) is

$$\left[4 \cdot \frac{1}{4} + 4 \cdot 3 \cdot \frac{1}{2} + 3 \cdot 3 \right] \pi r_{0.5}^2 = 16(2r_{0.5})^2 = 0.6021. \tag{16.36}$$

Therefore, only 39.79% of the state space is underrepresented by the L_∞-norm Gaussian membership functions. The difference is even more significant for lower values of desired representation. Consider the value of α at the boundaries of the discrete states used in the CAS algorithm:

$$r_\alpha = \frac{1}{(S)^{\frac{1}{m}} - 1} = 0.125, \tag{16.37}$$

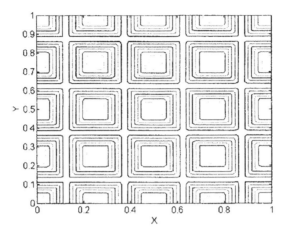

Fig. 16.3   Contour plot of a $[0, 1] \times [0, 1]$ state space into 25 fuzzy sets using Gaussian membership functions defined with L_∞-norm.

where

$$\alpha = \epsilon^{\frac{1}{4}}. \tag{16.38}$$

At this value of α, 21.46% of the state space is still underrepresented by the L_2-norm Gaussian membership functions, yet the entire state space is supported by the L_∞-norm Gaussian membership functions. The requirement that the existing fuzzy partition represent the state space S adequately implies that the L_∞-norm should be used over the L_2-norm. This issue is addressed in the sequel via the concept of a Δ-Modification Procedure.

The nature of the center-of-area defuzzification procedure is to provide a nonlinear weighted average interpolation to the learned actions for the discrete output of the CAS algorithm. In practice, the state that the GCS controller finally converges on, using σ^*, may not always be the s* due to the slight errors in the approximate dynamic programming estimates of $Q(i, a)$ for all i. Therefore, some type of modification to the learned control law is necessary.

For motivation, consider the three trajectories of θ plotted in Figure 16.4 for an inverted pendulum experiment. Trajectory 1 is the output of the CAS algorithm for $S = 49$. Note that the borders of the discrete states are represented by a horizontal dotted line and that the CAS algorithm output tends to bounce along the edge of the boundary of s*. This is a common occurrence when a discrete approximation to a continuous space is used. Trajectory 2 is the output of the fuzzy interpolation of the CAS output using σ^* and the L_∞-norm. The controller balances the pendulum at $\theta = 0.6°$ instead of 0 due to the slight errors in the estimated values of $Q(i, a)$. The theoretical convergence criteria of Q-learning requires an infinite number of visits to each state-action pair, which obviously cannot be achieved in practice. Therefore, some errors are always present. Trajectory 3 represents the modified output using the Δ-modification procedure explained in detail below. This method ensures that the pendulum is balanced at approximately $0°$.

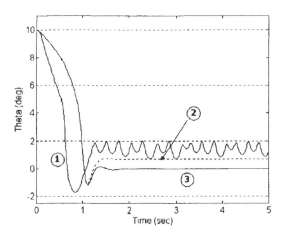

Fig. 16.4 Trajectories of θ for (1) the CAS algorithm, (2) the GCS algorithm, and (3) the GCS algorithm with Δ-modification, all with $S = 49$.

The essence of the Δ-modification is to determine an adjustment fuzzy state Δ with action a_Δ such that the modified control law

$$\hat{\pi}_\Delta(s) = \hat{\pi}(s) + \mu_\Delta a_\Delta \qquad (16.39)$$

drives the system to the set-point. Assume that the CAS algorithm has delivered a policy π^* such that the system does not fail for each state s during the test runs. Empirically, it has been determined that the center-of-area defuzzification with σ^* provides a response as shown in Trajectory 2 in Figure 16.4. During the first test run, denote the final state at time T using $a_\Delta = 0$ as $s(0,T)$. This state defines the location of the prototype member of Δ:

$$\mu_\Delta(s(0,T)) = 1, \qquad (16.40)$$

with $\sigma = \sigma^*$. Choosing an initial value $a_\Delta = a \in A$, a subsequent test run using $\hat{\pi}_\Delta$ provides the information necessary to update the estimate of a_Δ. Formulating the problem as:

$$f(a_\Delta) \equiv s(a_\Delta, T) - s^*, \qquad (16.41)$$

the secant algorithm may be used to determine the root of the function f.

Algorithm 16.3.2

> *1* Input: Controller from GCS algorithm.
> *2* Perform test run using $\hat{\pi}$.
> *3* $c \leftarrow 0$
> *4* $\Delta_c \leftarrow s(0, T)$
> *5* $a_{\Delta_c} \leftarrow a_0$
> *6* <u>while</u> $s(a_{\Delta_c}, T) \neq s^\star$ <u>do</u>
> *7*   Perform test run using $\hat{\pi}_{\Delta_c}$.
> *8* $\Delta_{c+1} \leftarrow s(a_{\Delta_c}, T)$
> *9* $a_{\Delta_{c+1}} \leftarrow a_{\Delta_c} - (\Delta_{c+1} - s^\star)(\Delta_{c+1} - \Delta_c)/(a_{\Delta_{c+1}} - a_{\Delta_c})\Delta_c$
> *10* <u>od</u>

16.4 EXPERIMENTAL INVESTIGATION OF THE GCS-Δ ALGORITHM

The experiments conducted in this section illustrate the GCS algorithm and identify the effects of the Δ modification procedure on the performance measure SSE. The factors that may have an impact are the cardinality of discrete state space, $|\mathbf{S}| = S$, the cardinality of reference action set, $|\mathbf{A}_A| = A$, and the number of expected iterations. A full factorial design of the 2^3 experiment estimates the effect of the main effects and small-order interactions on the computational savings for the 4-input/1-output inverted pendulum balancing problem [2]. The problem is to learn to balance an upright pole that is attached by a pivot to a cart, while keeping the cart at the center of the region it may move through. The discount factor γ was held constant at 0.99. The levels for each factor are given in Table 16.1. The performance improvements

Table 16.1 Levels of Variables in Full Factorial Experiment

Variable	0	1
State Space Cardinality	49	225
Action Space Cardinality	15	91
Expected Number of Iterations	100SA	500SA

for each of the 8 runs are given in Table 16.2. The *mean final θ value* is defined as the average value of θ over each of the 8 replications at 10.0 seconds in the test phase. The main effects and first order interaction are given in Table 16.3.

Each of the 8 replications of the 8 runs represents the application of the GCS-Δ algorithm. Each of the three factors is at either a high (1) or a low (0) level. From the performance improvements, an estimate of the effects and first-order interactions can be made. A plot of the main effects is shown in Figure 16.5, plotting the three factors: (1) State space cardinality, (2) Action space cardinality, (3) Number of expected iterations..

The experimental data provides several insights:

Table 16.2   Data for Full Factorial Experiment Showing the Performance Improvement

Factor Levels in Experimental Run	000	001	010	011	100	101	110	111
GCS Algorithm								
Mean Final Theta Value	0.51	-0.63	0.79	0.22	1.19	1.23	0.36	0.30
Std Deviation Final Theta Value	1.97	3.54	3.52	1.43	3.61	1.95	0.87	0.69
Mean SSE	15.81	28.15	27.04	13.62	25.77	17.92	9.11	10.78
Std SSE	9.81	17.88	23.55	6.74	29.97	13.47	6.63	6.54
Number of Failures	1	1	0	1	2	0	0	0
GCS-Delta Algorithm								
Mean Final Theta Value	-0.37	0.04	-0.41	-0.01	-0.18	0.34	-0.18	0.05
Std Deviation Final Theta Value	1.91	1.02	1.48	0.04	0.39	0.86	0.38	0.19
Mean SSE	11.06	11.80	9.78	2.80	6.13	9.48	4.77	4.15
Std SSE	16.13	11.13	12.85	0.47	4.02	8.45	4.56	1.75
Number of Failures	0	1	0	1	2	0	0	0

Table 16.3   Results of Analysis for Full Factorial Experiment

	LEVEL			MAIN EFFECTS			INTERACTIONS			
SSE Reduction	1	2	3	1	2	3	12	13	23	123
30.0%	-1	-1	-1	-30.0%	-30.0%	-30.0%	30.0%	30.0%	30.0%	-30.0%
58.1%	-1	-1	1	-58.1%	-58.1%	58.1%	58.1%	-58.1%	-58.1%	58.1%
63.8%	-1	1	-1	-63.8%	63.8%	-63.8%	-63.8%	63.8%	-63.8%	63.8%
79.5%	-1	1	1	-79.5%	79.5%	79.5%	-79.5%	-79.5%	79.5%	-79.5%
76.2%	1	-1	-1	76.2%	-76.2%	-76.2%	-76.2%	-76.2%	76.2%	76.2%
47.1%	1	-1	1	47.1%	-47.1%	47.1%	-47.1%	47.1%	-47.1%	-47.1%
47.6%	1	1	-1	47.6%	47.6%	-47.6%	47.6%	-47.6%	-47.6%	-47.6%
61.5%	1	1	1	61.5%	61.5%	61.5%	61.5%	61.5%	61.5%	61.5%
Mean 58.0%				0.3%	10.2%	7.1%	-17.4%	-14.7%	7.6%	13.9%

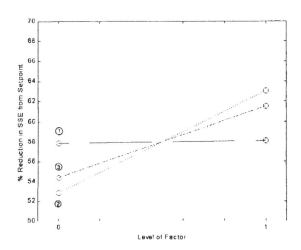

Fig. 16.5   Main effects of the three factors on the SSE performance.

1. The average performance improvement in SSE of 58% is a significant improvement. The Δ modification procedure reduces the approximation error of the GCS algorithm in an effective manner. The additional computations required are minimal since the procedure works during the testing phase of the learned control policy.

2. Though the main effect of increasing the expected number of iterations is positive, it does not guarantee improvement.

3. Similar to the GCS algorithm doing at least as well as the CAS, the GCS-Δ algorithm, with its basis in both, will perform at least as well as either.

Figures 16.6 and 16.7 show a sample trajectory for θ using the GCS algorithm and the GCS-Δ algorithm, respectively. The approximation error in the resulting near-optimal control policy is greatly reduced with the latter.

The CAS algorithm improves the performance of standard Q-learning by systematically reducing the action space to some arbitrarily small interval. This insures that an ϵ-optimal control policy is found under the strict quasi-convexity assumption. The GCS algorithm determines the value of a spread parameter, σ, which best interpolates the discrete state space in the CAS algorithm to the continuous state space. However, due to approximation error, some correction may be required by the Δ modification procedure. The performance of the proposed algorithms has been illustrated with the inverted pendulum balancing problem.

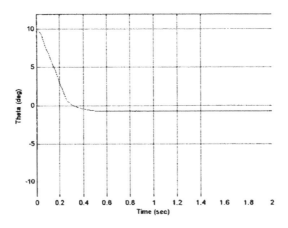

Fig. 16.6 Sample trajectory for the inverted pendulum balancing problem using the GCS algorithm.

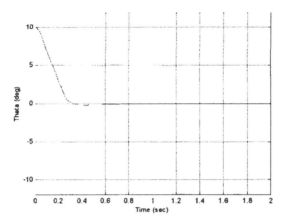

Fig. 16.7 Sample trajectory for the inverted pendulum balancing problem using the GCS-Δ algorithm.

16.5   DYNAMIC ALLOCATION OF CONTROLLER RESOURCES

A long-standing research question in dynamic programming is the allocation of resources for problems of high dimension [7, 30]. As state and action dimensions are increased, the resources used to approximate the state space and estimate the control action can increase exponentially. Fuzzy sets, through inference and defuzzification, can approximate any continuous function within any $\epsilon > 0$ accuracy [27]. However, the number of rules and membership functions required to achieve this may be prohibitive. Similar to the curse of dimensionality in dynamic programming, there exists a tradeoff in function approximation, as the number of nodes increase, the accuracy of the approximation may also increase, but the efficiency of the learning algorithm declines. This section presents a heuristic method to dynamically allocate nodes to the state space that achieve the desired approximation accuracy but does not burden the learning process in a prohibitive manner.

Additional state nodes require additional computations. Therefore, it is imperative to use a minimal amount of resources to successfully attack the problem at hand. As motivation, during a typical learning phase of 100,000 time steps in a testbed problem, the distribution of states visited in Figure 16.8 shows how little of the time is spent in the regions far from the set-point of [0,0]. With the majority of the time spent in the set-point region, it is conjectured that refining the approximation in that area will improve the performance. This conjecture is supported by the experiments in the sequel. Regions that are visited a disproportionately large number of times over the course of the learning phase of the controller are likely to be important regions where a finer resolution is beneficial.

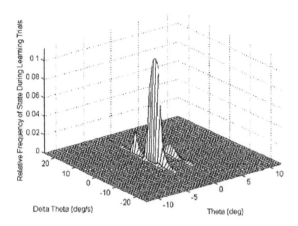

Fig. 16.8   Distribution of states visited during typical inverted pendulum learning phase.

Several exploration strategies may be used in Q-learning to identify the optimal action among the subset of actions at each state. Regions of high visitation under a uniform exploration strategy may not accurately depict regions of high visitation under the optimal control policy. To overcome this problem, the regions are determined using the learned control policy. This is easily accomplished during the policy update stage in the CAS algorithm. During that stage, the expected number of transitions until the set-point is reached is already estimated under the CAS, GCS, and GCS-Δ algorithms. Under a successful learned control policy π, it is expected that the region containing the set-point $s^\star$ will be visited with a high relative frequency and, therefore, will be one of the discrete regions where additional nodes are allocated.

If all $|\mathbf{S}| = S$ states are visited equally often then each state $\mathbf{s}_i$, $i = 1, \ldots, S$ has a relative frequency of $1/S$. The notion of a "highly visited state" is defined in this chapter as one that has been the visited with a relative frequency greater than p/S, where p is some integer in $1, \ldots, S$ such as 3 or 5. Denote the subset of regions that meet the criteria for high visitation as

$$\Omega = \{\mathbf{s}_i \mid \eta(\mathbf{s}_i) > p/S\}, \qquad (16.42)$$

where $\eta(\mathbf{s}_i)$ is a measure of the relative frequency $\mathbf{s}_i$ has been visited and $p/S \in (0, 1]$. The dynamic allocation of state nodes in the CAS algorithm is performed as outlined in Algorithm 16.5.1.

Algorithm 16.5.1

1 Set boundary condition: $V(\mathbf{s}^\star) = 0$.

2 Initialize $\mathbf{a}_j$ for all states.

```
 3   Set $Q_0(i, a_j) = M \gg 0$   $\forall i, j$.
 4   $n(i) \leftarrow 0 \; \forall i$
 5   while $n(i) < N \; \forall i$ do
 6       Perform an iteration of $Q$-learning.
 7       if Policy doesn't change for state $i$
 8           $n(i) \leftarrow n(i) + 1$
 9       else
10           $n(i) \leftarrow 0$
11       fi
12       if Reduction criterion is met for state $i$
13           Reduce IoU by $\beta < 1$ around $a_{j^*}$, $j^* = \mathrm{argmin}_j Q_n(i, a_j)$   $\forall i$ fi
14       for each state $s_i$ do
15           if $\eta(s_i) > p/S$
16               Divide state $s_i$ into $3^m$ sub-states. fi
17       od
```

Experimental data on the effects of adding nodes in regions of high visitation on SSE for the GCS-Δ algorithm provides the following insights [24]:

1. The average change in the SSE from using the standard uniform distribution of states to a distribution that clustered more states in regions of high visitation was $-44.7\% \pm 72.5\%$. The range contains 0 and no real effect can be interpreted.

2. The primary reason for the lack of visible improvement is that the GCS-Δ algorithm directly reduces approximation error in the learned control policy from the CAS and GCS algorithms. Adding nodes in regions of high visitation improves the output of the CAS, and therefore the GCS, algorithms but not necessarily the GCS-Δ. Other random factors are more significant.

3. The GCS-Δ algorithm can do no worse than the CAS and GCS algorithms, therefore even though there is no improvement, on the average, when adding the nodes in regions of high visitation, there may be some improvement in a particular instance.

Adding nodes dynamically during the learning phase, in selected regions of the state space, also provides significant improvements on the final SSE of the learned control policy. It allows the initial state space discretization to be rather sparse, with nodes added only where they are needed. This improves the computational efficiency of the algorithms by reducing the required number of iterations before each state is expected to be visited a prespecified number of times.

16.6   CONCLUSIONS AND FUTURE RESEARCH

In many control problems, the optimal policy is the more important part of the solution. The value of the functional equation when using an optimal policy is not necessarily required and, in some cases, model-free dynamic programming algorithms such as Q-learning expend additional computational effort to determine these values. This key property was exploited in the CAS and general continuous space (GCS) algorithms discussed in this chapter. The control policy derived by the CAS algorithm finds the near-optimal continuous action a^* for a discrete approximation of the state space S. The GCS algorithm generalized this policy to the entire continuous state space via a fuzzy additive model. Particular attention was paid to creating a fuzzy model which maintained integrity with the discrete approximation. Using an iterative procedure for determining the optimal σ value for the membership functions, the approximation of the continuous state space through a fuzzy partition and fuzzy additive model generates a control policy at least as good as the discrete approximation. Due to approximation error, some minor modifications to the control policy generated by GCS were required. The Δ-modification procedure accomplishes this task during the test phase of the controller. Approximation errors that affect the resulting learned control law of the GCS algorithm can be reduced significantly. The average improvement was a 58% reduction in the SSE after using the Δ-modification procedure.

Another area that generated considerable improvement in the learned control was the dynamic addition of state nodes during the learning phase in regions of high visitation. When dividing any region into smaller discrete parts, the new σ for the fuzzy membership functions must be chosen carefully so that the state space is not over- or underrepresented. The function defining this new σ was derived and implemented. The average improvements in the SSE was approximately 47% for the CAS algorithm and 51% for the GCS algorithm.

The theoretical aspects of the proposed algorithms are reinforced by experimental results conducted on several test problems [23, 24]. These include the classic inverted pendulum balancing problem, and real-world applications such as the DC servomotor positioning problem encountered in robotics, stabilizing a power generating system under a load [1], and retrieving a tethered satellite encountered in Space Shuttle missions [13]. Experience with these algorithms under the above environments has shown that the performance of the GCS-Δ controller is superior to several reinforcement learning controllers and conventional control methods.

Additional empirical studies relative to dynamically allocating resources to the fuzzy partition of the state space illustrate the advantages with regard to the performance improvements in the learned control policy as well as with computational efficiency. Future research problems along this line of investigation include the study of the optimal number of state nodes in the fuzzy discretization of the state space, cluster merging techniques for the elimination of unnecessary state nodes, efficiently and correctly handling randomly delayed reinforcements, and fuzzy termination states and times.

Bibliography

1. P. M. Anderson and A. A. Fouad, *Power System Control and Stability*, vol. 1, Iowa State University Press, Ames, IA, 1977.

2. C. W. Anderson and W. T. Miller III, Challenging control problems, *Neural Networks for Control*, pp. 474–509, MIT Press, Cambridge, MA, 1990.

3. A. G. Barto, S. J. Bradtke, and S. P. Singh, Learning to act using real-time dynamic programming, *Artificial Intelligence*, vol. 72, no. 1–2, pp. 81–138, 1995.

4. A. G. Barto, R. S. Sutton, and C. W. Anderson, Neuronlike adaptive elements that can solve difficult learning control problems, *IEEE Trans. Systems, Man, and Cybernetics*, vol. SMC-13, no. 5, pp. 834–846, 1983.

5. R. E. Bellman, *Dynamic Programming*, Princeton University Press, Princeton, NJ, 1957.

6. R. E. Bellman, *Adaptive Control Processes: A Guided Tour*, Princeton University Press, Princeton, NJ, 1961.

7. R. E. Bellman and S. E. Dreyfus, *Applied Dynamic Programming*, Princeton University Press, Princeton, NJ, 1962.

8. R. E. Bellman and S. E. Dreyfus, Functional approximations and dynamic programming, *Math Tables and Other Aids to Computations*, vol. 13, pp. 247–251, 1959.

9. R. E. Bellman, and R. E. Kalaba, *Dynamic Programming and Modern Control Theory*, Academic Press, New York, 1965.

10. H. R. Berenji, Fuzzy Q-learning: a new approach for fuzzy dynamic programming, *Proc. Third IEEE Conference on Fuzzy Systems*, pp. 486–491, Orlando, FL, 1994.

11. H. R. Berenji and S. K. Saraf. Competition and collaboration among fuzzy reinforcement learning agents, *Proc. 1998 IEEE International Conference on Fuzzy Systems*, vol. 1, pp. 622–627, Anchorage, AK, 1998.

12. D. P. Bertsekas and J. N. Tsitsiklis, *Neuro-Dynamic Programming*, Athena Scientific, Belmont, MA, 1996.

13. A. Boschitsch and O. O. Bendiksen, Nonlinear control laws for tethered satellites, *Advances in Astronautical Sciences,* vol. 62, pp. 257–276, 1986.

14. A. O. Esogbue, Neuro-fuzzy adaptive control: Structure, algorithms, and performance, *Applied Mathematical Reviews*, vol. 1, pp. 175–204, 2000.

15. A. O. Esogbue and W. E. Hearnes, Constructive experiments with a new fuzzy adaptive controller, *Proc. 1994 NAFIPS/IFIS/NASA Conference*, pp. 377–380, San Antonio, TX, 1994.

16. A. O. Esogbue and W. E. Hearnes, A learning algorithm for the control of continuous action space set-point regulator systems, *Journal of Computational Analysis and Applications*, vol. 1, no. 2, pp. 121–145, 1999.

17. A. O. Esogbue and J. A. Murrell, A fuzzy adaptive controller using reinforcement learning neural networks, *Proc. 2nd IEEE International Conference on Fuzzy Systems*, pp. 178–183, 1993.

18. A. O. Esogbue and J. A. Murrell, Optimization of a fuzzy adaptive network for control applications, *Proc. 5th International Fuzzy Systems Association World Congress*, pp. 1346–1349, Seoul, Korea, 1993.

19. A. O. Esogbue and J. A. Murrell, Advances in fuzzy adaptive control, *Computers & Mathematics with Applications*, vol. 27, no. 9–10, pp. 29–35, 1994.

20. A. O. Esogbue and Q. Song, On optimal defuzzification and learning algorithms: Theory and applications, *Journal of Fuzzy Optimization and Decision Making*, vol. 2, no. 4, pp. 283–296, 2003.

21. A. O. Esogbue, Q. Song, and W. E. Hearnes, Application of a self-learning fuzzy-neuro controller to the power system stabilization problem, *Proc. 1995 World Congress on Neural Networks*, vol. II, pp. 699–702, Washington, DC, 1995.

22. A. M. Geoffrion, Objective function approximation in mathematical programming, *Math. Programming*, vol. 3, pp. 23–37, 1977.

23. W. E. Hearnes, *Near-Optimal Intelligent Control for Continuous Set-Point Regulator Problems Via Approximate Dynamic Programming*, Ph.D. Thesis, School of Industrial and Systems Engineering, Georgia Institute of Technology, Atlanta, 1999.

24. W. E. Hearnes and A. O. Esogbue, Application of a near-optimal reinforcement learning controller to a robotics problem in manufacturing: A hybrid approach, *Journal of Fuzzy Optimization and Decision Making*, vol. 2, no. 3, pp. 183–213, 2003.

25. L. P. Kaelbling, M. L. Littman, and A. W. Moore, Reinforcement learning: A survey, *Journal of Artificial Intelligence Research*, vol. 4, 1996.

26. R. M. Kandadai and J. M. Tien, A knowledge-base generating heirarchical fuzzy-neural controller, *IEEE Trans. Neural Networks*, vol. 8, no. 6, pp. 1531–1541, 1997.

27. B. Kosko and J. A. Dickerson, Function approximation with additive fuzzy systems, in H. T. Nguyen, M. Sugeno, R. Tong, and R. Yager, (eds.), *Theoretical Aspects of Fuzzy Control*, chapter 12, pp. 313–347, Wiley, New York, 1995.

28. A. Y. Lew, *Approximation Techniques in Discrete Dynamic Programming*, Ph.D. Thesis, University of Southern California, Los Angeles, CA, 1970.

29. T. Miller III, R. S. Sutton, and P. J. Werbos, (eds.), *Neural Networks for Control*, MIT Press, Cambridge, MA, 1990.

30. T. L. Morin, Computational advances in dynamic programming, in M. L. Puterman, (ed.), *Dynamic Programming and Its Applications*, pp. 53–90, Academic Press, New York, 1978.

31. J. A. Murrell, *A Statistical Fuzzy Associative Learning Approach To Intelligent Control*. Ph.D. Thesis, School of Industrial and Systems Engineering, Georgia Institute of Technology, Atlanta, 1993.

32. C.-H. Oh, T. Nakashima, and H. Ishibuchi, Initialization of q-values by fuzzy rules for accelerating Q-learning, *Proc. 1998 IEEE International Joint Conference on Neural Networks*, pp. 2051–2056, Anchorage, AK, 1998.

33. J. Peng and R. Williams, Incremental multi-step Q-learning, *Machine Learning*, vol. 22, pp. 283–290, 1996.

34. R. S. Sutton, Learning to predict by the method of temporal differences, *Machine Learning*, vol. 3, pp. 9–44, 1988.

35. C. J. C. H. Watkins, *Learning From Delayed Rewards*, Ph.D. Thesis, Cambridge University, Cambridge, UK, 1989.

36. C. J. C. H. Watkins and P. Dayan, Q-learning, *Machine Learning*, vol. 8, pp. 279–292, 1992.

37. P. J. Werbos, An overview of neural networks for control, *IEEE Control Systems Magazine*, vol. 11, no. 1, pp. 40–41, 1991.

38. P. J. Werbos, Approximate dynamic programming for real-time control and neural modeling, in D. A. White and D. A. Sofge, (eds.), *Handbook of Intelligent Control: Neural, Fuzzy, and Adaptive Approaches*, pp. 493–525. Van Nostrand Reinhold, New York, 1992.

39. M. Wiering and J. Schmidhuber, Fast online $Q(\lambda)$, *Machine Learning*, vol. 33, no. 1, pp. 105–115, 1998.

40. R. J. Williams and L. C. Baird III, Analysis of some incremental variants of policy iteration: First steps toward understanding actor-critic learning systems, Technical Report NU-CCS-93-11, Boston, 1993.

17 Multiobjective Control Problems by Reinforcement Learning

DONG-OH KANG   ZEUNGNAM BIEN

Korea Advanced Institute of Science and Technology

Editor's Summary: Chapter 11 used hierarchical methods to solve multi-objective tasks. This chapter takes a different approach, using fuzzy control techniques. The mathematical background of multi-objective control and optimization is provided and a framework for an ADP algorithm with vector-valued rewards is introduced. Theoretical analyses are given to show certain convergence properties. A detailed algorithm implementation is presented, along with a cart-pole example.

17.1 INTRODUCTION

In our daily lives, we often confront decision-making situations in which more than one goal must be fulfilled. Difficulty arises when some of the objectives are conflicting with each other. In this case, inconsistency and lack of coherence among the objectives may prevent us from getting an optimal decision, and sometimes invoke sacrifice of at least one objective. Thus it is considered a challenging issue to investigate decision making problems with multiple objectives for efficient, satisfactory solutions. As for practical control problems, there are reported many examples with multiple objectives as can be seen in overhead crane control, automatic train operation system, and refuse incinerator plant control [1–4]. These kinds of control problems are called multiobjective control problems, where it is difficult to provide the desired performance with the control strategies based on single-objective optimization. In case of large-scale systems or ill-defined systems, the multiple objective control problem is more complicated to solve due to the uncertainties in the system models. The multiobjective control method based on the conventional multiobjective optimization techniques requires the exact model of the plant, which is often difficult to fulfill in large scale or ill-defined uncertain systems. On the other hand, reinforcement learning enables the control rule to be changed on the basis of the evaluative information about the control results rather than the exact information about the plant on the environment. Therefore, we propose to use reinforcement learning for the design of a multiobjective controller for large scale or ill-defined uncertain systems.

The multiobjective optimization problem has been intensively studied, especially in the area of economics or operational research area [5, 6]. The solution methods can be classified into two classes in terms of the form that is optimized. The first class is to optimize the scalarized value, which includes the weighted sum method, the sequential or ranking method, the constraint method, and the max-min method. In the methods, one scalar value is derived from the multiple objectives for the utility of the decision, and the conventional single objective optimization method is applied. The second class is to use the vector-valued utility. In this case, the complete ordering of the candidates is difficult, and the Pareto optimal concept is usually used. The Pareto optimal solutions are noninferior solutions among feasible solutions. The derivation of noninferior solutions is the major issue of the multiobjective optimization. The Multi-Objective Genetic Algorithm (MOGA) and the simulated annealing method are popular [7].

In the control area, there have been also several related studies, which include predictive control and fuzzy control. Yasunobu proposed a predictive fuzzy controller that uses rules based on skilled human operators' experience, and applied it to an automatic container crane and an automatic train operation system [1, 2]. The controller was designed to compute fuzzy performance indices to evaluate multiple objectives, and select a rule that produces the most desirable performance. K. Kim and J. Kim proposed a design method of fuzzy controller for multiple objectives in which some certainty factors are assigned in a heuristic manner to the obtained rules, and applied to calculate control inputs [8]. Lim and Bien proposed a rule modification scheme of fuzzy controller for multiobjective system via pre-determined satisfaction degree functions and rule sorting method of fuzzy c-means algorithm [9]. Recently, Yang and Bien proposed a programming approach using a fuzzy predictive model, and applied it to a MAGLEV ATO control problem [4]. To solve the multiobjective optimization problem, they assumed that a fuzzy predictive model of the plant is available and applied the max-min approach. But, their method is also dependent on a model, and thus it is hardly applicable for uncertain systems.

The method mentioned above usually requires some plant model or depends on human heuristics, and thus it is difficult to deal with conspicuous uncertainty. Reinforcement learning was applied to this situation in [10–12]. The reinforcement learning seems quite similar to the dynamic programming in the algorithmic form and a direct adaptive control technique in the content [11, 12]. It also uses the expected utility-like information about environment to decide the action, and update the information via interaction with the environment without using any model. In this sense, reinforcement learning can produce a potential solution to the multiobjective control problem for which information about the plant is not complete. Furthermore, different from the programming optimization process, the method can include heuristics and experience of human experts in its scheme by modifying its policy.

In this chapter, the multiobjective control problem is studied in which the plant with uncertain dynamics will be dealt with by a fuzzy logic-based decision making system, and reinforcement learning is adopted for design of an on-line multiobjective fuzzy logic controller. In Section 17.2, we review several important preliminary subjects including multiobjective optimization, and reinforcement learning. In Section 17.3, the policy improvement algorithm for Markov decision process with multiple reward is scrutinized. In Section 17.4, we apply the algorithm proposed in Section 17.3 to model-free multiple reward reinforcement learning for fuzzy control. Finally, some concluding remarks are given in the last section.

17.2 PRELIMINARY

17.2.1 Multiobjective Control and Optimization

Consider the MIMO plant with a controller which is dictated by multiple objective as shown in Figure 17.1.

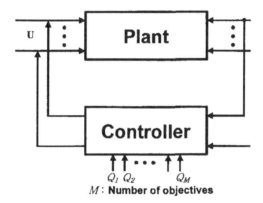

Fig. 17.1 Multiobjective control problem.

A general multiobjective control problem may be stated as a functional minimization problem as follows:

$$\begin{aligned}
\min \quad & \hat{Q}_1(\mathbf{x}(\cdot), \mathbf{u}(\cdot)), \hat{Q}_2(\mathbf{x}(\cdot), \mathbf{u}(\cdot)), \cdots, \hat{Q}_M(\mathbf{x}(\cdot), \mathbf{u}(\cdot)) \\
\text{s.t.} \quad & \hat{g}_i(\mathbf{x}(\cdot), \mathbf{u}(\cdot)) \le 0, i = 1, \cdots, I, \\
& \hat{h}_j(\mathbf{x}(\cdot), \mathbf{u}(\cdot)) = 0, j = 1, \cdots, J,
\end{aligned} \tag{17.1}$$

where $\mathbf{x}(\cdot)$ is a state function of the plant, and $\mathbf{u}(\cdot)$ is a control input. $\hat{Q}_k$ are objective functions, M is the number of objective functions, $\hat{g}_i \le 0$ are inequality constraints, I is the number of inequality constraints, $\hat{h}_j = 0$ are equality constraints, and J is the number of equality constraints. When the underlying variables $\mathbf{x}(\cdot)$ and $\mathbf{u}(\cdot)$ are real-valued, the problem is, in general, quite complicated to handle. In this chapter, we assume that the system under consideration can be discretized for proper modelling so that the problem can be handled as a discrete valued problem.

A general discrete valued multiobjective optimization problem can be mathematically defined as follows:

$$\min \quad Q_1(\mathbf{x}), Q_2(\mathbf{x}), \cdots, Q_M(\mathbf{x})$$
$$\text{s.t.} \quad \mathbf{x} \in \mathbf{X} \equiv \{\mathbf{x} \in R^n | g_i(\mathbf{x}) \leq 0, i = 1, \cdots, I, h_j(\mathbf{x}) = 0, j = 1, \cdots, J\},$$

$$(17.2)$$

where $\mathbf{x} \in R^n$ is a design variable, $Q_k(\mathbf{x})$ are objective functions, M is the number of objective functions, $g_i(\mathbf{x}) \leq 0$ are inequality constraints, I is the number of inequality constraints, $h_j(\mathbf{x}) = 0$ are equality constraints, and J is the number of equality constraints.

If we directly apply the concept of the optimality used in the single-objective problems to this multiobjective one, we arrive at the notion of a complete optimal solution [5, 6].

Definition 17.2.1 : Complete optimal solution [5]. *A solution* $\mathbf{x}^*$ *is called a complete optimal solution or ideal optimal solution to the problem in (17.2) , if , for all* $\mathbf{x} \in \mathbf{X}$, *there exists* $\mathbf{x}^* \in \mathbf{X}$ *such that* $Q_k(\mathbf{x}^*) \leq Q_k(\mathbf{x}), k = 1, 2, \cdots, M$.

In general, such a complete optimal solution that simultaneously minimizes all the multiple objective functions may not exist, especially when some of objective functions conflict with the others. Thus, instead of a complete optimal solution, another type of optimal solution, called a Pareto optimal solution, is often sought.

Definition 17.2.2 : Pareto optimal solution [5]. *A solution* $\mathbf{x}^*$ *is called a Pareto optimal solution or noninferior solution, if there does not exist* $\mathbf{x} \in \mathbf{X} - \{\mathbf{x}^*\}$ *such that* $Q_k(\mathbf{x}) \leq Q_k(\mathbf{x}^*), k = 1, 2, \cdots, M$, *and* $Q_k(\mathbf{x}) < Q_k(\mathbf{x}^*)$ *for at least one* k.

The concept of the Pareto optimal solution is depicted in Figure 17.2.

There are a number of methods to get a Pareto optimal solution in multiobjective optimization problems; the min-max method, the weighted sum method, the constraint method, and the goal programming are typical examples [5]. The methods derive one representative objective from multiple objectives and optimize the objective using single-objective optimization methods. The methods find the Pareto optimal solutions as depicted in Figure 17.3. Especially, the min-max method gives the single solution and is also applicable for non-convex problems [4, 5].

17.2.2   Markov Decision Process with Multiple Reward

In conventional reinforcement learning, the reward from the environment is a scalar value. In order to handle a multiobjective problem, the scalar reward may not be effective. The multiple rewards can be more suitable, corresponding to multiple objectives. In this case, a new scheme of reinforcement learning is required for the

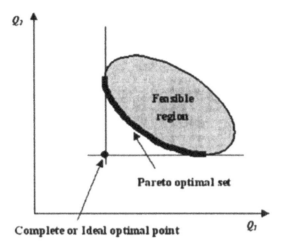

Fig. 17.2   Concept of Pareto optimal solutions.

solution to the multiobjective problem. In this subsection, some new concepts of Markov process with multiple rewards are introduced for the multiobjective problem.

As in the single objective problem, the policy for multiple reward case is expressed as the probability from the state space to the action space in case of the stochastic policy or as the mapping from the state space to the action space in case of the deterministic policy as in (17.3). But, in the multiobjective problem, the reward from the environment is a vector, not a scalar. The agent receives the rewards related with multiple objectives from the environment, and the rewards can be expressed as a vector as shown in (17.4). The value function of the multiobjective problem is also a vector because the value function is usually the discounted sum of the immediate reward over time as in (17.5). The action-value function is also vector-valued as in (17.6).

$$\pi(s,a) : S \times A \to [1,0] \ under \ stochastic \ policy,$$
$$\pi(s) : S \to A \ under \ deterministic \ policy. \tag{17.3}$$

$$\begin{aligned} \mathbf{r}_{t+1} &= \vec{\Re}^a_{t+1}(s,a) : S \times A \to R^M \\ &= \left[r^1_{t+1} r^2_{t+1} \cdots r^M_{t+1} \right]. \end{aligned} \tag{17.4}$$

$$\begin{aligned} \mathbf{V}(s) &= E(\sum_{k=0}^{\infty} \gamma^k \mathbf{r}_{t+k+1}|s_t = s) \ where \ 0 \le \gamma < 1 \\ &= \left[V_1 \ V_2 \cdots V_M \right]. \end{aligned} \tag{17.5}$$

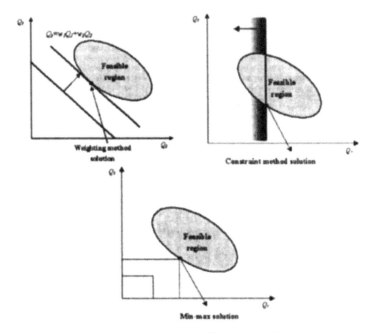

Fig. 17.3   Other examples of Pareto optimal solutions.

$$\mathbf{Q}(s, a) = [Q_1 \ Q_2 \cdots \ Q_M]^T. \tag{17.6}$$

Here, S is the set of possible states, A the set of available actions, M the number of objectives, and γ is the discount rate.

In case of the multiobjective problem, it is difficult to order the elements of a set of the solutions because of their multiple criteria. Instead, Pareto optimality often utilized to embrace the difficulty caused by multiplicity of criterion [5, 6]. The concept of Pareto optimality is applicable for multiple reward reinforcement learning. To be specific, the domination relation among the solutions is first defined as follows:

Definition 17.2.3 : Domination [5]. *A vector* $\mathbf{x} \in R^M$ *is said to dominate a vector* $\mathbf{y} \in R^M$ *if every element of* $\mathbf{x}$ *is larger than or equal to the corresponding element of* $\mathbf{y}$ *, and there exists at least one element of* $\mathbf{x}$ *that is larger than the corresponding element of* $\mathbf{y}$. *Formally, we write*

$$\mathbf{x} >_p \mathbf{y} \Leftrightarrow (\forall i, x_i \geq y_i) \ and \ (\exists i, x_i > y_i), \ i = 1, \cdots M, \tag{17.7}$$

where $\mathbf{x} = [x_1 x_2 \cdots x_M]^T$ *,* $\mathbf{y} = [y_1 y_2 \cdots y_M]^T$. Based on this concept, we can define Pareto optimal set of vector-valued state-value functions as follows:

Definition 17.2.4 : **Pareto optimal set of vector-valued state-value functions.**

$$Opt_p[\mathbf{V}] = \{\mathbf{V}^* \in \mathbf{V}_T | \; There \; exists \; no \; \mathbf{V} \in \mathbf{V}_T,$$
$$s.t. \; \mathbf{V}(s) >_p \mathbf{V}^*(s) \; for \; all \; s \in S\}, \qquad (17.8)$$

where $\mathbf{V}_T$ *is the set of all possible value functions and* $\mathbf{x} >_p \mathbf{y}$ *means* $\mathbf{x}$ *dominates* $\mathbf{y}$. Together with Pareto optimal set of vector-valued state-value functions, we can define Pareto optimal set of the vector-valued action-value functions as follows:

Definition 17.2.5 : **Pareto optimal set of vector-valued action-value functions.**

$$Opt_p[\mathbf{Q}] = \{\mathbf{Q}^* \in \mathbf{Q}_T | \; There \; exists \; no \; \mathbf{Q} \in \mathbf{Q}_T,$$
$$s.t. \; \max_a \mathbf{Q}(s,a) >_p \max_a \mathbf{Q}^*(s,a), \; for \; all \; s \in S\}, \qquad (17.9)$$

where $\mathbf{Q}_T$ *is the set of all possible vector-valued action-value functions.*

Definition 17.2.6 : **Pareto optimal policy.** *A policy* π^P *is called Pareto optimal if and only if the resulting vector-valued state-value function is a member of the Pareto optimal set of vector-valued state-value functions. Let* Π_p *denote Pareto optimal policies, that is,*

$$\Pi^p = \{\pi^p \in \Pi | \; There \; exists \; no \; \pi \in \Pi \,,$$
$$s.t. \; \mathbf{V}^\pi(s) >_p \mathbf{V}^{\pi^P}(s), \; for \; all \; s \in S\}, \qquad (17.10)$$

where $\mathbf{V}^\pi$ *is a vector-valued state-value function when a policy* π *is adopted, and* Π *is a set of possible policies.* It is remarked that the Pareto optimal set of the vector-valued state-value functions or the vector-valued action-value functions may contain more than one element. Different from the single reward case, thus, there can be multiple Pareto optimal candidates.

From the perspective of Pareto optimality, we can define the concept of multiple reward reinforcement learning, which, given the rewards from the environment, is concerned with derivation of the Pareto optimal policy and value function. From the multiobjective optimization technique explained in the previous subsection, we can say that if the max-min approach is applied to derive a policy of the multiple reward reinforcement learning problem, we may get the Pareto optimal policy. In fact, the method of the max-min multiobjective optimization technique is used for the solution of the multiple reward reinforcement learning in the present chapter.

We remark that the ordinary reinforcement learning can treat the multiobjective problem by letting the goal of the learning as the cost of the solution of each objective [13]. But, different from the ordinary reinforcement learning, the multiple reward reinforcement learning enables us to get the Pareto optimal policy more logically because it takes into account the trade-off information among the multiple objectives. Furthermore, the multiple reward reinforcement learning is expected to result in the fast convergence of the learning process by considering the multiple objectives.

17.3   POLICY IMPROVEMENT ALGORITHM FOR MDP WITH VECTOR-VALUED REWARD

17.3.1   Multiobjective Dynamic Programming

In this subsection, the method of multiobjective dynamic programming is proposed as a tool for the multiobjective optimization. The proposed multiobjective dynamic programming guarantees Pareto optimality of the solution. To proceed, we first define a concept called value difference, which is concerned with the scalar value function. Later, we shall consider the vector value case.

The value difference is the difference between the action-value of each possible action and the state-value of the current policy. $\pi_{t+1}(x)$ is a policy that has a different action from the current policy $\pi_t(x)$ at the current state s, and the same actions with $\pi_t(x)$ at the other states. Let $Q^{\pi_t}(x, \pi_{t+1}(x))$ be the action-value function when the action $\pi_{t+1}(x)$ is taken via the policy π_{t+1} at the state x , V^{π_t} the state-value function of the current policy π_t at the state x, and $\triangle(x, \pi_{t+1}(x))$ the difference between them. Then, we call the quantity $\triangle(x, \pi_{t+1}(x))$ the value difference.

Definition 17.3.1 : Value Difference.

$$
\begin{aligned}
\triangle(x, \pi_{t+1}(x)) &= Q^{\pi_t}(x, \pi_{t+1}(x)) - V^{\pi_t}(x) \\
&= \begin{cases} Q^{\pi_t}((x), \pi_t(x)) - V^{\pi_t}(x) = 0, & \text{if } x \neq s, \\ Q^{\pi_t}(s, \pi_{t+1}(s)) - V^{\pi_t}(s) \equiv \triangle(s, \pi_{t+1}(s)), & \text{if } x = s, \end{cases}
\end{aligned}
$$
$$(17.11)$$

where $Q^{\pi_t}(s, \pi_{t+1}(x)) = E_{\pi_t}\{R_t = \sum_{k=0}^{\infty} \gamma^k r_{t+k+1} | s_t = x, a_t = \pi_{t+1}(x)\}$, s is the current state.

Regarding the single value function, we choose the action as the new policy for the state, whose value difference is positive. The method increases the state value of the updated policy. Note that, if we choose the action whose value difference is negative, the state value of the new policy decreases. In case the value difference is zero, the state value does not change. From this perspective, the following Lemma 17.3.1 is derived.

Lemma 17.3.1 Upper/Lower Bound for $V^{\pi_{t+1}}(s)$.
If $\triangle(s, a) > 0$ and $\pi_{t+1}(s) = a$, then $V^{\pi_t}(s) + \frac{\triangle(s,a)}{1-\gamma} > V^{\pi_{t+1}}(s) > V^{\pi_t}(s) + \triangle(s, a) > V^{\pi_t}(s)$, and if $\triangle(s, a) < 0$, then $V^{\pi_t}(s) + \frac{\triangle(s,a)}{1-\gamma} < V^{\pi_{t+1}}(s) < V^{\pi_t}(s) + \triangle(s, a) < V^{\pi_t}(s)$. If $\triangle(s, a) = 0$, then $V^{\pi_{t+1}}(s) = V^{\pi_t}(s)$.

Proof: The proof can be found in the Appendix to this chapter.   ∎

As a result, if the value difference is kept positive by updating the policy, the overall state value of the policy increases. This means that we will get an optimal policy as Lemma 17.3.1 shows. From Lemma 17.3.1, we get the bound of the change of the state-value function by policy update.

In Lemma 17.3.1, the bound of the scalar state-value function is derived. In case of multiple objectives, the bound of the state-value function of each objective is derived from Lemma 17.3.1. By using these multiple bounds of multiple state-value functions, the dynamic programming algorithm is extended to the case of multiple objectives. Based on the value differences of the multiple objectives, the proposed multiobjective dynamic programming algorithm is derived as follows:

$$\pi_{t+1}(x) = \begin{cases} \pi_t(x), & x \neq s, \\ a^*, & x = s, \end{cases} \tag{17.12}$$

where

$$a^* = \begin{cases} \arg\max_{1 \leq i \leq M} [\Delta_i(s,a)], \\ \quad if \ \Delta_k(s,a) = 0 \ and \ \Delta_j(s,a) \geq 0, \\ \quad where \ k = \arg\min_{1 \leq i \leq M} \left[V_i^{\pi_t(s)} \right] \\ \quad for \ all \ a \ \in A(s), \ all \ j \neq k, 1 \leq j \leq M, \\ \arg\max_a \min_{1 \leq i \leq M} \left[V_i^{\pi_t(s)} + \Delta_i(s,a) \right] = \\ \quad \arg\max_a \min_{1 \leq i \leq M} [Q_i^{\pi_t}(s,a)], elsewhere. \end{cases}$$

Here, s is the current state, M is the number of objectives, and $\Delta_i(s,a)$ is the value difference of the i_{th} objective at the state s and action a.

The proposed method changes the policy incrementally. That is, the policy is updated by changing the action for the current state, and after the state transition, the action for the next transited state is updated. Therefore, the actions which are dedicated for the other states in the updated policy are the same as the actions in the current policy. This property enables us to update the policy on-line.

The proposed method is based on the max-min optimization technique, which updates the policy and gives the bigger minimum objective than the one which the original policy gives.

Lemma 17.3.2 *The policy updated via max-min optimization has the bigger minimum objective than the one which the original policy gives, that is,*

$$\max_a \min_{1 \leq i \leq M} [V_i^{\pi_t}(s) + \Delta_i(s,a)] \geq \min_{1 \leq j \leq M} V_j^{\pi_t}(s).$$

Proof: The proof can be found in the Appendix to this chapter. ∎

Based on Lemma 17.3.1 and Lemma 17.3.2, we can prove an important theorem in the dynamic programming method which is the policy improvement theorem. In the single objective case, the theorem guarantees the improvement of the state value of the policy updated by the dynamic programming. Similarly, in case of the multiple objectives, the minimum value among the multiple objectives is guaranteed to be nondecreasing. This guarantees that the proposed multiobjective dynamic programming method gives a nondecreasing minimum value among the multiple objectives.

Theorem 17.3.1 : Multiobjective Policy Improvement Theorem.

$$\min_{1 \le i \le M} V_i^{\pi_t}(s) \le \min_{1 \le i \le M} V_i^{\pi_{t+1}}(s)$$

if s is not visited again after time t. Here, M is the number of multiple objectives.
Proof: The proof can be found in the Appendix to this chapter. ∎

The theorem guarantees the nondecreasing property of the minimum value among multiple objectives. Figure 17.4 shows the resultant value function trends due to the proposed multiobjective dynamic programming. The state value of the present state is nondecreasing after update by the proposed method.

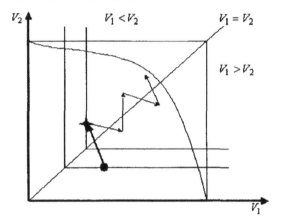

Fig. 17.4 State-value trend by the proposed mutiobjective dynamic programming.

In the theorem, we assume that the present state s should not be visited again after the present time t. In the realistic control problem, the state at the present time t is not revisited unless the state is a stable equilibrium state or the state in the limit cycle. Therefore, the assumption which is used in Theorem 17.3.1 is not too restrictive. In case of the stochastic environment, however, every state may be visited after the policy update is performed.

The following theorem shows the relation between the value difference and the Pareto optimal solution. If the value difference is zero for all states and objectives, the solution is Pareto optimal. This means that if the value difference is zero for all states and objectives, there is no improvement by the policy update. Therefore, the solution is optimal.

Theorem 17.3.2 *If $\triangle_i(s, \pi_{t+1}(s)) = 0$ for any $1 \le i \le M$, s, then π_{t+1} is a Pareto optimal solution.*

Proof: The proof can be found in the Appendix to this chapter. ∎

The following three corollaries show that the proposed method diminishes the value differences of the objectives, and finally gives a sub Pareto optimal solution.

Corollary 17.3.1 *The upper bound of* $\triangle_i^t(s, a)$ *for a state s is nonincreasing as time t increases, where* $\triangle_i^t(s, a)$ *is the value difference of the objective i at the time t.*
Proof: The proof can be found in the Appendix to this chapter.   ■

Corollary 17.3.2 *The lower bound of* $\triangle_i^t(s, a)$ *for a state s is nondecreasing as time t increases.*
Proof: The proof can be found in the Appendix to this chapter.   ■

Corollary 17.3.3 *The bound of the difference between*

$$\max_{1 \le i \le M} V_i^{\pi_t}(s) \text{ and } \min_{1 \le j \le M} V_j^{\pi_t}(s)$$

is nonincreasing for a state s as time t increases.
Proof: The proof can be found in the Appendix to this chapter.   ■

Because the upper bounds of the value differences diminish, the value differences will be nonincreasing, and remain within a certain value. This means the policy updated by the proposed method will be similar to the Pareto optimal solution within a certain bound.

17.4 MODEL-FREE MULTIPLE REWARD REINFORCEMENT LEARNING FOR FUZZY CONTROL

In the previous subsection, we have assumed that a model of the environment exists. If a model is not available, or inaccurate, the multiobjective dynamic programming may not be directly applicable for the mutiobjective control problem. In this section, a model-free reinforcement learning scheme is proposed. When the model is absent, the state value function is obtained using the temporal difference learning. In this section, the model-free multiple reward reinforcement learning schemes are applied for designing a fuzzy controller of a multiobjective control problem.

17.4.1 Multiple Reward Adaptive Critic Reinforcement Learning

The adaptive heuristic critic is popular in reinforcement learning [11, 12]. The adaptive critic is used to estimate the state value function. The adaptive critic and the policy are updated via the temporal difference learning. The same adaptive heuristic critic scheme is similarly applicable to the multiple reward reinforcement learning problem. In this chapter, the multiple adaptive critics are adopted for the estimation of the state-value functions for the multiple rewards. The policy is updated via the temporal difference which is selected among the temporal differences for the multiple rewards. The overall structure of the multiple reward adaptive critic reinforcement learning is depicted in Figure 17.5. For each reward, the state value for the reward is estimated. For every action, the temporal differences of the multiple rewards are obtained as follows:

$$\delta^j = r_t^j + \gamma \tilde{V}_j(s_t) - \tilde{V}_j(s_{t-1}), j = 1, \cdots, M, \qquad (17.13)$$

where s_t is the state at the time t, r_t^j is the reward for the j_{th} objective, δ^j is the temporal difference for the j_{th} reward, $\tilde{V}_j(s)$ is the estimated state value function of the state s for the j_{th} reward, and M is the number of the objective. So, for each adaptive critic, the state value function is updated using each temporal difference as follows:

$$\begin{cases} \tilde{V}_j(s) = \tilde{V}_j(s) + \alpha \delta^j, & if \ s = s_{t-1}, \\ \tilde{V}_j(s) = \tilde{V}_j(s), & otherwise, \end{cases} \qquad (17.14)$$

where α is the learning rate, and $j = 1, \cdots, M$. For the policy, one of the temporal differences for the rewards is selected. In this chapter, based on the multiobjective dynamic programming proposed in the previous subsection, the temporal difference for the update of the policy is selected as follows:

$$\delta^p = \delta^k, \qquad (17.15)$$

where

$$k = \begin{cases} \arg \max \delta^i, \\ \quad if \ \delta^m = 0 \ for \ m = \arg \min_i \{\tilde{V}^i(s_t)\}, \ or \ \forall j, \delta^j \geq 0 \ and \ \exists \ell, \delta^\ell > 0, \\ \arg \min_i \{\tilde{V}^i + \delta^i\}, \quad otherwise. \end{cases}$$

The policy has the quality parameters for each action. The selected temporal difference is used to update the quality parameter as follows:

$$Q(s, a) = Q(s, a) + \beta \delta^p, \qquad (17.16)$$

where $Q(s, a)$ is the quality of the action a for the state s, and β is the learning rate. In the policy, the quality parameters are used to modulate the probability of the action selection. There are various probability distribution of the selection of the action for the policy in the reinforcement learning area [14]. The probabilistic policy is adopted for both exploration and exploitation. In this chapter, the ϵ-greedy selection method is adopted [14].

17.4.2   Adaptive Fuzzy Inference System

As discussed in the previous subsection, the multiple reward reinforcement learning scheme is constructed on the bases of the discrete states and actions. The state-value function and the action-value function are used for the optimization. The policy is a rule which correlates the states with actions. In the multiple reward reinforcement learning for the multiobjective control problem, we need a function approximation structure for the state-value function, and controller for the policy for the continuous

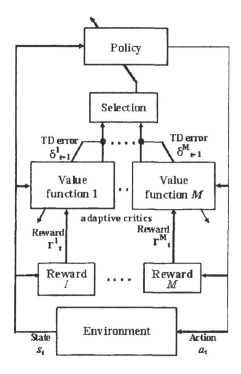

Fig. 17.5   Overall Structure of the multiple reward adaptive critic reinforcement learning.

state and action. That is, because the continuous state space requires the continuous output of the state value function, the function approximation is used for the state-value function prediction. The neural networks and neuro-fuzzy systems are popular methods for the purpose. Especially, the neuro-fuzzy system is suitable for using the expert knowledge. In this section, we adopt a fuzzy inference system to implement the critics and the policy of the multiobjective adaptive critic. The fuzzy inference system enables us to utilize the knowledge which is possibly obtained from human experts. Since the learning mechanism needs an adaptive structure, we use an adaptive fuzzy inference system proposed by Jouffe [14]. Jouffe proposed the multiple consequent term fuzzy controller and the fuzzy inference system with singleton consequent fuzzy sets for value function prediction. He applied the reinforcement learning for updating the controller [14]. The Jouffe's fuzzy inference system has multiple consequent terms for each fuzzy rule, which is equivalent to the rule base with inconsistent rules. As shown by Yu and Bien, the fuzzy logic controller with inconsistent rule-base can have a better performance [15]. As such, we use the Jouffe's adaptive fuzzy inference system structure for the multiobjective fuzzy logic controller.

The fuzzy inference system is composed of four layers as depicted in Figure 17.6. Each layer of the fuzzy inference system has the following function:

Layer 1) Input layer.

Layer 2) Membership calculation of each antecedent term.

Layer 3) Rule base node. It connects the input from the layer 2 to the output.

Layer 4) In the node, the defuzzification is performed and the final output of the fuzzy controller is determined.

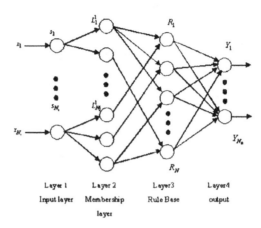

Layer 1   Layer 2   Layer 3   Layer 4

Input layer   Membership   Rule Base   output

layer

Fig. 17.6   Adaptive Fuzzy Inference System where N_i: number of the input variables, N: number of the rules, N_o: number of the outputs [14].

In this chapter, the singleton output fuzzy terms are used, and the center-average defuzzification is adopted as:

$$y = \frac{\sum_{i=1}^{N} y_i \mu_i(\mathbf{x})}{\sum_{i=1}^{N} \mu_i(\mathbf{x})}, \tag{17.17}$$

where N is the number of rules, y_i is the consequent singleton term of the rule i, $\mu_i(\mathbf{x})$ is the firing strength of the rule i and $\mathbf{x}$ is the input variable.

Each output of the adaptive critics is the predicted state-value function as:

$$\tilde{V}_{j,t}(\mathbf{S}(t)) = \mathbf{v}_{j,t} \Phi_t^T = \frac{\sum_{i=1}^{N} v_{i,j} \mu_i(\mathbf{S}(t))}{\sum_{i=1}^{N} \mu_i(\mathbf{S}(t))}, \tag{17.18}$$

where $j = 1, \cdots, M$ is the index for the objective, M is the number of objective, $\mathbf{v}_{j,t} = [v_{1,j} v_{2,j} \cdots v_{N,j}]$ is the vector of the output singleton terms for each critic, $\Phi_t = \left[\frac{\mu_1(\mathbf{S}(t))}{\sum_{i=1}^{N} \mu_i(\mathbf{S}(t))} \frac{\mu_2(\mathbf{S}(t))}{\sum_{i=1}^{N} \mu_i(\mathbf{S}(t))} \cdots \frac{\mu_N(\mathbf{S}(t))}{\sum_{i=1}^{N} \mu_i(\mathbf{S}(t))} \right]$ is the vector of the truth values of each rule, $\mathbf{S}(t)$ is the state of the environment at time t. In the system, the multiple adaptive critics share the common antecedent part with each other and the policy fuzzy controller as depicted in Figure 17.7.

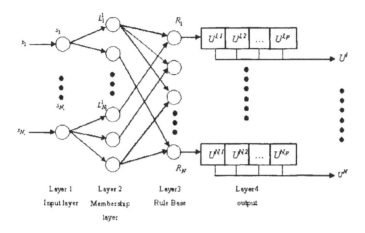

Fig. 17.7   Fuzzy controller using AFIS [14].

In the fuzzy controller which implements a policy, each rule has multiple consequent terms for each rule as follows:

$$R_i : If\ s_1\ is\ L_1^i\ and\cdots and\ s_N\ is\ L_N^i,\ then\ u\ is\ \{U^{i,1}, U^{i,2}, \cdots, U^{i,p}\},$$

$$(17.19)$$

where $\mathbf{S} = \{s_1, \cdots, s_N\}$ is a set of input variables, N is the number of input variables, $\{L_1^i, \cdots, L_N^i\}$ is a fuzzy set of antecedent of rule i, and u is the output variable, while $\{U^{i,1}, U^{i,2}, \cdots, U^{i,p}\}$ is a term set of consequent of rule i, p is the number of terms in the consequent. It is noted that, the output consequent in (17.19) is not just a single constant term as in usual Takagi-Sugeno-Kang representation but a p-tuple of numbers. At any instance, the only one output consequent term should be selected in the scheme. The concept of the multiple consequent terms is depicted in Figure 17.8. The utility $Q^{i,k}$ is assigned to each output consequent term $U^{i,k}$. For each rule, one output consequent term among a p-tuple of consequent terms is selected in consideration of the expected utilities $Q^{i,k}$ of actions $U^{i,k}$, $k = 1, \cdots, p$. The method to select one output consequent term for each rule is described in detail in [14]. After selecting one consequent term U^i for each rule, the scheme calculates the final controller output as follows:

$$u = \frac{\sum_{i=1}^{N} U^i \mu_i(\mathbf{S})}{\sum_{i=1}^{N} \mu_i(\mathbf{S})}.$$

$$(17.20)$$

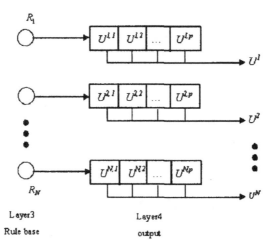

Fig. 17.8   Multiple output term case for each rule.

The utilities $Q^{i,k}$ are the parameters of the actor, that is, the parameters of the associative search element in an adaptive heuristic critic structure. They are updated at every action. Therefore, in the adaptive fuzzy inference system, the parameters to adjust are the consequent singletons of the multiple adaptive critics and the policy fuzzy controller : $\mathbf{v}_{j,t}, Q_t^{i,k}$.

17.4.3   Multiple Reward Adaptive Critic Reinforcement Learning for Fuzzy Control

We apply the proposed multiple reward adaptive critic reinforcement learning to the adaptive fuzzy inference system introduced earlier. The overall learning algorithm is described as follows:

Step 1) Calculate the state value function of each critic using the current evaluation function fuzzy system as follows:

$$\tilde{V}_{j,t-1}(\mathbf{S}_t) = \tilde{v}_{j,t-1}\Phi^T(\mathbf{S}_t), \tag{17.21}$$

where $j = 1, \cdots, M$ is the index for the objective, M is the number of objectives, and $\mathbf{S}_t$ is the state at time t.

Step 2) Compute the temporal differences of the critics. As the feedback value to the fuzzy controller, select one which has the minimum state-value among them as follows:

$$\delta_t^j = r_t^j + \gamma \tilde{V}_{j,t-1}(\mathbf{S}_t) - \tilde{V}_{j,t-1}(\mathbf{S}_{t-1}), j = 1, \cdots, M,$$
$$\delta_t^p = \delta_t^k,$$

where

$$
k = \begin{cases} \arg\max \delta_t^i, \\ \quad if\ \delta_t^m = 0\ for\ m = \arg\min_i\{\tilde{V}_{t-1}^i) \\ \quad and\ \forall j, \delta_t^j \geq 0\ and\ \exists \ell, \delta_t^\ell > 0, \\ \arg\min_i\{\tilde{V}_{t-1}^i + \delta_t^i\}, \\ \quad otherwise. \end{cases} \tag{17.22}
$$

Here, γ is the discount rate, and r_t^j is the reward of the j_{th} objective at time t.

Step 3) Update the parameter vectors of the fuzzy inference systems as follows:

$$
\bar{v}_{jt} = \bar{v}_{j,t-1} + \beta \delta_t^j \bar{\phi}_{t-1}, j = 1, \cdots, M,
$$
$$
Q_t^{U_{i,k}} = Q_{t-1}^{U_{i,k}} + \varphi \delta_t^p e_{t-1}^{U_{i,k}}, i = 1, \cdots, N, k = 1, \cdots, q, \tag{17.23}
$$

where β and φ are learning rates, $e_t^{U_{i,k}}$ and $\bar{\phi}_t = [\bar{\phi}_1\ \bar{\phi}_2 \cdots \bar{\phi}_N]$ are the eligibility traces of the utility of the action and the value function [14].

Step 4) Calculate the values of the state using the updated evaluation function networks as follows:

$$
\tilde{V}_{j,t}(\mathbf{S}_t) = \tilde{v}_{j,t}\Phi^T(\mathbf{S}_t), \tag{17.24}
$$

where $\tilde{V}_{j,t}$ means the output of each critic after the parameters are updated.

Step 5) Update the eligibility traces as follows:

$$
\bar{\Phi}_t = \Phi_t + \gamma\lambda\bar{\Phi}_{t-1},
$$
$$
e_t^{i,k}(U^i) = \begin{cases} \lambda' e_{t-1}^{i,k} + \phi_{R_i}(\mathbf{S}_t), & if\ U_t^i = U^{i,k}, \\ \lambda' e_{t-1}^{i,k}, & otherwise, \end{cases} \tag{17.25}
$$

where λ is the eligibility rate, λ' is the actor recency factor, and $\phi_{R_i}(\mathbf{S}_t) = \frac{\mu_i(\mathbf{S}_t)}{\sum_{j=1}^N \mu_j(\mathbf{S}_t)}$ is the truth value of rule i, and U_t^i is the selected fuzzy consequent singleton of fuzzy rule $i, i = 1, \cdots, N$.

Step 6) Select the new action U_t and take the action as follows:

$$
U_t = \frac{\sum_{i=1}^N U_t^i \mu_i(\mathbf{S}_t)}{\sum_{i=1}^N \mu_i(\mathbf{S}_t)}, \tag{17.26}
$$

where the selection process is the stochastic selection, that is, ϵ-greedy selection, which is explained in Jouffe [14].

17.4.4   Simulation

To show the effectiveness of the proposed method, some simulation is conducted for the inverted cart pole as depicted in Figure 17.10. It is fourth-order system, and the goals of the system are

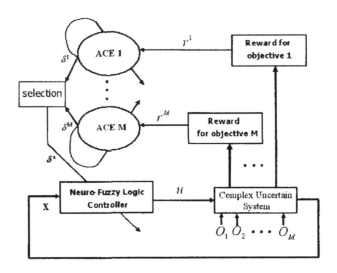

Fig. 17.9   Structure of multiple reward reinforcement learning.

1) Erect the pole upright and keep it upright.

2) The position of the cart should not be far away from the reference position.

As one can see, the two objectives can be conflicting in many situations. Thus, it is a good example of the multiobjective problem. The state equation of the system is as follows:

$$
\begin{aligned}
\mathbf{x}(t+1) &= \mathbf{x}(t) + T\{f(\mathbf{x}(t)) + g(\mathbf{x}(t))u(t)\}, \\
f(\mathbf{x}) &= [f_1(\mathbf{x})\ f_2(\mathbf{x})\ f_3(\mathbf{x})\ f_4(\mathbf{x})]^T, \\
f_1(\mathbf{x}) &= \dot{\theta}, \\
f_2(\mathbf{x}) &= \frac{g\sin\theta - \cos\theta\left[m\ell\dot\theta^2\sin\theta - \mu_c sgn(\dot x)\right]/(m_c+m) - \mu_p\dot\theta/m\ell}{\ell\left[3/4 - m\cos^2\theta/(m_c+m)\right]}, \\
f_3(\mathbf{x}) &= \dot{x}, \\
f_4(\mathbf{x}) &= \frac{m\ell\left[\dot\theta^2\sin\theta - \ddot\theta\cos\theta\right] - \mu_c sgn(\dot x)}{m_c+m}, \\
g(\mathbf{x}) &= [g_1(\mathbf{x})\ g_2(\mathbf{x})\ g_3(\mathbf{x})\ g_4(\mathbf{x})]^T, \\
g_1(\mathbf{x}) &= 0,\ g_2(\mathbf{x}) = \frac{-\cos\theta}{\ell\left[3/4 - m\cos^2\theta/(m_c+m)\right]}, \\
g_3(\mathbf{x}) &= 0,\ g_4(\mathbf{x}) = \frac{1}{m_c+m}, \\
\mathbf{x} &= \left[\theta\ \dot\theta\ x\ \dot x\right],
\end{aligned} \tag{17.27}
$$

where x is the position of the cart, $\dot{x}$ is the velocity of the cart, θ is the angle of the pole, and $\dot{\theta}$ is the angular velocity of the pole.

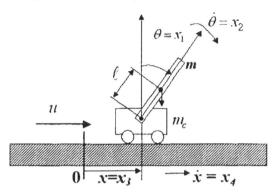

Fig. 17.10   The inverted cart pole system.

The parameters of the inverted cart are as follows:

$$\begin{cases} m_c = 1Kg & : mass\ of\ the\ cart. \\ m = 0.1Kg & : mass\ of\ the\ pole. \\ \ell = 0.5m & : length\ of\ the\ pole. \\ T_s = 0.02sec & : sampling\ time. \\ \mu_c = 0.0005 & : friction\ coefficient\ of\ the\ cart\ on\ the\ track. \\ \mu_p = 0.0005 & : friction\ coefficient\ of\ the\ pole\ on\ the\ cart. \\ g = 9.8m/sec^2 & : constant\ of\ gravity. \end{cases} \qquad (17.28)$$

The objectives of the system are that the system is kept in the area of $|\theta| \leq 12°$ and $|x| \leq 1m$. The following rewards are given as:

If $|\theta| \geq 12°$, then $r^1 = 0$, else $r^1 = -1$. If $|x| \geq 1m$, then $r^2 = 0$, else $r^2 = -1$.

The task of the inverted pendulum is an episodic task, and if the failure occurs, the state is reset to the initial state.

The parameters of the learning algorithm for the system are $\gamma = 0.95$, $\lambda = 0.9$, $\lambda' = 0.9$, $\beta = 0.01$, $\vartheta = 0.01$. We use $3 \times 3 \times 3 \times 3 = 81$ rules for input variables. The output of the controller is restricted in $|u| \leq 10N$.

Figure 17.11 compares several methods using the time step for success when the initial posture of the inverted pendulum is (0,0,0,0). The average failures over 100 runs are shown in Figure 17.11. As one can see, the proposed method is superior as the number of the consequent singleton increases. Barto's method uses two output choices such as bang-bang control, and shows somewhat poor results of 76.2 average failures [11]. Figure 17.12 shows the computation time for several methods. The Pentium II 450MHz processor is used to calculate the computation time of the methods. Barto uses the box system for the input space partition, and the box system requires little computation. In case of a fuzzy inference system, the calculation of

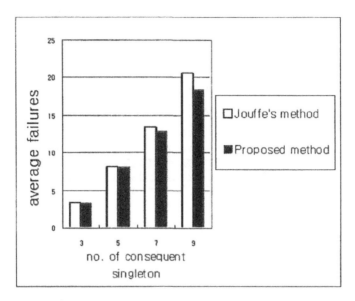

Fig. 17.11   Comparison with single objective case (average failures).

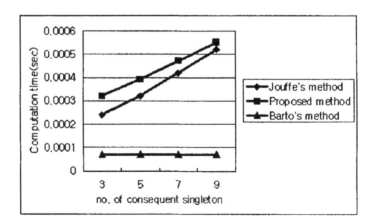

Fig. 17.12   Comparison with single objective case (computation time).

the fuzzy firing strength requires some computation time. Therefore, the proposed method takes more time than the Barto's method. The proposed method uses the multiobjective prediction, and therefore, the computation for the multiple objective is required. That means the computation for the proposed method requires more time than the single objective case. However, Figure 17.12 shows that this method incurs only the very modest worst-case computational penalty of 33% over the single objective case. As the number of consequent singleton increases, the relative computational penalty gets small, that is, 5% in case of 9 consequent singletons.

17.5   SUMMARY

In this chapter, we have proposed a reinforcement learning technique to be applied for multiobjective control problem. The multiobjective dynamic programming is considered for multiobjecitve optimization of the policy, and its max-min optimization is applied for multiple reward reinforcement learning in selecting the feedback temporal difference among multiple temporal differences of rewards. Using the adaptive fuzzy inference structure, the multiple reward reinforcement learning is extended to the fuzzy controller for a multiobjective control problem. Some simulation results are given to show the effectiveness of the proposed method.

We remark that the issues on the convergence of the proposed method to the Pareto optimality and the relation between the parameters of reinforcement learning and the control results remain to be further studied.

Appendix: Some Proofs of Theorems and Lemmas in Section 17.3

Definition 17.3.1: Value Difference.

$$
\begin{aligned}
\Delta(x, \pi_{t+1}(x)) &= Q^{\pi_t}(x, \pi_{t+1}(x)) - V^{\pi_t}(x) \\
&= \begin{cases} Q^{\pi_t}((x), \pi_t(x)) - V^{\pi_t}(x) = 0, & \text{if } x \neq s, \\ Q^{\pi_t}(s, \pi_{t+1}(s)) - V^{\pi_t}(s) \equiv \Delta(s, \pi_{t+1}(s)), & \text{if } x = s, \end{cases}
\end{aligned}
$$

$$(17.29)$$

where $Q^{\pi_t}(s, \pi_{t+1}(x)) = E_{\pi_t}\{R_t = \sum_{k=0}^{\infty} \gamma^k r_{t+k+1} | s_t = x, a_t = \pi_{t+1}(x)\}$.

Lemma 17.3.1: Upper/Lower Bound for $V^{\pi_{t+1}}(s)$.

If $\Delta(s, a) > 0$ and $\pi_{t+1}(s) = a$, then $V^{\pi_t}(s) + \frac{\Delta(s,a)}{1-\gamma} > V^{\pi_{t+1}}(s) > V^{\pi_t}(s) + \Delta(s, a) > V^{\pi_t}(s)$, and if $\Delta(s, a) < 0$, then $V^{\pi_t}(s) + \frac{\Delta(s,a)}{1-\gamma} < V^{\pi_{t+1}}(s) < V^{\pi_t}(s) + \Delta(s, a) < V^{\pi_t}(s)$. If $\Delta(s, a) = 0$, then $V^{\pi_{t+1}}(s) = V^{\pi_t}(s)$.

Proof: Let Q^{t+k} denote the action value of the state s if we choose the action for the state s by the new policy from time t to time $t + k - 1$, and follow the original policy π_t from time $t + k$, where $k = 1, 2, \cdots$. Then,

$$
\begin{aligned}
Q^t &= Q^{\pi_t}(s, \pi_t(s)) = V^{\pi_t}(s), \\
Q^{t+1} &= Q^{\pi_t}(s, \pi_{t+1}(s)) = \sum_{s_{t+1}} P_{ss_{t+1}}^{\pi_{t+1}(s)} \left[R_{ss_{t+1}}^{\pi_{t+1}(s)} + \gamma V^{\pi_t}(s_{t+1}) \right], \\
&= E_{\pi_{t+1}}\{r_{t+1} + \gamma V^{\pi_t}(s_{t+1}) | s_t = s\},
\end{aligned}
$$

where $P_{s_1 s_2}^a$ and $R_{s_1 s_2}^a$ are the probability and the reward value for the state to transit from s_1 to s_2 when the action a is taken. And

$$
\begin{aligned}
\Delta^{t+1} &= Q^{t+1} - Q^t = \Delta(s, \pi_{t+1}(s))\}, \\
Q^{t+2} &= E_{\pi_{t+1}}\{r_{t+1} + \gamma Q^{\pi_t}(s_{t+1}, \pi_{t+1}(s_{t+1}))) | s_t = s\}, \\
E_{\pi_{t+1}} &\{r_{t+1} + \gamma E_{\pi_{t+1}}\{r_{t+2} + \gamma V^{\pi_t}(s_{t+2})\} | s_t = s\} \\
&= E_{\pi_{t+1}}\{r_{t+1} + \gamma r_{t+2} + \gamma^2 V^{\pi_t}(s_{t+2}) | s_t = s\}, \\
\Delta^{t+2} &= Q^{t+2} - Q^{t+1} \\
&= E_{\pi_{t+1}}\{\gamma[Q^{\pi_t}(s_{t+1}, \pi_{t+1}(s_{t+1})) - V^{\pi_t}(s_{t+1})] | s_t = s\} \\
&= \gamma E_{\pi_{t+1}}\{\Delta(s_{t+1}) | s_t = s\} = \gamma \sum_{s_{t+1}} P_{ss_{t+1}}^{\pi_{t+1}(s)} \Delta(s_{t+1}) \\
&= \gamma P_{ss}^{\pi_{t+1}(s)} \Delta(s, \pi^{t+1}(s)), \\
\Delta^{t+2} &\propto \Delta(s, \pi_{t+1}(s)), |\Delta^{t+2}| \leq \gamma |\Delta(s, \pi^{t+1}(s))|.
\end{aligned}
$$

And

$$Q^{t+3} = E_{\pi_{t+1}}\{r_{t+1} + \gamma r_{t+2} + \gamma^2 Q^{\pi_t}(s_{t+2}, \pi_{t+1}(s_{t+2})))|s_t = s\},$$
$$\triangle^{t+3} = Q^{t+3} - Q^{t+2}$$
$$= E_{\pi_{t+1}}\{\gamma^2 [Q^{\pi_t}(s_{t+2}, \pi_{t+1}(s_{t+2})) - V^{\pi_t}(s_{t+2})]|s_t = s\}$$
$$= \gamma^2 E_{\pi_{t+1}}\{\triangle(s_{t+2}, \pi_{t+1}(s_{t+2}))|s_t = s\}$$
$$= \gamma^2 \sum_{s_{t+2}} P^{\pi_{t+1}(s)}_{ss_{t+2}} \triangle(s_{t+2}, \pi_{t+1}(s_{t+2}))$$
$$= \gamma^2 \triangle(s, \pi^{t+1}(s)) \sum_{s_{t+1}} P^{\pi_{t+1}(s)}_{ss_{t+1}} P^{\pi_{t+1}(s_{t+1})}_{s_{t+1}s},$$
$$\triangle^{t+3} \propto \triangle(s, \pi_{t+1}(s)), |\triangle^{t+3}| \le \gamma^2 |\triangle(s, \pi^{t+1}(s))|.$$

From the above, we have

$$\triangle^{t+k} \propto \triangle(s, \pi^{t+1}(s)), |\triangle^{t+k}| \le \gamma^{k-1}|\triangle(s, \pi^{t+1}(s))|. \tag{17.30}$$

Then, the following relation is derived:

$$|\triangle^t| = |\sum_{k=1}^{\infty} |\triangle^{t+k}| \le |\triangle(s, \pi^{t+1}(s))| \sum_{k=1}^{\infty} \gamma^{k-1} = \frac{|\triangle(s, \pi^{t+1}(s))|}{1 - \gamma},$$
$$|V^{\pi_{t+1}}(s) - V^{\pi_t}(s)| = |\triangle^t| \le \frac{|\triangle(s, \pi^{t+1}(s))|}{1 - \gamma}. \tag{17.31}$$

From (17.31) , the conclusion is straightforward.   ∎

Lemma 17.3.2:   The policy updated via max-min optimization has the bigger minimum objective than the one which the original policy gives, that is,

$$\max_a \min_{1 \le i \le M} [V_i^{\pi_t}(s) + \triangle_i(s, a)] \ge \min_{1 \le j \le M} V_j^{\pi_t}(s).$$

Proof:

Suppose

$$\max_a \min_{1 \le i \le M} [V_i^{\pi_t}(s) + \triangle_i(s, a)] < \min_{1 \le j \le M} V_j^{\pi_t}(s). \tag{17.32}$$

Naturally,

$$\min_{1 \le i \le M} [V_i^{\pi_t}(s) + \triangle_i(s, a)] < \max_{\bar{a}} \min_{1 \le j \le M} [V_j^{\pi_t}(s) + \triangle_j(s, \bar{a})] \quad for \; all \; a \in A(s). \tag{17.33}$$

That means

$$\min_{1 \le i \le M} [V_i^{\pi_t}(s) + \triangle_i(s, a)] < \min_{1 \le j \le M} [V_j^{\pi_t}(s) + \triangle_j(s, a^*)] \quad for \; all \; a \in A(s), \tag{17.34}$$

where

$$a^* = \arg\max_{a} \min_{1 \le j \le M} \left[V_j^{\pi_t}(s) + \triangle_j(s, a) \right].$$

From (17.32), (17.34), we can derive

$$\min_{1 \le i \le M} \left[V_i^{\pi_t}(s) + \triangle_i(s, a) \right] < \min_{1 \le j \le M} \left[V_j^{\pi_t}(s) + \triangle_j(s, a^*) \right]$$
$$< \min_{1 \le k \le M} V_k^{\pi_t}(s) \text{ for all } a \in A(s). \quad (17.35)$$

If we choose $a = \hat{a} = \pi_t(s)$, then

$$\triangle_i(s, \hat{a}) = \triangle_i(s, \pi_t(s)) = 0 \text{ for all } i,$$
$$\min_{1 \le i \le M} \left[V_i^{\pi_t}(s) + \triangle_i(s, \hat{a}) \right] = \min_{1 \le i \le M} V_i^{\pi_t}(s). \quad (17.36)$$

From (17.35), (17.36),

$$\min_{1 \le i \le M} V_i^{\pi_t}(s) < \min_{1 \le k \le M} V_k^{\pi_t}(s). \quad (17.37)$$

This is the contradiction. Therefore, we can conclude that

$$\max_{a} \min_{1 \le i \le M} \left[V_i^{\pi_t}(s) + \triangle_i(s, a) \right] \ge \min_{1 \le j \le M} V_j^{\pi_t}(s).$$

■

Theorem 17.3.1: Multiobjective Policy Improvement Theorem.

$$\min_{1 \le i \le M} V_i^{\pi_t}(s) \le \min_{1 \le i \le M} V_i^{\pi_{t+1}}(s)$$

if s is not visited again after time t. Here, M is the number of multiple objectives.
Proof:
Case 1)
If $\triangle_k(s, a) = 0$ and $\triangle_j(s, a) \ge 0$ for all j and $\exists \ell, \triangle_\ell(s, a) \ne 0$, where

$$k = \arg\min_{1 \le i \le M} \left[V_i^{\pi^*(s)} \right] \text{ for all } a \in A(s).$$

By Lemma 1, $V_k^{\pi_t}(s) \le V_j^{\pi_t}(s) \le V_j^{\pi_{t+1}}(s)$ and $V_k^{\pi_t}(s) = V_k^{\pi_{t+1}}(s)$, then $V_k^{\pi_{t+1}}(s) \le V_j^{\pi_{t+1}}(s)$.
Therefore,

$$\min_{1 \le k \le M} V_k^{\pi_t}(s) \le \min_{1 \le k \le M} V_k^{\pi_{t+1}}(s). \quad (17.38)$$

Case 2)

Let a^* such that

$$a^* = \arg\max_a \min_{1 \leq i \leq M} \left[V_i^{\pi^*(s)} + \Delta_i(s, a) \right].$$

Then, by Lemma 17.3.2,

$$\min_{1 \leq i \leq M} V_i^{\pi^*(s)} \leq \min_{1 \leq i \leq M} \left[V_i^{\pi^*(s)} + \Delta_i(s, a) \right].$$

Therefore, if $\Delta_m(s, a^*) \geq 0$ by Lemma 17.3.1,

$$V_m^{\pi_t}(s) + \frac{\Delta_m(s, a^*)}{1 - \gamma} \geq V_m^{\pi_{t+1}}(s) \geq V_m^{\pi_t}(s) + \Delta_m(s, a^*) \geq V_m^{\pi_t}(s) \geq \min_i V_i^{\pi_t}(s).$$
(17.39)

If $\Delta_m(s, a^*) < 0$, then $V_m^{\pi_{t+1}}(s) < V_m^{\pi_t}(s)$ by Lemma 17.3.1.

By assumption, $V_m^{\pi_{t+1}}(s) = V_m^{\pi_t}(s) + \Delta_m(s, a^*)$ for all s, m.

Therefore,

$$V_m^{\pi_t}(s) + \frac{\Delta_m(s, a^*)}{1 - \gamma} < V_m^{\pi_{t+1}}(s) = V_m^{\pi_t}(s) + \Delta_m(s, a^*) < V_m^{\pi_t}(s) \ for \ all \ s.$$
(17.40)

By Lemma 17.3.2,

$$\min_{1 \leq i \leq M} V_i^{\pi_t}(s) \leq \min_{1 \leq j \leq M} \left[V_j^{\pi_t}(s) + \Delta_j(s, a^*) \right] \leq V_m^{\pi_t}(s) + \Delta_m(s, a^*)(s)$$
$$= V_m^{\pi_{t+1}} \ for \ all \ s, m \ \ (17.41)$$

From (17.39), (17.41), we can conclude that

$$\min_{1 \leq i \leq M} V_i^{\pi_t}(s) \leq V_m^{\pi_{t+1}}(s) \ for \ all \ s, \ and \ for \ all \ m.$$

∎

Theorem 17.3.2: If $\Delta_i(s, \pi_{t+1}(s)) = 0$ for any $1 \leq i \leq M, s$, then π_{t+1} is a Pareto optimal solution.

Proof:

If $\Delta_i(s, \pi_{t+1}(s)) = 0$ for all $1 \leq i \leq M, s$, then $\pi_{t+1} = \pi_t$, which is straightforward. In other cases, we consider three following cases. Let $\bar{\pi}$ be the present policy π_t. Then, for $\pi_{t+1}(s) = \pi'(s) \neq \bar{\pi}(s)$, there exists at least one i, $\Delta_i(s, \pi_{t+1}(s)) \neq 0$.

Case 1)

For $1 \leq k \leq M$ such that

$$k = \arg\min_{1 \leq i \leq M} \left[V_i^{\pi_t(s)} + \Delta_i(s, \bar{\pi}(s)) \right] = \arg\min_{1 \leq i \leq M} \left[V_i^{\pi_t(s)} \right],$$

if $\Delta_k(s, \pi'(s)) = 0$, $\Delta_j(s, \pi'(s)) > 0$, and $k \neq j$, then $\pi'(s)$ can be the updated policy $\pi_{t+1}(s)$ at the state s.

However, according to (17.12), this is impossible, because if the present policy has the value as follows:

$$\bar{\pi}(s) = \max_a \max_{1 \leq i \leq M} [\Delta_i(s, a)] \neq \pi'(s), \text{then } V_i^{\bar{\pi}(s)} > V_i^{\pi'(s)} \text{ for some } i.$$

This means $\bar{\pi}(s)$ is not dominated by $\pi'(s)$.

Case 2)

If there exists any objective i such that $\Delta_i(s, \pi'(s)) < 0$, then $V_i^{\bar{\pi}(s)} > V_i^{\pi'(s)}$.
This means $\bar{\pi}(s)$ is nondominated by $\pi'(s)$.

Case 3)

If for all i, $\Delta_i(s, \pi'(s)) \geq 0$ and, there exists at least one j, $\Delta_j(s, \pi'(s)) > 0$, then then $\pi'(s)$ can be the updated policy $\pi_{t+1}(s)$ at the state s.

However, this is impossible for the same reason with case 1.

From case 1, case 2 and case 3, $\pi_{t+1} = \pi_t = \bar{\pi}$ is a Pareto optimal policy. ∎

Corollary 17.3.1: The upper bound of $\Delta_i^t(s, a)$ for a state s is nonincreasing as time t increases, where $\Delta_i^t(s, a)$ is the value difference of the objective i at the time t.

Proof:

From the definition, $Q_i^{\pi_t}(s, a) = V_i^{\pi_t}(s) + \Delta_i^t(s, a) \leq \frac{\gamma}{1-\gamma}$.

This means

$$\Delta_i^t(s, a) \leq \frac{\gamma}{1-\gamma} - V_i^{\pi_t}(s) \leq \frac{\gamma}{1-\gamma} - \min_{1 \leq i \leq M} V_i^{\pi_t}(s).$$

Because $V_i^{\pi_t}(s) \leq V_i^{\pi_{t+1}}(s)$, the upper bound of $\Delta_i^t(s, a)$ is nonincreasing for a state s. ∎

Corollary 17.3.2: The lower bound of $\Delta_i^t(s, a)$ for a state s is nondecreasing as time t increases.

Proof:

From Lemma 17.3.2, $V_i^{\pi_t}(s) + \Delta_i^t(s, a) \geq \min_j V_j^{\pi_t}(s)$.

This means

$$\Delta_i^t(s, a) \geq \min_{1 \leq j \leq M} V_j^{\pi_t}(s) - V_i^{\pi_t}(s) \geq \min_{1 \leq j \leq M} V_j^{\pi_t}(s) - \frac{\gamma}{1-\gamma}.$$

Because $V_i^{\pi_t}(s) \leq V_i^{\pi_{t+1}}(s)$, the lower bound of $\Delta_i^t(s, a)$ is non-decreasing for a state s. ∎

Corollary 17.3.3: The bound of the difference between

$$\max_{1 \leq i \leq M} V_i^{\pi_t}(s) \text{ and } \min_{1 \leq j \leq M} V_j^{\pi_t}(s) \text{ is nonincreasing for a state } s \text{ as time } t \text{ increases.}$$

Proof:

Let

$$d_i = V_i^{\pi_t}(s) - \min_{1 \le j \le M} V_j^{\pi_t}(s), j = 1, \cdots, M.$$

Because

$$d_i = V_i^{\pi_t}(s) - \min_{1 \le j \le M} V_j^{\pi_t}(s) \ge 0, 0 \le V_i^{\pi_t}(s) \le \frac{\gamma}{1 - \gamma},$$

then

$$\frac{\gamma}{1 - \gamma} - \min_{1 \le j \le M} V_j^{\pi_t}(s) \ge - \max_{1 \le i \le M} V_i^{\pi_t}(s) - \min_{1 \le j \le M} V_j^{\pi_t}(s) \ge d_i \ge 0. \quad (17.42)$$

As one can see from (17.42), the upper bound of d_i is nonincreasing for a state s, because $V_i^{\pi_t}(s) \le V_i^{\pi_{t+1}}(s)$. ∎

Bibliography

1. S. Yasunobu and T. Hasegawa, Evaluation of an automatic container crane operation system based on predictive fuzzy control, *Control Theory and Advanced Technology*, vol. 2, pp. 419–432, 1986.

2. S. Yasunobu and S. Miyamoto, Automatic train operation system by predictive fuzzy control, in M. Sugeno, (ed.), *Industrial Application of Fuzzy Control*, pp. 1–18, North-Holland: Elsevier Science Publishers, Amsterdam, 1985.

3. Y. S. Song, *Design of Fuzzy Sensor-Based Fuzzy Combustion Control System for Refuse Incinerator*, Master Thesis in KAIST, Dept. of Automation and Design Engineering, 1997.

4. Z. Bien, D. Kang, and S. Yang, Programming approach for fuzzy model-based multiobjective control systems, *International Journal on Fuzziness, Uncertainty, and Knowledge-Based Reasoning*, vol. 7, no. 4, pp. 289–292, 1999.

5. Y. J. Lai and C. L. Hwang, *Fuzzy Multiple Objective Decision Making*, Springer-Verlag, Berlin, 1994.

6. M. Sakawa, *Fuzzy Sets and Interactive Multiobjective Optimization*, Plenum Press, New York, 1993.

7. C. M. Fonseca and P. J. Fleming, Multiobjective genetic algorithms, *IEE Colloquium on Genetic Algorithms for Control Systems Engineering*, pp. 6/1–6/5, 1993.

8. K. Kim and J. Kim, Multicriteria fuzzy control, *Journal of Intelligent and Fuzzy Systems*, vol. 2, pp. 279–288, 1994.

9. T. Lim and Z. Bien, FLC design for multi-objective system, *Journal of Applied Mathematics and Computer Science*, vol. 6, no. 3, pp. 565–580, 1996.

10. R. S. Sutton, A. G. Barto, and R. J. Williams, Reinforcement learning is direct adaptive optimal control, *IEEE Control Systems Magazine*, vol. 12, no. 2, pp. 19–22, 1992.

11. G. Barto, R. S. Sutton, and C. W. Anderson, Neuronlike adaptive elements that can solve difficult learning control problems, *IEEE Trans. Systems, Man, and Cybernetics*, vol. SMC-13, no. 5, pp. 834–846, 1983.

12. L. P. Kaebling, M. L. Littman, and A. W. Moore, Reinforcement learning: A survey, *Journal of Artificial Intelligence Research,* vol. 4, pp. 237–285, 1996.

13. C.-T. Lin and I.-F. Chung, A reinforcement neuro-fuzzy combiner for multi-objective control, *IEEE Trans. Systems, Man and Cybernetics,* Part B, vol. 29, no. 6, pp. 726–744, 1999.

14. L. Jouffe, Fuzzy inference system learning by reinforcement methods, *IEEE Trans. Systems, Man, and Cybernetics,* Part C, vol. 28, no. 3, pp. 338–355, 1998.

15. Z. Bien and W. Yu, Extracting core information from inconsistent fuzzy control rules, *Fuzzy Sets and System,* vol. 71, no. 1, pp. 95–111, 1995.

Handbook of Learning and Approximate Dynamic Programming
Edited by Jennie Si, Andy Barto, Warren Powell and Donald Wunsch
Copyright © 2004 The Institute of Electrical and Electronics Engineers, Inc.

18 Adaptive Critic Based Neural Network for Control-Constrained Agile Missile

S. N. BALAKRISHNAN   DONGCHEN HAN

University of Missouri—Rolla

Editor's Summary: This chapter uses the adaptive critic approach, which was introduced in Chapters 3 and 4, to steer an agile missile with bounds on the angle of attack (control variable) from various initial Mach numbers to a given final Mach number in minimum time while completely reversing its flight path angle. While a typical adaptive critic consists of a critic and controller, the agile missile problem needs chunking in terms of the independent control variable and, therefore, cascades of critics and controllers. Detailed derivations of equations and conditions on the constraint boundary are provided. For numerical experiments, the authors consider vertical plane scenarios. Numerical results demonstrate some attractive features of the adaptive critic approach and show that this formulation works very well in guiding the missile to its final conditions for this state constrained optimization problem from an envelope of initial conditions.

18.1   INTRODUCTION

In order to explore and extend the range of operations of air-to-air missiles, there have been studies in recent years with a completely new concept. Launch the missile as usual from an aircraft but find a control that can guide it to intercept a target even if it is in the rear hemisphere (see Figure 18.1).

The best emerging alternative to execute this task is to use the aerodynamics and thrust to turn around the initial flight path angle of zero to a final flight path angle of 180 degrees. (Every missile-target scenario can be considered as a subset of this set of extremes in the flightpath angle.)   In this study, the problem is made more practical by limiting the missile to fly below a certain value of angle of attack. This sort of constraint is usually imposed in a problem due to the controller (actuator) or

463

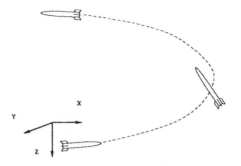

Fig. 18.1   Agile missile trajectory.

structural load limits, and so on. In calculus of variations, this problem is called the 'free final time and control-constrained' and it is very difficult to obtain solutions for it when the initial flight envelope is large. To our knowledge, there is no publication dealing with this difficult problem except with a single set of initial conditions.

In this study, however, solutions are obtained through the framework of optimal control. Optimization has been a field of interest to mathematicians, scientists and engineers for a long time. For optimal solutions which encompass perturbations to the assumed initial conditions or a family of initial conditions, a control system designer can use 'neighboring optimal' control [1] or dynamic programming [1]. Neighboring optimal control allows pointwise solutions to a (optimal) two-point boundary value problem to be used with a linearized approximation over a range of initial conditions. However, the neighboring optimal solution can fail outside the domain of validity of the linearization process. Dynamic programming can handle a family of initial conditions for linear as well as nonlinear problems. The usual method of solution, however, is computation-intensive. Furthermore, the solution is not generally available as a feedback that can be implemented in practice [2, 3].

Werbos [10] proposed a means to get around this numerical complexity by using 'approximate dynamic programming'(ADP) formulations. His methods approximate the original problem by a discrete formulation. The solution to the ADP formulation is obtained through the two-neural network adaptive critic approach. This ADP process, through the nonlinear function approximation capabilities of neural networks, overcomes the computational complexity that plagued the dynamic programming formulation of optimal control problems. More important, this solution can be implemented on-line, since the control computation requires only a few multiplications of the network weights which are trained off-line. This technique was used by Balakrishnan and Biega [6] to solve an aircraft control problem. Several authors have used neural networks to "optimally" solve nonlinear control problems [4–6]. For example, Kim and Calise [7] have proposed a neural network based control correction based on Lyapunov theory. A major difference between their approach and this study is that the development of guidance law/control is based on optimal control; hence, it is stabilizing and at the same time minimizing a cost.

In this study, a cascade of dual neural networks is used as a framework for the solutions of linear as well as nonlinear, *finite-time* optimal control problems with a special application to an agile missile. In a typical adaptive critic design, the controller output does not depend the current time but only the current states; in this problem, by contrast, the controller output has to be different for the same values of the state since the time left to complete the task plays a role in how much control has to be used. Hence, a cascade of controllers is synthesized by 'chunking' the independent variable. Rest of this paper is organized as follows: Approximate dynamic programming development in the context of a fairly general finite time optimal control problem presented in Section 18.2. Hamiltonian corresponding to the control constraints, features of constrained problems and optimal solutions on the constraint boundary are also discussed. Equations of motion for the agile missile are given in Section 18.3. Neural network solutions to minimum time bounded control of the agile missile are also discussed in Section 18.3. In Section 18.4, it is shown how to use the neurocontroller as a feedback controller. Simulation results and conclusions are presented in Section 18.5 and Section 18.6, respectively.

18.2   PROBLEM FORMULATION AND SOLUTION DEVELOPMENT

18.2.1   Approximate Dynamic Programming

In this section, general development on the optimal control of the nonlinear systems is presented in an ADP framework. Detailed derivations of these conditions may also be found in [6, 10] which are repeated here for the sake of clarity and completeness. The development in this section will subsequently be used in synthesizing the neural networks for optimal control of the agile missile. Note that the neural network solution consists of two parts — one solution when the control constraint is active and another solution when the constraint is not active.

Discrete description of a fairly general system model is given by

$$x_{i+1} = f_i(x_i, u_i), \tag{18.1}$$

where $f_i()$ can be either linear or nonlinear; i indicates the stage . The problem is to find a control sequence u_i to minimize the cost function J, where

$$J = \phi[x_N] + \sum_{i=0}^{N-1} L_i[x_i, u_i]. \tag{18.2}$$

Subject to a control inequality constraint:

$$C(u_i) \leq 0. \tag{18.3}$$

In Eq. (18.2), $L_i()$ can be a linear or nonlinear function of the states and/or control and $\phi()$ can be a linear or nonlinear function of terminal states.

Note that in an approximate dynamic programming formulation, Eq. (18.2) is rewritten as

$$J = \sum_{k=1}^{N-1} \Psi_k \left(x_k, u_k \right), \tag{18.4}$$

where x_k and u_k represent the $n \times 1$ state vector and $m \times 1$ control vector respectively at time step k. N represents the number of discrete time steps. By using Eq. (18.4), the *cost function from time step* k to $(N - 1)$ can be written as

$$J_k = \sum_{\tilde{k}=k}^{N-1} \Psi_{\tilde{k}} \left(x_{\tilde{k}}, u_{\tilde{k}} \right). \tag{18.5}$$

This cost can be split into the cost from $(k + 1)$ to $(N - 1)$, denoted by $J_{(k+1)}$ and the cost to go from k to $(k + 1)$ (called the utility function), Ψ_k as

$$J_k = \Psi_k + J_{k+1}. \tag{18.6}$$

We define the $n \times 1$ *costate* vector at time step k as

$$\lambda_k \equiv \frac{\partial J_k}{\partial x_k}. \tag{18.7}$$

Then the necessary condition for optimality for optimal control is

$$\frac{\partial J_k}{\partial u_k} = 0, \tag{18.8}$$

that is,

$$\frac{\partial J_k}{\partial u_k} = \left(\frac{\partial \Psi_k}{\partial u_k} \right) + \left(\frac{\partial J_{k+1}}{\partial u_k} \right) = \left(\frac{\partial \Psi_k}{\partial u_k} \right) + \left(\frac{\partial x_{k+1}}{\partial u_k} \right)^T \lambda_{k+1}. \tag{18.9}$$

Combining (18.8) and (18.9), we arrive at

$$\left(\frac{\partial \Psi_k}{\partial u_k} \right) + \left(\frac{\partial x_{k+1}}{\partial u_k} \right)^T \lambda_{k+1} = 0. \tag{18.10}$$

We derive the co-state propagation equation in the following way.

$$
\begin{aligned}
\lambda_k &= \frac{\partial J_k}{\partial x_k} \\
&= \left(\frac{\partial \Psi_k}{\partial x_k}\right) + \left(\frac{\partial J_{k+1}}{\partial x_k}\right) \\
&= \left[\left(\frac{\partial \Psi_k}{\partial x_k}\right) + \left(\frac{\partial u_k}{\partial x_k}\right)^T \left(\frac{\partial \Psi_k}{\partial u_k}\right)\right] \\
&\quad + \left[\left(\frac{\partial x_{k+1}}{\partial x_k}\right) + \left(\frac{\partial x_{k+1}}{\partial u_k}\right)\left(\frac{\partial u_k}{\partial x_k}\right)\right]^T \left(\frac{\partial J_{k+1}}{\partial x_{k+1}}\right) \\
&= \left[\left(\frac{\partial \Psi_k}{\partial x_k}\right) + \left(\frac{\partial X_{k+1}}{\partial x_k}\right)^T \lambda_{k+1}\right] \\
&\quad + \left(\frac{\partial u_k}{\partial x_k}\right)^T \left[\left(\frac{\partial \Psi_k}{\partial u_k}\right) + \left(\frac{\partial x_{k+1}}{\partial u_k}\right)^T \lambda_{k+1}\right],
\end{aligned}
\tag{18.11}
$$

with the boundary condition $\lambda_N = (\partial \phi(x_N)/\partial x_N)^T$.

The above problem formulation can be used to solve for control where no constraint exists. In order to accommodate bounds on control, a modified formulation is in order. For this purpose, define a quantity called the Hamiltonian [1] for this problem where the Hamiltonian, H_i, is given by

$$
H_i = L_i(x_i, u_i) + \lambda_{i+1}^T f_i(x_i, u_i) + \mu_{i+1}^T C(u_i).
\tag{18.12}
$$

In Eq. (18.12), μ_{i+1} is a time-invariant parameter. Now the additional requirement for optimal bounded control is that:

$$
\mu_i \begin{cases} \geq 0 & C = 0, \\ = 0 & C < 0. \end{cases}
\tag{18.13}
$$

For problem with control variable inequality constraints, the following equations hold at the junction between unconstrained and constrained part of the trajectory [1]:

$$
\begin{aligned}
\lambda_{i-} &= \lambda_{i+}, \\
H_{i-} &= H_{i+}, \\
\tfrac{\partial J_i}{\partial u_i}(-) &= \tfrac{\partial J_i}{\partial u_i}(+),
\end{aligned}
\tag{18.14}
$$

that is, the control inequality constraint will not form a discontinuity, that is the λ, H, u, μ are continuous across the junction points between the unconstrained control arc and constrained control arc. So, the control inequality constraint problem is different from unconstrained problem only in that μ needs to be calculated.

For $C < 0$ (constraint is not active), the constraint associate parameter $\mu_i = 0$ and Eq. (18.10) is used in calculating u_i. For $C = 0$ (the constraint is active), the control constraint equation is used to compute u_i and Eq. (18.12) along with Eq. (18.10) are used in the computation of μ_i.

With control solution described, a systematic process of using the time (or stage)-indexed neural networks to solve for control is developed next. Note that since the terminal conditions are given at the last stage, the solution proceeds backwards.

18.2.2 General Procedure for Finite Time Problems Using Adaptive Critics

For finite time (or finite-horizon) problems, a solution with neural networks evolves in two stages:

Synthesis of Last Network:

1) Note that $\lambda_N = \left(\frac{\partial \phi(x_N)}{\partial x_N}\right)^T$. For various random values of x_N, λ_N can be calculated.

2) Use the state-propagation Eq. (18.1) and optimality condition in Eq. (18.10) to calculate u_{N-1} for various x_{N-1} by randomly selecting x_N and the corresponding λ_N from step 1.

3) With u_{N-1} and λ_N, calculate λ_{N-1} for various x_{N-1} by using the costate propagation Eq. (18.11).

4) Train two neural networks. For different values of x_{N-1}, the u_{N-1} network outputs u_{N-1} and the λ_{N-1} network outputs λ_{N-1}. We have optimal control and costates for various values of the state at stage $(N - 1)$ now.

Other Networks:

1) Assume different values of states at x_{N-2} at stage $(N - 2)$ and pick a random network (or initialized with u_{N-1} network weights) called u_{N-2} network to output u_{N-2}. Use u_{N-2} and x_{N-2} in the state propagation equation to get x_{N-1}. Input x_{N-1} to the λ_{N-1} network to obtain λ_{N-1}. Use x_{N-2} and λ_{N-1} in the optimality condition in Eq. (18.10) to get target u_{N-2}. Use this to correct the u_{N-2} network. Continue this process until the network weights show little changes. This u_{N-2} network yields optimal u_{N-2}.

2) Using random x_{N-2}, output the control u_{N-2} from the u_{N-2} network. Use these x_{N-2} and u_{N-2} to get x_{N-1} and input x_{N-1} to generate λ_{N-1}. Use x_{N-2}, u_{N-2} and λ_{N-1} to obtain optimal λ_{N-2}. Train a λ_{N-2} network with x_{N-2} as input and obtain optimal λ_{N-2} as output.

3) Repeat the last two steps with $i = N - 1, N - 2, \ldots 0$, until u_o is obtained.

A schematic of the network development is presented in Figure 18.2.

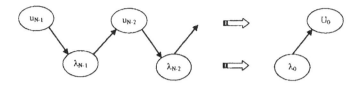

Fig. 18.2   Schematic of successive adaptive critic synthesis.

18.3   MINIMUM TIME HEADING REVERSAL PROBLEM IN A VERTICAL PLANE

The equations of motion in a vertical plane are presented and the minimum time problem is formulated this section. The main goal of this study is to find the control (angle-of-attack) history to minimize the time taken by the missile in reversing its flightpath angle while using an upper bound on the control, angle of attack.. In many engagements, most of the flight is dominated by two-dimensional motion-either in a horizontal or vertical plane. It should be noted that extension of this method to a three-dimensional engagements is straight-forward.

18.3.1   Equations of Motion of A Missile in a Vertical Plane

The non-dimensional equations of motion of a missile (represented as a point mass) in a vertical plane are:

$$M' = -S_w M^2 C_D - sin\gamma + T_w cos\alpha, \qquad (18.15)$$

$$\gamma' = \frac{1}{M}\left[S_w M^2 C_L + T_w sin\alpha - cos\gamma\right], \qquad (18.16)$$

$$X_E = M cos\gamma, \qquad (18.17)$$

$$Z_E = -M sin\gamma, \qquad (18.18)$$

where prime denotes differentiation with respect to nondimensional time, τ.

The non-dimensional parameters used in Eqs. (18.15) – (18.18) are:

$$\tau = g/at; \ T_w = T/mg; \ S_w = \rho a^2 S/2mg \text{ and } M = V/a.$$

In these equations, M is the flight Mach number, γ, the flightpath angle, α, the aerodynamic angle of attack, T, the solid rocket thrust, m, the mass of the missile, S, the reference aerodynamic area, V, the speed of the missile, C_L, the lift coefficient, C_D, the drag coefficient, g, the acceleration due to gravity, a, the speed of sound, ρ, the atmospheric density, and t is the flight time. Note that C_D and C_L are functions of angle of attack and flight Mach number and a neural network is trained to out put C_D and C_L with angle of attack and flight Mach number as inputs. X_E and Z_E are the non dimensional velocities in the horizontal and vertical directions, respectively.

18.3.2   Minimum Time Optimal Control Problem

The objective of the minimization process is to find the control (angle-of-attack) history to minimize the time taken by the missile to reverse its flightpath angle completely with limits on angle of attack while the Mach number changes from a given value from an envelope of Mach numbers to a final Mach number of 0.8. Mathematically, this problem is stated as to find the control minimizing J, the cost function where

$$J = \int_0^{t_f} dt, \tag{18.19}$$

with $\gamma(0) = 0deg$, $M(0) \equiv 0$ given, $\gamma(t_f) = 180deg.$ and $M(t_f) = 0.8$. This constrained optimization problem comes under the class of 'free final time' problems in calculus of variations and is difficult to solve. No general solution exists which generates optimal paths for flexible initial conditions.

In this study, such solutions are sought by using the ADP approach. In order to facilitate the solution using neural networks, the equations of motion are reformulated using the flightpath angle as the independent variable. This process enables us to have a fixed final condition as opposed to the 'free final time'. It should be observed that the independent variable in the transformed system should be monotonically increasing to allow proper indexing of the neural networks in the temporal domain. The transformed dynamic equations are:

$$\frac{dM}{d\gamma} = \frac{(-S_w M^2 C_D - sin\gamma + T_w \alpha)M}{S_w M^2 C_L - cos\gamma + T_w sin\alpha}, \tag{18.20}$$

$$\frac{dt}{d\gamma} = \frac{aM}{g(S_w M^2 C_L - cos\gamma + T_w sin\alpha)}, \tag{18.21}$$

and the transformed cost function is given by

$$J = \int_0^\pi \frac{aM}{g(S_w M^2 C_L cos\gamma + T_w sin\alpha)} d\gamma, \tag{18.22}$$

subject to the control variable inequality constraint:

$$\alpha \leq \alpha^*, \tag{18.23}$$

(here $\alpha^* = 120$)

$$C[u_i] = \alpha - \alpha^* \leq 0. \tag{18.24}$$

In this study, the final velocity is treated as a hard constraint; this means that the flight path angle and the velocity constraints are met *exactly* at the final point. The dynamics and associated optimal control equations are expressed in discrete form in order to use them with discrete feed-forward neural networks. The discrete system equations are given by

$$M_{k+1} = M_k + \frac{\left(-S_w M_k^2 C_{Dk} - \sin \gamma_k + T_{wk} \cos \alpha_k\right) M_k}{S_w M_k^2 C_{Lk} - \cos \gamma_k + T_{wk} \sin \alpha_k} \cdot \delta \gamma_k,$$

$$t_{k+1} = t_k + \frac{a_{M_k} \delta \gamma_k}{g\left(S_w M_k^2 C_{Lk} - \cos \gamma_k + T_{wk} \sin \alpha_k\right)}. \qquad (18.25)$$

The discrete Hamiltonian is

$$
\begin{aligned}
H_k &= \left(\frac{dt}{d\gamma}\right)_k \delta \gamma_k + \lambda_{k+1} M_{k+1} + \mu_{k+1} C[u_{i+1}] \\
&= \frac{a M_k \delta \gamma_k}{g\left(S_w M_k^2 C_{Lk} - \cos \gamma_k + T_{wk} \sin \alpha_k\right)} \\
&\quad + \lambda_{k+1} \left[M_k + \frac{\left(-S_w M_k C_{Dk} - \sin \gamma_k + T_{wk} \cos \alpha_k\right) M_k \delta \gamma_k}{S_w M_k^2 C_{Lk} - \cos \gamma_k + T_{wk} \sin \alpha_k} \right] \\
&\quad + \mu_{k+1} \alpha_k .
\end{aligned}
\qquad (18.26)
$$

The denominator of the first term on the right hand side sans g appears in many expressions and its derivatives are needed. Hence, for convenience, we define

$$den = S_w M_k C_{Lk} - \cos \gamma_k + T_{wk} \sin \alpha_k, \qquad (18.27)$$

$$\frac{\partial den}{\partial \alpha_k} = S_w M_k^2 \frac{\partial C_{Lk}}{\partial \alpha_k} + T_{wk} \cos \alpha_k, \qquad (18.28)$$

$$\frac{\partial den}{\partial M_k} = 2 S_w M_k^2 C_{Lk} + S_w M_k^2 \frac{\partial C_{Lk}}{\partial M_k}. \qquad (18.29)$$

Derivatives of lift and drag coefficients with respect to the angle of attack and the Mach number are obtained from a neural network which stores these coefficients. The costate equation is

$$
\begin{aligned}
\lambda_k &= \frac{a \cdot \delta \gamma_k}{g \cdot den} - \frac{a M_k \delta \gamma_k}{g \cdot den^2} \cdot \frac{\partial den}{\partial M_k} \\
&\quad + \lambda_{k+1} \cdot \delta \gamma_k \frac{\left(-3 S_w M_k^2 C_{Dk} - S_w M_k^3 \frac{\partial C_{Dk}}{\partial M_k} - \sin \gamma_k + T_{wk} \cos \alpha_k\right)}{den} \\
&\quad \lambda_{k+1} + \lambda_{k+1} \cdot \delta \gamma_k \frac{\left(S_w M_k^2 C_{Dk} + \sin \gamma_k - T_{wk} \cos \alpha_k\right) \cdot M_k}{den^2} \cdot \frac{\partial den}{\partial M_k}.
\end{aligned}
\qquad (18.30)
$$

Optimality condition is obtained in an expanded form as

$$
\begin{aligned}
&\frac{a}{g} \cdot \frac{\partial den}{\partial \alpha_k} + \lambda_{k+1} \cdot \left(S_w M_k^2 \frac{\partial C_{Dk}}{\partial \alpha_k} + T_{wk} \sin \alpha_k\right) \cdot den \\
&+ \lambda_{k+1} \left(- S_w M_k^2 C_{Dk} - \sin \lambda_k + T_{wk} \cos \alpha_k\right) \cdot \frac{\partial den}{\partial \alpha_k} + \mu_{k+1} = 0. \quad (18.31)
\end{aligned}
$$

18.3.3   Development of Neural Network Solutions

Neural network synthesis for a finite time problem consists of two steps. It begins from the last stage and proceeds backwards. Note that $\phi(\cdot)$ is zero in this formulation; however, the final state, M_N is specified.

18.3.3.1   Last Network Synthesis

1. Final Mach number, M_N, is fixed at 0.8. For random values of M_{N-1}, calculate α_{N-1} from the state propagation equation.

2. Use optimality condition to solve for appropriate λ_N.

3. From the costate propagation equation, calculate λ_{N-1}.

4. Train two neural networks: The α_{N-1} network outputs α_{N-1} for different values of M_{N-1} and the λ_{N-1} network outputs λ_{N-1} for different values of M_{N-1}. We have optimal α_{N-1} and λ_{N-1} now.

18.3.3.2   Synthesis of Other Networks

1. Assume different values of M_{N-2} and use a random neural network (or initialized with α_{N-1} network) called α_{N-2} network to output α_{N-2}. Use M_{N-2} and α_{N-2} to obtain M_{N-1}. Input M_{N-1} to λ_{N-1} network to get λ_{N-1}.

   Use M_{N-2}, λ_{N-1} in the optimality equation to solve for α_{N-2}. Use this α_{N-2} to correct the network. Continue this process until α_{N-2} network converges. This α_{N-2} network yields optimal α_{N-2}.

2. Using random M_{N-2} into α_{N-2} network obtains optimal $\forall_{N-2}$. Use M_{N-2} and α_{N-2} to obtain M_{N-1} and input to λ_{N-1} network to generate λ_{N-1}. Use M_{N-2}, α_{N-2} , and λ_{N-1} in costate equation to obtain optimal λ_{N-2}. Train λ_{N-2} network with M_{N-2} as input. We have λ_{N-2} network that yields optimal λ_{N-2}.

3. Repeat steps 1 and 2 with $k = N - 1, N - 2, \ldots, 0$, until we get α_o .

A schematic of the network development is presented in Figure 18.2.

Note that this procedure sweeps backwards [11]. At each step, we check whether the angle of attack constraint is violated. If it is, then the control target is made equal the limit value. Note that the μ value needs not to be calculated since it doesn't affect the procedure.

18.4   USE OF NETWORKS IN REAL-TIME AS FEEDBACK CONTROL

Assume any M_0 [within the trained range]. Use α_0 neural network to find optimal α and integrate until γ_1 for α_1 network is reached; use the M_1 values to find α_1 from the α_1 neural network and integrate until γ_2 is reached, and so on, until γ_f is reached.

Note that the forward integration is done in terms of time (which is available as an incidental variable as a function of the flightpath angle Eq. (18.25)). As a result, even though the network synthesis is done off-line, the control is a feedback process based on current states.

18.5 NUMERICAL RESULTS

Representative numerical results with the agile missile simulation are provided in this section. Tables of aerodynamic data of C_L and C_D variations with Mach numbers and angle of attack were provided in [12, 13]. Outputs from the neural network models of these tables were used for extracting these coefficients at different angles of attack. These data allow estimation of the aerodynamic coefficients at unusually high angles of attack that are encountered in such rear hemisphere flights. All the neural networks in this study are feedforward networks. A feedforward network was selected in order to facilitate numerical derivatives of the outputs with respect to the inputs. Each network has a three-layered structure with the first layer having a tangent sigmoidal activation function, the second layer having a logarithmic sigmoidal activation function and the third layer having a unit gain. Each layer consists of nine neurons. The results proved that the choices were adequate. There was no effort to optimize the structure of networks in this study. A Levenberg-Marquardt method is used to train the networks. Based on the authors' experience, many other training methods could have been applied, equally as well.

The control variable inequality constraint is chosen such that the angle of attack, $\alpha \leq 120°$. This value does not have any particular meaning, it is only chosen for testing the adaptive critic neural network technique for control constrained problems. The solution process as pointed out earlier proceeds backward from the last step. At each step, the control value was checked to see whether it exceeded the limit. Otherwise, the solution marched on as if the problem is unconstrained. Thirty-seven networks were needed to implement this optimal process. Time histories of Mach number, costates, angle of attack and the flightpath angle with an initial Mach number equal to 0.8 are presented in Figure 18.3.

Note that all these results are forward integration in terms of time. From Figure 18.3(a), it is clear that the ADP formulation met the final Mach number constraint exactly; the corresponding costate history is given in Figure 18.3(b). From Figure 18.3(c), it can be seen that the bound on control input was met exactly as well. Figure 18.3(d), presents the reversal of the flight path angle that was the objective of the problem. The real advantage of using the adaptive critic approach is clear from the Mach number history with flightpath angle in Figure 18.4(a).

For each trajectory with an initial Mach number varying from 0.6 to 0.8, the final Mach number is 0.8. That is, the same cascade of neurocontroller is used to generate optimal control for an envelope of initial conditions. In order to compare the control constrained solution with the unconstrained solution, we also plot the Mach number variations with the flight path angle and the angle of attack variations for both

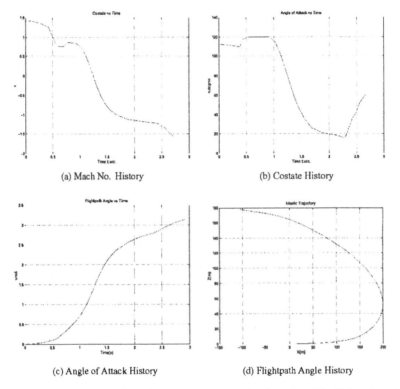

(a) Mach No. History

(b) Costate History

(c) Angle of Attack History

(d) Flightpath Angle History

Fig. 18.3   Time histories of Mach number, costates, angle of attack and the flightpath angle with $M_0 = 0.8$, $M_f = 0.8$.

the unconstrained and control constrained problem in Figures 18.4(b) and 18.4(c), respectively. From the Mach number history, we can see that the Mach numbers for the constrained problem are slightly higher than its corresponding part without the constraint. This is to be expected since smaller angle of attacks result in higher speeds. Further numerical experiments were conducted to test the robustness of the controller. Six controller in the mid-flightpath angle region (which means that some of the controls are held constant for longer periods) were removed and the trajectories were generated by using the remaining controllers. Mach number histories and the angle of attack of a 37-networks controller and a 31-networks controller are presented in Figure 18.4(a) and Figure 18.5(b).

From the simulation results, it can be seen that even though the trajectory is less optimal, the lesser network configuration still delivers the missile at the exact final Mach number of 0.8. Note that if further compactness is required, we can reduce the controller to be embedded in a single network by adding the flight path angle as an extra indexing input.

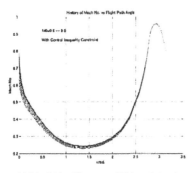

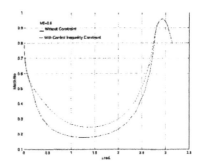

(a) Mach No. History vs. Flightpath Angle (M_0 varies from 0.6 to 0.8 and $M_f = 0.8$)

(b) Mach No. Comparison (Bounded Control v.s. Unconstraint Problem, $M_0 = 0.8, M_f = 0.8$)

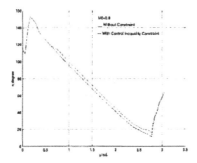

(c) Control History (Bounded Control v.s. Unconstraint Problem, $M_0 = 0.8, M_f = 0.8$)

Fig. 18.4   Mach number and control histories.

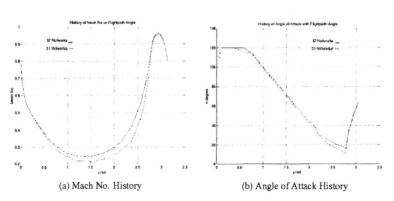

(a) Mach No. History

(b) Angle of Attack History

Fig. 18.5   Mach number and angle of attack histories ($M_0 = 0.8, M_f = 0.8$).

18.6   CONCLUSIONS

An approximate dynamic programming based formulation with an adaptive critic implementation was used to find optimal solution for a 'bounded control, free final time' problem associated with an agile missile control. The neural network controllers are able to provide (near) optimal control to the missile from an envelope of initial Mach numbers to a fixed final Mach number of 0.8 in minimum time. An added advantage in using these neurocontrollers is that they provide minimum time solutions even when we change the initial flight path angle from zero to any non zero (positive) value. To our knowledge, there has been no one tool (other than dynamic programming) which provides such solutions.

Acknowledgments

This research was supported partially by an Air Force Grant #F08630-96-1-0001 and a National Science Foundation Grant #ECS-9976588.

Bibliography

1. A. E. Bryson and Y. Ho, *Applied Optimal Control*, pp. 128–211, Hemisphere Publishing Co., New York, 1975.

2. A. Weinreb and A. E. Bryson, Optimal control system with hard control bounds, *IEEE Trans. Automatic Control*, vol. 30, no. 11, pp. 1135–1138, 1985.

3. B. Wie, C. H. Chuang and J. Sunkel, Minimum-time pointing control of two-link manipulator, *Journal of Guidance, Control and Dynamics*, vol. 13, no. 5, pp. 867–872, 1990.

4. K. J. Hunt, Neural networks for controller systems, A survey, *Automatica*, vol. 28, no. 6, pp. 1083–1112, 1992.

5. D. A. White and D. Sofge, *Handbook of Intelligent Control*, Chapters 3,5,8,12,13, Van Nostrand Reinhold, New York, 1992.

6. S. N. Balakrishnan and V. Biega, Adaptive critic based neural networks for aircraft optimal control, *Journal of Guidance, Control, and Dynamics*, vol. 19, no. 4, pp. 893–898, 1996.

7. B. S. Kim and A. J. Calise, Nonlinear flight control using neural networks, *AIAA Journal of Guidance, Control, and Dynamics*, vol. 20, no. 1, pp. 26–33, 1997.

8. A. G. Barto, R. S. Sutton, and C. W. Anderson, Neuronlike adaptive elements that can solve difficult learning control problems, *IEEE Trans. Systems, Man, and Cybernetics*, vol. SMC-13, pp. 834–846, 1983.

9. P. J. Werbos, Neurocontrol and supervised learning; An overview and evaluation, in D. A. White and D. A. Sofge, (eds.), *Handbook of Intelligent Control*, Van Nostrand Reinhold, New York, 1992.

10. P. J. Werbos, Approximate dynamic programming for real-time control and neural modeling, in D. A. White and D. A. Sofge, (eds.), *Handbook of Intelligent Control*, Van Nostrand Reinhold, New York, 1992.

11. D. Han and S. N. Balakrishnan, Adaptive critic based neural networks for agile missile control, *AIAA Guidance, Navigation, and Control Conference and Exhibit*, pp. 1803–1812, Boston, 1998.

12. K. D. Burns et al., Missile DATCOM, WL-TR-91-3039.

13. M. Innocenti, Integrated approach to guidance and control of alternate control technology flight vehicle, Annual Technical Report, prepared under Grant: F08630-94-1-001 for Department of the Air Force Wright Laboratory, Eglin Air Force Base, 1995.

19 Applications of Approximate Dynamic Programming in Power Systems Control

GANESH K VENAYAGAMOORTHY   RONALD G HARLEY
DONALD C WUNSCH
University of Missouri—Rolla   Georgia Institute of Tech.

Editor's Summary: This chapter presents application examples in the field of power systems control using adaptive-critic designs. These adaptive-critic techniques were introduced in Chapters 3 and 4, but this chapter does contain an in-depth review of the basic ideas and how they apply to the specific problems examined in this chapter. The primary systems examined here are the three-phase micro-alternators at the University of Natal in Durban, South Africa. The design and training of the procedure for the critic and action networks are described in detail. Both simulation results and actual hardware implementations are discussed. Detailed results are presented that compare how different adaptive-critic schemes perform and what type of performance can be expected with realistic problems. For another discussion about control of realistic power systems see Chapter 23.

19.1   INTRODUCTION

Power system control essentially requires a continuous balance between electrical power generation and a varying load demand, while maintaining system frequency, voltage levels and the power grid security. However, generator and grid disturbances can vary between minor and large imbalances in mechanical and electrical generated power, while the characteristics of a power system change significantly between heavy and light loading conditions, with varying numbers of generator units and transmission lines in operation at different times. The result is a highly complex and non-linear dynamic electric power grid with many operational levels made up of a wide range of energy sources with many interaction points. As the demand for electric power grows closer to the available sources, the complex systems that ensure the stability and security of the power grid are pushed closer to their edge.

Synchronous turbogenerators supply most of the electrical energy produced by mankind, and are largely responsible for maintaining the stability and the security of the electrical network. The effective control of these devices, is therefore, very important. However, a turbogenerator is a highly non-linear, non-stationary, fast acting, multi-input-multi-output (MIMO) device with a wide range of operating conditions and dynamic characteristics that depend on the power system to which the generator is connected too. Conventional automatic voltage regulators (AVRs) and turbine governors are designed based on some linearized power system model, to control the turbogenerator in some optimal fashion around one operating point. At any other operating points the conventional controller technology cannot cope well and the generator performance degrades, thus driving the power system into undesirable operating states [1]. Additionally, the tuning and integration of the large number of control loops typically found in a power station can prove to be a costly and time-consuming exercise.

Many novel control strategies have been suggested to deal with all these non-ideal plant properties as well as aging of plant components. Indeed significant research has demonstrated that adaptive control can improve the overall control in turbogenerator systems [2, 3], with the objective of extending operational stability margins [2]. These techniques have performed acceptably well in power stations [4], and commercial power system control manufacturers are beginning to apply self-tuning strategies.

Previous investigations have illustrated that self-tuning schemes will work well if the preconditions for stability and convergence are satisfied. However, the nonlinear nature of power systems implies that the estimated linear models are only valid for a small region about a given operating point. In practice, however, a power system is frequently subjected to various disturbances such as tap-changing, line and load switching and occasional major disturbances such as short circuits or lightning surges. All of these may cause excessive variations in plant outputs, leading to abrupt changes in operating conditions, and possibly hunting of the generators. Therefore, if a self-tuning controller is to work safely in practice, it must be reliable and robust. This can be very complex and unwieldy, with expert systems being suggested to fulfill this role [5].

Most adaptive control algorithms use linear models [6] with certain assumptions of types of noise and possible disturbances. Moreover, for the design of adaptive controllers, it has to be assumed that the number of system inputs equals the number of system outputs. Where necessary this is achieved by using a transformation to reduce the dimensions of the output space, with the drawback that this degrades the description of the system dynamics. While this is possible for SMIB systems, it becomes unwieldy and almost impossible in large multi-machine power systems. Consequently, the issues of unmodeled dynamics and robustness arise in practical applications of these adaptive control algorithms. To allow for all these uncertainties, the traditional controllers are typically designed with large safety margins. In the era of a deregulated electricity industry, and an emphasis on competitive pricing, it will become necessary to reduce these safety margins as much as possible while still maintaining a reliable service.

One of the most important problems arising from large-scale electric power system interconnection is the low frequency oscillation [7]. Power System Stabilizers (PSSs) are used to damp such oscillations, but the particular position and transfer function of a PSS is not a simple decision and is usually also based on some linearized system model, as initially proposed by deMello and Concordia [8] based on the single-machine-infinite-bus (SMIB) linearized model. A practical PSS must be robust over a wide range of operating conditions and capable of damping local, intra-area and inter-area system modes of oscillation [9, 10]. The participation of the generators in the different modes of the oscillation depends on the location of the generators in the system; some may participate in one mode only and others in more than one [11]. From this perspective, the conventional PSS design approach based on a SMIB linearized model in the normal operating condition has some drawbacks:

(a) There are uncertainties in the linearized model resulting from the variation in the operating condition, since the linearization coefficients are derived typically at some nominal operating condition.

(b) To implement the PSS for a multimachine power system, its parameters need to be tuned to coordinate with machines and utilities.

Consequently, a realistic solution for stabilizing the low frequency oscillations of a multimachine system is a stabilizer designed from a nonlinear multimachine model in the first place. Difficulties in a PSS design come from the handling of nonlinearities and interactions among generators. During the low frequency oscillation, the rotor oscillates due to the imbalance between mechanical and electrical powers. Thus handling the nonlinear power flow properly is the key to the PSS design for a multimachine power system. Unfortunately, it is not that simple to handle the nonlinear interacting variables by conventional analytical methods.

In recent years, renewed interest has been shown in power systems control using nonlinear control theory, particularly to improve system transient stability [12, 13]. Instead of using an approximate linear model, as in the design of the conventional power system stabilizer, nonlinear models are used and nonlinear feedback linearization techniques are employed on the power system models, thereby alleviating the operating point dependent nature of the linear designs. Nonlinear controllers significantly improve the power system's transient stability. However, nonlinear controllers have a more complicated structure and are difficult to implement relative to linear controllers. In addition, feedback linearization methods require exact system parameters to cancel the inherent system nonlinearities, and this contributes further to the complexity of stability analysis. The design of decentralized linear controllers to enhance the stability of interconnected nonlinear power systems within the whole operating region remains a challenging task [14].

However, the use of Computational Intelligence, especially Artificial Neural Networks (ANNs), offers a possibility to overcome the above-mentioned challenges and problems of conventional analytic methods. ANNs are good at identifying and controlling nonlinear systems [15, 16]. They are suitable for multi-variable applications, where they can easily identify the interactions between the system's inputs and outputs. It has been shown that a Multilayer Perceptron (MLP) neural network using deviation signals (e.g., deviation of terminal voltage from its steady value) as inputs, can identify [17] the complex and nonlinear dynamics of a single machine infinite bus configuration, with sufficient accuracy, and this information can then be used to design a nonlinear controller which will yield an optimal dynamic system response irrespective of the load and system configurations.

Previous publications have reported on the different aspects of neural network based control of generators. However, these neurocontrollers require continual online training of their neural networks after commissioning [18]. In most of the above results, an ANN is trained to approximate various nonlinear functions in the nonlinear system. The information is then used to adapt an ANN controller. Since an ANN identifier is only an approximation to the underlying nonlinear system, there is always residual error between the true plant and the ANN model of the plant. Stability issues arise when the ANN identifier is continually trained online and simultaneously used to control the system. Furthermore, to update weights of the ANN identifier online, gradient descent algorithms are commonly used. However, it is well known in adaptive control that a brute force correction of controller parameters, based on the gradients of output errors, can result in instability even for some classes of linear systems [19, 20]. Hence, to avoid the possibility of instability during online adaptation, some researchers proposed using ANNs such as radial basis functions, where variable network parameters occur linearly in the network outputs, such that a stable updating rule can be obtained [21]. To date, the development of nonlinear control using ANNs is similar to that of linear adaptive control because the ANNs are used only in linearized regions. Unfortunately, unlike linear adaptive control, where a general controller structure to stabilize a system can be obtained with only the knowledge of relative degrees, stabilizing controllers for nonlinear systems are difficult to design. As a result, most research on ANN based controllers has focused on nonlinear systems, whose stabilizing controllers are readily available once some unknown nonlinear parts are identified, such as

$$x^n = f(x^{n-1}, ..., x) + bu \qquad (19.1)$$

with full state feedback, where f is to be estimated by an ANN. Even though some methods have been suggested for using ANNs in the context of a general controller structure [22, 23], the stability implication of updating a network online is unknown. Furthermore, since an ANN controller can have many weights, it is questionable whether the network can converge fast enough to achieve good performance. Besides, in closed loop control systems with relatively short time constants, the computational time required by frequent online training could become the factor that limits the maximum bandwidth of the controller.

Adaptive critic designs (ACDs) are neural network designs capable of optimization over time, under conditions of noise and uncertainty. This family of ACDs is an addition to other existing neural network based techniques for control and optimization [32]. ACDs are considered to be the most powerful and complicated designs.

This chapter presents work by the authors [24–26] based on the adaptive critic's technique for designing a turbogenerator neurocontroller, which overcomes the risk of instability [27], the problem of residual error in the system identification [28], input uncertainties [29], and the computational load of online training. The neurocontroller augments/replaces the conventional automatic voltage regulator and turbine governor, and is trained in an offline mode prior to commissioning. Two different types of Adaptive Critics are discussed, namely the Heuristic Dynamic Programming (HDP) type and the Dual Heuristic programming (DHP) type. Results are presented for a single-machine-infinite-bus, as well as for a multimachine power system, showing that the DHP neurocontroller produces the best results in comparison to the HDP neurocontroller and conventional controllers even with the inclusion of a power system stabilizer.

Section 19.2 of this chapter describes adaptive critic designs and approximate dynamic programming. The mathematical equations are given for the HDP and DHP designs. Section 19.3 describes the general training procedure for the Critic and Action networks. Section 19.4 describes the power system models studied with adaptive critic designs. Section 19.5 presents some simulation and hardware implementation results on the single machine and multimachine power systems.

19.2 ADAPTIVE CRITIC DESIGNS AND APPROXIMATE DYNAMIC PROGRAMMING

19.2.1 Background

The adaptive critic designs (ACDs) are neural network designs capable of optimization over time under conditions of noise and uncertainty. The simplest adaptive critic designs learn slowly on large problems but they are successful on many real world difficult small problems. Complex adaptive critics may seem breathtaking, at first, but they are the only design approach that shows potential of replicating critical aspects of human intelligence: ability to cope with a large number of variables in parallel, in real time, in a noisy nonlinear non-stationary environment.

A family of ACDs was proposed by Werbos [32] as a new optimization technique combining concepts of reinforcement learning and approximate dynamic programming. For a given series of control actions that must be taken sequentially, and not knowing the effect of these actions until the end of the sequence, it is impossible to design an optimal controller using the traditional supervised learning neural network. The adaptive critic method determines optimal control laws for a system by successively adapting two ANNs, namely an *Action neural network* (which dispenses the control signals) and a *Critic neural network* (which 'learns' the desired performance index for some function associated with the performance index). These two neural networks approximate the Hamilton-Jacobi-Bellman equation associated with optimal control theory. The adaptation process starts with a non-optimal, arbitrarily chosen, control by the action network; the critic network then guides the action network toward the optimal solution at each successive adaptation. During the adaptations, neither of the networks need any 'information' of an optimal trajectory, only the desired cost needs to be known. Furthermore, this method determines optimal control policy for the entire range of initial conditions and needs no external training, unlike other neurocontrollers.

Dynamic programming prescribes a search which tracks backward from the final step, retaining in memory all suboptimal paths from any given point to the finish, until the starting point is reached. The result of this is that the procedure is too computationally expensive for most real problems. In supervised learning, an ANN training algorithm utilizes a desired output and, having compared it to the actual output, generates an error term to allow the network to learn. The backpropagation algorithm is typically used to obtain the necessary derivatives of the error term with respect to the training parameters and/or the inputs of the network. However, backpropagation can be linked to reinforcement learning via the critic network which has certain desirable attributes.

The technique of using a critic, removes the learning process one step from the control network (traditionally called the "action network" or "actor" in ACD literature), so the desired trajectory is not necessary. The critic network learns to approximate the *cost-to-go* or strategic utility function (the function J of Bellman's equation in dynamic programming) and uses the output of the action network as one of its inputs, directly or indirectly.

In the Dynamic Programming, or Markov Decision Process (MDP), literature, problems are described in terms of five essential characteristics:

1. Epochs

2. States

3. Actions

4. Rewards

5. Transition probabilities.

Most of the literature has focused on finite state spaces, where states are known with certainty. Actions may have probabilistic rules associated with them, and rewards and (obviously) transition probabilities may also be nondeterministic. Recent literature has addressed the extension to where states are also nondeterministic. These are known as Partially Observable Markov Decision Processes (POMDPs). It is straightforward to show that a POMDP can be transformed to a regular MDP with continuously-valued state variables [33]. Since Adaptive Critics typically have been used in control problems with continuous-valued state spaces, they trivially are applicable to POMDPs. This observation, by itself, is sufficient reason for the family of Adaptive Critic approaches to be known and utilized outside the confines of the Intelligent Control community.

Different types of critics have been proposed. For example, Watkins [34] developed a system known as Q-learning, explicitly based on dynamic programming. Werbos, on the other hand, developed a family of systems for approximating dynamic programming [32]; his approach subsumes other designs for continuous domains. For example, Q-learning becomes a special case of Action-Dependent Heuristic Dynamic Programming (ADHDP), which is a Critic approximating the J function (see Section 19.2.2 below), in Werbos' family of adaptive critics. A Critic which approximates only the derivatives of the function J with respect to its states, called the Dual Heuristic Programming (DHP), and a Critic approximating both J and its derivatives, called the Globalized Dual Heuristic Programming (GDHP), complete this ACD family. These systems do not require exclusively neural network implementations, since any differentiable structure is suitable as a building block. The interrelationships between members of the ACD family have been generalized and explained in detail by Prokhorov [35, 36].

19.2.2   Heuristic Dynamic Programming Neurocontroller

Figure 19.1 shows a model dependent HDP Critic/Action design. The HDP Critic neural network is connected to the Action neural network through a Model neural network of the plant. These three different neural networks are each described in Sections 19.2.2.1, 19.2.2.2 and 19.2.2.3 and are taken for the purposes of this study to be a three-layer feedforward neural network with a single hidden layer with sigmoid transfer function. The input and output layers have linear transfer functions.

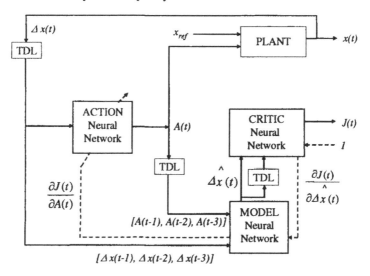

Fig. 19.1   A model dependent HDP Critic/Action design.

19.2.2.1   Model Neural Network

For model dependent designs it is assumed that there exists a Model neural network which is able to predict the states/outputs $x(t + 1)$, of the plant at time $t + 1$, given at time t, the states/outputs $x(t)$ and the action signals $A(t)$:

$$\hat{x}(t + 1) = f(x(t), A(t)). \tag{19.2}$$

In addition to signals at time t, delayed values of these signals can be used as well depending on the complexity of the plant dynamics [17]. For the purposes of this study the Model neural network predicts the *changes* in the outputs $\Delta x(t + 1)$, at time $t + 1$ [17]. In Figure 19.1, the inputs to the Model network are time-delayed values (TDL) of both the plant and the action network outputs.

A neural network based technique to develop a Model network using supervised learning is shown in Figure 19.1 and more details can be found in [15]. The conventional static backpropagation algorithm is used in training the neural network. This Model neural network can undergo *offline* or *online training*, as required by the application.

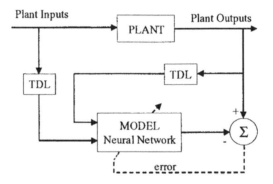

Fig. 19.2   Development of a neural network model of a plant.

The ACD controllers designed in this study use the model dependent designs. It is important for the description of the following sections to mention that the inputs to the Model neural network are the three time delayed values of the Plant and the action network outputs $(\Delta x(t - 1), \Delta x(t - 2), \Delta x(t - 3), A(t - 1), A(t - 2), A(t - 3))$ (Figure 19.1) in order to predict $\Delta \hat{x}(t)$. In other words, a third order neural network model is used. Explicit details on the development of a Model neural network are given in [17, 25, 37].

19.2.2.2   Critic Neural Network

Heuristic Dynamic Programming has a Critic neural network that estimates the function J (cost-to-go) in the Bellman equation of dynamic programming, expressed as follows:

$$J(x(t)) = \sum_{k=0}^{\infty} \gamma^k U(x(t+k)), \tag{19.3}$$

where γ is a discount factor for finite horizon problems ($0 < \gamma < 1$), $U(\cdot)$ is the utility function or the local cost and $x(t)$ is an input vector to the Critic. The Critic neural network is trained forward in time (multi-time steps ahead), which is of great importance for real-time operation.

Figure 19.3 shows the HDP Critic adaptation/training. The inputs to the Critic are outputs from the Model neural network and its time-delayed values (Figure 19.1). Two Critic neural networks are shown in Figure 19.3 having the same inputs and outputs but at different time instants. The first Critic neural network has inputs from time steps t, $t-1$ and $t-2$, and the second Critic neural network has inputs from time steps $t+1$, t and $t-1$. Their corresponding outputs are $J(t)$ and $\hat{J}(t+1)$ respectively. The second Critic neural network estimates the function $\hat{J}$ (cost-to-go) at time $t+1$ by using the Model neural network to get inputs one step ahead. As a result it is possible to know the Critic neural network output $\hat{J}(t+1)$ at time t.

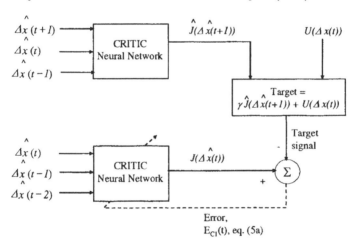

Fig. 19.3   HDP Critic neural network adaptation/training.

The Critic network tries to minimize the following error measure over time

$$\|E_1\| = \frac{1}{2} \sum_t E_{C1}^2(t), \tag{19.5}$$

$$E_{C1}(t) = J(\Delta \hat{x}(t)) - \gamma \hat{J}(\Delta \hat{x}(t+1)) - U(\Delta x(t)), \tag{19.5a}$$

where $\Delta x(t)$ is the changes in $x(t)$, a vector of observables of the plant (or the states, if available). The utility function U is dependent on the system controlled and a typical function is given in [25]. It should be noted that only for the purposes of this study, changes in the state variables are used rather than state variables. The necessary condition for (19.3) to be minimal is given by:

$$\frac{1}{2}\frac{\partial}{\partial W_{C1}}\left(E_{C1}^2(t)\right) = \left(E_{C1}\frac{\partial E_{C1}(t)}{\partial W_{C1}}\right) = 0. \tag{19.5b}$$

The weights' update for the Critic network using the backpropagation algorithm is given as follows:

$$\Delta W_{C1} = -\eta E_{C1}(t)\frac{\partial E_{C1}(t)}{\partial W_{C1}}, \tag{19.6a}$$

$$\Delta W_{C1} = -\eta\{J(\Delta\hat{x}(t)) - \gamma\hat{J}(\Delta\hat{x}(t+1)) - U(\Delta x(t))\} \tag{19.6b}$$
$$\times\frac{\partial\{J(\Delta\hat{x}(t)) - \gamma\hat{J}(\Delta\hat{x}(t+1)) - U(\Delta x(t))\}}{\partial W_{C1}},$$

where η is a positive learning rate and W_{C1} are the weights of the Critic neural network. The same Critic network is shown in two consecutive moments in time in Figure 19.3. The critic network's output $\hat{J}[\Delta\hat{x}(t+1)]$ is necessary in order to provide the training signal $\gamma\hat{J}[\Delta\hat{x}(t+1)] + U(\Delta x(t))$, which is the desired/target value for $\hat{J}[\Delta\hat{x}(t)]$.

19.2.2.3   Action Neural Network

The objective of the Action neural network in Figure 19.1, is to minimize $J(\Delta x(t))$ in the immediate future, thereby optimizing the overall cost expressed as a sum of all $U(\Delta x(t))$ over the horizon of the problem. This is achieved by training the Action neural network with an error signal $\partial J/\partial A$. The gradient of the cost function J, with respect to the outputs A, of the Action neural network, is obtained by backpropagating $\partial J/\partial J$ (i.e., the constant 1) through the Critic neural network and then through the pretrained Model neural network to the Action neural network. This gives $\partial J/\partial A$ and $\partial J/\partial W_A$ for all the outputs of the Action neural network, and all the Action neural network's weights W_A, respectively. The weights' update in the Action neural network using backpropagation algorithm is given as follows:

$$\|E_2\| = \frac{1}{2}\sum_t E_{A1}^2(t), \tag{19.7a}$$

where

$$E_{A1} = \frac{\partial J(t)}{\partial A(t)}, \tag{19.7b}$$

and

$$\frac{\partial J(t)}{\partial A(t)} = \frac{\partial J(t)}{\partial\Delta\hat{x}(t)}\frac{\partial\Delta\hat{x}(t)}{\partial A(t)}. \tag{19.7c}$$

the weight change in the Action network ΔW_{A1} can be written as:

$$\Delta W_{A1} \propto \frac{\partial E_2}{\partial W_A}. \tag{19.8a}$$

Equation (19.8a) can be further written as:

$$\Delta W_{A1} = -\alpha E_{A1}(t)\frac{\partial E_{A1}(t)}{\partial W_{A1}}, \tag{19.8b}$$

$$\Delta W_{A1} = -\alpha \frac{\partial J(t)}{\partial A(t)} \frac{\partial}{\partial W_{A1}} \left(\frac{\partial J(t)}{\partial A(t)}\right), \tag{19.8c}$$

where α is a positive learning rate.

With (19.6b) and (19.8c), the training of the Critic and the Action networks can be carried out. The general training procedure for the Critic and the Action networks are described in Section 19.3.

19.2.3 Dual Heuristic Programming Neurocontroller

The Critic neural network in the DHP scheme shown in Figure 19.4, estimates the derivatives of J with respect to the vector $\Delta\hat{x}$ (outputs of the Model neural network) and learns minimization of the following error measure over time:

$$\|E_3\| = \sum E_{C2}^T(t)E_{C2}(t)s, \tag{19.9}$$

where

$$E_{C2}(t) = \frac{\partial J[\Delta\hat{x}(t)]}{\partial \Delta\hat{x}} - \gamma\frac{\partial \hat{J}[\Delta\hat{x}(t+1)]}{\partial \Delta x(t)} - \frac{\partial U[\Delta x(t)]}{\partial \Delta x(t)}, \tag{19.10}$$

where $\partial(\cdot)/\partial\Delta x(t)$ is a vector containing partial derivatives of the scalar $(\cdot)$ with respect to the components of the vector Δx. The Critic neural network's training is more complicated than in HDP, since there is a need to take into account all relevant pathways of backpropagation as shown in Figure 19.4, where the paths of derivatives and adaptation of the Critic are depicted by dashed lines. In Figure 19.4, the dashed lines mean the first backpropagation and the dotted-dashed lines mean the second backpropagation.

The Model neural network in the design of DHP Critic and Action neural networks are obtained in a similar manner to that described in Section 19.2.2.1.

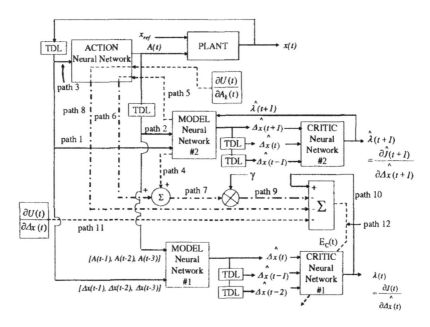

Fig. 19.4   DHP Critic neural network adaptation.

In the DHP scheme, application of the chain rule for derivatives yields:

$$
\frac{\partial \hat{J}[\Delta \hat{x}(t+1)]}{\partial \Delta x_j(t)} = \sum_{i=1}^{n} \hat{\lambda}_i(t+1) \frac{\partial \Delta \hat{x}_i(t+1)}{\partial \Delta x_j(t)}
$$

$$
+ \sum_{k=1}^{m} \sum_{i=1}^{n} \hat{\lambda}_i(t+1) \frac{\partial \Delta \hat{x}_i(t+1)}{\partial A_k(t)} \frac{\partial A_k(t)}{\partial \Delta x_j(t)}, \quad (19.11)
$$

where $\hat{\lambda}_i(t+1) = \partial \hat{J}[\Delta \hat{x}(t+1)]/\partial \Delta \hat{x}_i(t+1)$, and n, m, j are the numbers of outputs of the Model, Action and Critic neural networks respectively. By exploiting (19.11), each of n components of the vector $E_{C2}(t)$ from (19.10) is determined by

$$
E_{C2j}(t) = \frac{\partial J[\Delta \hat{x}(t)]}{\partial \Delta \hat{x}_j(t)} - \gamma \frac{\partial \hat{J}[\Delta \hat{x}(t+1)]}{\partial \Delta x_j(t)}
$$

$$
- \frac{\partial U[\Delta x(t)]}{\partial \Delta x_j(t)} - \sum_{k=1}^{m} \frac{\partial U(t)}{\partial A_k(t)} \frac{\partial A_k(t)}{\partial \Delta x_j(t)}. \quad (19.12)
$$

The signals in Figure 19.4 which are labeled with a path number, represent the following:

(i). Path 1 represents the outputs of the plant fed into the Model neural network #2. These outputs are $\Delta x(t)$, $\Delta x(t-1)$ and $\Delta x(t-2)$.

(ii). Path 2 represents the outputs of the Action neural network fed into the Model neural network #2. These outputs are $A(t)$, $A(t-1)$ and $A(t-2)$.

(iii). Path 3 represents the outputs of the plant fed into the Action neural network. These outputs are $\Delta x(t)$, $\Delta x(t-1)$ and $\Delta x(t-2)$.

(iv). Path 4 represents a backpropagated signal of the output of the Critic neural network #2 through the Model neural network with respect to path 1 inputs. The backpropagated signal on path 4 is

$$\sum_{i=1}^{n} \hat{\lambda}_i(t+1) \frac{\partial \Delta \hat{x}_i(t+1)}{\partial \Delta x_j(t)} \text{ in (19.11)}.$$

(v). Path 5 represents a backpropagated signal of the output of the Critic neural network #2 through the Model neural network with respect to path 2 inputs. The backpropagated signal on path 3 is

$$\sum_{i=1}^{n} \hat{\lambda}_i(t+1) \frac{\partial \Delta \hat{x}_i(t+1)}{\partial \Delta A_k(t)} \text{ in (19.11)}.$$

(vi). Path 6 represents a backpropagation output of path 5 signal ((v) above) with respect to path 3. The signal on path 6 is

$$\sum_{k=1}^{m}\sum_{i=1}^{n} \hat{\lambda}_i(t+1) \frac{\partial \Delta \hat{x}_i(t+1)}{\partial A_k(t)} \text{ in (19.11)}.$$

(vii). Path 7 is the sum of the path 4 and path 6 signals resulting in

$$\partial \hat{J}[\Delta \hat{x}(t+1)]/\partial \Delta x_j(t), \text{ given in (19.11)}.$$

(viii). Path 8 is the backpropagated signal of the term $\partial U(t)/\partial A_k(t)$ (Figure 19.5) with respect to path 3 and is

$$\sum_{k=1}^{m} \frac{\partial U(t)}{\partial A_k(t)} \frac{\partial A_k(t)}{\partial \Delta x_j(t)} \text{ in (19.12)}.$$

(ix). Path 9 is a product of the discount factor γ and the path 7 signal, resulting in term $\gamma \partial \hat{J}[\Delta \hat{x}(t+1)]/\partial \Delta x_j(t)$ in (19.12).

(x). Path 10 represents the output of the Critic neural network #1,
$$\partial J[\Delta \hat{x}(t)]/\partial \Delta \hat{x}(t).$$

(xi). Path 11 represents the term $\partial U(t)/\partial \Delta x(t)$ (Figure 19.5).

(xii). Path 12 represents $E_{C2j}(t)$ given in (19.12) and as follows:
$$Path12 = E_{C2j}(t) = Path10 - Path9 - Path11 - Path8.$$

The partial derivatives of the utility function $U(t)$ with respect to $A_k(t)$, and $x(t)$, $\partial U(t)/\partial A_k(t)$ and $\partial U(t)/\partial \Delta x(t)$ respectively, are obtained by backpropagating the utility function, $U(t)$ through the Model network as shown in Figure 19.5.

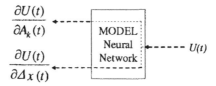

Fig. 19.5   Backpropagation of $U(t)$ through the Model neural network.

The adaptation of the Action network in Figure 19.4, is illustrated in Figure 19.6 which propagates $\lambda(t + 1)$ back through the Model network to the Action network. The goal of such adaptation can be expressed as follows [35, 36]:

$$\frac{\partial U[\Delta x(t)]}{\partial A(t)} + \gamma \frac{\partial \hat{J}[\Delta \hat{x}(t + 1)]}{\partial A(t)} = 0 \qquad \forall t.$$

The error signal for the Action network adaptation is therefore given as follows:

$$E_{A2}(t) = \frac{\partial U[\Delta x(t)]}{\partial A(t)} + \gamma \frac{\partial \hat{J}[\Delta \hat{x}(t + 1)]}{\partial A(t)}. \tag{19.13}$$

The weights' update expression [36], when applying backpropagation, is as follows:

$$\Delta W_{A2} = -\alpha \left[\frac{\partial U[\Delta x(t)]}{\partial A(t)} + \gamma \frac{\partial \hat{J}[\Delta \hat{x}(t + 1)]}{\partial A(t)} \right]^T \frac{\partial A(t)}{\partial W_{A2}}, \tag{19.14}$$

where α is a positive learning rate and W_{A2} are weights of the DHP Action neural network.

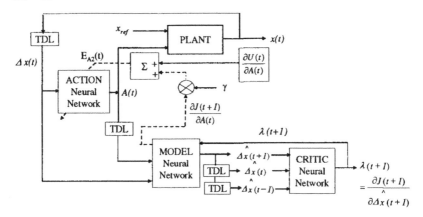

Fig. 19.6   DHP Action neural network adaptation.

19.3   GENERAL TRAINING PROCEDURE FOR CRITIC AND ACTION NETWORKS

The training procedure is that suggested in [36] and it is applicable to any ACD. It consists of two separate training cycles: one for the Critic, and the other for the Action. An important measure is that the Action neural network is pretrained with conventional controllers (Proportional Integral Derivative, PID) controlling the plant in a linear region. The Critic's adaptation is done initially with the pretrained Action network, to ensure that the whole system, consisting of the ACD and the plant remains stable. Then the Action network is trained further while keeping the Critic neural network weights fixed. This process of training the Critic and the Action one after the other, is repeated until an acceptable performance is reached. It is assumed that there is no concurrent adaptation of the pretrained Model neural network, and W_C is initialized to small random values.

In the Critic's training cycle, an incremental optimization of (19.5) and/or (19.9) is carried out using a suitable optimization technique (e.g., backpropagation). The following operations are repeated N_C times:

1. Initialize $t = 0$ and $\Delta x(0)$

2. Compute the output of the Critic neural network at time t,
   $J(t)$ or $\lambda(t) = f_C(\Delta x(t), W_C)$

3. Compute the output of the Action neural network at time t,
   $A(t) = f_A(\Delta x(t), W_A)$

4. Compute the output of the Model neural network at time $t + 1$,
   $\Delta x(t + 1) = f_M(\Delta x(t), A(t), W_M)$

5. Compute the output of the Critic neural network at time $t + 1$,
   $\hat{J}(t + 1)$ or $\hat{\lambda}(t + 1) = f_C(\Delta x(t + 1), W_C)$

6. Compute the Critic neural network error at time t,
   $E_{C1}(t)$ from (19.5a) or $E_{C2}(t)$ from (19.10).

7. Update the Critic neural network's weights using the backpropagation algorithm.

8. Repeat steps 2 to 7.

The functions $f_C(\Delta x(t), W_C)$, $f_A(\Delta x(t), W_A)$ and $f_M(\Delta x(t), A(t), W_M)$ represent the Critic, the Action and the Model neural networks with their weights W_i, respectively.

In the Action neural network's training cycle, an incremental learning is also carried out using the backpropagation algorithm, as in the Critic neural network's training cycle above, and the list of operations for the Action neural network's training cycle is almost the same as that for the Critic neural network's cycle above (steps 1 to 7). However, (19.7b) or (19.13) are used for updating the Action neural network's weights instead of using (19.5a) or (19.10). The Action's training cycle is repeated N_A times while keeping the Critic's weights W_C fixed. N_C and N_A are the lengths of the corresponding training cycles. It is important that the whole system consisting of the ACD and the plant remains stable while both of the Critic and Action networks undergo adaptation.

19.4   POWER SYSTEM

The micro-machine laboratory at the University of Natal in Durban, South Africa, has two 3 kW, 220 V, three-phase micro-alternators, and each one represents both the electrical and mechanical aspects of a typical 1000 MW alternator. The laboratory power system is simulated in the MATLAB/SIMULINK environment and simulation studies with neurocontrollers are carried out prior to hardware implementations. The laboratory single machine infinite bus power system in Figure 19.7 consists of a micro-alternator, driven by a dc motor whose torque — speed characteristics are controlled by a power electronic converter to act as a micro-turbine, and a single short transmission line which links the micro-alternator to a voltage source which has a constant voltage and frequency, called an infinite bus. The parameters of the micro-alternators, determined by the IEEE standards are given in Tables 19.1 and 19.2 [30]. A time constant regulator is used to insert negative resistance in series with the field winding circuit [30], in order to reduce the actual field winding resistance to the correct per-unit value.

Table 19.1   Micro-Alternator #1 Parameters

$T'_{d0} = 4.50$ s	$X'_d = 0.205$ pu	$R_s = 0.006$ pu
$T''_{d0} = 33$ ms	$X''_d = 0.164$ pu	$H = 5.68$ s
$T''_{q0} = 0.25$ s	$X_q = 1.98$ pu	$F = 0$
$X_d = 2.09$ pu	$X''_q = 0.213$ pu	$p = 2$ pole pairs

Table 19.2   Micro-Alternator #2 Parameters

$T'_{d0} = 3.72$ s	$X'_d = 0.205$ pu	$R_s = 0.006$ pu
$T''_{d0} = 33$ ms	$X''_d = 0.164$ pu	$H = 5.68$ s
$T''_{q0} = 0.25$ s	$X_q = 1.98$ pu	$F = 0$
$X_d = 2.09$ pu	$X''_q = 0.213$ pu	$p = 2$ pole pairs

The practical system uses a conventional AVR and exciter combination of which the transfer function block diagram is shown in Figure 19.8, and the time constants

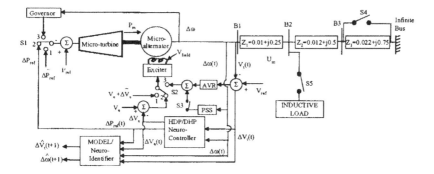

Fig. 19.7 The single machine infinite bus configuration with the conventional AVR and governor controllers, and neurocontroller.

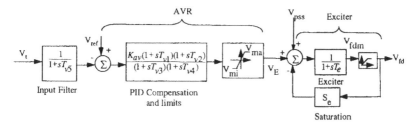

Fig. 19.8 Block diagram of the AVR and exciter combination.

Table 19.3   AVR and Exciter Time Constants

T_{v1}	0.616 s	T_{v5}	0.0235 s
T_{v2}	2.266 s	T_e	0.47 s
T_{v3}	0.189 s	K_{av}	0.003
T_{v4}	0.039 s		

and gain are given in Table 19.3 [30]. The exciter saturation factor S_e is given by

$$S_e = 0.6093 \ \exp(0.2165 V_{fd}). \tag{19.15}$$

$T_{v1}, T_{v2}, T_{v3}, T_{v4}$ are the time constants of the PID voltage regulator compensator; T_{v5} is the input filter time constant; T_e is the exciter time constant; K_{av} is the AVR gain; V_{fdm} is the exciter ceiling voltage; and, V_{ma} and V_{mi} are the AVR maximum and minimum ceiling voltages.

The block diagram of the power system stabilizer (PSS) used to achieve damping of the system oscillations is shown in Figure 19.9 [7]. The considerations and

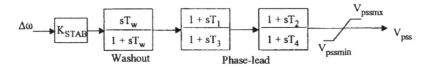

Fig. 19.9   Block diagram of the power system stabilizer.

Table 19.4   PSS Time Constants and Gain

T_w	3 s	T_3	0.045 s
T_1	0.2 s	T_4	0.045 s
T_2	0.2 s	K_{STAB}	33.93

procedures used in the selection of the PSS parameters are similar to that found in [7] and these parameters are given in Table 19.4.

A separately excited 5.6 kW thyristor controlled dc motor is used as a prime mover, called the micro-turbine, to drive the micro-alternator. The torque-speed characteristic of the dc motor is controlled to follow a family of rectangular hyperbola to emulate the different positions of a steam valve, as would occur in a real typical high pressure (HP) cylinder turbine. The three low pressure (LP) cylinders' inertia are represented by appropriately scaled flywheels attached to the micro-turbine shaft. The micro-turbine and governor combination transfer function block diagram is shown in Figure 19.10, where, P_{ref} is the turbine input power set point value, P_m is the turbine output power, and $\Delta\omega$ is the speed deviation from the synchronous speed. The turbine and governor time constants and gain are given in Table 19.5 [30].

The gain K_{av} (0.003) of the AVR in Table 19.3 is obtained by suitable choice of the gain and phase margin as described in [31]. The gain K_g (0.05) of the governor in Tables 19.5 is chosen to give a drop of 5%. Transmission lines are represented by using banks of lumped inductors and capacitors.

A three-machine power system shown in Figure 19.11 is set up by using the two micro-alternators and the infinite bus as the third machine.

19.5   SIMULATION AND HARDWARE IMPLEMENTATION RESULTS

19.5.1   Simulation Studies

This section describes the simulation studies carried out on the use of feedforward neural networks to implement Adaptive Critic Designs (ACDs) based nonlinear adaptive controllers for turbogenerator(s)/micro-alternator(s) in a single-machine-infinite-bus (SMIB) power system first and then in a multimachine power system.

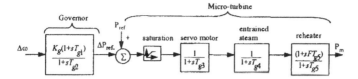

Fig. 19.10   Block diagram of the micro-turbine and governor combination.

Table 19.5   Governor and Micro-Turbine Constants

Phase advance compensation, T_{g1}	0.264 s
Phase advance compensation, T_{g2}	0.0264 s
Servo time constant, T_{g3}	0.15 s
Entrained steam delay, T_{g4}	0.594 s
Steam reheat time constant, T_{g5}	2.662 s
pu shaft output ahead of reheater, F	0.322
Gain K_g	0.05

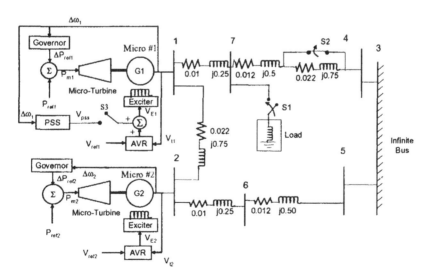

Fig. 19.11   Multimachine power system consisting of two micro-alternators G1 and G2 which are conventionally controlled by the AVRs, governors and a PSS.

19.5.1.1   Single Machine Infinite Bus Power System

The single-machine-infinite-bus power system is shown in Figure 19.7. Although a micro-alternator and a micro-turbine are studied here, they nevertheless represent on a per unit (pu) basis the correct parameters of large multi-megawatt systems as explained in Section 19.4. The conventional controllers, the Automatic Voltage Regulator (AVR) and governor are described in Section 19.4 with their parameters given in Tables 19.3 and 19.4. For the SMIB power system, the objectives of the controllers are to damp out local oscillations and keep the terminal voltage constant during transients. The presence of inter-area oscillations in the SMIB power system is minimal and as such a Power System Stabilizer (PSS) is not necessary and is therefore not included. However, the PSS is subsequently added for the multimachine power system in Section 19.5.2, where the inter-area oscillations do exist.

The dynamic and transient operation of the HDP and DHP neurocontrollers are compared with the operation of the conventional controller (AVR and turbine governor, excluding the PSS) under two different conditions: ±5% step changes in the terminal voltage setpoint and a temporary three-phase short circuit on the infinite bus. At this point, the training of the HDP and the DHP neurocontrollers have been completed and terminated before the following evaluation tests are carried out. The performance of the HDP/DHP neurocontroller in Figure 19.7 (switches S1 and S2 in position '2') is compared with that of the conventional AVR and governor controllers (switches S1 and S2 in position '3'), by evaluating how quickly they respond and damp out oscillations in the terminal voltage and rotor angle. Restoring terminal voltage and rotor angle to steady state after any changes is important for the stability of the power system. The results showed in this section have been partly published [24, 38].

Step changes in the terminal voltage reference V_{ref} or V_e (Figure 19.7)
Figures 19.12 and 19.13 show the terminal voltage and the rotor angle of the micro-alternator for ±5% step changes in the terminal voltage with the micro-alternator operating at 1 pu power and 0.85 lagging power factor, and line impedance $Z_1 = 0.02 + j0.4$ pu. The neurocontrollers clearly outperform the conventional controllers. The response of the continually online trained neurocontroller (COT) [18] is also plotted in solid dashed line for comparison.

From Figures 19.12 and 19.13, it can be seen that the COT controller has a faster rise time compared with the HDP and DHP neurocontrollers. The DHP neurocontroller has a better damping than the COT and conventional controller, and a faster rise time than the HDP. The damping factor and rise time are influenced by the choice of the coefficients in the local utility function [25] and the discount factor, and the inherent characteristics of the HDP and DHP schemes.

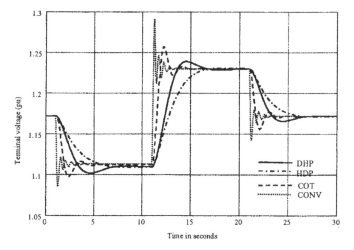

Fig. 19.12 Terminal voltage of the micro-alternator for $\pm 5\%$ step changes in the terminal voltage reference (transmission line impedance $Z_1 + Z_2$).

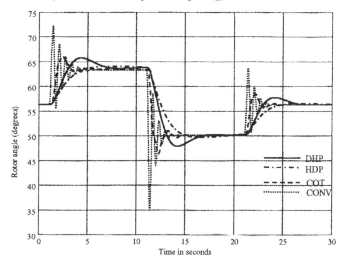

Fig. 19.13 Rotor angle of the micro-alternator for $\pm 5\%$ step changes in the terminal voltage reference (transmission line impedance $Z_1 + Z_2$).

Short circuit test at new transmission line impedance $Z_1 + Z_2 + Z_3$

In power systems faults such as three-phase short circuits occur from time to time, and because they prevent energy from the generator reaching the infinite bus, it means that most of the turbine shaft power goes into accelerating the generator during the fault. This represents a severe transient test for the controller performance. Figures 19.14

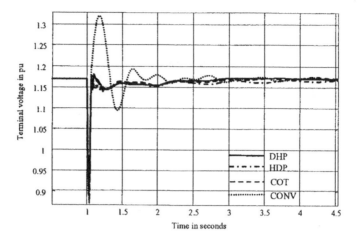

Fig. 19.14   Terminal voltage of the micro-alternator for a temporary 50 ms three-phase short circuit (transmission line impedance $Z_1 + Z_2 + Z_3$).

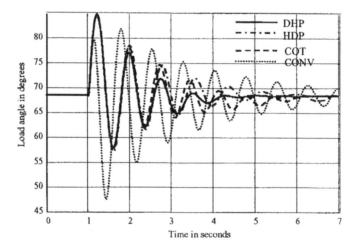

Fig. 19.15   Rotor angle of the micro-alternator for a temporary 50 ms three-phase short circuit (transmission line impedance $Z_1 + Z_2 + Z_3$).

and 19.15 show the response of all four controllers for the three-phase temporary short circuit for 50 ms with the new transmission line impedance $Z_1 + Z_2 + Z_3$. Here, it is obvious that the DHP controller clearly beats the other three controllers in terms of offering the greatest oscillation damping especially in the rotor angle. The DHP controller proves its robustness to changes in the system configurations.

19.5.1.2   Multimachine Power System

The multimachine power system is shown in Figure 19.11. The parameters of the micro-alternators and conventional controllers are given in section 19.4 of this chapter. Based on the results for the SMIB power system in section 19.5.1.1, the DHP controller has the best performance, hence, the DHP neurocontroller is the only one that is now implemented on the multimachine power system. The performance of the DHP neurocontroller is now compared with that of the conventional controllers, one of which is equipped with a power system stabilizer. Sections 19.5.1.2.1 and 19.5.1.2.2 describe the multimachine power system with one DHP neurocontroller and two DHP neurocontrollers respectively.

Multimachine power system with one DHP neurocontroller and one conventional controller

The multimachine power system with one DHP neurocontroller and one conventional AVR and governor is shown in Figure 19.16. The design procedure of the DHP neurocontroller is similar to that of the SMIB power system described in sections 19.2 and 19.3. The Model neural network training is carried out on a multimachine power system as described in [25, 26] and used with fixed weights in the design of the DHP neurocontroller in this section.

At different operating conditions and disturbances, the transient performance of the DHP neurocontroller is compared with that of the conventional controllers (AVR and governor), as well as with that of the AVR equipped with a PSS. For the operating condition ($P_1 = 0.20$ pu, $Q_1 = -0.02$ pu and $P_2 = 0.20$ pu, $Q_2 = -0.02$ pu), the conventional automatic voltage regulators, governors and the power system stabilizer are designed to give their best performance [7, 31].

Multimachine power system with two DHP neurocontrollers

The multimachine power system with DHP neurocontroller #1 and #2 controlling micro-alternator #1 and #2 respectively is shown in Figure 19.17.

Three different controller combination studies are carried out, namely

- Case a — conventional controller on G1 and DHP neurocontroller on G2.

- Case b — conventional controller with a PSS on G1 and DHP neurocontroller on G2.

- Case c — DHP neurocontrollers on both G1 and G2.

4% step change at the first operating condition

At the operating condition ($P_1 = 0.20$ pu, $Q_1 = -0.02$ pu and $P_2 = 0.20$ pu, $Q_2 = -0.02$ pu), a 4% step increase occurs at $t = 1$s in the desired terminal voltage (in V_{ref1} in Figure 19.11 and V_{e1} in Figures 19.16 and 19.17) of micro-alternator #1. Figures 19.18 and 19.19 show that the DHP neurocontrollers on the two micro-alternators (case c) ensure no overshoot on the terminal voltage unlike with cases a and b, and that the DHP neurocontroller also provides superior speed deviation damping. For

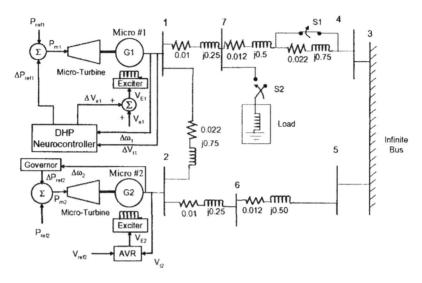

Fig. 19.16   Multimachine power system with one DHP neurocontroller, and one conventional AVR and governor.

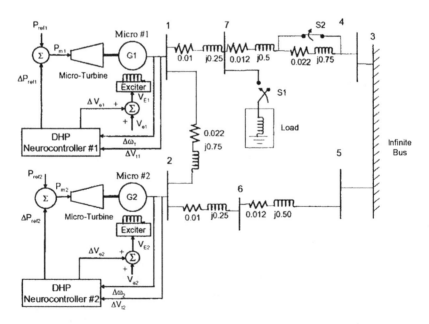

Fig. 19.17   Multimachine power system with two DHP neurocontrollers.

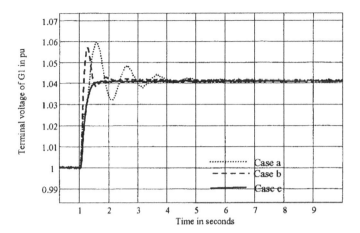

Fig. 19.18   Terminal voltage of micro-alternator #1 for a 4% step change in its terminal voltage reference V_{ref1} or V_{e1}.

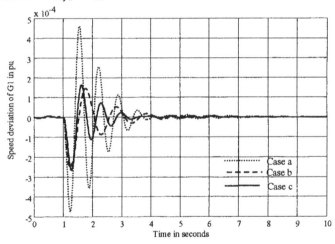

Fig. 19.19   Speed deviation of micro-alternator #1 for a 4% step change in its terminal voltage reference V_{ref1} or V_{e1}.

this same disturbance, Figure 19.20 shows the speed deviation of micro-alternator #2 and it is clear that with a DHP neurocontroller on micro-alternator #1 (case c), the speed deviation damping of micro-alternator #2 is also much improved.

Three-phase short circuit at the second operating condition

At another operating condition ($P_1 = 0.50$ pu, $Q_1 = -0.1493$ pu and $P_2 = 0.65$ pu, $Q_2 = -0.1341$ pu), a 100 ms short circuit occurs at t = 1s at bus 6 (Figures 19.11,

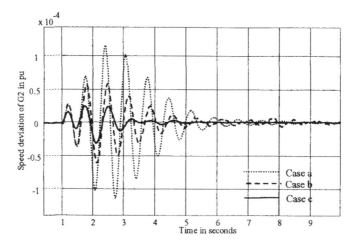

Fig. 19.20   Speed deviation of micro-alternator #2 for a 4% step change in its terminal voltage reference V_{ref1} or V_{e1} of micro-alternator #1.

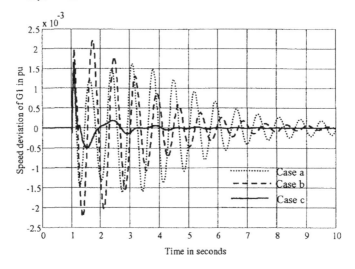

Fig. 19.21   Speed deviation of micro-alternator #1 for a 100ms three-phase temporary short circuit at bus 6.

19.16 and 19.17). Figure 19.21 shows the speed deviation of micro-alternator #1 and it is clear that with a DHP neurocontroller on micro-alternator #1, the speed deviation damping of the micro-alternators has improved.

In all the above tests, the DHP neurocontrollers (case *c*) have performed excellently compared with the conventional controllers (case *a*) and even with the inclusion of a power system stabilizer (case *b*) under different operating conditions. The results clearly prove the robustness of the DHP neurocontrollers which is as a result of the powerful adaptive critic design algorithms. Many more tests were done to confirm this. The DHP neurocontrollers' performances can be improved further by using different discount factors and local utility functions.

19.5.2   Hardware Implementation

This section describes the laboratory practical implementation and testing of the DHP type ACD controllers. The purpose of these tests is to confirm via practical measurements the potential of adaptive critic based neurocontrollers which have been demonstrated during the simulation studies above, both for a single machine and a multimachine power system. However, the laboratory implementation on micro-machines is also intended to form a basis for possible future investigations into the use of such neurocontrollers on large multi-megawatt sized power plants in a real-world power station.

19.5.2.1   Single Machine Infinite Bus Power System
A temporary 125 ms three-phase short circuit at the first operating condition (P = 0.2 pu, Q = 0 pu)
At the first operating condition, a temporary 125 ms duration three-phase short circuit at bus B2 (Figure 19.7) is carried out at $t = 10$ s. Figures 19.22 and 19.23 show the terminal voltage and the load angle response for this test. DHP, PSS+CONV and CONV indicate for the response with the neurocontroller, the power system stabilizer plus the conventional controller, and the conventional controller, respectively.

The DHP neurocontroller again provides superior damping compared with the conventional controllers. In the terminal voltage response the DHP neurocontroller responds fast and as a result the dip in the terminal voltage is lower than that experienced when the conventional controllers control the micro-alternator. This is of great benefit in not only restoring the micro-alternator stability, but also minimizing poor power quality.

A temporary 125 ms three-phase short circuit at the second operating condition (P = 0.3 pu, Q = 0 pu)
At the second operating condition, a temporary 125 ms duration three-phase short circuit at bus B2 (Figure 19.7) is carried out at $t = 10$ s. Figures 19.24 and 19.25 show the terminal voltage and the load angle response for this test. The DHP neurocontroller again performs superior to the conventional controllers at this operating point. The damping of the terminal voltage and the load angle is again better with the DHP neurocontroller. For the terminal voltage response in particular, the performances of the PSS and the conventional controllers have degraded compared with the terminal voltage response in Figure 19.22.

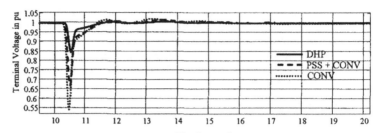

Fig. 19.22 Terminal voltage response for a temporary 125 ms three-phase short circuit at bus B2 (Figure 19.7) for $P = 0.2$ pu and $Q = 0$ pu.

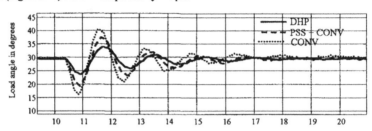

Fig. 19.23 Load angle response for a temporary 125ms three-phase short circuit at bus B2 (Figure 19.7) for $P = 0.2$ pu and $Q = 0$ pu.

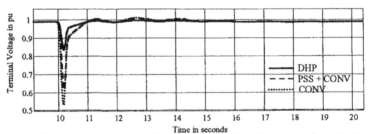

Fig. 19.24 Terminal voltage response for a temporary 125 ms three-phase short circuit at bus B2 (Figure 19.7) for $P = 0.3$ pu and $Q = 0$ pu.

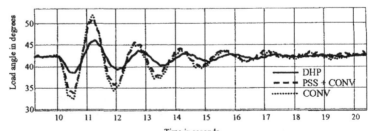

Fig. 19.25 Load angle response for a temporary 125ms three-phase short circuit at bus B2 (Figure 19.7) for $P = 0.3$ pu and $Q = 0$ pu.

The following section describes the implementation and the performance evaluation of the DHP neurocontrollers on a multimachine power system.

19.5.2.2   Multimachine Power System

As in the simulation studies for the multimachine power system, first a DHP neurocontroller is developed for G1 and then a DHP neurocontroller is developed for G2. The DHP neurocontrollers are tested for dynamic and transient operation for the following two disturbances:

(a)   An inductive load addition along the transmission line by closing switch S1.

(b)   An increase in the transmission line impedance by opening switch S2.

Four different controller combination studies are carried out for the above disturbances and at different operating conditions for G1 and G2:

- Case a — conventional controller on both G1 and G2.

- Case b — conventional controller with a PSS on G1 and conventional controller on G2.

- Case c — DHP neurocontroller on G1 and conventional controller on G2.

- Case d — DHP neurocontrollers on both G1 and G2.

An inductive load addition

At the operating condition ($P = 0.2\ pu$, $Q = 0\ pu$), an inductive load, $P = 0.8$ pu at power factor (pf) of 0.85 lagging, is added to the transmission line at bus 7 by closing switch S1 at time $t = 10$ s. Figure 19.26 shows load angle response of micro-alternator #2 for the four different controller combinations (cases a to d). The DHP neurocontrollers (cases c and d) ensure minimal overshoot on the load angle compared with the conventional controllers. This is to be expected since the AVR and the governor parameters have been tuned for only small disturbances at this operating point. The terminal voltage response of micro-alternator #2 is not shown, because relatively little disturbance and improvement are experienced since the fault is closer to micro-alternator #1. For the same disturbance, the load angle response of micro-alternator #1 is shown in Figure 19.27 The PSS (case b) on micro-alternator #1 improves the performance of the conventional controllers but still the DHP neurocontroller #1 (case c) performs better. The influence of the DHP neurocontroller #2 (case d) on micro-alternator #2 does not improve the load angle response on micro-alternator #1, this is because the DHP neurocontroller #1 does not undergo further training after the DHP neurocontroller #2 has been installed. However, with further training of the DHP neurocontrollers, it might be possible to improve even more on the performances. It is clear that the two DHP neurocontrollers (case d) give the best performance of the four different controller combinations (case a to d).

Addition of a series transmission line

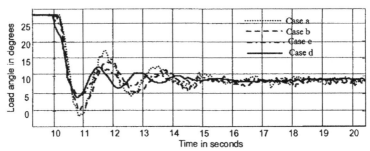

Fig. 19.26   Load angle response of micro-alternator #2 for an inductive load addition at bus 7 for P = 0.2 pu and Q = 0 pu.

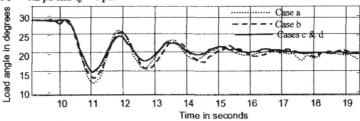

Fig. 19.27   Load angle response of micro-alternator #1 for an inductive load addition at bus 7 for P = 0.2 pu and Q = 0 pu.

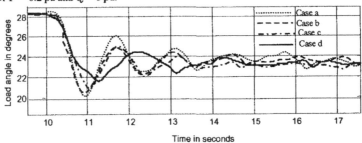

Fig. 19.28   Load angle response of micro-alternator #2 for series transmission line impedance increase by opening switch S2 for P = 0.2 pu and Q = 0 pu.

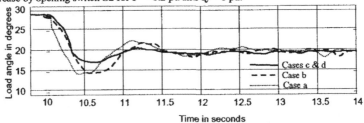

Fig. 19.29   Load angle response of micro-alternator #1 for series transmission line impedance increase by opening switch S2 for P = 0.2 pu and Q = 0 pu.

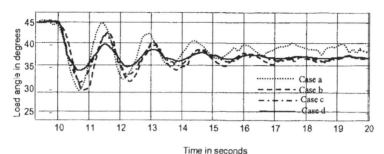

Fig. 19.30   Load angle response of micro-alternator #2 for an inductive load addition at bus 7 for $P = 0.3$ pu and $Q = 0$ pu.

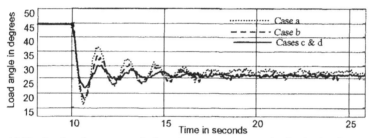

Fig. 19.31   Load angle response of micro-alternator #1 for an inductive load addition at bus 7 for $P = 0.3$ pu and $Q = 0$ pu.

At the operating condition ($P = 0.2pu, Q = 0pu$), the series transmission line impedance is increased at time $t = 10$ s from $Z = 0.022 + j0.75$ pu to $Z = 0.044 + j1.50$ pu by opening switch S2. Figure 19.28 shows the load angle response of micro-alternator #2 for this test with the four different controller combinations. Clearly the DHP neurocontrollers (case d) again exhibit superior damping and ensure lesser overshoots compared with the performance of the conventional controllers even when equipped with a PSS.

The load angle response of micro-alternator #1 for the same disturbance is shown in Figure 19.29. It is clear the DHP neurocontrollers exhibit the best damping of the controllers. Again, as in the case of the load addition disturbance result of Figure 19.27, the addition of a DHP neurocontroller on micro-alternator #2 does not improve the performance on micro-alternator #1. However, the performance could possibly be improved if further training of the DHP neurocontrollers were to be carried out.

Inductive load addition at a different operating condition

At the operating condition ($P = 0.3pu, Q = 0pu$), an inductive load, $P = 0.8$ pu at power factor (pf) of 0.85, is added to the transmission line at bus 7 by closing switch S1 at time $t = 10$ s. Figure 19.30 shows the load angle response of micro-alternator #2 for the four different controller combinations. The two DHP neurocontrollers (case d) ensure minimal overshoot and better damping of the load angle compared with the other controller combinations. This is to be expected since the conventional AVR and the governor parameters have been tuned for only small disturbances at the operating point ($P = 0.2pu, Q = 0pu$). For the same disturbance, the load angle response of micro-alternator #1 is shown in Figure 19.31. The PSS (case b) on micro-alternator #1 improves the performance of the conventional controllers but still the DHP neurocontroller (case c) performs better. The influence of a DHP neurocontroller #2 on micro-alternator #2 does not improve the load angle response on micro-alternator #1, because the DHP neurocontroller #1 does not undergo further training after DHP neurocontroller has been installed. However, with further training of the DHP neurocontrollers, it might be possible to improve more on the performances. It is clear that the two DHP neurocontrollers give the best performance of the four different controller combinations.

Additional tests, comparisons, and details on training and implementations are given in [24–26].

19.6   SUMMARY

This chapter has presented the investigations on the design and implementation of Adaptive Critic based neurocontrollers to replace the conventional automatic voltage regulators (AVRs) and governors on generators, in both a single-machine-infinite-bus and a multimachine power system. It has presented the concept for Adaptive Critic Designs (ACDs) for the design of optimal controllers for practical power systems and validated the concept on a laboratory prototype system.

For turbogenerators in a single-machine-infinite-bus power system, the results have shown that it is possible to design and apply ACD based neurocontrollers based on Heuristic Dynamic Programming and Dual Heuristic Programming neurocontrollers [24] that operate successfully. In addition the performances of ACD based neurocontrollers have been compared with the indirect adaptive neural controller based on continual online training, and also with the conventional automatic voltage regulator and governor [18]. From these comparisons it can be concluded that the DHP neurocontroller exhibits the best performance and robustness under many different operating conditions and system configurations.

The *neurocontrol* of turbogenerators in a SMIB power system has been extended to DHP based neurocontrol of turbogenerators in a multi-machine power system. The performance and robustness of these DHP neurocontrollers have been compared against the conventional AVRs and governors [25, 39] and in addition, with an AVR equipped with a conventional power system stabilizer [40]. Again, from these comparisons it is concluded that the DHP neurocontroller exhibits the best performance and robustness under different operating conditions and system configurations in the multi-machine power system.

The *Adaptive Critic Design* based neurocontrollers have the great advantage that once trained, their weights/parameters remain fixed and therefore avoid the risk of instability associated with continual online training. The convergence guarantee of the Critic and Action neural networks during offline training was shown in [27, 41]. In addition, the heavy computational load of online training only arises during the offline training phase and therefore makes the online real time implementation cost of the neurocontrollers less expensive. The processing hardware cost is a small fraction of the cost of turbogenerators and therefore this is not a big issue.

The *Adaptive Critic Design* based nonlinear neurocontrollers designed in this chapter are all based on approximate models obtained by neuroidentifiers, but nevertheless exhibit superior performance in comparison to the conventional linear controllers which use more extensive linearized models. This benefit of a neuroidentifier agrees with the conclusions on the comparison of using approximate and exact models in adaptive critic designs which was explicitly shown in [28].

All these features are desirable and important for industrial applications which require a neurocontroller technology that is nonlinear, robust and stable.

Acknowledgments

The authors acknowledge the University of Natal, Durban, South Africa, for allowing the usage of the micro-machines laboratory. The financial support from the National Science Foundation, USA and National Research Foundation, South Africa, are also gratefully acknowledged.

Bibliography

1. B. Adkins and R. G. Harley, *The general theory of alternating current machines,* Chapman and Hall, London, 1975.

2. Q. H. Wu and B. W. Hogg, Laboratory evaluation of adaptive controllers for synchronous generators, *Automatica,* vol. 27, no. 5, pp. 845–852, 1991.

3. Q. H. Wu and B. W. Hogg, Adaptive controller for a turbogenerator system, *IEE Proc.,* vol. 135, pt D, no. 1, pp. 35–42, 1988.

4. O. P. Malik, C. X. Mao, K. Prakash, G. Hope, and G. Hancock, Tests with a microcomputer based adaptive synchronous machine stabilizer on a 400 MW thermal unit, *IEEE Trans. Electronic Computation,* vol. 8, no. 1, pp. 6–12, 1992.

5. D. Flynn, B. W. Hogg, E. Swidenbank, and K. J. Zachariah, Expert control of a self-tuning automatic voltage regulator, *Control Engineering Practice,* vol. 3, no. 11, pp. 1571–1579, 1995.

6. D. J. G. Morrell and B. W. Hogg, Identification and validation of turbogenerators models, *Automatica,* vol. 26, no. 1, pp. 135–156, 1990.

7. P. Kundur, M. Klein, G. J. Rogers, and M. S. Zywno, Application of power system stabilizers for enhancement of overall system stability, *IEEE Trans. Power Systems,* vol. 4, no. 2, pp. 614–626, May, 1989.

8. F. P. deMello and C. Concordia, Concepts of synchronous machine stability as affected by excitation control, *IEEE Trans. Power Apparatus and Systems,* vol. PAS-87, pp. 1426–1434, 1968.

9. M. R. Khaldi, A. K. Sarkar, K. Y. Lee, and Y. M. Park, The modal performance measure of parameter optimization of power system stabilizers, *IEEE Trans. Energy Conversion,* vol. 8, no. 4, pp. 660–666, 1993.

10. P. M. Anderson and A. A. Fouad, *Power System Control and Stability,* IEEE Press, New York, 1994.

11. Y. Kitauchi and H. Taniguchi, Experimental verification of fuzzy excitation control system for multi-machine power system, *IEEE Trans. Energy Conversion,* vol. 12, no. 1, pp. 94–99, March 1997.

12. J. W. Chapman, M. D. Ilic, C. A. King, L. Eng, and H. Kaufman, Stabilizing a multimachine power system via decentralized feedback linearizing excitation control, *IEEE Trans. Power System*, vol. 8, no. 3, pp. 830–839, 1993.

13. Y. Wang, D. J. Hill, L. Gao, and R. H. Middleton, Transient stability enhancement and voltage regulation of power system, *IEEE Trans. Power System*, vol. 8, pp. 620–627, 1993.

14. Z. Qiu, J. F. Dorsey, J. Bond, and J. D. McCalley, Application of robust control to sustained oscillations in power systems, *IEEE Trans. Circuits System I: Fundamental Theory and Applications*, vol. 39, no. 6, pp. 470–476, 1992.

15. K. S. Narendra and K. Parthasarathy, Identification and control of dynamical systems using neural networks, *IEEE Trans. Neural Networks*, vol. 1, no. 1, pp. 4–27, 1990.

16. K. J. Hunt, D. Sbarbaro, R. Zbikowski, and R. J. Gawthrop, Neural networks for control systems — a survey, *Automatica*, vol. 28, no. 6, pp. 1083–1112, 1992.

17. G. K. Venayagamoorthy and R. G. Harley, A continually online trained artificial neural network identifier for a turbogenerator, *Proc. IEEE International Electrical Machine and Drives Conference (IEMDC)*, pp. 404–406, Seattle, WA, 1999.

18. G. K. Venayagamoorthy and R. G. Harley, A continually online trained neurocontroller for excitation and turbine control of a turbogenerator, *IEEE Trans. Energy Conversion*, vol. 16, no. 3, pp. 261–269, 2001.

19. P. Parks, Liapunov redesign of model reference adaptive control systems, *IEEE Trans. Automatic Control*, vol. 11, pp. 362–367, 1966.

20. P. Osburn, H. Whitaker, and A. Kezer, New developments in the design of model reference adaptive control systems, *Proc. IAS 29th Annual Meeting*, New York, 1961.

21. R. M. Sanner and J. E. Slotine, Gaussian networks for direct adaptive control, *IEEE Trans. Neural Networks*, vol. 3, no. 6, pp. 837–863, 1992.

22. M. I. Jordan and D. E. Rumelhart, Forward models: Supervised learning with a distal teacher, *Cognitive Science*, vol. 16, pp. 307–354, 1992.

23. A. U. Levin and K. S. Narenda, Control of nonlinear dynamical systems using neural networks: Controllability and stabilization, *IEEE Trans. Neural Networks*, vol. 4, no. 2, pp. 192–206, March 1993.

24. G. K. Venayagamoorthy, R. G. Harley, and D. C. Wunsch, Comparison of heuristic dynamic programming and dual heuristic programming adaptive critics for neurocontrol of a turbogenerator, *IEEE Trans. Neural Networks*, vol. 13, no. 3, pp. 764–773, 2002.

25. G. K. Venayagamoorthy, R. G. Harley, and D. C. Wunsch, Dual heuristic programming excitation neurocontrol for generators in a multimachine power system, *IEEE Trans. Industry Applications,* vol. 39, no. 2, pp. 382–384, 2003.

26. G. K. Venayagamoorthy, R. G. Harley, and D. C. Wunsch, Implementation of adaptive critic based neurocontrollers for turbogenerators in a multimachine power system, *IEEE Trans. Neural Networks,* vol. 14, no. 5, pp. 1047–1064, 2003.

27. D. Prokhorov and L. A. Feldkamp, Analyzing for Lyapunov stability with adaptive critics, *Proc. International Conference on Systems, Man and Cybernetics,* vol. 2, pp. 1658–1161, 1998.

28. T. T. Shannon and G. G. Lendaris, Qualitative models for adaptive critic neurocontrol, *Proc. International Joint Conference on Neural Networks,* vol. 1, pp. 455-460, IJCNN 1999, Washington, DC, 1999.

29. Z. Huang and S. N. Balakrishnan, Robust adaptive critic based neurocontrollers for systems with input uncertainties, *Proc. International Joint Conference on Neural Networks, (IJCNN 2000),* vol. 3, pp. 67-72, Como, Italy, 2000.

30. D. J. Limebeer, R. G. Harley, and S. M. Schuck, Subsychronous resonance of the Koeberg turbogenerators and of a laboratory micro-alternator system, *Trans. SA Institute of Electrical Engineers,* pp. 278–297, 1979.

31. W. K. Ho, C. C. Hang, and L. S. Cao, Tuning of PID controllers based on gain and phase margin specifications, *Proc. 12th Triennial World Congress on Automatic Control,* pp. 199–202, 1993.

32. P. J. Werbos, Approximate dynamic programming for real time control and neural modelling," in D. A. Whit and D. A. Sofge, (eds.), *Handbook of Intelligent Control,* pp. 493–525, Van Nostrand Reinhold, New York, 1992.

33. A. R. Cassandra, L. P. Kaelbling, and M. L. Littman, Acting optimally in partially observable stochastic domains, *Proc. Twelfth National Conference on Artificial Intelligence (AAAI-94),* vol. 2, pp. 1023–1028, 1994.

34. C. J. Watkins and P. Dayan, Q-Learning, *Machine Learning,* vol. 8, pp. 279–292, 1992.

35. D. V. Prokhorov, *Adaptive Critic Designs and Their Applications,* Ph.D. Thesis, Texas Tech University, 1997.

36. D. V. Prokhorov and D. C. Wunsch, Adaptive critic designs, *IEEE Trans. Neural Networks,* vol. 8, no. 5, pp. 997–1007, 1997.

37. G. K. Venayagamoorthy, R. G. Harley, and D. C. Wunsch, Adaptive neural network identifiers for effective control of turbogenerators in a multimachine

power system, *Proc. IEEE Power Engineering Society Winter Meeting 2001*, pp. 1293–1298, Columbus, OH, 2001.

38. G. K. Venayagamoorthy, D. C. Wunsch, and R. G. Harley, Neurocontrol of turbogenerators with adaptive critic designs, *Proc. IEEE Africon*, pp. 489–494, Capetown, South Africa, 1999.

39. G. K. Venayagamoorthy, R. G. Harley, and D. C. Wunsch, Excitation and turbine neurocontrol with derivative adaptive critics of multiple generators on the power grid, *IEEE-INNS International Joint Conference on Neural Networks*, pp. 984–989, Washington, DC, 2001.

40. G. K. Venayagamoorthy, R. G. Harley, and D. C. Wunsch, A nonlinear voltage controller with derivative adaptive critics for multimachine power systems, *IEEE Power Industry Computer Applications Conference (PICA)*, pp. 324-329, Sydney, Australia, 2001.

41. D. V. Prokhorov and D. C. Wunsch, Convergence of critic-based training, *Proc. IEEE International Conference on Systems, Man, and Cybernetics*, vol. 4, pp. 3057–3060, 1997.

Handbook of Learning and Approximate Dynamic Programming
Edited by Jennie Si, Andy Barto, Warren Powell and Donald Wunsch
Copyright © 2004 The Institute of Electrical and Electronics Engineers, Inc.

20 Robust Reinforcement Learning for Heating, Ventilation, and Air Conditioning Control of Buildings

CHARLES W. ANDERSON   DOUGLAS HITTLE
MATT KRETCHMAR   PETER YOUNG

Colorado State University

Editor's Summary: This chapter is a case study, implementing the technique presented in Chapter 13. A detailed problem formulation is presented for a heating and cooling system and a step-by-step solution is discussed. A combined PI and reinforcement learning controller is designed within a robust control framework and detailed simulation results are presented. An internet link is provided to a website where information on experiments is provided.

20.1 INTRODUCTION

Typical methods for designing fixed feedback controllers result in sub-optimal control performance. In many situations, the degree of uncertainty in the model of the system being controlled limits the utility of optimal control design. Controllers can be manually tuned in the field, but it is very difficult to determine manual adjustments that result in overall improvement. Building energy systems are particularly troublesome since the process gain is highly variable, depending non-linearly on the load on components such as heating and cooling coils and on inlet conditions such as air temperature and air volume flow rate.

  With the development and implementation of high-speed digital control hardware, researchers in the HVAC industry have begun to explore self-tuning control systems for heating and air-conditioning processes. Heating and air-conditioning processes pose especially difficult control tuning issues because the gain of the system can vary with load and vary seasonally. For example, Figure 20.1 shows the percent

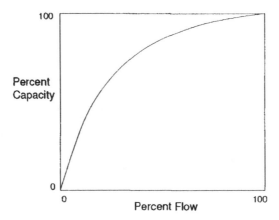

Fig. 20.1   Percent of capacity of a heating coil versus hot water flow rate through the coil.

of capacity of a heating coil versus hot water flow rate through the coil (data from manufacturer's software). Notice how much more rapidly capacity changes with changes in flow at low flow rates than at high flow. Even if a compensating non-linear valve is used to control the coil, the performance of the combination may have variable gain, depending on how much pressure drop occurs in the valve compared to the coil. The heating coil/valve system gain also changes with air flow rate and inlet water and air temperatures. In many HVAC subsystems individual loop gains change significantly with the state of the overall system.

Some of these issues have been addressed by Nesler and Stoecker [27] and Bekker, Meckel and Hittle [8]. These authors proposed various manual tuning strategies for heating and air-conditioning processes, which consequently require skilled human operators.

Self-tuning techniques have also been proposed. Brandt [9] summarizes several approaches to adaptive control. He makes the important point that so-called "jacket" software often needs to be implemented along with the adaptive control schemes to avoid trying to tune when the system is uncontrollable (e.g., when the heating coil is off) or when variations are so slow as to make adaptation inappropriate. Pinella et al. [30] and Dexter and Haves [15] describe some implementations of self-tuning procedures. More recently, Seem [32] proposed a new pattern recognition adaptive controller (PRAC) that adjusts the proportional and integral gain to stabilize unstable PI controllers.

Expert systems are also being considered. For example, Haberl et al. [20] discuss expert systems used to diagnose operating malfunctions in heating and air-conditioning systems. Norford et al. [28] discuss the implementation of a similar expert system to help building operators in determining proper operating strategies.

Most recently, researchers have begun to consider using neural networks to control the highly non-linear and time-varying HVAC processes. Ferrano and Wong [19], for example, use neural networks as a weather predicting strategy in applications involving thermal storage. Other examples include Seem and Braun's [33] use of neural networks to predict electrical demand and Miller and Seem's [25] prediction of time required to return a room to a desired temperature. Hepworth and Dexter [21] used a neural network of radial basis functions in parallel with a proportional controller on an HVAC system. They showed, by simulation and experiment, that their controller can compensate for plant non-linearities and adapt to plant degradation over time. Curtiss et al. [12, 13] used a neural network controller to gradually take over from a PID controller. Ahmed et al. [1] applied general regression neural networks to HVAC process control. Shavit, in Curtiss et al. [14], used a neural network to augment the output of a PI controller. The network attempts to modify the output of a PI controller to provide first-order response of the process while minimizing the actuator motion. The network is trained off-line, using simulation models of the various systems to which it is to be applied. This scheme has been field tested with better than expected results.

Concepts developed by Shavit and Seems [34] have been field-tested, indicating that they are more fully developed than others. Seem's PRAC controller uses the temporal pattern of the controlled variable to adjust the proportional and integral gain constants of a PI controller to stabilize and improve the performance of the system. Note that while this scheme is adaptive, the plant is still controlled by a traditional single-input–single-output PI controller.

Shavit uses a fixed (trained off line) neural network in parallel with a PI controller as a replacement for a traditional PI controller. The neural network plays the role of a non-linear controller that is well suited to non-linear HVAC systems.

In summary, both Seem and Shavit use single-input–single-output (SISO) feedback controllers, which is typical of current practice in the HVAC industry. The enormous potential of powerful techniques for MIMO control has not been explored in HVAC systems. Seem's scheme is linear but adaptive, while Shavit's is non-linear but fixed.

The neural network control schemes enumerated above have all used supervised learning, generally with back-propagation as the method for weight adjustment. The potential of reinforcement learning is largely unexplored for HVAC control problems.

An alternate approach to designing a controller that attempts to learn about the unknown dynamics of a system is to design one that is insensitive to such model uncertainties. This is the subject of robust control. This area has seen a great deal of research activity in recent years, with powerful new tools for system analysis and controller design emerging.

Early researchers in this area, including Doyle [16] and Safonov [31], developed stability analysis tools for multivariable linear time invariant (LTI) systems subject to structured dynamic LTI uncertainties. This work was extended to H_∞ performance analysis and controller synthesis in Doyle [17]. Subsequent work in this area has led to algorithms for parametric uncertainty in Young [40] and Barmish and Kang [6], and non-linear or time-varying uncertainty in Shamma [35] and Megretsky [24]. Tools are now available that use alternative norms for the performance measure, such as H_2 in Paganini [29], L_1 in Khammash and Pearson [23], and extensions to arbitrary L_p-induced norms for the LTI case in Bamieh and Dahleh [5] and the non-linear/time-varying case in Young and Dahleh [41]. More recently, some tools have started to emerge for performance measures other than norms, and in particular fixed inputs are considered in Elia et al. [18].

In tandem with this theoretical work, researchers have developed advanced computational software, such as the μ-Tools Matlab toolbox (Balas et al. [4]), which has facilitated the use of this work by practicing engineers in industry. Robust designs have been successfully applied to many interesting practical problems (Balas and Young [3]; Buschek and Calise [10]), though their utility for HVAC systems has yet to be considered.

A natural question that arises from this is to consider whether reinforcement learning and robust control can be usefully combined. Indeed, in his summary paper Werbos [39] remarks, "Most practical applications probably involve a mix of known and unknown dynamics, so that the ultimate optimum is to combine these two approaches." Some researchers have started to consider this possibility. These ideas have been exploited to address the well known stability problems of classical adaptive control schemes. Although this innovation is fairly recent, there has already been considerable success, and the ideas are now mainstream enough to find texts on the subject, such as Ioannou and Sun [22]. However, for learning controllers, only a very limited amount of work has been carried out in this direction. Suykens et al. [37] use a neural net to identify a plant model in the canonical linear fractional transformation (LFT) form with associated uncertainty description, so that this identified model set can then be used for a robust design. Bass and Lee [7] use a neural net to learn the dynamic inversion of a non-linear plant, and implement feedback linearization, so that a robust design can then be carried out for the linearized plant. It should be noted, however, that efforts in this regard are still in their infancy, and none of the approaches developed so far really exploits the power of both approaches. For instance, in both of the above examples one can only ever get the performance that the robust controller can deliver, and the neural net is essentially helping with the identification/linearization. However, one would like to exploit the power of a neural net controller to precisely tune the controller, so that the final delivered performance exceeds that of a robust controller based only on *a priori* information.

In this chapter, a reinforcement-learning (RL) agent is added in parallel to an existing feedback controller and trained to minimize the sum of the squared error of the control variable over time. First, results are shown of RL learning performance when applied to a non-linear simulation of a heating coil. Then, the system consisting of the heating coil model, a proportional-integral (PI) feedback controller, and RL agent are placed in the robust control framework described in Chapter 13. The heating coil model is linearized and integral-quadratic constraints (IQC) are used to characterize the non-linear and time-varying parts of the RL agent. This results in a synthesis of feedback control and reinforcement learning with guaranteed static and dynamic stability, even while learning.

20.2   HEATING COIL MODEL AND PI CONTROL

Underwood and Crawford [38] developed a model of an existing heating coil by fitting a set of second-order, non-linear equations to measurements of air and water temperatures and flow rates obtained from the heating coil. A diagram of the model with a PI controller is shown in Figure 20.2. The state of the modeled system is defined by the air and water input and output temperatures, T_{ai}, T_{ao}, T_{wi}, T_{wo}, and the flow rates of air and water, f_a, and f_w. The control signal, c, affects the water flow rate. Its value ranges from 670 for the maximally open position to 1400 for the maximally closed position. By changing the valve setting, the flow rate of the water in the heating coil can be increased or decreased. The valve setting, an input to the plant, is the output from the controllers. The flow rate of water, f_w, is directly affected by the valve setting. In turn, this affects the temperature of the water leaving the coil, T_{wo}, and ultimately the temperature of the air leaving the coil, T_{ao}, because the flow rate of the water determines how much thermodynamic energy can be delivered from the boiler to the heating coil in the duct. T_{ao} is the state variable that is to be controlled. Control performance is determined by how closely the output air temperature tracks the reference signal, which, in the HVAC case, is the desired thermostat setting or the set point.

The model is given by the following equations with constants determined by a least-square fit to data from an actual heating coil:

$$f_w(t) = 0.008 + 0.00703\big(-41.29 + 0.30932c(t-1)$$
$$- 3.2681 \times 10^{-4}c(t-1)^2 + 9.56 \times 10^{-8}c(t-1)^3\big), \qquad (20.1)$$
$$T_{wo}(t) = T_{wo}(t-1) + 0.64908 f_w(t-1)\big(T_{wi}(t-1) \qquad (20.2)$$
$$- T_{wo}(t-1)\big) + \big(0.02319 + 0.10357 f_w(t-1)$$
$$+ 0.02806 f_a(t-1)\big)\left(T_{ai}(t-1) - \frac{(T_{wi}(t-1) + T_{wo}(t-1))}{2}\right),$$
$$T_{ao}(t) = T_{ao}(t-1) + 0.19739 f_a(t-1)\big(T_{ai}(t-1)$$
$$- T_{ao}(t-1)\big) - \big(0.03184 + 0.15440 f_w(t-1)$$
$$+ 0.04468 f_a(t-1)\big)\left(T_{ai}(t-1) - \frac{(T_{wi}(t-1) + T_{wo}(t-1))}{2}\right)$$
$$+ 0.20569\big(T_{ai}(t) - T_{ai}(t-1)\big). \qquad (20.3)$$

For the experiments reported here, the variables T_{ai}, T_{wi}, and f_a were modified by random walks to model the disturbances and changing conditions that would occur in actual heating and air conditioning systems. The bounds on the random walks are $4 \le T_{ai} \le 10°C$, $73 \le T_{wi} \le 81°C$, and $0.7 \le f_a \le 0.9$ kg/s.

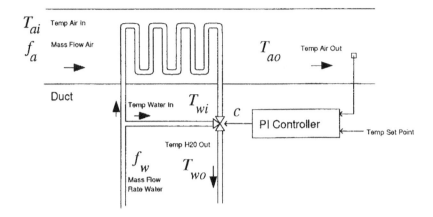

Fig. 20.2   The simulated heating coil under control of a PI feedback controller.

A PI controller was tuned to control the simulated heating coil. The best proportional and integral gains were determined by measuring the sum of the absolute value of the difference between the set point and the actual exiting air temperature under a variety of disturbances. The PI control law is

$$c'(t) = k_p e(t) + k_i \int e(t) \, dt, \tag{20.4}$$

where $e(t)$ is the difference between the set point and the actual output air temperature at time t, k_p is the proportional gain, and k_i is the integral gain. $c'(t)$ is a normalized control value that ranges from 0 to 1 to specify flows rates from the minimum to the maximum allowed values. The best control law constants were determined to be $k_p = 0.185$ and $k_i = 0.178$. This normalized control signal is converted to the control signal for the model by

$$c(t) = 1400 - 730c'(t). \tag{20.5}$$

20.3   COMBINED PI AND REINFORCEMENT LEARNING CONTROL

A reinforcement learning agent trained using the Q-learning algorithm [36, 42] was combined with a PI controller as shown in Figure 20.3. Inputs to the reinforcement-learning agent were T_{ai}, T_{ao}, T_{wi}, T_{wo}, f_a, f_w, and the set point, all at time t. The output of the reinforcement learning agent is directly added to the output of the PI controller. The allowed output actions of the reinforcement learning agent was the set of discrete actions $A = \{-100, -50, -20, -10, 0, 10, 20, 50, 100\}$. The selected action is added to the PI control signal. The Q function is implemented using a quantized, or table look-up, method. Each of the seven input variables is divided

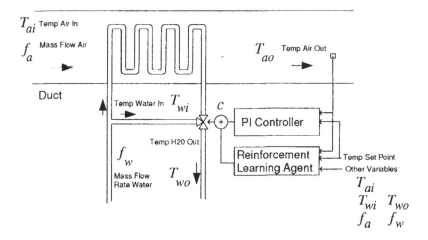

Fig. 20.3 The heating coil controlled by the sum of a PI controller and the action taken by the reinforcement learning agent.

into six intervals, which quantize the 7-dimensional input space into 6^7 hypercubes. Unique Q values for the nine possible actions are stored in each hypercube.

Reinforcement R at time step t is determined by the squared error between the controlled variable and its set point plus a term proportional to the square of the action change from one time step to the next:

$$R(t) = (T_{ao}(t)^* - T_{ao}(t))^2 + \beta(a_t - a_{t-1})^2, \qquad (20.6)$$

where T_{ao}^* is the set point. The action change term is introduced to reduce the fluctuation in the control signal to minimize the stress on the control mechanism. The values of a_t in this equation are indices of the set A, rather than actual output values.

The reinforcement-learning agent is trained for 1,000 repetitions, called *trials*, of a 500 time-step interaction between the simulated heating coil and the combination of the reinforcement learning agent and the PI controller, hereafter referred to as the RL/PI controller. The performance of the RL/PI controller was compared with the performance of just the PI controller. Figure 20.4 shows the set point and actual temperatures and the PI control signal over these 500 steps. The RMS error between the set point and actual output air temperature over these 500 steps for the PI controller acting alone is 0.93.

The training algorithm for the reinforcement learning agent depends on the parameters α_t, λ, γ, and β. After trying a small number of values, γ and β were fixed to $\gamma = 0.95$ and $\beta = 0.1$. A number of combinations of values for α and λ were tested, where α is held constant during a training run rather than setting it to values that decrease with time as required for the convergence proofs. Figure 20.5 shows

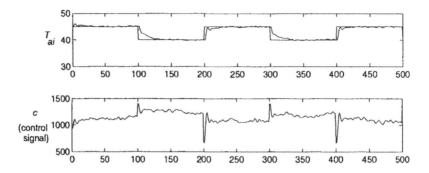

Fig. 20.4 Performance of PI controller. The top graph shows the set point and the actual output air temperature over time. Over this sequence, the RMS error is 0.93. The bottom graph shows the PI output control signal.

how performance depends on these parameter values. Performance was measured by the average RMS error over the final 30 trials and also by the average RMS error over all 1,000 trials. The first measure is of final performance and the second is an indication of the speed with which the error was reduced. The best parameter values were $\alpha = 0.1$ and $\lambda = 0.995$. These values were used to obtain the following results.

The reduction in RMS error while training the reinforcement-learning agent is shown in Figure 20.6. After approximately 80 training trials, the average RMS error between the set point and actual temperature was reduced below the level achieved by the PI controller alone. With further training the reinforcement learning agent converges on a deterministic policy that consistently achieves an RMS error of 0.85, about an 8.5% reduction in error from that achieved by the PI controller alone.

The resulting behavior of the controlled air temperature is shown in the first graph of Figure 20.7. The second graph shows the output of the reinforcement-learning agent. It has learned to be silent (an output of 0) for most time steps. It produces large outputs at set point changes and at several other time steps. The combined output of the RL/PI controller is shown in the third graph. The reinforcement-learning agent learns to augment the PI controller's output during the initial steps and at most of the set point changes. From time step 340 through about 480 the reinforcement-learning agent injects negative and positive values into the control signal. This is the pattern of outputs to which the controller converged on in most of the successful training runs of the reinforcement-learning agent. This behavior is due to the particular trajectories followed by the disturbance variables. The last three graphs of Figure 20.7 shows the disturbance trajectories for T_{ai}, T_{wi}, and f_a. At step 340 the value of f_a exceeds its previous values and the reinforcement learning agent compensates by injecting a pattern of negative and positive changes to the controller's output. Additional details of these experiments are described by Anderson et al. [2].

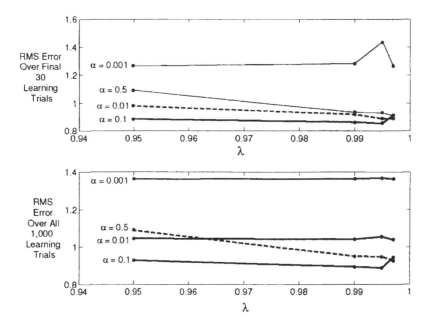

Fig. 20.5 Performance of combined reinforcement-learning agent and PI controller for different parameter values. Top graph shows performance averaged over final 30 learning trials. Bottom graph shows performance averaged over all 1,000 learning trials. Best parameter values are $\alpha = 0.1$ and $\lambda = 0.995$.

20.4   ROBUST CONTROL FRAMEWORK FOR COMBINED PI AND RL CONTROL

Guarantees of stability of any adaptive control scheme greatly increase its acceptability in a real control application. An approach providing such guarantees is presented in Chapter 13. Here that approach is applied to the heating coil model discussed in the previous section.

To construct traditional and robust controllers for the heating coil, it is recast as a linear, time-invariant (LTI) system with integral-quadratic constraints (IQCs). Due primarily to the complex dynamics of HVAC systems, a single LTI model is not adequate for approximating the dynamics of the non-linear system. Consequently, an LTI model is constructed that is reasonably accurate for only a limited operating range (around a set point temperature with static environmental variables). A Taylor series expansion about the desired operating point is used to construct the LTI model of the system. The LTI model is used for designing controllers and then the original, non-linear model is used for testing the performance of the controllers. The following

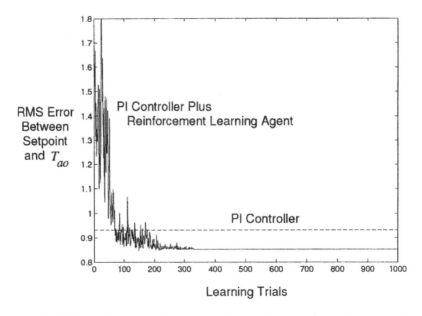

Fig. 20.6   Reduction in error with multiple training epochs. The amount of exploration is reduced during training.

parameters specify the operating point for the Taylor series expansion:

$$u = 972.9 \quad f_W = 0.2785, \tag{20.7}$$

$$T_{wo} = 55.45 \quad T_{wi} = 78.0, \tag{20.8}$$

$$T_{ao} = 45.0 \quad T_{ai} = 12.0. \tag{20.9}$$

The resulting linear model is

$$
\begin{aligned}
f_w(t) &= 0.2785 - (3.863\text{x}10^{-4}(u(t-1) - 972.9)), & (20.10)\\
T_{wo}(t) &= 93.5445(f_w(t-1) - 0.2785) + 0.792016(T_{wo}(t-1) - 55.45)\\
& \quad + 55.45, & (20.11)\\
T_{ao}(t) &= 0.8208(T_{ao}(t-1) - 45.0) + 45.0 + 0.0553(T_{wo}(t-1) - 55.45)\\
& \quad + 7.9887(f_w(t-1) - 0.2785). & (20.12)
\end{aligned}
$$

Since PI control is a dominant trend in the HVAC industry, here a PI controller is constructed using state-of-the-art tuning laws [11]. The PI controller and parameters are defined in Section 20.2. The tracking performance of the PI controller when implemented on the non-linear model is shown in the top graph of Figure 20.8. The

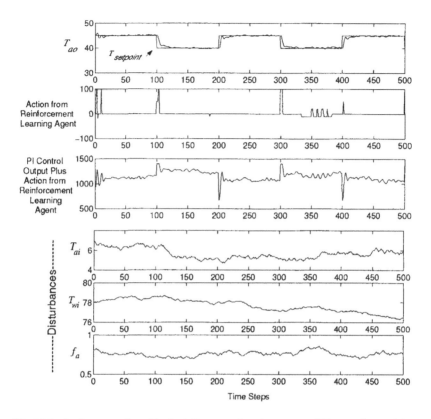

Fig. 20.7 Performance of combined reinforcement-learning agent and PI controller resulting in an RMS error of 0.85. Top graph shows the set point and the actual output air temperature over time. The second graph shows the output of the reinforcement-learning agent. The third graph shows the combined PI output and reinforcement-learning agent output. The bottom three graphs show the trajectories of the disturbance variables.

control performance is quite good as the PI controller has been finely tuned to suit this particular non-linear model.

Now a reinforcement learning component is added in parallel to the PI controller. The learning component consists of an actor network and a critic. The critic is trained via Q-learning [42] and the actor network is trained to output the action deemed optimal by the critic for each state. An LTI model with IQCs of the PI controller, neuro-controller, and plant is used to analyze the stability of the system while learning. The weights of the actor network are prevented from taking values for which the system becomes unstable by the robust reinforcement learning procedure defined in Chapter 13.

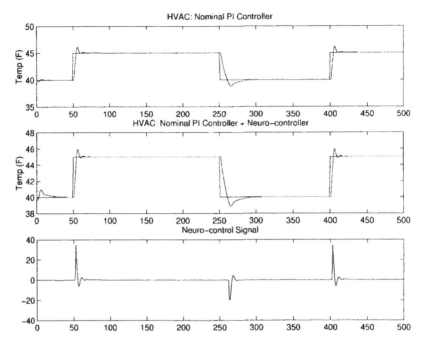

Fig. 20.8 Tracking performance of nominal controller and nominal plus trained neuro-controller. Bottom graph shows output of neuro-controller.

Inputs to both learning components are the tracking error and the three internal state variables of the non-linear model: F_w, T_{ao}, and T_{wo}. To choose the number of hidden units in the actor network, several different configurations were tested. A network with a single hidden layer of 10 units was found to perform well. The Q function is approximated by a table indexed by F_w, T_{ao}, and T_{wo} and the control action. Each of the four input dimensions is quantized into 10 intervals, resulting in a Q table of 10^4 entries.

The middle graph of Figure 20.8 shows the step responses of the combined PI/RL controllers. The output of the actor network is also shown in the bottom graph. The actor network learns to produce no output control signal for most cases. Only during a step transition in the reference input, the action network outputs a very small control signal to be added to the PI control signal.

The squared tracking error summed over the sequence shown in Figure 20.8 is 0.526 when just the PI controller is used. The error is decreased to 0.518 when the RL controller is added. This is a slight, 1.5%, decrease in error. The nominal PI controller provides fairly good tracking performance, leaving little room for improvement by the RL controller. This result is significant not for the small improvement achieved by the RL agent, but for the fact the system is guaranteed to be stable throughout the entire learning phase. Details of these results are presented by Kretchmar [26].

The RL controller uses state information that is not available to the PI controller. By exploiting this extra information, the RL controller may be able to implement better control performance. However, the addition of more state variables might not always be the best solution; it may be the case that a better performing controller can be found by using fewer state variables. The reason for this counter-intuitive relationship is that the added state information increases the complexity of the feedback loop which, in turn, allows more possibilities for unstable control. This may limit the flexibility inherent in the RL algorithm in order to guarantee control.

If the PI controller provides excellent control performance, then why consider applying this robust RL scheme to this task? The primary reason involves the difference between the real physical plant and the plant model. The dynamics of the plant model will be different from the unknown dynamics of the physical system. The PI controller may still be "tuned" for the physical plant, but the performance of this controller is expected to be substantially less than the performance of the PI controller on the plant model. Essentially, there is likely to be more room for improved control performance when the physical plant is involved.

20.5 CONCLUSIONS

Two experiments are described in augmenting PI controllers with reinforcement learning (RL) agents to learn improved control of a non-linear model of a heating coil. In the first experiment, the combined PI and RL controllers were applied to a the heating coil. The RL agent was trained using a Q-learning algorithm to form a tabular approximation to the optimal Q function. Results show that the reinforcement learning agent learns to modify the PI controller's output only at times when the PI controller is not adequate to meet performance objectives.

For the second experiment, the heating coil model is linearized to enable the application of robust stability analysis to the control system, including the RL agent. This required the addition of an artificial neural network as an action network to the RL agent. The resulting system was then trained with stability constraints in effect. The results show that a small amount of performance improvement occurred with the addition of the RL agent. More importantly, learning occurred in the context of guaranteed stability.

It is hypothesized that performance improvements achieved with RL agents would be much more substantial in a real heating coil due to uncertainties and non-linearities that are present in real systems. These uncertainties and non-linearities result in conservative robust controllers, whose performance can be improved by RL agents being fine-tuned through interactions with the real system. A physical heating coil with many sensors and actuators is being constructed at Colorado State University to research this question. The status of this project is summarized at the web site http://www.engr.colostate.edu/nnhvac.

Acknowledgments

This work was supported by the National Science Foundation through grants CMS-9401249 and CISE-9422007.

Bibliography

1. O. Ahmed, J. W. Mitchell, and S. A. Klein, Application of general regression neural network (GRNN) in HVAC process identification and control, *ASHRAE Trans.*, vol. 102, no. 1, 1996.

2. C. W. Anderson, D. C. Hittle, A. D. Katz, and R. M. Kretchmar, Synthesis of reinforcement learning, neural networks, and pi control applied to a simulated heating coil, *Artificial Intelligence in Engineering*, vol. 11, no. 4, pp. 423–431, 1997.

3. G. J. Balas, and P. M. Young, Control design for variations in structural natural frequencies, *AIAA Journal of Guidance, Dynamics and Control*, vol. 18, pp. 325–332, 1995.

4. G. J. Balas, A. K. Packard, J. C. Doyle, K. Glover, and R. S. Smith, Development of advanced control design software for researchers and engineers, *Proc. American Control Conference*, pp. 996–1001, Boston, MA, 1991.

5. B. Bamieh and M. Dahleh, Robust stability with structured time-invariant perturbations, *Proc. 31st IEEE Conference on Decision and Control*, pp. 1987–1990, Tuscon, AZ, 1992.

6. B. R. Barmish, and H. I. Kang, A survey of extreme point results for robustness of control systems, *Automatica*, vol. 29, pp. 13–35, 1993.

7. E. Bass, and K. Y. Lee, Robust control of nonlinear systems using norm-bounded neural networks, *Proc. 1994 IEEE International Conference on Neural Networks*, pp. 2524–2529, Orlando, FL, 1994.

8. J. Bekker, P. K. Meckel, and D. Hittle, A tuning method for first order processes with PI controllers, *ASHRAE Trans.*, vol. 97, no. 1, 1991.

9. S. Brandt, Adaptive control implementation issues, *ASHRAE Trans.*, vol. 92, no. 2B, pp. 211–219, 1986.

10. H. Buschek, and A. J. Calise, Hypersonic flight control system design using fixed order robust controllers, *Proc. 6th International Aerospace Planes and Hypersonic Technologies Conference*, Nashville, TN, 1995.

11. K. D. Cock, B. D. Moor, W. Minten, W. V. Brempt, and H. Verrelst, A tutorial on PID-control, Technical Report ESAT-SIST/TR 1997-08, Katholieke Universiteit, Leuven, 1997.

12. P. S. Curtiss, Experimental results from a network-assisted PID controller, *ASHRAE Trans.*, vol. 102, no. 1, 1996.

13. P. S. Curtiss, J. K. Kreider, and M. J. Brandemuehl, Adaptive control of HVAC processes using predictive neural networks, *ASHRAE Trans.*, vol. 99, no. 1, 1993.

14. P. S. Curtiss, G. Shavit, and J. K. Kreider, Neural networks applied to buildings—a tutorial and case studies in prediction and adaptive control, *ASHRAE Trans.*, vol. 102, no. 1, 1996.

15. A. Dexter, and P. Haves, A robust self-tuning predictive controller for HVAC applications, *ASHRAE Journal*, vol. 32, no. 12, pp. 12–22, 1990.

16. J. C. Doyle, Analysis of feedback systems with structured uncertainty, *IEEE Proc.*, vol. 129, no. D, pp. 242–250, 1982.

17. J. C. Doyle, Structured uncertainty in control system design, *Proc. 24th Conference on Decision and Control*, pp. 260–265, Kyoto, Japan, 1985.

18. N. Elia, P. M. Young, and M. A. Dahleh, Robust performance for both fixed and worst case inputs, *Proc. 34th IEEE Conference on Decision and Control*, pp. 3170–3175, New Orleans, LA, 1995.

19. F. Ferrano, and K. Wong, Prediction of thermal storage loads using a neural network, *ASHRAE Trans.*, vol. 96, no. 2, pp. 723–726, 1990.

20. J. S. Haberl, L. K. Norford, and J. V. Spadaro, Diagnosing building operational problems, *ASHRAE Journal*, vol. 31, no. 6, pp. 20–30, 1989.

21. S. J. Hepworth, and A. L. Dexter, Neural control of non-linear HVAC plant, *Proc. IEEE Conference on Control Applications*, vol. 3, no. 3, pp. 1849–1854, 1994.

22. P. A. Ioannou, and J. Sun, *Robust Adaptive Control*, Prentice-Hall, Englewood Cliffs, NJ, 1996.

23. M. Khammash, and J. B. Pearson, Performance robustness of discrete-time systems with structured uncertainty, *IEEE Trans. Automatic Control*, vol. 36, pp. 398–412, 1991.

24. A. Megretsky, Necessary and sufficient conditions of stability: A multiloop generalization of the circle criterion, *IEEE Trans. Automatic Control*, vol. 38, pp. 753–760, 1993.

25. R. Miller, and J. Seem, Comparison of artificial neural networks with traditional methods of predicting return time from night or weekend setback, *ASHRAE Trans.*, vol. 97, 1991.

26. R. M. Kretchmar, *A Synthesis of Reinforcement Learning and Robust Control Theory*, Ph.D. Thesis, Colorado State University, Department of Computer Science, Fort Collins, CO, 2000.

27. C. G. Nesler and W. F. Stoecker, Selecting the proportional and integral constants in the direct digital control of discharge air temperature, *ASHRAE Trans.*, vol. 90, no. 2B, pp. 834–845, 1984.

28. L. K. Norford, A. Allgeier, and J. V. Spadaro, Improved energy information for a building operator—exploring the possibilities of a quasi-real-time knowledge-based system, *ASHRAE Trans.*, vol. 96, no. 1, pp. 1515–1523, 1990.

29. F. Paganini, Necessary and sufficient conditions for robust H_2 performance, *American Control Conference*, pp. 1970–1975, Seattle, WA, 1995.

30. M. Pinella, D. C. Hittle, E. Wechselberger, and C. Pederson, Self-tuning digital integral control, *ASHRAE Trans.*, vol. 92, no. 2B, pp. 202–209, 1986.

31. M. G. Safonov, Stability margins for diagonally perturbed multivariable feedback systems, *IEEE Proc.*, vol. 129, no. D, pp. 251–256, 1982.

32. J. E. Seem, A new pattern recognition adaptive controller, Presented at 13th International Federation of Automatic Control World Congress, San Francisco, CA, 1996.

33. J. Seem, and J. Braun, Adaptive methods for real-time forecasting of building electrical demand, *ASHRAE Trans.*, vol. 97, 1991.

34. J. E. Seem, and H. J. Haugstad, Field and laboratory results for a new pattern recognition adaptive controller, *Proc. CLIMA 2000 International Conference*, p. 77, Brussels, Belgium, 1997.

35. J. S. Shamma, Robust stability with time-varying structured uncertainty, *IEEE Trans. Automatic Control*, vol. 39, pp. 714–724, 1994.

36. R. S. Sutton, and A. G. Barto, *Reinforcement Learning: An Introduction*, MIT Press, Cambridge, MA, 1998.

37. J. A. K. Suykens, B. L. R. De Moor, and J. Vandewalle, Nonlinear system identification using neural state space models, applicable to robust control design, *International Journal of Control*, vol. 64, pp. 129–152, 1995.

38. D. M. Underwood, and R. R. Crawford, Dynamic nonlinear modeling of a hot-water-to-air heat exchanger for control applications, *ASHRAE Trans.*, vol. 97, no. 1, pp. 149–155, 1991.

39. P. J. Werbos, An overview of neural networks for control, *Control Systems Magazine*, vol. 11, no. 1, pp. 40–41, 1991.

40. P. M. Young, Controller Design with Real Parametric Uncertainty, *International Journal of Control,* vol. 65, pp. 469–509, 1996.

41. P. M. Young, and M. A. Dahleh, Robust L_p Stability and Performance, *Systems & Control Letters,* vol. 26, pp. 305–312, 1995.

42. C. Watkins, *Learning with Delayed Rewards,* Ph.D. Thesis, Cambridge University Psychology Department, Cambridge, UK, 1989.

21 Helicopter Flight Control Using Direct Neural Dynamic Programming

RUSSELL ENNS   JENNIE SI
Boeing Company   Arizona State University

Editor's Summary: This chapter presents a complex continuous state system control problem using the direct NDP method introduced in Chapter 5. The system examined is an Apache helicopter modeled using the full-scale industrial simulator FLYRT. The focus is on how direct NDP can be used to control complex, realistic, and high-dimensional systems. The problem formulation, objectives, implementation, results, and insights that can be gained from doing direct NDP designs are discussed in detail. This case study is one of the applications examined in this book that addresses the challenging generalization issue.

21.1 INTRODUCTION

Generalization or scalability of an approximate dynamic programming methodology to large and complex problems is one of the major challenges to date. This chapter focuses on demonstrating that the direct neural dynamic programming (direct NDP) method introduced in Chapter 5 can be successfully applied to complex, realistic, higher dimensional systems. Furthermore it shows the promise of direct NDP as a robust approximate dynamic programming control system design technique based on measures such as learning statistics, problem scalability, and the range of problems handled.

In this chapter the original direct NDP mechanism presented in [11] is expanded on in order to improve its learning ability. Then, a comprehensive case study of this new design is presented. The example used is helicopter flight control, where the control goals are to perform both helicopter stabilization and maneuvering for a full-scale industrial helicopter model. The model is quite complex, being both highly nonlinear and having multiple inputs and multiple outputs (MIMO).

535

Direct NDP is a control methodology that provides an approximate solution to an optimal control problem that is often solved by dynamic programming. Thus, direct NDP is an approximate dynamic programming (ADP) method. Direct NDP was perceived as a strong candidate for a learning system for helicopter flight control for a number of reasons. First, it can be applied to complex systems such as helicopters without the need to decouple the control system into simpler subsystems. Therefore, it can learn to take advantage of any of the system's cross coupling characteristics when generating its control solution, including coupling benefits that may not be apparent to a control systems design engineer. Second, direct NDP and other ADP methods can deal with both explicitly and implicitly defined system performance measures which are usually a function of the system states and control actions. Third, direct NDP and other ADP methods avoid the "curse of dimensionality" that dynamic programming methods suffer from by providing approximate solutions. This, however, may also be considered as the down side when true "optimality" is demanded. Finally, direct NDP is an approximate dynamic programming method that does not require explicitly building a system model prior to learning to improve system performance. This has provided advantages for systems that are not well understood or difficult to model.

Balakrishnan was one of the first to use a form of reinforcement learning (adaptive critic based networks) for aircraft flight controls [1]. However, the research limited itself to the longitudinal axis and, as a result, the system only had a single control. Prokhorov et al. [9] have demonstrated their adaptive critic designs (ACD) in an auto-lander application, which takes in altitude, vertical speed, and horizontal position (three states), and in some cases pitch and horizontal speed (for five states), and computes the required pitch command. Thus, their demonstration was limited to scalar control. Their system was tested on a linearized two-dimensional model of a commercial aircraft. These reinforcement learning methods have taken a model-based approach, as opposed to the model-independent approach that can be used by direct NDP.

A limited amount of research has been done in the area of reinforcement learning or ADP in general for helicopter control. Buskey et al. [3] tested three learning architectures, single layer neural networks (SLNN), multi-layer neural networks (MLNN) and fuzzy associative memories (FAM) on a helicopter simulator. Simulation showed the possibilities of FAM as an on-line learning algorithm for autonomous helicopter control. The experiments were preliminary in terms of model scale and controller capacities. Simulations were based on a second order scaled-size model (linearized within the hover envelope). The controllers were only trained to maintain a stable hover.

Shim et al. [10] also compared three different control methodologies for autonomous helicopter control: linear robust multi-variable control, fuzzy logic control with evolutionary tuning and nonlinear tracking control. A simplified thrust-torque generation nonlinear model implemented in MATLAB Simulink was used for the control system design; however, only hover and low velocity regime were considered. The experiment showed the capability of the robust and fuzzy controller to handle uncertainties and disturbances with less robustness and a substantially wider range of flight envelopes covered by nonlinear control with more accurate system knowledge needed. However, the research was limited to near hover conditions and simplified system uncertainties and disturbances.

Bagnell and Schneider [2] used policy search methods to fly an autonomous helicopter. After the dynamical system was sampled to build a LWBR state-space model, the policy search method was then applied to evaluate and synthesize simple one-layer neural network controllers under distributions of the identified Markovian models toward a pre-defined cost criterion. The method was validated for the CIFER system developed by the U.S. Army, although only the so-called "core dynamics" of the helicopter were considered. The control was limited to the longitudinal and lateral axis and, as a result, the system only had two controls. The controller was trained to regulate the helicopter hovering about a point or a slowly varying trajectory.

Wan et al. [12] developed a model predictive neural control (MPNC) system which combines the conventional neural network (NN) control and the nonlinear state-dependent Riccati equation (SDRE) control for aggressive helicopter maneuvers. The SDRE controller design, repeated at every sample time, provided initial local asymptotic stability. The neural network controller was optimized by learning the weights to minimize the receding horizon MPC cost function. The NN controller was pre-trained to behave similarly to the SDRE controller, updated on-line for every N time steps where N is the receding horizon time with weights from the previous horizon used as the initialization. During the training, the SDRE controller provided stable tracking and good conditions, while the NN decreased the tracking error. The MPNC controller augmented the SDRE controller with the NN controllers by a relative weighting constant. As training progressed, the SDRE control reduced, which in turn gave more authority to the neural controller. Simulation was based on a helicopter model built via a simulation package called FlightLab. A twelve dimensional state vector was used with four control outputs. Various maneuvers such as rapid take-off and landings and "elliptic" maneuvers were simulated. Simulation results showed the improved performance over traditional SDRE and LQ control. However, their system was still constrained in the following two ways. First, their system took a model-based approach, as opposed to the model-independent approach used by other on-line learning algorithms. The SDRE controller required an analytical approximation of the system dynamics which meant a system model needed to be identified to generate the SDRE control. Therefore, issues like model mismatch would certainly affect the final controller performance. Second, the computational effort was extensive. A Riccati equation needed to be solved at every iteration. The Jacobians for the flight system also needed to be evaluated via perturbation during backward simulation for training the NN network. With current CPU speeds, this approach is precluded from use in real-time control.

The purpose of this research is not to compare the direct NDP method to any of the existing methods, be they reinforcement learning based, neural network based or otherwise. Rather, the purpose is to demonstrate the power of direct NDP as an approximate dynamic programming control methodology on a challenging controls problem that other approximate dynamic programming algorithms may not be able to handle. This is done by first showing how direct NDP can be used to stabilize a helicopter for five flight conditions (hover, 30, 60, 90 and 120 knots), and then showing how it can learn to perform a number of aircraft maneuvers (hover to 50 ft/s at various accelerations, up to the aircraft upper limits of 0.25 g ($8 ft/s^2$); and deceleration maneuvers from 100 ft/s to 50 ft/s at various decelerations). Simulations are performed in both clear air and in the presence of turbulence and step gusts using an industrial Dryden model. It is then shown how direct NDP can be used for helicopter flight control reconfiguration, demonstrating the possibility of using direct NDP to control the helicopter even in the presence of an actuator failure. This shows the promise of direct NDP as a useful reinforcement learning method in the failure accommodation realm. Unlike many results which are based on linearized models and corresponding assumptions, these direct NDP designs and simulations are conducted using FLYRT, a very realistic nonlinear system model.

The chapter is organized as follows. Section 21.2 briefly describes the helicopter model used for evaluating the direct NDP designs. Section 21.3 then expands on the direct NDP methodology developed in Chapter 5 and applies it to the helicopter flight control stabilization problem. Key to this expanded direct NDP methodology is the trim network which is crucial to the success of direct NDP for complex nonlinear systems. The direct NDP methodology is then further refined for the more complicated helicopter command tracking and reconfigurable flight control problems in Sections 21.4 and 21.5 and simulation results are presented. Section 21.6 then provides some conclusions.

21.2   HELICOPTER MODEL

A helicopter is a sophisticated system with multiple inputs used to control a significant number of states. There exists a large amount of cross coupling between control inputs and states. Further, the system is highly non-linear and changes significantly as a function of operating condition. For these reasons the helicopter serves as an excellent and challenging platform for testing approximate dynamic programming systems.

The helicopter's states are controlled by a main rotor and a tail rotor. There are three main rotor actuators whose positions, z_A, z_B, and z_C, control the position and orientation of a swash plate which in turn controls the main rotor's blade angles as a function of rotational azimuth. There is a single tail rotor actuator position (z_D) which controls the tail rotor's blade angles. The aircraft states are numerous. For flight control purposes the states of interest are the aircraft translational (u, v, w) and rotational (p, q, r) velocities and the aircraft orientation (θ, ϕ, ψ) for a total of nine states. The helicopter's longitudinal (u), lateral (v), and vertical velocities (w) are in ft/s. The helicopter's roll rate (p), pitch rate (q), and yaw rate (r) are in degrees/s. The helicopter's Euler angles, pitch (θ), roll (ϕ) and yaw (ψ) are in degrees. The states can be written in vector form as $\mathbf{x} = [u, v, w, p, q, r, \theta, \phi, \psi]$. The controls can be written in vector form as $\mathbf{u} = [z_A, z_B, z_C, z_D]$.

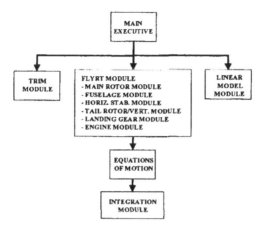

Fig. 21.1   Major components of helicopter model.

For simulation purposes a detailed helicopter model run at 50 Hz is used to evaluate the direct NDP controller's performance. The model, named FLYRT and shown in Figure 21.1, is a sophisticated nonlinear helicopter flight simulation model developed over the past two decades [8]. FLYRT models all the forces and moments acting on the helicopter. The rotor is modeled using a blade element model. FLYRT dynamically couples the six-degrees-of-freedom rigid body of the helicopter to the main rotor through Euler equations. The drive train is represented as a single degree of freedom model and is coupled to the main rotor, tail rotor and engine. The engine is modeled in sufficient detail to cover performance over all phases of flight, including ground modes. The landing gear is modeled as three independent units interfacing with a rigid airframe. Quaternions are used during state integration to accommodate large attitude maneuvers.

FLYRT also models the mechanical geometry between the actuators and the helicopter blades as well as the dynamics of the actuators. Each actuator is modeled as a first order lag with time constant $\tau = 0.03$, reflective of a typical actuator. Actuator rate and position limits are also modeled.

The Apache helicopter is the helicopter modeled for the simulation studies. The operating conditions for which the studies are performed are shown in Table 21.1. The center of gravity (C.G.) is listed in the standard Apache FS/WL/BL coordinate frame [8].

Table 21.1   Helicopter Operating Conditions

Weight	16324 lb
C.G. - FS/BL/WL	201.6 in, 0.2 in, 144.3 in
Temperature	59° F
Altitude	1770 ft

21.3   DIRECT NDP MECHANISM APPLIED TO HELICOPTER STABILITY CONTROL

Attention is now turned toward applying the direct NDP framework to the helicopter stabilization flight control problem. The objective of a direct NDP controller is to optimize a desired performance measure by learning to create appropriate control actions through interaction with the environment. The controller is designed to learn to perform better over time using only sampled measurements, without explicitly identifying a system model for control design purposes. In this application, the objective is to learn to create appropriate control actions solely by observing the helicopter states, evaluating the controller performance and adjusting the neural networks accordingly.

This section describes the implementation details of a direct NDP controller for helicopter stabilization and the rationale behind them. Figure 21.2 outlines the direct NDP control structure applied to helicopter stabilization. It consists of an action network, a critic network, and a trim network. The action network provides the controls required to drive the helicopter to the desired system state. The critic network approximates the cost function if an explicit cost function does not exist. The trim network, a major addition to prior direct NDP designs, provides nominal trim control positions as a function of the desired operating condition.

Note that in the next section an even more sophisticated direct NDP control structure shown in Figure 21.6 is developed. Such a control structure can be used to solve command tracking and other control problems more sophisticated than stabilization.

21.3.1   Critic Network

The critic network is used to approximate a cost function should an explicit cost function not be convenient or possible to represent. In this application, the network output $J(t)$ approximates the discounted total reward-to-go function,

$$R(t) = r(t+1) + \alpha r(t+2) + \alpha^2 r(t+3) + \cdots , \qquad (21.1)$$

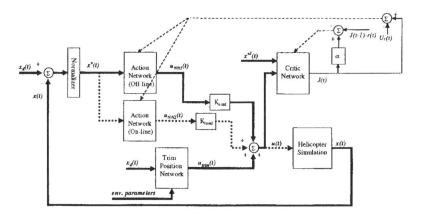

Fig. 21.2   Direct NDP based helicopter controller for stabilization, featuring simple action network, trim network, and critic network.

where $R(t)$ is the future accumulative reward-to-go value at time t, α is a discount factor for the infinite-horizon problem ($0 < \alpha < 1$), and $r(t)$ is the external reinforcement value at time t.

In prior direct NDP research, $r(t)$ was simply a binary reinforcement signal given by $r(t) = 0$ if successful, or $r(t) = -1$ if failure occurs at the end of an event. For this application, a more informative quadratic reinforcement signal at each sampling time is formulated

$$r(t) = -\sum_{i=1}^{n} \left(\frac{(x_i - x_{i,d})}{x_{i,max}} \right)^2 , \tag{21.2}$$

where x_i is the ith state of the state vector $\mathbf{x}$, $x_{i,d}$ is the desired reference state and $x_{i,max}$ is the nominal maximum state value.

The critic network is implemented with a standard multi-layer nonlinear feed-forward neural network as shown in Chapter 5. A sigmoid function is used for the nonlinearity. The inputs to the critic network are all states, normalized and squared. The output is the approximate cost function $J(t)$. The technique for training this critic network is shown in Chapter 5.

21.3.2   Action Network

The action network generates the desired plant control given measurements of the plant states. As with the critic network, the action network is implemented with a standard multi-layer nonlinear feed-forward neural network. The number of network outputs equals the control space dimension.

The principle in adapting the action network is to back-propagate the error between the desired ultimate objective, denoted by U_c, and the total reward-to-go $R(t)$. U_c is usually set to 0 since success was defined to be 0. Either the actual total reward function $R(t)$, or an approximation to it $J(t)$, is used depending on whether an explicit cost function or a critic network is available. In the latter case back-propagation is done through the critic network. Training of the action network is discussed in Chapter 5.

If desired, a second action network can be implemented to perform on-line learning to adapt to local flight conditions while the first action network's weights are frozen after having been trained off-line under specific common flight conditions. The second network's on-line weight adaptations based on its experiences should improve the controller's performance. It can be authority limited as required by the application. The results in this paper do not include the second aritifical neural network (ANN).

21.3.3   Trim Network

The trim (position) network is a neural network, or lookup table, that is trained, or programmed, to schedule the aircraft's nominal actuator position and aircraft orientation as a function of operating conditions. This section describes the concept of trim, explains why it is a critical element in nonlinear flight control system design, and provides an overview of a method for determining trim. Further details of the development of the trim network can be found in [5]. This trim method, though applied to helicopters here, can be extended to any general control system.

Controlling the helicopter is a nonlinear control system design problem. An important part of nonlinear control system design, often not discussed, is the ability to determine the trim states for the system (e.g., helicopter) over all operating conditions. The trim states are the positions of the controls and the dependent system states associated with achieving a desired steady state condition. When the system is trimmed, the change in state derivative (e.g., acceleration) is either zero or minimized.

For example, with an aircraft one tries to adjust (trim) the controls to balance the aerodynamic, inertial and gravitational forces and moments in all axes at all times. The aircraft is trimmed when the desired balance is achieved or the aircraft enters a desired steady state. In the case of a helicopter, the controls to be trimmed are the four actuators and the dependent states to be trimmed are the pitch, roll and yaw, that is, seven trim state variables in total. These states are trimmed for the desired specified steady state translational velocities (u, v, w) and angular rates (p, q, r), that is, six variables representing desired steady states.

When flying a conventional mechanically controlled helicopter, a pilot continually trims the helicopter via his closed loop control. Similarly, traditional PID based control techniques inherently trim the helicopter, the integrators serve as the trim component. The ability to incorporate a trim ability into direct NDP is paramount to successfully applying direct NDP to the helicopter control problem. It serves as a good first guess at the required control solution, turning the global optimization problem into a local optimization problem. Previous direct NDP control designs were successful because the systems that were tested (e.g., the inverted pendulum) had a zero trim requirement [11]. Similarly, many flight control papers have assumed linear models in which case there is also a zero trim requirement since the model is linearized about a trim condition. In both cases, the problem definition implicitly makes it a local optimization problem.

Trimming requires determining the seven trim state variables to include four control positions (u) and three body angles (defined below to be x^2) for a given flight condition. For this reason, the original state vector x is divided into two parts, $x^1 = [u, v, w, p, q, r]$ and $x^2 = [\theta, \phi, \psi]$, where x^1 represents the desired steady state variables for which the seven trim positions, $[u_{trim} \ x^2_{trim}]$, are to be determined.

Figure 21.3 shows the neural network structure used to determine the seven aircraft trim positions u_{trim} and x^2_{trim} for any desired trim steady state x^1_{trim}. The trim network is a one-layer weight feed-forward network with seven biases (b) that correspond to the seven trim states. The network has a nonlinear sigmoid function fanning out the outputs. In using the action network for trim, the inputs to the trim network are zeroed and there are seven outputs from the network including four controls and three body angles.

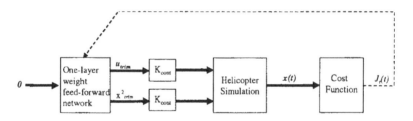

Fig. 21.3   Neural network structure for determining trim. The objective is to mininimize the specified cost function $J_t(t)$.

State equations associated with Figure 21.3 are as follows:

$$[u_{trim} \ x^2_{trim}] = f_{sig}(b), \tag{21.3}$$

where

$$f_{sig}(b) = [f_{sig}(b_1), \cdots, f_{sig}(b_7)], \tag{21.4}$$

and

$$f_{sig}(\xi) = 2(1 - e^{-\xi})/(1 + e^{-\xi}). \tag{21.5}$$

The cost function used is $J_t = 1/2\bar{x}^T\bar{x}$, where $\bar{x}_i = \frac{(x^1_i - x^1_{trim,i})}{x^1_{max,i}}$, $i = 1, \cdots, 6$.

The basic technique for determining the trim positions is to train the network, whose inputs are zeroed and that has biases, to minimize the cost function. After training is complete, the resulting biases then provide the aircraft trim positions. The weights in this network are of no importance other than the bias vector. This trim network is trained over a number of epochs where in each epoch, m, the plant model is initialized, the plant dynamics are evolved for a specified time t_f, and back-propagation is performed to adjust the trim network biases to minimize the objective at time t_f for that epoch, m. In general, the variable, v at time t_f for epoch m, is denoted as $v(m)$. For example, the objective function at time t_f for epoch m is denoted $J_t(m)$.

The network biases, b, are trained using the same gradient descent training method described for direct NDP. That is,

$$b(m+1) = b(m) + \Delta b(m), \tag{21.6}$$

$$\Delta b_i(m) = \lambda(m) \left[-\frac{\partial J_t(m)}{\partial b_i(m)} \right], i = 1 \ldots 7, \tag{21.7}$$

where λ is the learning rate. For the four biases corresponding to the trim controls

$$\frac{\partial J_t(m)}{\partial b_i(m)} = \frac{\partial J_t(m)}{\partial \mathbf{x}^1(m)} \frac{\partial \mathbf{x}^1(m)}{\partial u(m)} \frac{\partial u(m)}{\partial b_i(m)}, i = 1 \ldots 4. \tag{21.8}$$

For the three biases corresponding to the trim body angles

$$\frac{\partial J_t(m)}{\partial b_i(m)} = \frac{\partial J_t(m)}{\partial \mathbf{x}^1(m)} \frac{\partial \mathbf{x}^1(m)}{\partial \mathbf{x}^2(m)} \frac{\partial \mathbf{x}^2(m)}{\partial b_i(m)}, i = 5 \ldots 7. \tag{21.9}$$

In both cases the first partial is evaluated in the manner described earlier in the direct NDP mechanism. For the trim angles bias updates, the partials $\frac{\partial \mathbf{x}^1(m)}{\partial \mathbf{x}^2(m)}$ and $\frac{\partial \mathbf{x}^1(m)}{\partial u(m)}$ can be calculated numerically via perturbations to the system. Alternatively, the partial $\frac{\partial \mathbf{x}^1(m)}{\partial \mathbf{x}^2(m)}$ can be approximated analytically at hover from simple physics (rotate the gravity vector into the body frame and use appropriate small angle approximations). The partials are computed at hover only and used for all flight conditions for the results presented.

Once training is complete the trim positions are then determined from the network biases via the sigmoid function. Note that the trim network is trained off-line first, independent of the action and the critic networks in the direct NDP controller.

21.3.4 Other Considerations

Several other considerations need to be made in order to implement the direct NDP controller for control of sophisticated systems such as a helicopter. First, the action network is implemented as a traditional two-layer feed-forward network. However, since the action neural network's output (control) is typically limited to ± 1 by the sigmoidal non-linearity present in the last stage of the network, a control scaling factor is used for each control. The value chosen is typically $K_{cont} = u_{max}$ where u_{max} is the maximum control authority of the actuators. It is necessary to incorporate this scaling into the back-propagation when training the action network.

Second, the quadratic reinforcement signal described previously is used. Not only does this provide better information than the binary reinforcement signal, it is requisite for the command tracking control problem. Additionally, the normalization factor used in the reinforcement function (for a critic network), or cost function (for no critic), is decreased as a function of time at a specified rate until it reaches a lower limit. This allows the relative importance of each state to change with time as required by the application.

Third, more sophisticated failure criteria are implemented. Failure criteria are used to bound each state's allowed error during the controller's training phase. The allowed errors, shown in Table 21.2, are initially large and decrease as a function of time to an acceptable minimum. These failure criteria were chosen judiciously but no claims are made to their optimality. The results do show that these criteria create a control system that can control the helicopter well both in nominal conditions and when subjected to disturbances.

Heuristic failure criteria is one of the advantages of direct NDP if one does not have an accurate account of the performance measure. This is also one characteristic of the direct NDP design that differs from other neural control designs. The critic network plays the role of working out a more precise account of the performance measure for credit/blame assignment derived from the heuristic criteria. If the networks have converged, an explicitly desired state has been achieved which is reflected in the U_c term in the direct NDP structure.

Table 21.2   Failure Criterion for Helicopter Control

Aircraft State	Initial Allowed Error	Final Allowed Error	Error Rate
u, v, w	$20\,ft/s$	$4\,ft/s$	$-0.8\,ft/s/s$
p, q, r	$30°/s$	$6°/s$	$-1.2°/s/s$
θ, ϕ, ψ	$30°$	$6°$	$-1.2°/s$

Fourth, the input to the action network, x^*, needs to be a normalization of $(x - x_d)$ rather than simply a normalization of x. This provides non-zero state stabilization and command tracking capabilities. The input to the critic is $(x^*)^2$ which helps shape $J(x)$ and has resulted in significant performance improvements over results in prior research.

Fifth, network biases are added to the action network to accommodate control biases and disturbances in the plant (much like integrators are added to linear quadratic controllers).

21.3.5   Helicopter Stabilization Control Results

The first helicopter flight control goal to consider is that of stabilization. In stabilization, the goal is to maintain the aircraft's desired longitudinal velocity while maintaining zero lateral and vertical velocity and zero angular rates. To show robustness, each control goal is tested in three wind conditions: case (A) no wind, case (B) 10 ft/s step gust for 5 seconds, and case (C) turbulence simulated using a Dryden model with a spatial turbulence intensity of $\sigma = 5ft/s$ and a turbulence scale length of $L_W = 1750ft$. Characteristic to prior direct NDP research, the performance of direct NDP is summarized statistically. The statistical success of the direct NDP controller's ability to learn to control the helicopter is evaluated for a number of specified flight conditions.

For each flight condition 100 runs were performed to evaluate direct NDP's performance, where for each run the neural networks' initial weights were set randomly. Each run consisted of up to N attempts (trials) to learn to successfully control the system. An attempt was deemed successful if the helicopter stays within the failure criteria bounds described in Table 21.2 for the entire flight duration (1 minute). If the controller successfully controled the helicopter within N trials, the run was considered successful, if not, the run was considered a failure. For stabilization, N was 500.

For the 100 runs, the following statistic were collected: the successful run percentage, the average number of trials, and (for command tracking in the following section) the learning deviation. The success percentage reflects the percentage of runs for which the direct NDP system successfully learns to control the helicopter. The average number of trials is what it takes the direct NDP system to learn to control the helicopter. The learning deviation is the standard deviation of the number of trials from run to run. It is used to demonstrate the (in)consistency of the learning control performance.

Five flight conditions were considered for direct NDP based helicopter stabilization, the stabilization of the helicopter at hover and at 30, 60, 90 and 120 knots. The statistical success of the direct NDP controller's ability to learn to control the helicopter was evaluated for each flight condition. The results are shown in Table 21.3.

The neural network parameters used during training are provided in Table 21.4. The learning rates, β, for the action network and critic network were scheduled to decrease linearly with time (typically over a few seconds). In every time frame the weight equations were updated until either the error had sufficiently converged ($E < E_{tol}$) or N_{cyc} internal update cycles of the weights had occurred. N_h was the number of hidden nodes in the neural networks. Note that these parameters were chosen based on experience but were not tuned to optimize the results.

Two observations can be made about the results in Table 21.3. First, the results indicate that a large number of trials must be made before successful stabilization. This is not suprising for a learning system that learns from experience without any *a priori* system knowledge. The ramification is that this training needs to be done off-

Table 21.3   Direct NDP Learning Statistics for Helicopter Stabilization for Three Different Wind Conditions

	Condition (knots)	Hover	30	60	90	120
Case A	Percentage of Successful Runs	100%	100%	100%	100%	100%
	Average No. of Trials	18	47	36	30	70
Case B	Percentage of Successful Runs	28%	72%	82%	66%	30%
	Average No. of Trials	201	148	162	176	191
Case C	Percentage of Successful Runs	88%	63%	74%	77%	45%
	Average No. of Trials	128	214	206	186	200

Table 21.4   Neural Network Parameter Values Used for Stabilization

Parameter	α	$\beta_a(t_0)$	$\beta_a(t_f)$	$\beta_c(t_0)$	$\beta_c(t_f)$	$N_{cyc,a}$
Value	0.95	0.1	0.1	0.1	0.01	200
Parameter	$N_{cyc,c}$	$E_{tol,a}$	$E_{tol,c}$	N_h	K_{cont}	
Value	100	0.005	0.1	6	5.0	

line (i.e., not in a real helicopter), where failures can be afforded, until the controller is successfully trained. Once trained, the neural network weights are frozen and the controller structure shown in Figure 21.6 can be implemented in a helicopter. Limited authority on-line training can then be performed to improve system performance.

Second, direct NDP's learning ability appears to be slightly poorer at hover. This is speculated to be a phenomenon due to the lack of "persistence of excitation". At hover, the system states tend to evolve slower than at higher speeds and so there is less state variation (dynamics information) over the trials from which direct NDP can learn.

Once the system is successfully trained, the action network's weights are frozen and the system can be tested. Typical and statistical time histories of the system performance when tested are shown for a flight at 30 knots in turbulence (Figure 21.4) and a flight at 90 knots exposed to a step gust (Figure 21.5). The typical time history was a randomly chosen simulation selected from the repertoire of test runs for that flight condition. The average error and error deviation reflect the Monte Carlo sample mean and sample variance over time when testing all of the successfully trained networks reflected in the statistics shown in Table 21.3 for the specified flight condition. The results show that once the direct NDP controller has been successfully trained, it can reliably and consistently drive the system to its desired states. Time history plots for the other flight conditions are similar.

21.4   DIRECT NDP MECHANISM APPLIED TO HELICOPTER TRACKING CONTROL

Attention is now turned toward refining the direct NDP framework developed in the previous section for the helicopter flight control command tracking problem. Figure 21.6 outlines the direct NDP control structure applied to helicopter tracking control. It is the same as the design in the prior section except that the action network consists of a structured cascade of artificial neural networks. The explicit structure embedded in this action network, lacking in the prior direct NDP designs, allows the direct NDP controller to more easily learn and take advantage of the physical relationships and dependencies of the system.

Such structure in the action network is similar to classic controllers for helicopters, providing for inner loop body rate control, attitude control and outer loop velocity control. In this way, the explicit relationships between body angular rates, attitudes and translational velocities are taken advantage of. The potential advantage of the structured ANN over classic design methodologies is that it permits full cross-axes control coupling that many single-input-single-output (SISO) PID controller designs do not. However, the structured ANN does introduce a level of human knowledge/expertise to the direct NDP implementation that is not transparent to non-experts.

It is possible, but rather cumbersome, to show that a classic proportional controller can be equated to one instance (one set of weights) of the structured ANN if the network nonlinearities are removed (or linearized about the network operating point). Such a relationship between the two designs can be used to provide a good first guess of the action network weights should one want to apply this "expert" knowledge to the learning system. However, all results presented here were obtained for a learning system that was trained from scratch without using any expert knowledge.

To perform command tracking, a desired state vector $\mathbf{x}_d$ to be tracked needs to be specified. For helicopters, it is well established that only four of the states in the state vector $\mathbf{x}$ are explicitly controllable. In this experiment, the velocities u, v and w and the aircraft's yaw, ψ are the four states to be controlled. The rotational velocities and remaining Euler angles, pitch and roll, are determined by direct NDP to achieve the specified tracking goal. This new desired tracking vector, $\mathbf{x}_d^0 = [u_d, v_d, w_d, \psi_d]$, is a subset of the original desired state vector $\mathbf{x}_d$.

Referring to Figure 21.6, the inputs to the first ANN are the longitudinal and lateral velocity errors, $u_{err} = u_d - u$ and $v_{err} = v_d - v$, respectively. The first ANN outputs are the resulting desired pitch and roll of the helicopter, θ_d and ϕ_d. The inputs to the second ANN are the errors in the aircraft attitudes, $\theta_{err} = \theta_d + \theta_{trim} - \theta$, $\phi_{err} = $

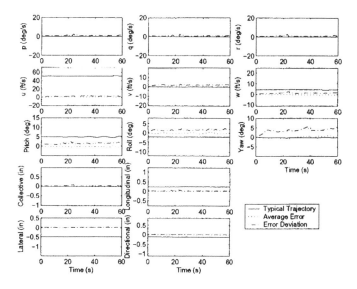

Fig. 21.4   Typical and statistical state and control trajectories for helicopter stabilization at 30 kts in turbulence.

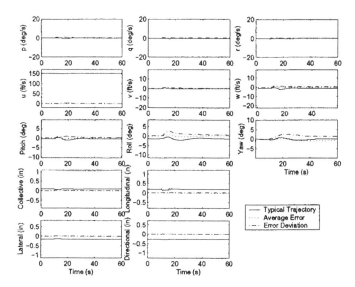

Fig. 21.5   Typical and statistical state and control trajectories for helicopter stabilization at 90 kts with a step gust.

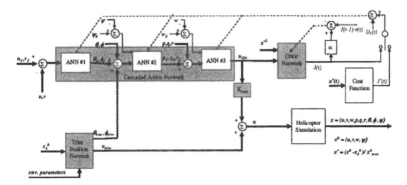

Fig. 21.6   Direct NDP based helicopter controller integrating three cascaded artificial neural networks in one action network, a trim network, and a critic network.

$\phi_d + \phi_{trim} - \phi$ and $\psi_{err} = \psi_d - \psi$. The second ANN outputs are the desired roll, pitch and yaw rates (p_d, q_d, r_d) of the helicopter, as a function of the attitude errors. These outputs are then summed with the actual angular rates to obtain the angular rate errors, $p_{err} = p_d - p$, $q_{err} = q_d - q$ and $r_{err} = r_d - r$. The angular rate errors and the vertical velocity error $w_{err} = w_d - w$ then form the inputs to the third ANN. The third ANN then computes the controls $u_{NN} = [u_{NN,1}, u_{NN,2}, u_{NN,3}, u_{NN,4}]$ as a function of the angular rate and vertical velocity errors. The resulting u_{NN}, which is normalized because of the ANN structure, is then scaled by the controller scaling gain K_{cont} and summed with the nominal trim control from the trim network. That is, $u = K_{cont}u_{NN} + u_{trim}$.

As described in Section 21.3, the objective of the direct NDP controller is to create a series of control actions to optimize a discounted total reward-to-go function. For the helicopter tracking problem we need to define a slightly different quadratic reinforcement signal

$$r(t) = -\sum_{i=1}^{n} \left(\frac{(x_i^0 - x_{d,i}^0)}{x_{max,i}^0} \right)^2 , \tag{21.10}$$

where in this application $n = 4$, x_i^0 is the i-th state variable of $\mathbf{x}^0 = [u, v, w, \psi]$, and $\mathbf{x}_d^0 = [u_d, v_d, w_d, \psi_d]$ is the desired state and $x_{max,i}^0$ is a normalization term.

The critic and action networks were then trained per Chapter 5. However, the equations governing the training of the action network were significantly more complex than prior work since back-propagation must be performed through ANNs 2 and 3 for the training of ANN 1. In essence, it is equivalent to training a six layer network if each ANN has two-layers of weights. The failure criteria used were the same as those for stabilization.

21.4.1   Helicopter Tracking Control Results

This section presents results showing the direct NDP controller's performance controlling the helicopter for a variety of maneuvers. For each maneuver, the control goal is for the helicopter to track the desired longitudinal velocity trajectory while maintaining zero lateral and vertical velocity and zero steady state angular rates. Statistics were then collected in the same manner as for the helicopter stabilization control results. The exception was the maximum numer of trials, N, which was 5000 for tracking. The parameter N was increased for command tracking because of the increased complexity of the task compared to stabilization. Since this is a complex, nonlinear, MIMO control system design problem, it provides a good idea of how well this ADP algorithm can generalize.

Fourteen maneuvers were considered: seven accelerations from hover to 50 ft/s at various accelerations and seven decelerations from 100 ft/s to 50 ft/s at various decelerations. Table 21.5 statistically summarizes the learning ability of the direct NDP controller to perform the seven acceleration maneuvers. Results for the seven deceleration maneuvers are provided in Table 21.6. All maneuvers are conducted for each of the three wind conditions cited above. The neural network parameters used during training for command tracking are provided in Table 21.7.

Table 21.5   Learning Statistics for Hover to 50 ft/s Maneuver at Various Accelerations for Three Different Wind Conditions

Cond.	Acceleration (ft/s^2)	2	3	4	5	6	7	8
Case A	Success Percentage	94%	62%	67%	65%	66%	66%	74%
	Average No. of Trials	1600	2019	2115	1950	1983	2028	1870
	Learning Deviation	214	339	324	307	293	306	252
Case B	Success Percentage	96%	66%	76%	70%	50%	57%	53%
	Average No. of Trials	1367	1720	1770	1874	2173	1970	2419
	Learning Deviation	191	280	255	275	381	337	400
Case C	Success Percentage	95%	98%	97%	85%	60%	56%	58%
	Average No. of Trials	642	824	1126	1843	1842	2145	2379
	Learning Deviation	115	128	165	263	313	333	403

In addition to the tabular learning statistics, both statistical and typical time history plots of the aircraft states are provided for two cases in the testing phase. Figure 21.7 shows both the statistical average state error and error deviation over all successful runs and a typical plot of the controller performance for a hover to 50 ft/s maneuver at an aggressive $5 ft/s^2$ acceleration in the presence of turbulence. Figure 21.8 shows both the statistical average state error and error deviation and a typical plot of the

Table 21.6   Learning Statistics for 100 ft/s to 50 ft/s Maneuver at Various Decelerations for Three Different Wind Conditions

Cond.	Acceleration (ft/s^2)	-2	-3	-4	-5	-6	-7	-8
Case A	Success Percentage	98%	90%	85%	80%	84%	76%	73%
	Average No. of Trials	759	1700	1610	2114	1516	1800	2045
	Learning Deviation	105	227	226	291	219	248	292
Case B	Success Percentage	99%	85%	74%	71%	77%	76%	78%
	Average No. of Trials	1260	1460	1650	1979	2030	1950	1737
	Learning Deviation	181	215	229	298	295	264	239
Case C	Success Percentage	100%	98%	93%	93%	97%	89%	91%
	Average No. of Trials	778	1258	1373	1489	1236	1350	1677
	Learning Deviation	105	180	183	200	162	186	220

controller performance for a 100 ft/s to 50 ft/s maneuver at $4ft/s^2$ deceleration in the presence of a step gust. Helicopter and control dynamics are similar for the other maneuvers once the learning controller becomes stabilized in learning.

It is worth mentioning that a comprehensive analysis on the convergence performance of an entire direct NDP system in general does not exist, nor does an analytical framework on the relationship between the performance of the direct NDP learning controller vs. the learning parameters. It has been argued [11] that updating individual networks alone, action or critic for example, may be viewed as a stochastic approximation problem and therefore, conditions similar to the Robbins-Monro algorithm may be used as guidelines in scheduling the learning parameters. Quantitatively, it is observed that the direct NDP learning parameters do impact the learning ability of the learning controller. For example, the learning rate for action networks can be tuned to perform different maneuvers with different system outcomes. This is illustrated by Table 21.8 which shows the direct NDP system performance for learning rates (β_a) of 0.2 and 0.02 for both more aggressive and less aggressive maneuvers. Lower learning rates improve the success for more aggressive maneuvers but decrease the learning ability (increase the number of trials) required to learn for less aggressive maneuvers.

It is interesting to note that despite what one may expect, overall the direct NDP controller more reliably and more quickly learns to control the helicopter in the presence of turbulence. This is clearly evident in Tables 21.5 and 21.6. The learning performance improvement can be attributed to the sustained larger excitation, due to turbulence, to both the neural networks inputs and the cost function evaluation. As a result, the network weights change more in the weight update Eq. (5.10) and thus the learning system explores more of the solution space and is less likely to become trapped in local minima that do not provide adequate control solutions. This suggests that in applications where turbulence or other excitation is not natural, it may be prudent to create an artificial equivalent in order to improve learning performance.

Also note that, as with the stabilization results, a large number of trials must occur to successfully learn to perform the maneuver. Further, the more aggressive the maneuver (the higher the acceleration), the more trials that are required. Again, the ramification is that this training needs to be done done off-line (i.e., not in a real helicopter), where failures can be afforded, until the controller is successfully trained. Once trained, the neural network weights are frozen and the controller can be implemented in a helicopter. Limited authority on-line training can then be performed to improve system performance.

Table 21.7   Neural Network Parameter Values Used for Command Tracking

Parameter	α	$\beta_a(t_0)$	$\beta_a(t_f)$	$\beta_c(t_0)$	$\beta_c(t_f)$	$N_{cyc,a}$
Value	0.95	0.02	0.02	0.1	0.01	200

Parameter	$N_{cyc,c}$	$E_{tol,a}$	$E_{tol,c}$	N_h	K_{cont}	
Value	100	0.005	0.1	6	2.5	

Table 21.8   Learning Statistics for Hover to 50 ft/s Maneuvers as a Function of Learning Rate

β_a	Acceleration (ft/s^2)	2	6	8
0.02	Success Percentage	94%	66%	74%
	Average No. of Trials	1600	1983	1870
	Learning Variance	214	293	252
0.2	Success Percentage	100%	80%	20%
	Average No. of Trials	248	1708	2809
	Learning Variance	71	421	439

21.5   RECONFIGURABLE FLIGHT CONTROL

Reconfigurable flight control (RFC) is the ability of a helicopter flight control system to reconfigure itself so that when a failure occurs, for example an actuator is dam-

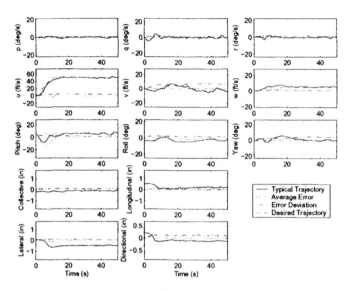

Fig. 21.7   Statistical and typical state and control trajectories of the helicopter for a hover to 50 ft/s maneuver at $5 ft/s^2$ acceleration in turbulence. Turbulence is simulated using a Dryden model with a spatial turbulence intensity of $\sigma = 5 ft/s$ and a turbulence scale length of $L_W = 1750 ft$.

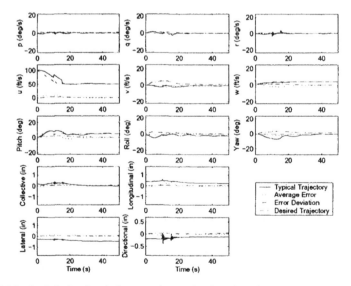

Fig. 21.8   Statistical and typical state and control trajectories of the helicopter for a 100 ft/s to 50 ft/s maneuver at $-4 ft/s^2$ acceleration exposed to a step gust. The step gust is 10 ft/s in magnitude and has a 5 second duration.

aged or fails, control of the helicopter remains. Such reconfigurable flight control has been widely demonstrated on fixed-wing aircraft, taking advantage of control redundancies. However, on any existing helicopter the loss of a main rotor actuator is catastrophic. This section describes how direct NDP can be be used as a reconfigurable flight control method for helicopter actuator failures, and demonstrates the method on a (rather benign) actuator failure.

Direct NDP already has been successfully demonstrated as a learning control scheme for helicopter flight control. From a neural control perspective this is a major accomplishment as it illustrates that neural networks can be taught to control very sophisticated unstable systems. It is now shown that direct NDP also has the capability to learn to control these sophisticated systems even when one of the underlying controls fails. One implication is that direct NDP may be applicable to failure accommodation for other physical systems where the failure accommodation solution is not apparent using known underlying design principles.

To perform reconfigurable flight control, the direct NDP controller simply needs to be trained to learn to recover from the failure in the same manner that it was trained to control the aircraft in the prior sections. The control goal is to maintain the desired aircraft states. If needed, weighting factors can be used in the reinforcement signal to place different emphasis on the various aircraft states. This allows certain less important states to be sacrificed while emphasis is placed on the more important aircraft states. For example, the control goal may permit some variation in longitudinal velocity, while maintaining emphasis on zero steady state angular rates (i.e., keep the aircraft upright).

The implementation of direct NDP for reconfigurable flight control can be identical to that presented earlier. However, it was found that direct NDP performs much better for reconfigurable flight control if one change is made. This change is to have the direct NDP controller "control" blade angles, rather than actuator position as shown in Figure 21.6. The performance of the new implementation is improved because it minimizes the amount of terms required to be computed during back-propagation when training the controller (and eliminates the difficulty in computing one of the terms). However, in doing so, this adds one piece of expert knowledge, the mechanical relationship that exists between the actuators and the blade angles in the upper flight control system (e.g., the swashplate and pitch links). While this relationship is well known, adding this "expert knowledge" does remove some of the purity of direct NDP as a generic learning system.

The ability of direct NDP to learn to control the helicopter when a control actuator fails is now demonstrated. Consider a helicopter flying at 60 knots that suffers a main rotor actuator failure at time $t = 5.0s$. Figure 21.9 shows the recovery from the failure when using direct NDP, in contrast to a classically designed PID controller which is unable to retain control of the helicopter.

Though the failure was more benign than what may be typically expected, in the sense that the failure position was very close to the trim position, it does illustrate that direct NDP is able to learn to accommodate such failure without any expert knowledge of the system. Such learning-based failure accommodation may have promise

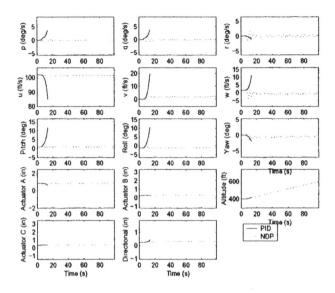

Fig. 21.9 Direct NDP is shown to be able to control the helicopter despite the presence of an actuator failure.

for failure recovery in other systems where the dynamics are not well understood. However, it should be mentioned that direct NDP for RFC does not find solutions for all failure cases in the work done to date. Thus, further work in making this method more reliable is required.

Note that in the above example, neither direct NDP nor the PID controller used any knowledge-based reconfigurable flight control method. However, direct NDP can be combined with knowledge-based reconfigurable flight control technique should they be available [6]. This permits the system to take advantage of known methodologies while using direct NDP to improve performance or handle cases that the known methodology may have difficulty with.

21.6 CONCLUSIONS

This chapter has advanced neural dynamic programming control research by introducing the direct NDP control structure to sophisticated helicopter stabilization and tracking control problems. Paramount to this was the development of a method for trimming the helicopter system and a structured approach to implementing the action network. A sophisticated nonlinear validated helicopter model was used to test the controller and its ability to learn to perform stabilization and a number of difficult maneuvers. The research has shown that the direct NDP is able to successfully stabilize a helicopter over a wide range of flight conditions and subject to various disturbances. The research has also shown that the direct NDP can control a helicopter for a wide range of realistic maneuvers and over a wide range of flight conditions, a few examples of which were illustrated. Thus, it appears that direct NDP is a viable candidate for controlling complex MIMO systems and is suited particularly well for on-line and complex multi-axes coupling control applications. The results have demonstrated the generalization capability of a learning control system, namely, direct NDP, to a large, continuous state problem. The same principles can also be directly applied to large, discrete state/action problems.

This chapter has presented a general overview of a complex design using direct NDP. The emphasis has been on the procedure of formulating the problem, setting up the objectives, implementation details, results and insights that can be gained from doing direct NDP designs. More detailed examinations of the helicopter problem including the helicopter model, the trim system, flight control system requirements, helicopter geometry and optimization for reconfiguration consideration, and many other technical details can be found in [7], or [4–6] for helicopter stabilization, tracking, and reconfiguration control, respectively.

Acknowledgments

The research was supported by the National Science Foundation under grants ECS-0002098 and ECS-0233529. The first author was also supported by an academic leave provided by the Boeing Company.

Bibliography

1. S. Balakrishnan and V. Biega, Adaptive-critic-based neural networks for aircraft optimal control, *AIAA Journal of Guidance, Control, and Dynamics,* vol. 19, no. 4, pp. 731–739, 1996.

2. J. A. Bagnell and J. Schneider, Autonomous helicopter control using reinforcement learning policy search methods, *Proc. International Conference on Robotics and Automation,* vol. 2, pp. 1615–1620, 2001.

3. G. Buskey, J. M. Roberts, and G. Wyeth, Online learning of autonomous helicopter control, *Australasian Conference on Robotics and Automation,* pp. 19–24, 2002.

4. R. Enns and J. Si, Apache helicopter stabilization using neural dynamic programming, *AIAA Journal of Guidance, Control, and Dynamics,* vol. 25, no. 1, pp. 19–25, 2002.

5. R. Enns and J. Si, Helicopter trimming and tracking control using direct neural dynamic programming, *IEEE Trans. Neural Networks,* vol. 14, no. 4, pp. 929–939, 2003.

6. R. Enns, and J. Si, Helicopter flight-control reconfiguration for main rotor actuator failures, *AIAA Journal of Guidance, Control, and Dynamics,* vol. 26, no. 4, pp. 572–584, 2003.

7. R. Enns, *Neural Dynamic Programming Applied to Rotorcraft Flight Control and Reconfiguration,* Ph.D. Thesis, Arizona State University, Tempe, AZ, 2001.

8. S. Kumar, J. Harding, and S. Bass, AH-64 apache engineering simulation non-real time validation manual, *Technical Report USAAVSCOM-TR-90-A-010,* 1990.

9. D. Prokhorov and D. Wunsch II, Adaptive critic designs, *IEEE Trans. Neural Networks,* vol. 8, no. 5, pp. 997–1007, 1997.

10. H. Shim, T. J. Koo, F. Hoffmann, and S. Sastry, A comprehensive study of control design for an autonomous helicopter, *Proc. 37th IEEE Conference on Decision and Control,* vol. 4 , pp. 3653–3658, 1998.

11. J. Si and Y. Wang, On-line learning by association and reinforcement, *IEEE Trans. Neural Networks,* vol. 12, no. 2, pp. 264–276, 2001.

12. E. A. Wan, A. A. Bogdanov, R. Kieburtz, A. Baptista, M. Carlsson, Y. Zhang, and M. Zulauf, Model predictive neural control for aggressive helicopter maneuvers, *Software Enabled Control: Information Technologies for Dynamical Systems,* IEEE Press, 2003.

Handbook of Learning and Approximate Dynamic Programming
Edited by Jennie Si, Andy Barto, Warren Powell and Donald Wunsch
Copyright © 2004 The Institute of Electrical and Electronics Engineers, Inc.

22 Toward Dynamic Stochastic Optimal Power Flow

JAMES A. MOMOH

National Science Foundation

Editor's Summary: This chapter deals with optimal power flow tools, which must be capable of handling multiple objectives while simultaneously satisfying operational constraints. This chapter contains an overview of the generalized optimal power flow problem, descriptions of the traditional techniques used for optimal power flow and their limitations, and discussion of the current state-of-the-art in robust optimal power flow. Finally, the author presents a generic framework for fitting ADP techniques into the optimal power flow framework. This is presented as an open problem for ADP, and the framework is given so that researchers can test and compare different techniques.

22.1   GRAND OVERVIEW OF THE PLAN FOR THE FUTURE OPTIMAL POWER FLOW (OPF)

Optimal Power Flow (OPF) is one of the most important operational functions of the modern day energy management system. There are several objectives of optimal power flow and among them are: finding the optimum generation amongst the existing generating units, such that the total generation cost is minimized, while simultaneously minimizing the voltage deviation of the system to improve the security and loss minimization of the entire network and furthermore attempting to satisfy the power balance equations and various other constraints in the system. The constraints imposed on the system include voltage magnitudes of the buses and active and reactive power constraints as well as transformer or phase shifter taps constraints, which are discrete in nature. The optimal power flow tool must be capable of handling multiple objectives while simultaneously satisfying operational constraints.

   A power flow may have any number of operating limit violations. When such conditions occur, the operator may wish to determine if the troubles can be alleviated by appropriate corrective actions. This analytical process is known as Optimal Power

Flow. The optimal actions analyses are formulated as an optimization problem with the objective of minimizing load curtailment, MW generation redispatched, and transformer phase angle adjustment. It includes a standard ac power flow solution with local automatic adjustments, power system network linearization, and a linear programming solution to relieve the overload and voltage limit violations. The corrective action algorithm must recognize several types of constraints and controls. Constraints are operating limits imposed on bus voltages, branch flows, and power transfers over interfaces, and so on. The system troubles in contingency analysis are violations of such operating constraints. Controls include generator real and reactive power generation, phase shifter angles, and busload curtailment schedule voltages of generators and ULTC-transformers, reference values of HVDC systems and Flexible AC Transmission System devices (Unified Power Flow Controller UPFC, Static VAR Compensators SVC, Solid State Synchronous Compensator STATCOM, and Thyristor Controlled Series Capacitor (TCSC)).

The objective of the corrective action algorithm is to observe all constraints while optimizing the weighted sum of the control movement. The priorities of control actions are phase shifter control followed by generator real power shifts, and if required, load curtailment. This priority order is maintained by assigning the higher value of weight to the lower priority controls. There are several approaches adopted in the analytical process, which involves the application of intelligent systems, heuristic methods and hybridized methods incorporating both heuristic and computational intelligence techniques.

Optimal power flows have been widely used in planning and real-time operation of power systems for active and reactive power dispatch to minimize generation costs and system losses and to improve system voltage profiles. Typically, problems of voltage constraints and stability issues have been assumed decoupled and thus treated independently. However, as the system operates closer to its stability limits, such as its voltage collapse point, due to market pressures, this assumption does not apply any longer, and hence there is a need to consider these limits within the OPF. By including these stability limits in the OPF problem, optimization procedures can also be used to improve the overall system security while accounting at the same time for the costs associated with it, which is becoming an important issue in open electricity markets. The voltage stability problem in power systems has been widely studied, and the basic mechanisms that lead to a network voltage collapse have been identified and are now clearly understood. It has been demonstrated that the overall stability of the system is closely associated with the proximity of a system to a voltage collapse condition, i.e., as the system approaches a voltage collapse point, its stability region becomes smaller, resulting in a system that is less likely to survive contingencies hence, as a first approximation, one can use voltage stability criteria to account for the overall system stability.

Some current research includes neural network optimal power flow when seen in its entirety. It is probably the case that the operation and control of an interconnected power system is the largest and most involved control problem of any kind in any field. As usual, the cost side of the optimization problem so formulated is revenue generation, which are among the largest of any single body or organization of unified structure and on the supply side is the life-blood of every modern and developed economy.

In common with current usage, the on-line computer-based facilities on which system operation, control and management now draw so extensively are referred to as Energy Management Systems (EMSs). Modern EMSs provide extensive on-line computer-based control and communication facilities for meeting the central requirement in electricity supply of achieving the greatest possible system utilization at the lowest possible running costs subject to very many constraints both technical and non-technical.

The optimal power flow function is a key one in the suite of software modules that make up an EMS. At the present time, the on-line control and management facilities that OPF can achieve are limited by the processing capabilities of the computer system that supports it. Because OPF is one of the most involved and extensive of all EMS functions, its computing system overheads are high.

With the current research and state-of-the-art technology in OPF tools, there are pertinent issues that have not been adequately tackled. The traditional OPF involves the determination of the instantaneous optimal steady state solution of the power grid without addressing the need for time dependency and to some extent handling of discrete as well as continuous parameters of the system components. The drawbacks are:

- Lack of Foresight: the capability of the existing OPF to predict the future in terms of asset valuation and economic rate of return on investment in power system infrastructure subject to various system dynamics and network constraints.

- Lack of explicit optimization technique to handle perturbation and noise.

Since foresight is critical to the power system as far as network expansion, right-of-way procurement for transmission corridors and overall rate of return on investment are concerned, new tools are needed for the operation of the deregulated power system markets.

With these features fully integrated into traditional OPF, the deregulated power system markets can fully deal with both foresight and stochastic dynamics of the optimal power flow problems of utility companies worldwide.

New OPF tools capable of handling the traditional OPF problems such as Unit Commitment, Economic dispatch, loss minimization, VAR Planning, congestion management through the evaluation of Available Transfer Capability, market designs and price caps as well as voltage stability problems are needed. The tools must have, in addition to handling the traditional OPF problems, a futuristic and time dependence module for evaluation of the impact of investment for the future while taking into account the stochastic nature of the power system network. These will therefore require new models of the existing system to be developed and incorporated into the new OPF.

22.1.1   Achievements of OPF and Documented Limitations

From the general class of OPF problems solved to date and the methodologies employed, there are significant achievements worthy of mention. Among them are:

Optimal power flow using successive sparse and non-sparse quadratic programming, successive non-sparse separable programming, Newton's method and most recently Interior point methods. Some of these implementation techniques, with appropriate modeling and formulation of the OPF problem, resulted in commercial software programs. These programs are being maintained regularly to achieve high performance and give better, accurate results compared to their earlier versions due to improvements made in the optimization theories and significant research advances over the years. These improvements include high flexibility and modeling of new objective functions and constraints, reliability of solutions obtained, good user friendly interface and visualizations.

Research has led to the development of an ultra-high-speed computing architecture with the aim of allowing all aspects of OPF to be implemented. The architecture is one based on the massively parallel-processing structure of arrays of neural networks. The main features of OPF in EMS are:

- Operating cost minimization;

- Network active-power loss minimization and restoration control strategies;

- Control of generator active-power loading levels, off-nominal tap-positions of transformers, target voltage settings of on-load tap-changing transformers, generator excitation controllers, and static VAR compensators, selection for service of fixed-value reactors and capacitors used for reactive-power and voltage control purposes;

- Inclusion of contingency constraints relating to any number of postulated outage conditions and power network configurations;

- Post-contingency rescheduling so that optimal operation is achieved without violating any security constraint;

- Ultra-high-speed computation in which the computing time is independent of size of the power network and the number of postulated contingency cases of which it is required to take account;

- Bidding analysis and strategizing of competitive bidding;

- Market oriented dispatch and demand side bidding;

- Bidding and contracting rules formulation;

- Ancillary services/reactive VAR support;

- Load curtailment analysis and post-optimal analysis;

- Dispute resolution and litigation.

22.1.1.1 Some Extended Application of OPF

- Capacitor Installation

  Optimal sizing and location of new capacitors in the network to ensure a defined level of steady-state security. This is solved as Security Constrained Optimal Power Flow to minimize the investment and operating costs. The current challenge is to embed this OPF solution in a multi-year planning process that considers variation in system load, generation and network topology changes.

- Transmission Service Pricing

  Transmission constrained marginal costs can be obtained for factors including binding power system operating constraints, production capacities, equipment regulation limits, area interchanges, losses and bus powers. The cost of MVAr supply and delivery is being addressed by utilities in the deregulated power industry.

  Through the use of OPF in EMS, utilities tend to benefit in system management and control in achieving the greatest possible network system utilization at the lowest possible operating costs.

22.1.1.2 Challenges and Limitations to On-Line OPF Implementation

There are numerous challenges faced in implementing OPF in the deregulated power system, including the following specific operational requirements and challenges.

One of the operational requirements is robustness. OPF solutions should not be sensitive to the starting point used, and changes in solution point should be consistent with that in the operating constraints.

The three main problems faced are: the existence of multiple local, weak convergence, or non-convergence of the solution method used; poor input data; and inadequate models of the power system to be able to solve real life problems. For local minima problems, the nature of actual power system is responsible for getting the solution stacked. However proper models and operating information is needed to fully specify the problem and eliminate the non-uniqueness. The convergence problem arises as a result of an inherent limitation of the methodology employed for the solution and hence more serious. This calls for new methods and tools from other fields of research to address such limitations.

Accurate data to mimic the existing power system must be modeled and fine tuned to meet the network standards and operating constraints before attempting to run the program. Choice of initial starting point of the solution algorithm is necessary to avoid entrapment in local extrema instead of global extrema of the operating regime of the power system network.

In the new deregulated power system, the capabilities of conventional optimal power flow program itemized above must be enhanced to handle new components, discrete and continuous parameters and accurate modeling of system components as well as extensive data involving many power systems interconnections. There is the need for expansion of the scope of OPF problem to embrace:

- More flexible controls and constraint priority strategies;

- Incorporation of control and load dynamics;

- Inclusion of start-up and shut-down constraints of certain controls and other operating constraints that meet specific practical requirements;

- Hydro modeling;

- Incorporating voltage stability and other dynamic constraints relevant in the on-line environment;

- Modeling of non-linear and voltage dependent loads;

- Coordination of controls with different response characteristics;

- Modeling of prohibited operating zones on the cost curves;

- Effective modeling of external system suitable for optimization applications;

- Time restrictions on control constraint violations;

- Cost penalty thresholds for constraints reinforcements;

- Effective branch switching modeling.

Conventional and existing OPF tools use all possible control actions for getting the optimal solution, but it may not be feasible to execute more than a limited number of control actions. The remedies could be (1) to limit number of control action of each type, (2) assign initiation cost for each control action, (3) utilize fuzzy based method to reduce ineffective controls, though further work is needed in this area to lead to robust solutions, and (4) compromise in the short run by reliance on near-optimal solutions that incorporate sound engineering rules.

For modeling discrete quantities of the power system, the step-size of the discrete variables should be sufficiently small to warrant continuous approximations. The possible approaches for accurate results and modeling of discrete variables are:

- Simulations of conventional power flow program with all discrete variables fixed on their steps;

- Introduction of penalty function for discrete controls;

- Implementing a trajectory of OPF control shifts that does not exacerbate existing violations or cause additional ones.

Consistency of OPF with other online functions is vital to the overall benefit of the system. This required consistency in the study mode and research as well as closed loop implementation of OPF is a current subject of research. The design of interface and of OPF in closed-loop mode with other functions, namely, Unit Commitment, Economic Dispatch, Security analysis, and Automatic Generation Control AGC presents formidable tasks. The central aspect is the effective coordination of OPF-ED-AGC control hierarchy.

22.2   GENERALIZED FORMULATION OF THE OPF PROBLEM

Minimize $F(x, u)$,

subject to $\begin{array}{ll} g(x, u) = 0 & \text{Equality constraints,} \\ h(x, u) \leq 0 & \text{Inequality constraints,} \end{array}$

where $x's$ are the state variables of the power system and $u's$ are the control parameters of the power system.

A. OBJECTIVE FUNCTION DEFINITION

1. Minimum fuel cost

The cost function of the generators is assumed to be quadratic and the objective function can be stated as:

$$\text{Minimize } F = \sum_{i=1}^{NG} (a_i P_{gi}^2 + b_i P_{gi} + c_i), \qquad (22.1)$$

where

$$\begin{array}{rl} P_{gi} & : \quad \text{Real power output of the } i^{th} \text{ generator;} \\ a_i, b_i \text{ and } c_i & : \quad \text{Cost coefficients of the } i^{th} \text{ generator;} \\ NG & : \quad \text{Total number of generator in the system.} \end{array}$$

2. Minimum line flow overloads

The objective function for minimal line overload on the transmission system is:

$$\text{Minimize } F = \sum_{i=1}^{NL} \left(P_{ij}(t) - P_{ij}^{\max}(t) \right)^2, \qquad (22.2)$$

where

$$\begin{array}{rl} P_{ij}(t) & : \quad \text{Real power flow through the } ij^{th} \text{ transmission line at time stage } t; \\ P_{ij}^{\max}(t) & : \quad \text{Maximum power flow through the } ij^{th} \text{ transmission line at time} \\ & \qquad \text{stage } t; \\ NL & : \quad \text{Total number of overloaded lines.} \end{array}$$

3. Minimum transmission losses

The losses in a power system are directly related to the slack bus power. The optimal power flow model for loss minimization problem in a typical power system is:

$$\text{Minimize } F_L = F(P_{\text{slack}}), \qquad (22.3)$$

where

P_{slack} : Power at the swing or reference bus;

F_L : System Loss function;

P_{slack} : Active power being delivered at the slack bus, and it is a function of the system state variables such that $P_{slack} = F_{slack}(V, \theta, T)$.

4. Minimal adjustment of phase shifters

Here we seek to minimize the number of phase shifters that are adjusted in a give time frame such that the objective function is:

$$\text{Minimize } F = \sum_{i=1}^{NS} W_i \phi_i, \tag{22.4}$$

where

W_i : Weighting factor of the phase shifter;

ϕ_i : Phase shift angle of the ith phase shifter;

NS : Total number of phase shifters in the network.

5: Minimal generation impact

The least generation shift is desired for the control action that is taken to relieve or reduce the power system congestion. The mathematical formulation of the objective function is stated as:

$$\text{Minimize } F = \sum_{i=1}^{NG} \left(P_{gi}(t) - P_{gi}^o(t) \right)^2, \tag{22.5}$$

where

$P_{gi}^o(t)$: Real power output of the ith generator before installing the phase shifter at time stage t;

ϕ_i : Actual real power output of the ith generator after installing the phase shifter at time stage t;

NG : Total number of overloaded lines.

B. CONSTRAINTS MODELING IN OPTIMAL POWER FLOW

In addition to the general linear/nonlinear constraints, the constraints relating to phase shifter variables such as phase shifter angle and maximal adjustment numbers should be included in the OPF formulation with phase shifter. It is important to select suitable constraints for OPF with phase shifter and the candidate constraints are as follows:

1. Upper/Lower limits of real power output of generators

$$P_{gi}^{\min} \leq P_{gi} \leq P_{gi}^{\max}, \tag{22.6}$$

where

$$\begin{aligned}
P_{gi} &: \text{Active power of generator } i; \\
P_{gi}^{min}, P_{gi}^{max} &: \text{Lower and upper real generation limits of unit } i, \text{ respectively.}
\end{aligned}$$

2. Power balance equation

$$\sum_{i=1}^{NG} P_{gi} = \sum_{k=1}^{ND} P_{dk} + P_L, \tag{22.7}$$

where

$$\begin{aligned}
P_{gi} &: \text{Active power of generator } i; \\
P_{dk} &: \text{Active power at load bus } k; \\
P_L &: \text{Total system losses;} \\
NG &: \text{Total number of generator buses in the network;} \\
ND &: \text{Total number of load buses in the network.}
\end{aligned}$$

3. Upper/Lower limits of reactive power output of generator

$$Q_{gi}^{\min} \leq Q_{gi} \leq Q_{gi}^{\max}, \tag{22.8}$$

where

$$\begin{aligned}
Q_{gi} &: \text{Reactive power of generator } i; \\
Q_{gi}^{min}, Q_{gi}^{max} &: \text{Lower \& upper reactive generation limits of unit } i, \text{ respectively.}
\end{aligned}$$

4. Upper/Lower limits of node voltage at PV buses

$$V_{gi}^{\min} \leq V_{gi} \leq V_{gi}^{\max}, \tag{22.9}$$

where

$$\begin{aligned}
V_{gi} &: \text{Node voltage of generator } i; \\
V_{gi}^{min}, V_{gi}^{max} &: \text{Lower and upper voltage limits of generator bus } i, \text{ respectively.}
\end{aligned}$$

5. Upper/Lower limits of node voltage at PQ buses

$$V_{di}^{\min} \leq V_{di} \leq V_{di}^{\max}, \tag{22.10}$$

where

$$
\begin{aligned}
V_{gi} &: \text{Node voltage at load bus } i; \\
V_{di}^{min}, V_{di}^{max} &: \text{Lower and upper limits of the node voltage at load bus } i.
\end{aligned}
$$

6. Power flow constraints of transmission lines

$$
P_{ij}^{\min} \leq P_{ij} \leq P_{ij}^{\max}, \tag{22.11}
$$

where

$$
\begin{aligned}
P_{ij} &: \text{Active power flow through line } L_{ij}; \\
P_{ij}^{min}, P_{ij}^{max} &: \text{Lower and upper power limits of line } L_{ij}.
\end{aligned}
$$

7. Upper/Lower limits of transfer capacity of the network

$$
\left[V_i^2 + V_j^2 - 2V_i V_j \cos(\theta_i - \theta_j) \right] / Z_L(l)^2 I_{L\max}^2(l) \leq 0, \tag{22.12}
$$

where

$$
\begin{aligned}
V_i &: \text{Voltage magnitude at bus } i; \\
V_j &: \text{Voltage magnitude at bus } j; \\
I_L &: \text{Thermal or current limit of line, } L_{ij}; \\
Z_L &: \text{Impedance of transmission line, } L_{ij}.
\end{aligned}
$$

8. Upper/Lower transformers limits (ULTCs)

$$
T_i^{\min} \leq T_i \leq T_i^{\max}, \qquad i \in NT, \tag{22.13}
$$

where

$$
\begin{aligned}
T_i &: \text{Tap setting of transformer } i; \\
T_i^{min}, T_i^{max} &: \text{Lower and upper tap limits of transformer } i, \\
& \text{respectively.}
\end{aligned}
$$

9. Step length of phase shift angle

$$
\phi_i^{\min} \leq \phi_i \leq \phi_i^{\max} \quad \forall \quad i \in \{1, NPS\}, \tag{22.14}
$$

where

$$
\begin{aligned}
\phi_i &: \text{Phase angle of the } i^{th} \text{ transformer;} \\
NPS &: \text{Total number of phase shifters in the network.}
\end{aligned}
$$

10. The maximal adjustment times of phase shifter per day

$$\sum_{i=1}^{N^{sch}} \phi_i^{\min} \leq \sum_{i=1}^{N^{sch}} \phi_i(t) \leq \sum_{i=1}^{N^{sch}} \phi_i^{\max}, \qquad (22.15)$$

where

$\phi_i(t)$: Angle of phase shifter transformer i at time t;

ϕ_i^{max} : Upper phase angle limit of the ith phase shifter;

ϕ_i^{min} : Lower phase angle limit of the ith phase shifter;

N^{sch} : Set of scheduled phase shifter transformers.

For the above-mentioned constraint set, it is necessary to analyze whether all these constraints are active or suitable for optimal power flow calculations with phase.

22.3 GENERAL OPTIMIZATION TECHNIQUES USED IN SOLVING THE OPF PROBLEM

Various optimization methods are utilized in the solution of the general OPF problem with multiple objectives and with equality and inequality constraints. These methods include:

- Linear programming;

- Nonlinear programming;

- Quadratic programming;

- Interior point (including Dual Affine algorithm, Primal Affine algorithm and Barrier algorithm);

- Hybridized method incorporating any of the conventional approaches and computational intelligence method;

- Computational intelligence techniques including evolutionary programming, artificial neural networks and adaptive critics networks.

These methods are suitable for the evaluation of the generalized optimal power flow problem. Each is suitable in its own domain with peculiar strengths and weakness. Based on the level of accuracy, faster convergence and storage required any of the methods could be used for the requirements needed.

In the evaluation of multi-objective functions coupled with both equality and inequality constraints analytical hierarchical process and pareto-optimal analysis could be used in conjunction with any of the above methods to assign priority and ranking to controls options used in the general formulation of the Optimal Power Flow problem.

Description of Optimization Techniques

Optimization techniques are characterized by the efficiency of the techniques employed for obtaining a 'good' starting or initial point, the robustness and completeness of the iterative process, and the termination conditions. Many approaches currently exist for solving the optimization problems in power system such as unit commitment, resource allocation, economic dispatch or generation rescheduling, network reconfiguration, and load shedding. These optimization approaches, some of which have been reviewed earlier, are characterized by the certain attributes/criteria that are used to evaluate optimal power flow tools. These attributes and criteria for evaluation include:

Adaptability

This is an evaluation measure of the capability to add new constraints, limits, and optimization function, which remain solvable by the optimization technique that is used in the optimal power flow tool.

Practicality

This involves evaluation of the method's ability to handle practical systems as well as physical system constraints. Further, it involves the ability of the method to display or convey the correct answers constantly to the user under different system conditions and inputs.

Efficiency / Reliability

This involves the method's ability to compute and convey the correct solution to the optimization problem in a minimum or reasonable time.

Sensitivity

This involves the measure of the method's reaction to practical changes such as the effect of parameter or network changes in the system. The sensitivity of the optimization method can be a useful attribute for quickly computing global optimal solution in an environment where computational time is costly.

A brief review and evaluation of the mathematical programming methods commonly used in optimal power flow is discussed below.

22.3.1   Linear Programming (LP)

Linear Programming, introduced by G.B. Dantzig (1947), exploits the linear structure of many problems in a wide variety of applications, including power system operation and planning. Typically, Linear Programming utilizes two method of computation. They are (1) the Simplex Method and (2) the Revised Simplex Method. The revised simplex method is more commonly used in practical applications. Variants of Interior

Point methods are also used for Linear Programming problems and several OPF tools exist using LP formulations.

There exist several solution techniques for solving the optimal power flow problem that are based on linear programming models. The objective function in the OPF problem is approximated to a linear relationship relative to the state and control variables. A first order Taylor series approximation is used which assumes continuity and convexity of the objective function.

Linear Programming has been used extensively to solve real and reactive generation costs in OPF as well as transmission loss calculations. The method, typically the revised simplex technique, assumes a linear model of the problem, which introduces some degree of inaccuracy in the final results. Its ability to handle discrete controls is done using several variants of Integer programming such as the branch and bound method.

22.3.2 Nonlinear Programming

In power system applications and analysis, the Newtonian method in nonlinear programming is among the most popular approaches. Other techniques in classical optimization include Quadratic Programming (QP), Lagrangian Multiplier and Reduced Gradient Methods. These methods are applicable to solving typical optimal power flow problems of electric power system. This is because most problems can be sufficiently modeled as piece-wise nonlinear functions suitable for the power system which is inherently nonlinear.

Nonlinear Programming (NLP) typically employs Lagrangian or Newtonian Techniques for the constrained and unconstrained optimization problems. The approach assumes that all objective functions are modeled as smooth and continuous functions. However, the mathematical response of the power system diverts from this assumption in many instances. Static devices such as the phase shifter transformers, series and shunt components (Capacitor and Inductor banks), Flexible AC Transmission Devices (FACTs), and Under Load Tap Changers (ULTCs) are devices with discrete transfer and control functions. Thus, nonlinear programming methods often suffer if the approximations to the actual models for these devices are not sufficiently accurate. The λ lambda parameter utilized in this approach does not reflect the future of the power system and therefore lacks foresight. Integrating foresight or dynamics and stochasticity into this approach will ensure the correct response and behavior of the power system in the future and can help in power system planning.

In power system applications, the Newtonian method takes advantage of the convexity and continuity conditions of the load flow model and is used in many powerful optimization tools where solution accuracy of AC power flow calculations is to be preserved. The commonly used Newton, Gradient Search, and MINOS optimization packages are available in the form of commercial-grade programs for solving the general optimal power flow problems.

22.3.3   Interior Point Method

Interior Point methods represent a special case for solving optimization problems using the linear programming model. The basic approach was developed and benchmarked by Karmakar et al. 1986. This has led to many variants of the approach, which can be adapted to many common real-time problems. These variants include the Primal Affine method, the Dual Affine method, and the Projection or Karmakar's technique and an *extended* Interior Point algorithm that was developed at Howard University.

The projection scaling method or Karmakar's algorithm is characterized by its ability to converge with any feasible initial point. This makes the method attractive and powerful for engineers and software developers. But the method depends on the location of a good initial point inside the polytope or region that is bounded by the constraint set of the optimization problem. Increasing the dimension of the problem does not degrade the computational speed of the algorithm.

The primal affine method is faster than the projection scaling method. It is a special case of the projection scaling method with improved computation of the step length of the translation vectors in the search space. The method solves linear objective function subject to linear *equality* constraints.

The dual affine method is the fastest of the three commonly used variants of Interior Point being reviewed in this report. Similar to the primal affine method, it is an enhanced version of Karmakar's scaling algorithm. The method utilizes a "barrier" or Logarithmic approach in computing the step lengths of the scaling vectors. This method is applicable to optimization problems where the objective function is linear subject to linear inequality constraints. A logarithmic penalty function is used in the formulation of the Lagrange function of the constrained optimization problem. The method employs minimal inverse calculations to increase the speed of computation.

Finally, in the past decade, primal-dual algorithms have emerged as the most important and useful algorithms from the interior-point class. Primal-dual algorithms, which include path-following algorithms (short-step, long-step, and predictor-corrector), potential-reduction algorithms, and infeasible-interior-point algorithms, can be extended to more general classes of problems such as semi-definite programming, and general problems in convex programming.

22.3.4   Barrier Optimization Method

The Barrier approach to a constrained optimization problem is the process of applying a logarithmic or concave function as a shadow price or special purpose Lagrangian multiplier. This process has let to an improvement in the solution of the Karush-Kuhn-Tucker (KKT) conditions of a linear programming problem. The method of bounding the interior of the solution space, which is also used to analyze the efficiency of the conjugate-gradient method in solving a system of linear equations, can also be also extended to the interior point method. The relationships between the primal-dual Newton interior-point method and the logarithmic barrier function method has led to more robust optimization algorithms in this class that are less prone to infeasibility unless the problem is ill-conditioned.

A special relationship between the primal-dual Newton interior-point method and the logarithmic barrier function method has also led to solution of LP problems with a primal-dual Logarithmic Barrier method. The solution method is based on solving the primal-dual penalty-barrier equations for a sequence of decreasing values of the scalar penalty-barrier parameter. An augmented penalty-barrier function is used as an objective function for both the primal and dual variables.

22.4 STATE-OF-THE-ART TECHNOLOGY IN OPF PROGRAMS: THE QUADRATIC INTERIOR POINT (QIP) METHOD

The linear programming model that incorporates linear objectives and constraints is often not suitable for adequate results to an OPF problem. As such, without compromising the speed advantage of the projection of scaling methods of interior point, a Quadratic Interior Point (QIP) method has been developed.

In the QIP formulation used in power system applications, the objective function is quadratic and the constraints set of the network and the devices on the power system have been linearized. The Quadratic Interior Point technique is by far the fastest of the variants of the interior point methods. Also, it is better at converging to an acceptable solution giving a wide range of initial point solutions. The QIP technique is the backbone of the computational tool, Robust Interior point Optimal Power Flow (EPRI/Howard University) under contract No. 3788-01. The research thrust therein is aimed at extending its capability to handle discrete variable such as phase shifter.

The QIP method is used in the state of the are in Optimal Power Flow Problems. In the previous section, we provided an analytical formulation and modeling of the OPF-based phase shifter problem. The solution strategy involves decomposing the problem into a multi-stage procedure. The first stage involves solving the optimal power flow problem using the QIP method. This method is incorporated in the Robust Interior Point Optimal Power Flow Program (RIOPF) and is used to obtain the optimal operating points without phase shifters. The second stage of the optimization process involves application of the rule-base scheme to optimize the discrete phase shifter tap setting in order to solve the line overload problem. The result of the OPF calculations serves as input to the rule base system. The firings of the rules were done to compute the discrete phase shifter tap setting under faulted or contingent scenarios.

Next, perform Cost-Benefit Analysis on the selected control strategies. It provides an economic rationale for using phase shifter control. Finally, the overall implementation strategy contains a post-optimal analysis for shadow price calculation in order to determine the impact of the phase shifters on the system. The final recommendations for phase shifter tap settings for optimal overload relief is now based on technical as well as economic judgments.

Interior Point Algorithm

The above-mentioned OPF model with phase shifter is a nonlinear mathematical programming problem. It can be reduced by an elimination procedure. The reduction of the OPF model is based on the linearized load flow around base load flow solution for small perturbation. The reduced OPF model has the format:

$$\text{Min } F = \frac{1}{2}X^T Q X + G^T X + C, \tag{22.16}$$

subject to

$$AX = B, \text{ with } X \geq 0. \tag{22.17}$$

The model given in Eq. 22.16 - 22.17 has a quadratic objective function subject to the linear constraints that satisfy the basic requirements of quadratic interior point (QIP) scheme. Generally, the effectiveness of interior point methods depends on a good starting point. The extended quadratic interior point (EQIP) method with improved initial conditions is presented in this paper. It features the improved starting point and faster convergence.

Integrated OPF Design with IP Application

This section describes the overall optimal power flow design scheme that incorporates the phase shifter control. The scheme features an interior point-based optimal power flow model and a rule-based scheme that is used to calculate the optimal adjustment schemes of the phase shifter tap settings, limit checking of constraints, line overloading state checking, sensitivity analysis for ranking phase shifters, rule-based system to handshake the adjustment of phase shifters, and a CBA and Shadow Price module.

22.5 STRATEGY FOR FUTURE OPF DEVELOPMENT

New extensions of RIOPF are needed for enhanced performance and improved capability for general-purpose OPF problems embodying formulation of new power system constraints and the introduction of new power system components such as Distributed Generation resources and FACTS device technology. This new extensions should be able to handle the optimal location and sizing of DGs and other system components

The RIOPF is a special purpose OPF tool that can be extended to integrate computational intelligence techniques for enhanced capability and performance in a dynamical changing network topology of the power system and stochasticity of input data for running the program.

For a dynamically changing network, the static optimization techniques cannot suffice hence the integration of Approximate Dynamic Programming (ADP) and other computational intelligence into RIOPF is a big challenge and requires substantial research effort in that direction.

This needed research for effective integration of RIOPF from power system perspective and ADP from applied mathematics can be of great help for the utility industry that currently depends upon static optimization techniques with fixed topological networks of the power grid.

The main goal of the review of the conventional OPF tools is to enhance the existing OPF tools and make them more capable in handling decision making for large optimization problems involving power system networks with changing topology and network congestion constraints. The idea is to infuse computational intelligence techniques to be able to accommodate foresight into its decision making capability as well as ensure that the stochastic nature of the state of the system network are adequately catered for.

The conventional (traditional) OPF tools lack *two basic ingredients* that are essential for the smooth operation of the power system.

These ingredients are:

- *Lack of foresight:* The capability of the existing OPF to predict the future in terms of asset valuation and economic rate of return on investment in power system infrastructure subject to various system dynamics and network constraints.

- *Lack of explicit optimization technique to handle perturbation and noise.* The power system is not static but changing with respect to load demands at any time period, hence these stochastic activities need to be modeled to reflect the practical network and hence obtain better results which have real physical implications. Probabilistic models for the power system change in load, control options and state estimations parameters has to be developed to enhance the future direction of OPF.

Since foresight is critical to the power system as far as network expansion, right of way procurement for transmission corridors and overall rate of return on investment are concerned, new tools are needed for the operation of the deregulated power system markets.

With these features fully integrated into the traditional OPF, the deregulated power system markets can fully deal with both foresight and stochastic features required for the optimal power flow problems of utility companies worldwide.

From the review of available of power system literature on OPF, their formulations and solution methodologies, the following methods have been utilized to solve the OPF problem.

Approximate Dynamic Programming (ADP) a computational intelligence technique can incorporate the time feature required for the future. Dual Heuristic Programming (DHP), a class of ADP can be used with the appropriate modeling of the power system to incorporate foresight and to enhance the stochastic description and prediction of the power network.

A unified OPF capable of handling dynamics and stochasticity will have to appropriately model (1) the system state variables and controls options which may be continuous or discrete and (2) integration with other computational intelligence such as GA or Evolutionary Programming of differing capabilities to handle continuous and discrete variables as well as differentiability of functions of state variables and controls. Genetic Algorithms can achieve combinatorial best solution when the ADP utilizes its strength in formulating the fitness function for global optimization. GA ensures global optimization and well suited for handling discontinuous or discrete functions due to its bitwise operation and manipulation of genetic operators of the fitness function.

22.5.1 Approximate Dynamic Programming (ADP)

Dynamic programming prescribes a search which tracks backward from the final step, retaining in memory all suboptimal paths from any given point to the finish, until the starting point is reached. The result of this is that the procedure is too computationally expensive for most real world problems [3].

There are three levels of computational intelligence, namely:

(1) Model based Adaptive Critics Designs using Approximate Dynamic Programming (ADP);

(2) Basic mammalian brain level;

(3) Levels beyond basic mammalian brain (Symbolic, semiotic, multi-modular or distributed and quantum neural nets).

Adaptive critic designs are neural network designs capable of optimization over time under conditions of noise and uncertainty. A family of ACDs was proposed by Werbos [2] as a new optimization technique combining concepts of reinforcement learning and approximates dynamic programming. For a given series of control actions that must be taken sequentially, and not knowing the effect of these actions until the end of the sequence, it is impossible to design an optimal controller using the traditional supervised learning ANN.

The adaptive critic method determines optimal control laws for a system by successively adapting two ANNs, namely, an *action neural network* (which dispenses the control signals) and a *critic neural network* (which 'learns' the desired performance index for some function associated with the performance index). These two neural networks approximate the Hamilton-Jacobi-Bellman equation associated with optimal control theory. The adaptation process starts with a non-optimal, arbitrarily chosen control by the action network; the critic network then guides the action network toward the optimal solution at each successive adaptation. During the adaptations, neither of the networks needs any 'information' of an optimal trajectory, only the desired cost needs to be known. Furthermore, this method determines optimal control policy for the entire range of initial conditions and needs no external training.

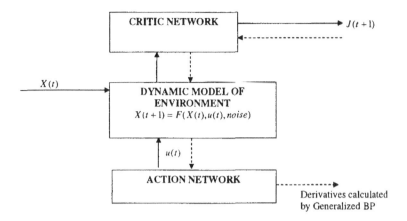

Fig. 22.1   Constituents of a general purpose-learning machine.

Table 22.1   Nomenclature for Adaptive Dynamic Programming

u	:	Action vectors
U	:	The utility which the system is to maximize or minimum
X	:	Complete state description of the plant to be controlled
γ	:	Usual discount rate or interest rate that is needed only in infinite-time-horizon problems
J	:	Secondary or strategic utility function
A	:	Action network
E	:	Error measure or function
W	:	Neural network weights
λ	:	Output of the critic network and the approximation to the J function
η	:	Learning rate

Action-Dependent Heuristic Dynamic Programming (ADHDP) is a critic approximating the J function, in Werbos' family of adaptive critics. A critic which approximates only the derivatives of the function J with respect to its states is called the Dual Heuristic Programming (DHP), and a critic approximating both J and its derivatives, called the Globalized Dual Heuristic Programming (GDHP), complete this ACD family. These systems do not require exclusively neural network implementations, since any differentiable structure is suitable as a building block.

The nomenclature used in this section for the Adaptive Dynamic Programming is summarized in Table 22.1. Additional terminologies are defined when necessary.

Consider

$$X(t+1) = F[X(t), u(t), t], \tag{22.18}$$

where $X \in \underline{X} \subset R^n$ is the vector of state variables of the power system, which includes voltages and angles, active reactive power flows.

$u \in \underline{u} \subset R^m$ is the vector of control actions encountered in power systems which include phase shifters, and transformer taps, reactive VAR support, excitation system of generators, and so on.

The **Critic Network** is the offline ANN trainer. It is needed for evaluation of policy formulated and is needed for long range futuristic regimes and stability margin prediction of systems as well as taking current issues into considerations. The critic neural network in the DHP scheme in Figure 22.2, estimates the derivatives of J with respect to the vector ΔX, and learns minimization of the following error measure over time:

$$J(t) = \sum_{k=0}^{\infty} \gamma^k U(t+k), \tag{22.19}$$

$$\|E\| = \sum_t E^T(t)E(t), \tag{22.20}$$

where

$$E(t) = \frac{\partial J[\Delta X(t)]}{\partial \Delta X(t)} - \gamma \frac{\partial J[\Delta X(t+1)]}{\partial \Delta X(t)} - \frac{\partial U(t)}{\partial \Delta X(t)}, \tag{22.21}$$

where $\partial(.)/\partial \Delta X(t)$ is a vector containing partial derivatives of the scalar $(.)$ with respect to the components of the vector ΔX. The DHP critic neural network structure has two linear output neurons. The critic neural network's training is more complicated than in HDP, since there is a need to take into account all relevant pathways of back-propagation as shown in Figure 22.2, where the paths of derivatives and adaptation of the critic are depicted by dashed lines.

This diagram shows the implementation of (22.22). The same critic network is shown for two consecutive times, t and $t+1$. Discount factor $\gamma = 0.5$. Dashed lines show BP paths. The output of the critic network $\lambda(t+1)$ is back propagated through the Model from its outputs to its inputs, yielding the first term of (22.22) and $\partial J(t+1)/\partial A(t)$. The latter is back propagated through the Action from its output to its input forming the second term of (22.21). BP of the vector $\partial U(t)/\partial A(t)$ through the Action results in a vector with components computed as the last term of (22.23). The summation produces the error vector $E(t)$ for critic training.

In the DHP scheme, application of the chain rule for derivatives yields

$$\frac{\partial J[\Delta X(t+1)]}{\partial \Delta X_i(t)} = \sum_{i=1}^{n} \lambda_i(t+1) \frac{\partial X_i(t+1)}{\partial \Delta X_i(t)}$$

$$+ \sum_{k=1}^{m} \sum_{i=1}^{n} \lambda_i(t+1) \frac{\partial \Delta X_i(t+1)}{\partial A_k(t)} \frac{\partial A_k(t)}{\partial \Delta X_i(t)}, \tag{22.22}$$

where $\lambda_i(t+1) = \partial J[\Delta X(t+1)]/\partial \Delta X_i(t+1)$, and n, m are the numbers of outputs of the model and the action neural networks, respectively. By exploiting Eq.

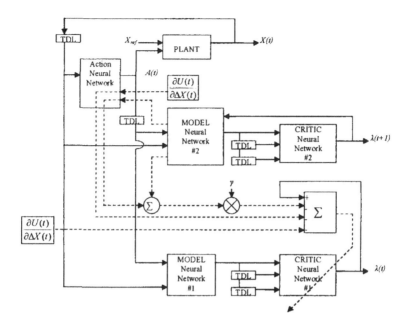

Fig. 22.2   DHP critic neural network adaptation.

(22.22), each of n components of the vector $E(t)$ from Eq. (22.21) is determined by

$$E(t) = \frac{\partial J[\Delta X(t)]}{\partial \Delta X_i(t)} - \gamma \frac{\partial J[\Delta X(t+1)]}{\partial \Delta X_i(t)} - \frac{\partial U(t)}{\partial \Delta X_i(t)} - \sum_{k=1}^{m} \frac{\partial U(t)}{\partial A_k(t)} \frac{\partial A_k(t)}{\partial \Delta X_i(t)}.$$

(22.23)

The adaptation of the action neural network in Figure 22.1 is illustrated in Figure 22.2 which propagates $\lambda(t+1)$ back through the model network to the action network. The goal of such adaptation can be expressed as follows [4, 5]:

$$\frac{\partial U(t)}{\partial A(t)} + \gamma \frac{\partial J(t+1)}{\partial A(t)} = 0 \quad \text{for all } t.$$

(22.24)

The weights' update expression, when applying back-propagation, is as follows:

$$\Delta W_A = -\eta \left[\frac{\partial U(t)}{\partial A(t)} + \frac{\partial J(t+1)}{\partial A(t)} \right]^T \frac{\partial A(t)}{\partial W_A},$$

(22.25)

where η is a positive learning rate and W_A is the weights of the action neural network in the DHP scheme. The structure of the action neural network is identical to that of the action network in the HDP scheme. The general derivations of the equations in this section are shown in [4] in detail. The word "Dual" is used to describe the fact that the target outputs for the DHP critic training are calculated using back-propagation in

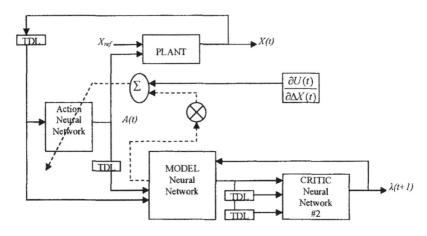

Fig. 22.3   DHP action network adaptation.

a generalized sense; more precisely, it does use dual subroutines (states and co-states) to back propagate derivatives through the model and action neural networks,

The **Action Network** acts as the controller and adjust the control set to meet the objective of the problem formulation. It implements the undersigned policy generated based on the model formulated and is taken back for evaluation by comparing it with real life problems and solutions. With these requirements any policy formulation based on these converges to an optimal solution.

BP paths are shown with dashed lines. The output of the critic $\lambda(t + 1)$ at time $(t + 1)$ is back propagated through the Model from its outputs to its inputs, and the resulting vector is multiplied by the discount factor γ and added to $\partial U(t)/\partial A(t)$. Then an incremental adaptation of the action network is carried out in accordance with Eq. (22.25).

Power systems networks for optimization consist of generators and a number of Distributed Generation resources and FACTS devices transmission networks, loads and area to be served, and so on.

The challenge is to define different performance measure and criteria that will minimize or optimize dynamically the state and controls of the power systems variables such as voltages, system losses and speed deviation of generator units. The impact of contingency and different load management under emergency and restorations stages will be sought using a new OPF scheme proposed below.

Given the power systems with dynamic nonlinear time varying multi-variable complex processes. Any optimization method must therefore be carried out in real time and adapt to continuously changing conditions. Using the recent development in power electronics such as FACTS and software strategies such as hybrid of interior point method and adaptive critics software program for optimization we are able to provide a global intelligent controller which will provide a global dynamic OPF. Specifically, it will coordinate the controls intelligently and ensure handling of the dynamically changing conditions, adapt, optimize, communicate and negotiate appropriately the optimum strategy for the networks.

With the time domain feature of both the critic network and the action network, the future could easily be analyzed, using object nets based intelligent controllers.

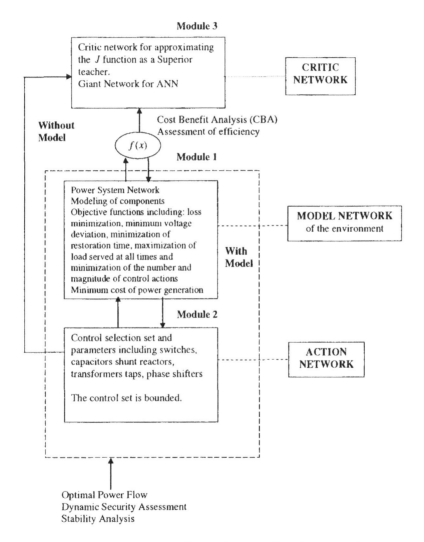

Fig. 22.4   General framework for application of Adaptive dynamic programming to power systems.

Module 1

Read power system parameters and obtain distribution function for state estimation of measurement errors inherent in data, ascertain and improve accuracy of data. Infer relationships between the past data and future ones of unknown period using time series and dynamical systems and in all cases determine the time dependent model approximation behavior of the systems generating the data. Define the model and with the uncertainties, this step includes defining the problem objective and constraint functions for each problem.

Module 2

Determine the feasibility region of operation of the power systems and the emergency state with corresponding violations under different contingencies. Enumerate and schedule different control options over time for different contingency scenarios. Coordinate the controls and perform post optimizations of additional changes. Evaluate results and perform sensitivity analysis studies.

Module 3

For post-optimization process, evaluation and assessment of control options during contingencies are necessary. This module handles the post optimization process by through cost benefit analysis to evaluate the various controls. (How cost effective and efficient). In the power system parlance, a big network, which will perform this evaluation, is essential and indispensable). The critic network from ADP techniques will help realize the dual goals of cost effectiveness and efficiency of the solution via the optimization process.

22.5.2   Stochastic Modeling of the System Using ADP

To extend the meaning of the stochastic dynamic modeling from adaptive dynamic programming to power stem optimization problem, we partitioned the discussion into two parts:

1. Dynamics (anticipatory) nature: In power systems, dynamics comes as a result of the time dependence of the control set required for the optimization process. For any contingency such as changes in loading due to weather conditions within any time interval, adjustment of capacitor switching, transformer taps, minimizing voltage deviation using FACTS devices technology must be coordinated optimally for the efficient and cost effective operation of the power systems.

2. Stochastic: Probabilistic modeling of faults for reliability studies, and loading, as well as available transfer capability (ATC) of transmission lines are needed to deal with unpredictable events such as faults on a line, equipment failure etc. these models have been addressed in power system literature over the years. Moreover, weather variations effects such as solar radiation storms being stochastic in nature affects thermal rating of transmission line. New and efficient computatitional techniques are therefore required for stochastic modeling of the new power system with new devices such as distributed generation

sources and FACTS devices to ensure that priority of load, market structures and planning of generation, transmission and distribution systems are well taken care of.

In assessing the dynamic security of the system and identifying bottlenecks on transmission grids, these two qualities must be assessed.

Dynamic model of the environment refers to the actual power system to be optimized. In this case refers to model of the objective function, which could be either loss minimization, minimum voltage deviation, or cost function comparison. This leads to improvement of the conventional solution by taking it a step further and assessing the cost benefit analysis of the solution using the strategic utility function.

Assuming a stochastic model of the power system network due to change in load and other system dynamics given as:

$$X(t+1) = F(X(t), u(t)) + \text{noise},$$

where the only form of noise is error in the forecast $X(t+1)$.

There are two main methods of handling the noise vector in F network namely the simulated approach and the imputed noise approach.

Further assumptions are:

A set of vectors for $X(t)$ for $t = 1$ through T, the model network F and the critic network J are kept constant. For the simulation approach, noise vectors are simulated for each value of and for each component of the vector noise a random value is picked based on the assumed probability distribution of the noise component. This procedure yields a set of vectors of $noise(t)$ for all t. This set of noise data are held constant, a derivative is then calculated through one pass of the data with respect to a weight W_i',

$$\frac{\partial}{\partial W_i'} \sum_t J(F(X(t), u(X(t), W_i'), \text{noise}(t))).$$

The probability distribution utilized must come from a substantive knowledge and behavior of the power system network.

Development of such a model can be achieved by building a network which inputs $R(t)$ and $u(t)$ and aims at $X(t+1)$ as the targets for its output.

22.5.3 Candidate Power System Problems for DSOPF

22.5.3.1 Example 1: Unit Commitment Application

The problem of generator unit commitment is to meet load demands over a particular time interval and it fits between economic dispatch, maintenance and production scheduling of generation resources while meeting network constraints. Traditionally, unit commitment problems have been stated in the standard format as an optimization problem with the following objective function and constraints:

$$\text{Minimize } F = \sum_{t=1}^{T} \sum_{i=1}^{N} [u_i(t) \cdot F_i(E_i(t)) + S_i(t) + f_i(t)].$$

The constraint models for the unit commitment optimization problem are as follows:

- System energy balance

$$\sum_{i=1}^{N} \frac{1}{2} [u_i(t)P_{g_i}(t) + u_i(t-1)P_{g_i}(t-1)] = P_D(t) \quad \text{for } t \in \{1, T\}.$$

- System spinning reserve requirements

$$\sum_{i=1}^{N} u_i(t)\tilde{P}_{g_i}(t) \geq P_D(t) + P_R(t) \quad \text{for } t \in \{1, T\}.$$

- Unit generation limits

$$P_{g_i}^{\max} \leq P_{g_i}(t) \leq \tilde{P}_{g_i}(t) \quad \text{for } t \in \{1, T\} \text{and } i \in \{1, N\}.$$

- Energy and power exchange

$$E_i(t) = \frac{1}{2} [P_{g_i}(t) + P_{g_i}(t-1)] \quad \text{for } t \in \{1, T\},$$

$$\sum_{i=1}^{N} P_{g_i}(t) \geq (\text{Net Demand} + \text{Reserve}).$$

In case the units should maintain a given amount of reserve, its upper bounds must be modified accordingly. Therefore, we have:

$$P_{g_i}^{\max} = P_{g_i}^{\max} - P_{g_i}^{reserve},$$

$$\text{Demand} + \text{Losses} \leq \sum_{i=1}^{N} P_{g_i} - \sum_{i=1}^{N} P_{g_i}^{reserve},$$

$$C_{cold} = C_O(1 - e^{\alpha t}) + C_L,$$

where

$$
\begin{aligned}
C_{cold} &: \quad \text{Cost to start an off-line boiler,} \\
\alpha &: \quad \text{Unit's Thermal time constant,} \\
t &: \quad \text{Time in seconds,}
\end{aligned}
$$

$$
\begin{aligned}
C_L &: \quad \text{Labor Cost to up the units,} \\
C_O &: \quad \text{Cost to start up a cold boiler,} \\
C_{banked} &= \quad C_B t + C_L,
\end{aligned}
$$

where

$$
\begin{aligned}
C_B &: \quad \text{Cost to start up a banked boiler,} \\
t &: \quad \text{Time in seconds.}
\end{aligned}
$$

Modeling in Unit Commitment
Nomenclature

F : Total operation cost on the power system;

$E_i(t)$: Energy output of the i_{th} unit at hour t;

$F_i(E_i(t))$: Fuel cost of the i_{th} unit at hour t when the generated power is equivalent to $E_i(t)$;

N : Total number of units in the power system;

T : Total time under which Unit Commitment is performed;

$P_{gi}(t)$: Power output of the i_{th} unit at hour t;

$\tilde{P}_{gi}(t)$: Constrained generating capability of the i_{th} unit at hour t;

$P_{gi}^{max}(t)$: Maximum power output of the i_{th} unit;

$P_{gi}^{min}(t)$: Minimum power output of the i_{th} unit;

$S_i(t)$: Start-up Cost of the i_{th} unit at hour t;

$F_i(t)$: Ramping Cost of the i_{th} unit at hour t;

$P_D(t)$: Net system power demand at hour t;

$P_R(t)$: Net system spinning reserve hour t;

$\lambda(t)$: Lagrangian Multiplier for the system *power balance constraint* at hour t;

$\mu(t)$: Lagrangian Multiplier for the system *spinning reserve* constraint at hour t;

$u_i(t)$: Commitment state of the i_{th} unit at hour t.

The following optimization approaches are useful for dealing with the unit commitment problems:

- Dynamic Programming,
- Lagrangian Relaxation methods.

Due to the computational burden required for dynamic programming and the subsequent curse of dimensionality, the unit commitment problem becomes intractable hence approximate methods of computation are preferred for near-optimal or optimal solutions.

Adaptive Dynamic Programming Reinforces the Learning Process

First we present an example for the unit commitment problem and then advance to apply the adaptive dynamic programming approach to the unit commitment problem based on the general framework of application to power systems.

Problem Statement

Find the optimal operation cost of a four units plant using dynamic programming to find the optimum unit commitment schedules covering an 8-hour period.

Table 22.2 gives each unit characteristics including the maximum and minimum power output of each unit, the incremental heat rate of fuels used, no load energy specifying the amount of energy derivable from the unit when (on spinning), start up energy required for boiler thermal systems.

Table 22.3 represents the load being served by the units over 2-hour period.

Table 22.2   Load Data (All Time Periods = 2h)

UNIT	LIMITS OF THE UNIT $P_{g_i}^{max}$ (MW)	$P_{g_i}^{min}$ (MW)	INCREMENTAL HEAT RATE (BTU/KWH)	NO LOAD ENERGY (MBTU/H)	START-UP ENERGY (MBTU/H)
1	500	70	9950	300	800
2	250	40	10200	210	380
3	150	30	11000	120	110
4	150	30	11000	120	110

Table 22.3   Load Duration Curve

TIME t	LOAD, $P_D(t)$ (MW)
1	600
2	800
3	700
4	950

Table 22.4   Start Up and Shut Down

UNIT	MINIMUM UP TIME (HR)	MINIMUM DOWN TIME (HR)
1	2	2
2	2	2
3	2	4
4	2	4

Note: Fuel cost = 1 R/MBTU

Table 22.5   Operating Cost for Various Load Levels

Combination	Units				Operating cost for various load levels PD (MW)			
	1	2	3	4	600	700	800	950
A	1	1	0	0	6505	7525	*	*
B	1	1	1	0	6649	7669	8705	*
C	1	1	1	1	6793	7813	8833	10475

Table 22.4 represents the start up and down time required for the units.

The following Table 22.5 gives the characteristic of all combinations needed as well as the operating cost for each at the loads in the load data. An "*" indicates that a combination can not supply the load. The starting conditions are (a) at the beginning of the 1st period units 1 and 2 are up, (b) units 3 and 4 are down and have been down for 8 hours.

Solution Using Dynamic Programming
STARTING CONDITION

At the starting conditions, u#1 and Unit #2 are up and Unit #3 and Unit #4 are down for 8 hours,

$$\text{Load} = 600\text{MW}, l = 1, \ K = A.$$

All the states A, B, and C will be considered.

For state A, we have

$$F(1, A) = \min \left\{ C^{\min}(1, A) + S(0, A : 1, A) + F^{\min}(0, A) \right\}.$$

The operating cost is $C^{\min}(1, A) = 6505(\$/h)$, the transition cost is $S(0, A : 1, A) = 0$, but because we have two hours, the total cost should be

$$F(1, A) = (2 \times 6505) + 0 + 0 = 13010(\$).$$

For state B, we have

$$F(1, B) = \min \left\{ C^{\min}(1, B) + S(0, A : 1, B) + F^{\min}(0, A) \right\}.$$

The operating cost is $C^{\min}(1, B) = 6649(\$/h)$. In the transition from state A to state B only unit #3 came on line, hence the transition cost will be the start up cost of unit 3 coming on line, which is

$$
\begin{aligned}
S(0, A : 1, B) &= \text{ start up cost of unit #3} \\
&= 110 MBtu \times 1.00\$/MBtu \\
&= \$110. \\
TotalCost = F(1, B) &= (2 \times 6649) + 110 + 0 \\
&= \$13408.
\end{aligned}
$$

For state C,

$$F(1, C) = \min \left\{ C^{\min}(1, C) + S(0, A : 1, C) + F^{\min}(0, A) \right\}.$$

The operating cost is $C^{\min}(1, C) = 6793(\$/h)$. In the transition from state A to state C only unit #3 and unit #4 came on line. Hence the transition cost consists of the start up cost for the two units,

$$
\begin{aligned}
\Rightarrow S(0, A : 1, B) &= (110 MBtu + 110 MBtu) \times 1.00\$/MBtu \\
&= \$220, \\
\Rightarrow F(1, C) &= (2 \times 6793) + 220 + 0 \\
&= \$13806.
\end{aligned}
$$

NEXT INTERVAL

$l = 2$, maximum load level = 800MW.

In this period, state A will not be considered. We will consider states B and C and save strategies A and B in period 1.

For state A and B, $L = A$ and B,

$$F(2, B) = \min_{(A, B)} \left\{ C^{\min}(2, B) + S \left[\begin{array}{c} 1, A : 2, B \\ 1, B : 2, B \end{array} \right] + F^{\min} \left[\begin{array}{c} 1, A \\ 1, B \end{array} \right] \right\}.$$

Operating cost:

$$C^{\min}(2, B) = 8705(\$/h).$$

Transition cost:

$$S(1, A : 2, B) = 110MBtu \times 1.00\$/MBtu = \$110,$$
$$S(1, B : 2, B) = \$0.$$

Total Cost:

$$F(2, B) = (2 \times 8705) + \min \begin{bmatrix} 110 + 13010 \\ 0 + 13408 \end{bmatrix}$$
$$= \$30530.$$

Considering State C:

$$l = 2, K = C, L = A\&B \text{ (saved strategies)}.$$

From the forward DP,

$$F(2, C) = \min_{(A,B)} \left\{ C^{\min}(2, C) + 5 \begin{bmatrix} 1, A : & 2, C \\ 1, B : & 2, C \end{bmatrix} + F^{\min} \begin{bmatrix} 1, & A \\ 1, & B \end{bmatrix} \right\}.$$

Operating cost:

$$C^{\min}(2, C) = 8833(\$/h).$$

Transition cost:

$$S(1, A : 2, C) = (110MBtu + 110Mbtu) \times 1.00\$/MBtu = \$220,$$
$$S(1, B : 2, C) = (110Mbtu + 110Mbtu) * 1.00\$/Mbtu = \$110.$$

Total Cost:

$$F(2, C) = (2 \times 8833) + \min \begin{bmatrix} 220 + 13010 \\ 110 + 13408 \end{bmatrix}$$
$$= 17666 + min \begin{bmatrix} 13230 \\ 13518 \end{bmatrix}$$
$$= \$30896.$$

NEXT PERIOD

Maximum load level is 700MW. In this period we cannot shut down Unit #3, since if we do (because it is possible to shut it down due to the minimum up/down time rules) that we cannot supply the load of 950 MW in the last 6–8 hours. This is because Unit 3 will be needed but if it is down, then it has to be down for 4 hours to obey the minimum down time period. Hence, we consider states B and C only in this period.

State B:

$$K = B; \quad l = 3; \quad L = B\&C.$$

Using the Forward DP,

$$F(3, B) = \min_{(B,C)} \left\{ C^{\min}(3, B) + 5 \begin{bmatrix} 2, B & 3, B \\ 2, C & 3, B \end{bmatrix} + F^{\min} \begin{bmatrix} 2, & B \\ 2, & C \end{bmatrix} \right\}.$$

Operation Cost:

$$
\begin{aligned}
C_{\min}(3, B) &= 7669\$/h. \quad \text{- from table,} \\
S(2, B : 3, B) &= \$0, \\
S(2, C : 3, B) &= \$0.
\end{aligned}
$$

Total Cost:

$$
\begin{aligned}
F(3, B) &= (2 * 7669) + \min \begin{bmatrix} 30530 \\ 30896 \end{bmatrix} \\
&= \$45868.
\end{aligned}
$$

Considering State C,

$$
\begin{aligned}
F(3, C) &= \min \left\{ C^{\min}(3, C) + 5 \begin{bmatrix} 2, B : & 3, C \\ 2, C : & 3, C \end{bmatrix} + F^{\min} \begin{bmatrix} 2, B \\ 2, C \end{bmatrix} \right\} \\
&= (2 * 7813) + \min \begin{bmatrix} 110 + 30530 \\ 0 + 30896 \end{bmatrix} \\
&= \$46266.
\end{aligned}
$$

NEXT INTERVAL

$l = 4$, maximum load = 950 MW, hence only state C will be considered because it is the only state that can supply the load. Saving the two strategies B and C we have,

$$
\begin{aligned}
F(4, C) &= \min_{(B\&C)} \left\{ C^{\min}(4, C) + 5 \begin{bmatrix} 3, B : & 4, C \\ 3, C : & 4, C \end{bmatrix} + F^{\min} \begin{bmatrix} 3, & B \\ 3, & C \end{bmatrix} \right\} \\
&= (10475 * 2) + \min \begin{bmatrix} 110+ & 45868 \\ 0+ & 46266 \end{bmatrix} \\
&= 20950 + \min \begin{bmatrix} 45978 \\ 46266 \end{bmatrix} \\
&= \$66928
\end{aligned}
$$

In a table form, the optimum unit commitment schedule will be as Table 22.6. Figure 22.5 illustrates the computational process for the 8 hours period.

Table 22.6   Optimum Unit Commitment Schedule

Time Period	Load	Combination	Units			
			1	2	3	4
1	600	A	1	1	0	0
2	800	B	1	1	1	0
3	700	B	1	1	1	0
4	950	C	1	1	1	1

Complete Solution: Up/Down Time minimum rules obeyed.

State Combination	Units				Total Capacity
	1	2	3	4	
C	1	1	1	1	1050
B	1	1	1	0	900
A	1	1	0	0	750

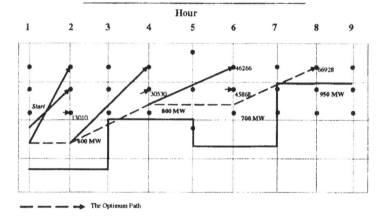

Fig. 22.5   Computational process for the 8 hours period.

Using Figure 22.4 to develop and enhance the unit commitment example presented. We develop the interface of the Adaptive Dynamic Programming (ADP) concept to enhance the unit commitment scheme to yield a global dynamic optimal power flow.

The action network module provides the necessary controls and their corresponding limits for meeting an optimal schedule over time. These are ramping rates, unit generation, downtimes and uptimes, spinning reserve requirements, energy and power exchange, unit's availability etc. The purpose of the action network is to provide the control variables needed for the optimization of the model objective functions as it is traditional done in ADP process.

At the model level in the frame work for the unit commitment problem consisting of the objective function and the constraints, the unit commitment problem is typically optimized using Lagrange relaxation method or the well known dynamic programming or recently evolutionary programming and particle swarm programming techniques. The corresponding ADP approach either keeps the method or solves the model free approach using recurrent neural network techniques. The optimized generation schedules are presented to the critic network to assess the achievement of the multi-objectives formulation in the model such as cost effectiveness, reliability, security and efficiency.

This evaluator feeds back the set of controls to the action network for readjustment and aims at improving the overall cost function and performance of the unit commitment schedule. To account for the stochastic nature of load variation and network changes, the model is refined by the giant critic network to learn from these changes.

22.5.3.2   Example 2: Application to Distribution System Reconfiguration

Distribution networks are generally configured radially for effective and non-complicated protection schemes. Under normal operation conditions, distribution feeders may be reconfigured to satisfy the following objectives:

- Minimum distribution line losses.

- Optimum voltage profile.

- Relieve the overloads in the network.

While all load requirements and maintaining the radial structure of the network. Distribution reconfiguration application schemes are used to control switching operations to clear the abnormal conditions or to improve service quality for a stressed distribution system.

Minimize $\sum Z_b I_b$,

subject to $[A]_i = I$,
where

$$
\begin{array}{rcl}
Z_b & : & \text{Impedance of the branch,} \\
I_b & : & \text{Complex current flow in the branch } b, \\
i & : & m\text{-vector of complex branch currents,} \\
m & : & \text{Total number of the branches,}
\end{array}
$$

I : n-vector of complex nodal injection currents,

n :   Total number of network nodes,

A :   Network incidence matrix, whose entries are

$$(ap, b) = \begin{cases} = 1 & : \text{if branch } b \text{ starts from the node } p, \\ = -1 & : \text{if the branch } b \text{ starts from the node } b, \\ = 0 & : \text{if the branch is not connected to the node } p. \end{cases}$$

The problem of feeder reconfiguration for voltage deviation minimization could be studied using the 32-bus distribution network. This class of problem is solved using integer programming or heuristic programming. The problem can be cast in ADP formulation and using the framework in Figure 22.4 to determine the optimum switching sequences while minimizing losses and voltage deviation.

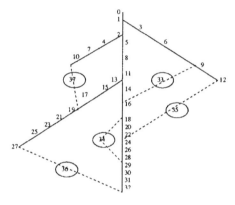

Fig. 22.6   Bus Distribution System with 5 Normally Open Switches.

$u(t)$:   Action vectors, open/close status for 37 switches.

$U(t)$:   The objective function.

r :   Usual discount rate or interest rate that is needed only in infinite-time horizon problems.

J :   Secondary or strategic utility function.

$X(t)$:   Real and active loads to the 32 buses and 5 normally open switches.

To solve the reconfiguration problem we use the framework of ADP application to power systems distribution system reconfiguration problem is modeled as combinatorial optimization with variables representing the switching status of equipment. The controls sequences in the action networks include capacitors, reactors, transformer taps and phase shifters which are options available for reconfiguration of the system and amount of control offered by each equipment, which may be discrete or continuous.

As before the critic network evaluates the control set adopted for any contingency and assigns cost and benefits derivable from such actions. At convergence using the classical or ADP based OPF we obtain the optimal load flow pattern results in minimum voltage deviation for the optimum reconfiguration under different co loading conditions and uncertainties

22.5.3.3 *Example 3: Dynamic security assessment of the power systems*

The increasing complexity of power systems due to their growth and operation close to designed limits, economic constraints and new regulations required for restructuring calls for the incorporation of dynamic security assessment into Supervisory Control and Data Acquisition (SCADA) and Energy Management Systems (EMS). Current research employed methodologies of hybridized nature to handle the security assessment of the power systems thereby improve the accuracy of existing tools used in analyzing the system. These hybrid methods consists of time domain simulations and transient energy functions as well as computational intelligence techniques such as neural networks, fuzzy logic and genetic algorithms. Significant progress had been made in this area of research making the analysis of security assessment online.

The classical method involves computation of critical energy, total energy system at the instant of fault and then determine the stability margin which is based on the difference between the critical energy and energy at fault clearing. The index varies with topology changes and uncertainties in load. To achieve the desired stability control actions are taken to reduce the system instability. This is a classic problem, which has been done using ANN for training and to assess the system conditions. The incorporation of this stability index as a constraint in OPF is being a research investigation for now. The treads of SDOPF framework in Figure 22.4 is a candidate problem for obtain assessment and profitability.

Finally, as we gain more experience and confidence in the application of ADP to power systems optimization problems several other challenging OPF problems will be tested as part of an ongoing research investigation.

22.6 CONCLUSION

Optimal Power Flow (OPF) is one of the most important operational functions of the modern day energy management system. There are several objectives of optimal power flow that must be considered simultaneously when determining an optimal control stratrgy. The main drawbacks of conventional OPF tools is that they lack foresight, the ability to predict the future in terms of asset valuation and economic rate of return on investment; and lack of explicit optimization techniques to handle perturbation and noise, the power system is not static, but changing with respect to load demands at any given time period.

Approximate Dynamic Programming (ADP) is a computational intelligence technique that can incorporate the time features required for the future. Dual Heuristic Dynamic Programming (DHP), a class of ADP, can be used, with appropriate modeling of the power system, to incorporate foresight and to enhance stochastic description and prediction of the power network. This chapter has presented a method for incorporating DHP into an extension of OPF.

Bibliography

1. T. Miller III, R. S. Sutton, and P. J. Werbos, *Neural Networks for Control, a Bradford Book,* MIT Press, Cambridge, MA, London.

2. P. J. Werbos, Approximate dynamic programming for real-time control and neural modeling, in D. A. White and D. A. Sofge, (eds.) *Handbook of Intelligent Control,* pp. 493–525, Van Nostrand Reinhold, New York, 1992.

3. G. K. Venayagamoorthy, R. G. Harley, and D. C. Wunsch, Intelligent control of turbogenerator exciter/turbine on the electric power grid to improve power generation and stability, *Proc., ICPSOP 2002,* Abuja, Nigeria.

4. D. Prokhorov and D. C. Wunsch, Adaptive critic designs, *IEEE Trans. Neural Networks,* vol. 8, no. 5, pp. 997–1007.

5. J. A. Momoh and J. Z. Zhu, Improved interior point method to OPF problems, *IEEE Trans. PES,* vol. 14, pp. 1114–1130, 1999.

23 Control, Optimization, Security, and Self-healing of Benchmark Power Systems [†]

JAMES A. MOMOH   EDWIN ZIVI
National Science Foundation   U.S. Naval Academy

Editor's Summary: This chapter presents several challenging and benchmark problems from the field of power systems. The first benchmark is the IEEE 118 Bus commercial terrestrial Electrical Power System (EPS). The second benchmark represents a finite inertia hybrid ac/dc shipboard Integrated Power System (IPS). The analytic utility and Navy benchmark models and their respective simulations have been experimentally validated and have been used to determine system reliability, reconfigurability, stability, and security. The challenge is to provide novel control and optimization methods and tools to improve the quality of service despite natural and hostile disruptions under uncertain operating conditions. Along with these problems several smaller problems are also presented which demonstrate different aspects of the challenges of power system control. The purpose of this chapter is formulation of problems that ADP methods could be applied to, therefore the emphasis is more on detailed problem description and simulation, not on any particular solution.

23.1 INTRODUCTION

Market-driven power system restructuring, along with advances in distributed generation, energy conversion and protection technologies, requires new control, stabilization and optimization techniques. Moreover, these systems are composed of dynamically interdependent subsystems for which static and isolated design and optimization may fail. For example, the need for robust, distributed infrastructures is articulated by the Architectures for Secure and Robust Distributed Infrastructures

---

[†] The views expressed here are those of the authors, Momoh (on leave from Howard University), and Zivi and not the official views of NSF and U.S. Naval Academy.

initiative [1]:

> *"The major barrier constraining the successful management and design of large-scale distributed infrastructures is the conspicuous lack of knowledge about their dynamical features and behaviors. Up until very recently analysis of systems such as the Internet, or the national air traffic system, have primarily relied on the use of non-dynamical models, which neglect their complex, and frequently subtle, inherent dynamical properties. These traditional approaches have enjoyed considerable success while systems are run in predominantly cooperative and "friendly" environments, and provided that their performance boundaries are not approached. With the current proliferation of applications using and relying on such infrastructures, these infrastructures are becoming increasingly stressed, and as a result the incentives for malicious attacks are heightening. The stunning fact is that the fundamental assumptions under which all significant large-scale distributed infrastructures have been constructed and analyzed no longer hold; the invalidity of these non-dynamical assumptions is witnessed with the greater frequency of catastrophic failures in major infrastructures such as the Internet, the power grid, the air traffic system, and national-scale telecommunication systems."*

The power industry is currently developing new technologies, architectures and control strategies to achieve adaptable, efficient and secure power systems. In support of these efforts, two benchmark problems are presented herein. The first benchmark is a commercial terrestrial Electrical Power System (EPS). The second benchmark represents a finite inertia hybrid ac/dc shipboard Integrated Power System (IPS). The analytic utility and Navy benchmark models and simulations have been experimentally validated and used to determine system reliability, reconfigurability, stability and security. The challenge is to provide novel control and optimization methods and tools to improve the quality of service despite natural and hostile disruptions under uncertain operating conditions.

This chapter presents two reference benchmark systems addressing two challenging problems:

1. Utility power system optimization for scheduling unit commitment subject to unit availability, dynamic response and transmission constraints.

2. Maximizing continuity of service despite temporal and spatially clustered "bursts" of natural or hostile disruption.

Many of the control strategies, optimization techniques and learning agents presented in this book have the potential to significantly improve the efficiency and dependability of the terrestrial Electric Power System and the shipboard Integrated Power System.

23.2   DESCRIPTION OF THE BENCHMARK SYSTEMS

Since objective benchmark problems with realistic constraints are often difficult to obtain, challenge problems are presented in terms of the two benchmark systems. The *IEEE 118 Bus* EPS benchmark is representative of large commercial ac terrestrial systems. The shipboard IPS benchmark provides a complementary isolated micro-grid containing solid-state power conversion, regulation and dc distribution. These benchmarks provide reliable, validated system models, assumptions and simulations. Moreover, model complexity, convergence and numerical issues have been suppressed in favor of facilitating the investigation of innovative control, optimization design and planning strategies for robust, high performance, self-healing power systems.

23.2.1   General Introduction to Utility Power Systems

A typical power system consists of generation, transmission and distribution elements. The commercial benchmark system is the *IEEE 118 Bus* electric power system shown in Figure 23.1. Data for this benchmark is available from
http://www.ee.washington.edu/research/pstca/.
Generally, utility systems share common fundamental characteristics including:

- Electric power is generated using synchronous machines that are driven by steam turbines, gas turbines, hydro-turbines or internal combustion engines.

- Generated power is transmitted from generating sites over long distances to load centers that are spread over wide areas.

- Three phase ac systems comprise the main means of generation, transmission and distribution of electric power.

- Tight voltage and frequency regulation is required to maintain high-quality product.

Electric power is produced at generating stations and transmitted to consumers through an intricate network involving transmission lines, transformers and switching devices. The following hierarchy is used to classify the transmission network: (1) Transmission system, (2) Sub transmission system, and (3) Distribution system.

The transmission system interconnects all major generating stations and main load centers in the system. It forms the backbone of the integrated power system and operates at the highest voltage levels (typically, 230 kV and above). The generator voltages are usually in the range of 11 to 35 kV. These voltages are stepped up to the transmission voltage level, and power is transmitted to transmission sub-stations where the voltages are stepped down to the subtransmission level at substations (typically, 69 kV to 138 kV). The generation and transmission subsystems are often referred to as the bulk power system. The subtransmission system transmits power

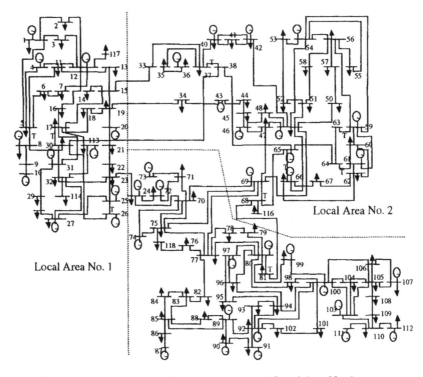

Fig. 23.1   *IEEE 118 Bus* Terrestrial Benchmark System.

at a lower voltage and in smaller quantities from the transmission substation to the distribution substations. Large industrial customers are commonly supplied directly from the subtransmission system. The distribution system is the final stage in the transfer of power to the individual customers. The primary distribution voltage is typically between 4.0 kV and 34.5 kV. Primary feeders at this voltage level supply small industrial customers. The secondary distribution feeders supply residential and commercial customers at 120/240 V. Unlike typical terrestrial power systems, the Navy benchmark system is an isolated finite inertia micro-grid. A brief description of the Navy system follows.

23.2.2   Navy Power System Testbed

The Navy shipboard Integrated Power Systems will distribute approximately 100MW of power for propulsion, ship service power and pulsed loads. Gas turbine prime movers drive 3-phase ac power for propulsion, pulsed loads and ship service distribution subsystems. In the benchmark system, ac to dc power supplies feed port and starboard longitudinal dc distribution buses which feed dc Zonal Electrical Distribution System (DCZEDS) subnets. Figure 23.2 presents the Navy reduced scale benchmark system. The hardware prototype for the ac portion of the Navy benchmark system is installed at Purdue University. Ac power is obtained from a 59 kW wound-rotor synchronous generator (560 V, 60 Hz) driven by a 150 Hp four-quadrant dynamometer. The primary ac load is a 37 kW propulsion drive, consisting of a dc-link based propulsion power converter driving an induction motor and four-quadrant load emulator. The ac bus includes a passive harmonic filter, a pulsed load, and a 15 kW dc power supply which feeds the dc zonal electrical distribution system. The hardware prototype for the 3 zone dc distribution portion of the Navy benchmark system is installed at the University of Missouri at Rolla. The port and starboard bus voltages are obtained through 15 kW ac to dc rectifier-based power supplies (PS). In zone regulation is achieved through 5 kW converter modules (CMs) that convert the primary dc bus voltages from 500V dc to approximately 420V dc within each zone. The inverter module (IM) converts 420 V dc zonal bus power to 5kW of 230 V, 3-phase ac power. The Motor Controller (MC) drives a 5 kW ac motor. Finally, the Constant Power Load (CPL) represents a 5 kW constant power wide-bandwidth power electronic based load.

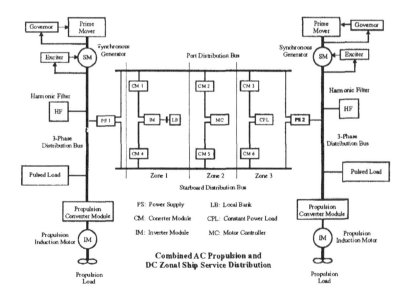

Fig. 23.2   Navy Power System Reference System.

23.3   ILLUSTRATIVE TERRESTRIAL POWER SYSTEM CHALLENGE PROBLEMS

The following power systems engineering problems have been chosen to illustrate the nature of problems encountered in the power industry and the limitations of the presently available tools for analyzing the networks. Some of the referred texts for this exercise are [2, 24–28]. The analyses of various problems feature the system description, the topology, state variables, network constraints, the static and dynamic equations and network parameters.

Generalizing the standard power flow equations, allows static optimization strategies to be developed solving the Optimal Power Flow (OPF) problem. Optimization is an essential tool to obtain large power systems that are: (1) affordable, (2) reliable, (3) secure and (4) economically sustainable. The ability to achieve this essential Optimal Power Flow is limited by solution: (1) accuracy, (2) speed and (3) robustness. Optimization of system operation requires:

- Hourly "unit commitment" decisions regarding whether a unit is on or off at a given hour.

- Hourly "hydrothermal problem" decisions that take advantage of hydroelectric plant generation flexibility to manage water reserve levels and improve system performance.

- Longer-term "maintenance-scheduling problem" decisions that minimize the production cost without violating system reserve requirements.

In the following, scaled down versions of 3-bus and 5-bus systems, representing realistic systems, are presented to facilitate the discussions and the understanding of power system concepts. The three examples below address issues on (1) load flow (2) economic dispatch of generating units (3) transient stability of power system.

23.3.1   Load Flow Problem and Solution

The power flow equations are expressed as follows:

$$S_i^* = P_i - jQ_i = V_i^* \sum_{j=1}^{n} Y_{ij} V_j, \tag{23.1}$$

where S_i^*, P and Q represent the conjugate of the apparent power, the real power and the reactive power, respectively. The real and reactive powers are computed from standard power flow equations given by:

$$P_i = |V_i| \sum_{j=1}^{n} |Y_{ij}||V_j| \cos(\theta_i - \theta_j - \Psi_{ij}), \tag{23.2}$$

$$Q_i = |V_i| \sum_{j=1}^{n} |Y_{ij}||V_j| \sin(\theta_i - \theta_j - \Psi_{ij}). \tag{23.3}$$

Using rectangular coordinates in the form of $V_i = e_i + j\, f_i$ and $Y_{ij} = G_{ij} + j\, B_{ij}$, we obtain:

$$P_i = e_i \left(\sum_{j=1}^{n} (G_{ij}e_j - B_{ij}f_j) \right) + f_i \left(\sum_{j=1}^{n} (G_{ij}f_j + B_{ij}e_j) \right), \tag{23.4}$$

$$Q_i = f_i \left(\sum_{j=1}^{n} (G_{ij}e_j - B_{ij}f_j) \right) - e_i \left(\sum_{j=1}^{n} (G_{ij}f_j + B_{ij}e_j) \right). \tag{23.5}$$

$|Y_{ij}|$ — Magnitude of Admittance of line from bus i to bus j.

Ψ_{ij} — Phase angle of the admittance of the line from bus i to bus j.

e_i, f_i — Real and imaginary components of the complex voltage at node i.

e_j, f_j — Same as e_i, f_i for node j.

G_{ij}, B_{ij} — Real and imaginary components of the complex admittance matrix elements.

θ_i, θ_j — Phase angle of voltages at bus i and bus j, respectively.

V_i, V_j — Magnitude of voltages at bus i and bus j, respectively.

The input data requirements for the power system analysis are:

- Y-Bus matrix,

- System load,

- Slack bus Voltage and angle,

- System (network) limits -line flows, bus generation (active and reactive).

Once the power system problem has been formulated and operating conditions specified, steady state solutions can be obtained. However, since the power system is composed of **dynamically** interdependent subsystems with various *load perturbations* and *system contingencies*, solving the power flow for every conceivable scenario and control strategy poses an exhausting computational challenge. Moreover, the parameters and input data for both the utility and navy systems have *inherent stochasticity*, which must be addressed efficiently. Allowance for *stochasticity* calls for variations in system parameters and inputs of approximately ±10%. Computational solutions must also be capable of predicting future behavior in support of *forecasting* control strategies. Predictions are also required for economic planning and investment. In conclusion, traditional simulation methodologies must be enhanced to incorporate stochastic modeling and selection of appropriate statistical distributions.

New methods for simulating emerging power systems from the computational intelligence and stochastic programming communities could have considerable impact on system security and efficiency.

Consider a 3-bus utility power system given as Figure 23.3:

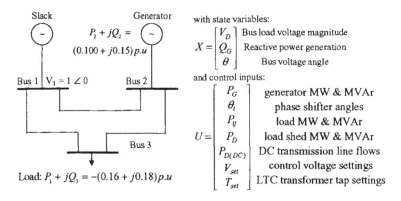

with state variables:

$$X = \begin{bmatrix} V_D \\ Q_G \\ \theta \end{bmatrix} \begin{matrix} \text{Bus load voltage magnitude} \\ \text{Reactive power generation} \\ \text{Bus voltage angle} \end{matrix}$$

and control inputs:

$$U = \begin{bmatrix} P_G \\ \theta_t \\ P_y \\ P_D \\ P_{D(DC)} \\ V_{set} \\ T_{set} \end{bmatrix} \begin{matrix} \text{generator MW \& MVAr} \\ \text{phase shifter angles} \\ \text{load MW \& MVAr} \\ \text{load shed MW \& MVAr} \\ \text{DC transmission line flows} \\ \text{control voltage settings} \\ \text{LTC transformer tap settings} \end{matrix}$$

Fig. 23.3 3-Bus Utility System.

With control constraits:		Operating constraints:
C1.	Generator output in MW.	C4.	Line and transformer flows (MVA, Amps, MW, MVAr).
C2.	Transformer tap limits.	C5.	MW and MVAr interchanges.
C3.	Shunt capacitor range.	C6.	MW and MVAr reserve margins (fixed/dynamic).
C7.	Voltage, angle (magnitude, difference).		

Performance objective functions:

P1. Active power objectives.
P2. Economic dispatch
     (minimum cost losses, MW generation or transmission losses).
P3. Environmental dispatch.
P4. Maximum power transfer.

Reactive power objectives (MW and MVAr Loss minimization)

P5. Minimum deviation from a target schedule.
P6. Minimum control shifts to alleviate violations.
P7. Least absolute shift approximation of control shift.

With the following initial conditions:

IC1. $V_{slack} = 1\angle 0 p.u.$: Slack bus voltage.
IC2. p_D : Total system load.
IC3. p_G : Generator (PV) active power generation.
IC4. $|V_G|$: Magnitude of generator (PV) bus voltage.

23.3.2   Optimization Problem (Minimization of Cost of Generation)

Optimization problems involve minimization of functions with both equality and inequality constraints. Consider the following power system optimization problem general formulation:

$$\text{Minimize the function } f(x, u),$$
$$\text{subject to:} \qquad g(x, u) = 0, \qquad (23.6)$$
$$h(x, u) \leq 0,$$

where x and u are the state and control variables of the power system. In solving this problem, the Lagrangian is formulated and the Kuhn-Tucker conditions of optimality for first order derivatives as well as second order derivatives are invoked to obtain the general solution.

Power Systems Example on Optimization [2]

The parameters of a 5-bus system are given in Table 23.1 and the cost functions in dollars per hour are:

$$F_1 = 0.0060 P_{1g}^2 + 2.0 P_{1g} + 140, \qquad (23.7)$$
$$F_2 = 0.0075 P_{2g}^2 + 1.5 P_{2g} + 120, \qquad (23.8)$$
$$F_3 = 0.0070 P_{3g}^2 + 1.8 P_{3g} + 80, \qquad (23.9)$$

assuming that $0.95\,p.u. \leq V_i \leq 1.05\,p.u.;\ i = 1,\ \dots N bus$. Assuming all generators are rated at 200 MW, you can use the initial generation schedule with:

- The OPF program to obtain the absolute minimum cost of this system as well as the real and reactive generation schedule.

- The OPF program to obtain the loss minimum of this system, the reactive power of generation, and the optimal voltage profile.

Table 23.1   5-Bus System Impedance and Line Charging Data

Bus Code From Bus i to Bus j		Line Impedance pu	½ Line Charging Susceptance (pu)	Line Limits (MW)
1	2	$0.02 + j0.06$	J0.030	30
1	3	$0.08 + j0.24$	J0.025	40
2	3	$0.06 + j0.18$	J0.020	50
2	4	$0.06 + j0.18$	J0.020	80
2	5	$0.04 + j0.12$	J0.015	40
3	4	$0.01 + j0.03$	J0.010	180
4	5	$0.08 + j0.24$	J0.025	120

Table 23.2   5-Bus Initial Generation Schedule

Bus i	Bus Voltage, V_i $|V_i|V\angle\theta_i$		Power Generation		Load Level	
	Magnitude (pu)	Angle (degrees)	P_{gi} (MW)	Q_{gi} (MVAr)	P_{load} (MW)	Q_{load} (MVAr)
1	1.060	0.0	98.4	-	0	0
2	1.056	-2.27	40.0	23.2	20	10
3	1.044	-3.69	30.0	30.0	45	15
4	1.041	-4.16	0.0	10.0	40	5
5	1.030	-5.35	0.0	0.0	60	10

Solution to a Deterministic Optimization Problem

The optimization problem is solved using a Quadratic sensitivity program based on Lagrangian approach. From the OPF program, we obtain the following optimal values of P, Q and V:

- Absolute minimum cost = 2.7403,
  P_{g1} = 97.48MW, P_{g2} = 40.00MW and P_{g3} = 30.00MW,
  Q_{g1} = -17.86MVAr, Q_{g2} = -0.260MVAr and Q_{g3} = 33.94MVAr,

- Loss minimum = 0.024763, $|v_4|$ = 1.04535 pu, $|v_5|$ = 1.02052 pu, Q_{g1} = -18.87MVAr and Q_{g2} = 1.38MVAr.

The objective of power system optimization problems is to minimize the economic dispatch (minimum cost losses); MW generation or transmission losses; environmental dispatch; maximum power transfer; minimum deviation from a target schedule; minimum control shifts to alleviate violations; and/or least absolute shift approximation of control shift. Power system optimization problems have been solved using a variety of methods with various efficiencies measured in terms of reliability, speed of operation, flexibility, and maintainability. However, some methodologies are only suitable for particular objective functions. Robust, multi-objective optimization methodologies are needed.

To date, Lagrangian and other optimization approaches have focused on the solution to steady-state problems in which the load and network parameters are assumed constant. In reality, the system is a complex dynamically interdependent system subjected to uncertainty due to time varying loads, parameter changes, disturbances and noise. Increased use of solid-state based technologies, such as Flexible AC Transmissions Systems (FACTS) will continue to inject additional noise. The present optimal power flow methods perform creditably in addressing the problems posed so far. However, simplifying assumptions to ease the computational burden of existing solutions limits the applicability of present work to solve the ever-increasing complexity in power systems. One possible extension of present methods would be the introduction of stochastic optimal power flow tools that are capable of handling future uncertainties over time.

23.3.3   Transient Stability Problem

Transient stability in power systems [3, 4] is concerned with the system's ability to remain in synchronism following a major disturbance such as a line outage. Protective relays are placed strategically throughout the system to detect faults and to trigger the opening of circuit breakers to isolate the fault. Therefore, the power system can be considered as going through changes in configuration in three stages, from pre-fault, fault-on, to post-fault systems. There are several approaches to solving the angle stability problem, including: (1) the Transient Energy function method, (2) Partial Energy function approach, and (3) the Hybrid method. In the Transient Energy function method, the energy margin is obtained as the difference between the system's transient energies, for example, the difference between the potential energy of the system at the controlling unstable equilibrium point (UEP) and the kinetic energy gained as a result of the disturbance. In the Transient Energy function method, the normalized difference between the system energy at fault clearing and the energy at the controlling UEP (the normalized energy margin [EM]) is used as the index of stability. The system is stable if EM is positive and it is unstable if EM is negative. EM analysis also provides useful information leading to the effective application of control actions.

This section provides the mathematical formulation of Transient Energy function method. Generally, a multi-machine system is represented by a set of differential equations of the form:

$$\dot{\delta} = \omega, \tag{23.10}$$

$$M\dot{\omega} = P_m - P_e, \tag{23.11}$$

$$T'_{do}\dot{E}'_q = E_{fd} - E'_q + I_d(X_d - X'_d), \tag{23.12}$$

$$T'_{do}\dot{E}_d = -E_d - I_q(X_q - X'_q), \tag{23.13}$$

$$T_{ex}\dot{E}_{fd} = -E_{fd} + K_{ex}(V_{ref} - V_t), \tag{23.14}$$

$$E'_q - V_q = I_dX'_d + I_qr_s, \tag{23.15}$$

$$E'_d - V_d = -I_qX'_q + I_dr_s, \tag{23.16}$$

where

δ :   Rotor or torque angle,

$\dot{\delta}$:   First derivative of the rotor angle w.r.t time,

ω :   Angular speed of rotor,

$\dot{\omega}$:   First derivative of the angular speed w.r.t time (angular acceleration),

X_l :   Leakage reactance,

E'_d :   d-axis transient generated voltage,

E'_{fd} : d-axis transient field voltage,

r_a : Armature resistance,

V_t : Terminal voltage of the generator,

X_d : d-axis synchronous reactance,

X'_d : d-axis transient reactance,

I_d : d-axis current,

I_q : q-axis current,

T'_{do} : d-axis open circuit transient time constant,

X_q : q-axis synchronous reactance,

X'_q : q-axis transient reactance,

T'_{qo} : q-axis open circuit transient time constant,

T_{ex} : Excitation system time constant,

V_{ref} : Referenced voltage.

The electrical power transmitted from each generator to the rest of the system is calculated using the following equations:

$$P_{aj} = P_{mj} - P_{ej} - \frac{M_j}{M_T} P_{COI}, \qquad (23.17)$$

where

$$P_{COI} = \sum_{j=1}^{NR} P_{mj} - P_{ej}; \quad M_T = \sum_{j=1}^{NR} M_j,$$

and

P_{mj} = Mechanical power of the j-th generator,

P_{ej} = Electrical power output of the j-th generator,

M = Rotor Inertia; NR = number of Rotors.

The expression can be re-written as:

$$P_{aj} = P_j - \sum_{\substack{k=1 \\ k \neq j}}^{NR} (C_{jk} \sin \theta_{jk} + D_{jk} \cos \theta_{jk}) - \sum_{l=1}^{ND} (C_{jl} \sin \theta_{jl} + D_{jl} \cos \theta_{jl}),$$

$$(23.18)$$

where

$P_j = P_{mj} - E_j^2 G_{ij}, \quad C_{jk} = E_j E_k B_{jk}, \quad ND$ = number dc bu,

$D_{jk} = E_j E_k G_{jk}, \quad C_{jl} = E_j V_l B_{il}, \quad D_{jl} = E_j E_l G_{jl},$

$\theta_{jk} = \theta_j - \theta_k, \quad \theta_{jl} = \theta_j - \theta_l,$

θ_i = Generator rotor angle, $\quad \theta_j$ = Angle of converter ac bus,

j, k = Generator internal nodes, $\quad l$ = Converter bus (ac side).

Computation of the Systems Energy Margin

The EM computation involves calculating the kinetic energy (KE) and potential energy (PE) terms. The total KE is computed from the speeds of all machines at clearing:

$$KE = \frac{1}{2} \sum_i M_i \omega_i^2. \tag{23.19}$$

To account for the fact that not all the kinetic energy at the end of the disturbance contributes to system separation, the concept of corrected kinetic energy KE_{co} has been introduced as:

$$KE_{corr} = \frac{1}{2} M_{eq} \omega_{eq}^2, \tag{23.20}$$

where

$$M_{eq} = \frac{M_{cr} M_{sys}}{M_{cr} + M_{sys}}; \text{ and } \omega_{eq} = \omega_{cr} \omega_{sys}.$$

The inertial center of the group of machines whose rotor angles have advanced beyond 90 degrees at the controlling UEP has inertia M_{cr} and angular speed ω_{cr}. The corresponding terms for the rest of the generators are M_{sys} and ω_{sys}. The Potential Energy margin (PE) between clearing, θ^{cl}, and the unstable equilibrium point, θ^u, is given by

$$PE = \sum_i \int_{\theta_i^{cl}}^{\theta_i^u} (Pm_i - Pe_i)\, d\theta_i. \tag{23.21}$$

In the reduced formulation, PE can be divided into three components: (i) the position energy associated with the constant mechanical power output and shunt admittances at the machine internal buses; (ii) the magnetic energy associated with the line susceptances of the network; and (iii) the dissipation energy associated with the transfer conductance of the reduced network. The dissipation term consists of a path dependent integral, for which a closed form expression has been previously obtained using a linear angle trajectory between θ^{cl} and θ^u [24]. The computation of PE is formulated as an ordinary differential equation problem of the form:

$$\dot{y}_i = (Pm_i - Pe_i)(\theta_i^u - \theta_i^{cl}), \tag{23.22}$$
$$\dot{\theta}_i = (\theta_i^u - \theta_i^{cl}). \tag{23.23}$$

This formulation is subject to the network power flow equations. The set of differential-algebraic equations is solved in the time interval 0.0 to 1.0, using the same variable-step, variable order integration routine used to determine the fault-on trajectory, to obtain the potential energy margin to any desired degree of accuracy.

At the end of the integration process, PE is computed as:

$$PE = \sum_i y_i, \tag{23.24}$$

where the EM with KE correction is obtained from:

$$\Delta V = \frac{PE - KE_{corr}}{KE_{corr}}. \tag{23.25}$$

The energy margin determines the stability of the system as follows:

If $\Delta V > 0.0$ then, the system is stable;

If $\Delta V = 0.0$ Undetermined;

If $\Delta V < 0.0$ System is Unstable.

Using the above equations, the stability analysis is computed for each of the machines and if the energy margin is positive for all the generators, then the system is stable. Otherwise the system is unstable. In the TEF computation, the critical clearing time (T_{cl}) used is 0.1 seconds; this gives the speed and angle at a considerable operating level.

Example on Transient Stability Analysis

A one-line diagram of the previous 3-Bus power system is as presented below as Figure 23.4. The system is subjected to a 3-phase fault on bus 2. The fault is cleared after 0.1 seconds. Determine whether or not the system is stable after the fault is cleared.

Transient Stability Analysis Solution

The power generation at bus 2 and slack bus voltage magnitude and angle are given in Figure 23.4. The line data and results are given below. Since PE is bigger than KE throughout the cycle, the system is "transient" stable.

The main concern in transient stability is the need for synchronism of ac system synchronous machines. If a system becomes unstable, at least one rotor angle becomes unbounded with respect to the rest of the system. In transient stability, primary component to be modeled is the synchronous generators and the typical time frame of concern 1 to 30 seconds. The objectives of power system transient stability are:

- Minimize of disturbance severity and duration,

- Increase forces restoring synchronism,

- Reduce the accelerating torque by reducing input mechanical power,

- Reduce the accelerating torque by applying artificial load.

Accurate analytical transient analysis results are critical for optimal operation of the power system.

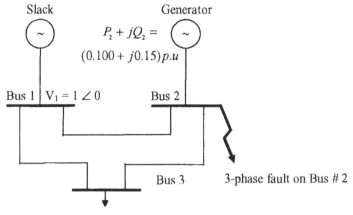

Load: $P_3 + jQ_3 = -(0.16 + j0.18)p.u$

Fig. 23.4 A 3-Bus Utility System with Bus 2 under 3-phase faulted condition.

3-Bus Transient Stability Line Data					
Line #	From	To	R	X	Shunt
1	1	2	.05000	.22000	.080
2	2	3	.02500	.25000	060
3	1	3	.10000	.18000	.110

Pre-fault Bus Voltage Results		
Bus #	Voltage Magnitude	Bus Angle
1	1.000	.0000
2	1.046	-.0092
3	1.024	-.0291

Unstable & Stable Equilibrium Points		
Bus #	S.E.P	U.E.P
1	.8747	2.2669
2	.5009	2.6406

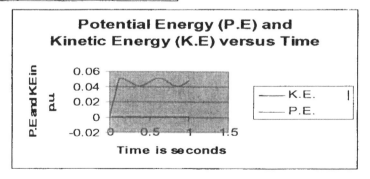

Fig. 23.5 3-Bus Energy Margin simulation results.

Relevance

Transient stability analysis using the energy margin approach provides an excellent analysis of the power system following a disturbance. The results can be applied to stability constrained optimal power flow. This type of OPF can be used in the enhancement of the power systems operation depending on the type of objective function and controls being considered. There are several computational tools for analyzing the transient stability of the system including EMTP (Electromagnetic Transient Program), HUTSP (Howard University Transient Stability Program), and EPRI-based PSAP packages. Most, if not all, of these programs do not handle the problems of noise and complexity introduced due to the introduction of new technology such as FACTS devices and load regulation devices.

Extensions

The use of the Energy Margin approach for solving transient stability problems has been very successful in transient stability analysis. However, future of electric power systems require new computational tools such as ADP for achieving improved performance that can incorporate increased non-linearity, noise, complexity and uncertainty.

The scalability of the new tools developed using ADP concepts will enhance working models for analysis and controls for optimum performance of the navy ship systems as well. We discuss herein the solved examples of the benchmark systems and discuss areas of research grand challenges for future research.

23.4   ILLUSTRATIVE NAVY POWER SYSTEM CHALLENGE PROBLEMS

Isolated power electronics based electrical distribution systems are becoming increasingly common for renewable and mobile applications including air, land, and sea vehicles. In this section, three methods for analyzing the stability of power electronics based power distribution systems are reviewed and applied to the zonal dc distribution portion of the Navy benchmark system originally presented in Figure 23.2 and documented in [5–7]. Each of these methods has practical limitations, which limit their applicability to real world systems. The challenge involves using ADP methods for online system stability monitoring and augmentation despite uncertain and disruptive operating conditions.

As shown in Figure 23.6, this dc zonal distribution testbed consists of two ac to dc power supplies feeding port and starboard dc distribution buses.

Each of three load distribution zones is fed from both the port and starboard dc distribution buses via load sharing dc to dc converters. Note that diodes prevent a fault in one bus from being fed by the opposite bus. Zonal loads are represented by an Inverter Module (IM), a motor controller (MC), and a Constant Power Load (CPL). The dc zonal distribution hardware testbed was developed by the U.S. Navy and the

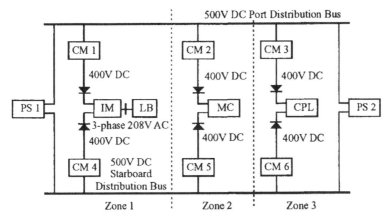

PS: Power Supply   IM: Inverter Module   LB: Local Bank
SSCM: Converter Module   MC: Motor Controller   CPL: Constant Power Load

Fig. 23.6 Navy Benchmark dc Distribution System.

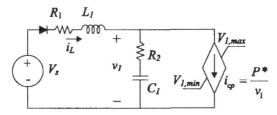

Fig. 23.7 Example dc/dc Power Converter.

Energy Systems Analysis Consortium (ESAC) and is located at the University of Missouri at Rolla.

These power electronics based systems facilitate a high degree of automation and offer nearly instantaneous reconfiguration capabilities. Moreover, power electronic converters can provide excellent regulation of their output objectives. For example, the dc to dc converter, shown in Figure 23.7, can maintain an essentially constant output voltage regardless of input disturbances. From a load regulation perspective, this property is highly desirable. However, Eqs. (23.26) and (23.27) show that this constant power demand causes the converter incremental input impedance with respect to input voltage to be negative. This negative incremental input resistance has an undesirable destabilizing effect on the interdependent overall power system dynamics,

$$P^*(t) = i_{cp}(t)\, v_l(t) \quad \Rightarrow \quad i_{cp}(t) = \frac{P^*(t)}{v_l(t)}, \qquad (23.26)$$

$$\frac{\partial\, i_{cp}(t)}{\partial\, v_l(t)} = -\frac{P^*(t)}{v_l^2(t)} = \frac{1}{Z(t)} \quad \Rightarrow \quad Z(t) = -\frac{v_l^2(t)}{P^*(t)}. \qquad (23.27)$$

Three stability analysis methods are discussed in the following sections: time-domain simulation, generalized immittance analysis, and the direct method of Lyapunov. The predictions of the time-domain simulation and the generalized immittance analysis are compared with experimentally measured results. In particular, these two methods have recently been shown to predict the stability (or lack thereof) of the hardware test system [8]. A non-linear Lyapunov method using polytopic modeling techniques is also presented. The dc stability analysis concludes with consideration of the limitations of existing methods and opportunities for new ADP based analysis.

23.4.1   Preliminaries

Dc Stability Definitions

Although the definitions of an equilibrium point, an operating point, and of stability are crisply defined mathematically, some clarification is required for physical systems. In a model detailed enough to portray the switching action of the power semiconductors, many of the state variables continue to cyclically change even under steady state conditions. Herein, an equilibrium point is a point at which the fast or dynamic average of the derivatives of the state variables is zero [9]. An operating point is defined as an equilibrium point about which the system is being studied. If conditions are such that there is only one possible equilibrium point then these terms become synonymous. An operating point of the system model is said to be locally stable if, when perturbed from an operating point by a small amount, the system model returns to that operating point. An operating point of the system model is said to be globally stable if the operating point can be perturbed by any amount and still return to that operating point. A dc power system is said to be locally stable about an operating point if the system voltages and currents vary only at the forcing frequencies associated with the switching of the power semiconductors and that the average values of these variables is such that all power converters are operating properly. In other words, the system is said to be stable if, neglecting switching induced ripple, the voltages and currents are constant in the steady-state and the level of these voltages and currents is such that all converters are operating in their intended modes of operation. These comments with regard to stability are intended for informal discussion only. For a thorough and rigorous discussion, the reader is referred to [10] and [11].

Dc System Operation

Returning to the system, diagrammed in Figure 23.6, the two ac to dc power supplies (PS1 and PS2), independently feed the port and starboard dc distribution buses. Each zone is fed by port and starboard converter modules (CM), which use voltage droop to share power. Diodes prevent a fault in one bus from being fed by the opposite bus. The three characteristic loads consist of an inverter module (IM) that in turn feeds an ac load bank (LB), a motor controller (MC), and a generic constant power load (CPL). Robustness in this system is achieved as follows. First, in the event that either a power supply fails, or a distribution bus is lost, the other bus will pick up full system load without interruption in service. Imposing current limits on the converter modules mitigates faults between the converter module and diode and as before, the opposite bus can supply the zone. Finally, faults within the components are mitigated through the converter module controls.

Operating the PS1 and PS2 power supplies in an uncontrolled rectifier mode results in the configuration given in Figure 23.8. The primary side ac voltage is a nearly ideal 480 V l-l rms source at 60 Hz. The transformer parameters are: primary leakage inductance: 1.05 mH, primary winding resistance: 191 mΩ, secondary leakage inductance: 1.05 mH, secondary winding resistance: 191 mΩ, magnetizing inductance: 10.3 H and primary to secondary turns ratio of 1.30. All of these parameters apply to the wye-equivalent T-equivalent transformer model and are referred to the primary winding. Finally, the dc link inductance, L_{dc}, is 9.93 mH, the resistance of this inductance, $r_{L_{dc}}$, is 273 mΩ, the dc capacitor, C_{dc}, is 461 mF, and the effective series resistance of the dc capacitor, $r_{C_{dc}}$, is 70 mΩ. This capacitor is removed for some studies as noted.

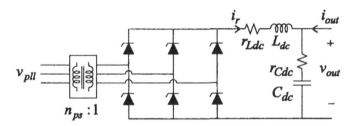

Fig. 23.8   Ac to dc power supply.

The converter modules regulate the flow of power from the dc distribution buses into each zone. The circuit diagram is provided in Figure 23.9. Although the parameters vary from converter module to converter module, typical converter module parameters are: $r_{C_{in}} = 1078$ mΩ, $C_{out} = 454 \mu f$, $r_{C_{out}} = 70$ mΩ, $r_{L_{out}} = 101$ mΩ.

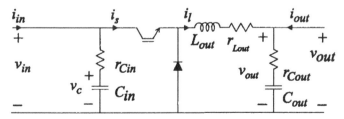

Fig. 23.9   Dc to dc converter module circuit diagram.

The experimentally validated converter module controller is presented in Figure 23.10 and Table 23.3. The commanded output voltage is v_{out}^* and the commanded inductor current is i_l^*. This current command, in conjunction with the measured current i_l is used by a hysteresis modulator so that the actual current closely tracks the measured current. The undesirable constant power destabilizing effect of negative incremental converted module input resistance is stabilized by a feedback compensater $H_{sf}(s)$, given by

$$H_{sf}(s) = K_{sf} \frac{\tau_{sf1}s}{(\tau_{sf1}s + 1)(\tau_{sf2}s + 1)}. \tag{23.28}$$

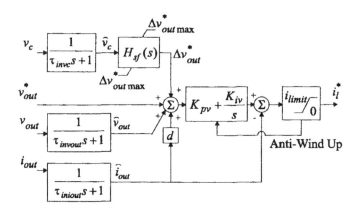

Fig. 23.10   Converter module controls.

Table 23.3   Dc to Dc Converter Module Control Parameters

$\tau_{invc} = 7.96\mu s$	$\tau_{invout} = 7.96\mu s$	$\tau_{iniout} = 7.96\mu s$
$v_{out}^* = 420$ V	$d = 0.8$ A/V	$K_{pv} = 0.628$ AV
$K_{sf} = 0.1$	$\tau_{sf1} = 20$ ms	$\tau_{sf2} = 5$ ms
$\Delta v_{outmax}^* = 20$ V	$i_{limit} = 20$ A	$K_{iv} = 219$ A/Vs

For the sake of brevity, the inverter module (IM), motor controller (MC), and constant power load (CPL) will not be discussed in detail herein. The salient dynamics of these components may be represented by a capacitor with capacitance C_x and effective series resistance r_x in parallel with an ideal constant power load of P_x.

Parameters for these equivalent circuits are listed in Table 23.4. While this simplistic description can be used to a first approximation, a more detailed analysis was used in the actual studies presented. The reader is referred to [5–7] for a more detailed description of these components.

Table 23.4   Equivalent Load Parameters

Component	$C_x, \mu F$	r_x, mΩ	P_x, kW
IM	590	127	4.69
MC	877	105	2.93
CPL	374	189	5.46

Using this test system, two scenarios are studied. For each case, it is assumed that the starboard bus is out of service due to a fault and that the remainder of the system is being fed from the port power supply. All loads are operating at the capacities listed in Table 23.4. The difference between the two cases is the converter parameters. For Case 1, all parameters are as listed thus far. For Case 2, the power supply output capacitance is removed, and the input capacitance to all the converter modules is reduced from $C_{in} = 459 \mu F$ and $r_{C_{in}} = 1078$ mΩ to average values of $C_{in} = 101 \mu F$ and $r_{C_{in}} = 213$ mΩ.

23.4.2   Stability Analysis Method 1: Time-Domain Simulation

Perhaps the most straightforward means to examine system performance is through the use of time-domain simulation. There are fundamentally two types of simulations that are typically used in this class of systems: so-called 'detailed' model based simulations and non-linear average value model (NLAM) based simulations. In this analysis, 'detailed' refers to a simulation in which the switching action of each semiconductor is included, even if only on an idealized 'on' or 'off' basis. Non-linear average value based models refer to simulations where the switching is represented on an average value basis. As a result, state variables are constant in the steady state as is still true for ac systems expressed in a synchronous reference frame [9]. Figure 23.11 depicts the performance of the detailed simulation of the test system for the two cases described in the previous section. Initially, the parameters are those for Case 1. As can be seen, the waveforms are constant, aside from the switching induced ripple. Approximately one-half through the study, the parameters are changed to match Case 2. The port bus voltage now contains a low-frequency oscillation, which is not related to any of the semiconductor switching frequencies. The conclusion that Case 2 is unstable has been experimentally validated.

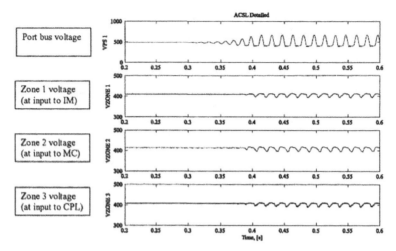

Fig. 23.11   Test system performance, detailed simulation.

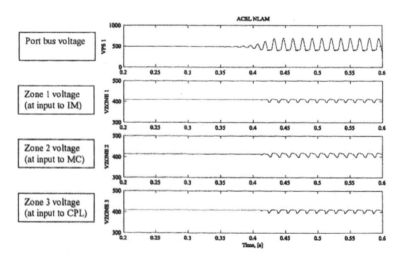

Fig. 23.12   Test system performance, NLAM simulation.

Figure 23.12 the results of the same experiment conducted with non-linear average value model (NLAM) based simulation. One difference between Figure 23.11 and Figure 23.12 is the absence of the switching induced ripple in the waveforms. Otherwise, the simulations are compatible in their predictions, although the exact details of the waveforms vary once the instability has commenced. This variation is due to the chaotic behavior unstable nonlinear systems tend to exhibit. The observation that the two models are in good agreement during transients involving stable conditions [13] supports this observation. Although the conclusions of each time-domain study are similar, the NLAM has a computational advantage in that the switching of the semiconductors does not periodically excite the dynamics. As a result, stiff integration algorithms can be used more effectively than for detailed model simulations. An additional feature of NLAMs is that they can be automatically linearized using many simulation languages including MATLAB [14] and ACSL [15]. As a tool for examining stability, however, the use of time-domain simulation has drawbacks. The primary drawback is that a given study only predicts the stability of a single operating point for a particular perturbation. One valid approach to gain confidence in system behavior is to run massive numbers of studies. However, there is always a possibility that an unstable operating point or scenario can be overlooked. Exploiting ADP methods to determine the boundaries of stable operation would be of particular value.

23.4.3   Stability Analysis Method 2: Generalized Immittance Analysis

An alternative stability analysis is to use the method of generalized immittance analysis [16–19]. This is a frequency domain based technique, which has two important characteristics. First, in a single analysis it can be used to test the local stability of all operating points of interest. Second, unlike eigenanalysis, it can be used to set forth design specifications that ensure stability. For example, given a source characterization this method can be used to deduce properties that the load must satisfy in order to ensure the local stability of all operating points of interest.

To illustrate this method, consider the simple source-load system of Figure 23.13. Let the small-signal impedance characteristic of the source at an operating point x be denoted Z_x, and let the small-signal admittance characteristic of the load be denoted Y_x. Let the set Z represent the generalized impedance and the set Y represent the generalized admittance. Thus, $Z_x \in Z$ and $Y_x \in Y$ for all operating points of interest. The variation of values stems both from nonlinearities as well as parameter uncertainties.

After selecting a stability criterion, the generalized admittance analysis can be used to ensure a Nyquist gain margin GM and phase margin PM. The resulting stability constraint is best viewed in the immittance space. To this end, consider Figure 23.14. The x-axis of this figure is log of frequency, the y-axis is real part in hybrid dB [19], and the z-axis is imaginary part in hybrid dB. The volume to the right is a forbidden region of the load admittance. The forbidden region is obtained using the stability criteria as well as the generalized source impedance. The volume

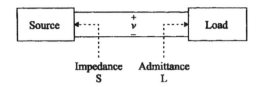

Fig. 23.13   Simple source – load system.

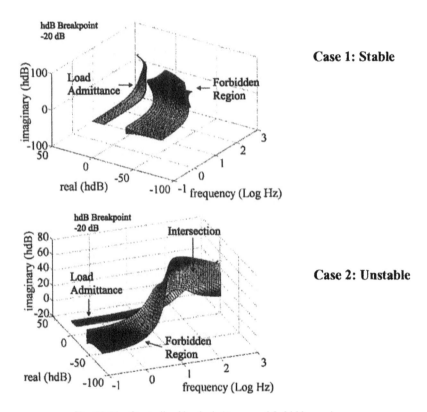

Fig. 23.14   Generalized load admittance and forbidden regions.

to the left represents a generalized load admittance. If the load admittance does not intersect the forbidden region, the system is stable in a small signal sense with the specified phase margin, gain margin and parameter uncertainties.

The basic ideas of generalized immittance analysis are set forth in [16–18]. These papers are primarily concerned with simple source-load systems. The extension of the method to large-scale systems is considered in [19] where a series of mapping functions is used to reduce any given system to a single source load equivalent. The steps to analyze the system are illustrated in Figure 23.15 and proceed as:

- Identify system configuration,

- Combine each conversion model and load,

- Combine the resulting three parallel aggregate L-converters into a single source and a single effective load.

Details on converter types, mapping operations, and intermediate stability tests along with more complete examples including other systems are set forth in [19].

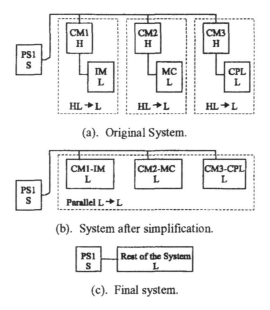

(a). Original System.

(b). System after simplification.

(c). Final system.

Fig. 23.15 Immittance based reduction process.

Experimental Validation

One of the key measures in evaluating these methods is how well they predict measured performance. Figure 23.16 and Figure 23.17 depict the measured time domain performance for Case 1 and Case 2 conditions, respectively. As can seen, the experimental results are consistent with both the time domain simulation and the Generalized Immittance Analysis. Once again, ADP methods could provide important new ways to explore the boundaries of system stability and to extend the treatment of system nonlinearities.

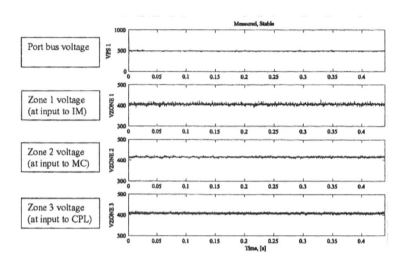

Fig. 23.16 Measured system performance for case 1.

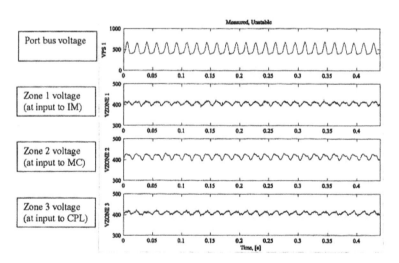

Fig. 23.17 Measure system performance for case 2.

23.4.4 Stability Analysis Method 3: Polytopic Stability Analysis

In the preceding section, stability analysis of a power electronics based distribution system using time domain simulation and generalized immittance analyses was demonstrated, with excellent results. However, there are shortcomings inherent to either method. In the case of time domain simulation, the results are limited to a very narrow range of conditions. Massive numbers of trajectories must be evaluated to gain confidence in the system performance. In the case of generalized immittance analysis, in some sense a more powerful result is obtained. Using a single analysis, an entire range of operating points can be proven to be locally stable. Furthermore, this approach can be used as a design synthesis tool by providing a method to formulate component specifications. However, the generalized immittance design approach does not guarantee a bounded response in the presence of large disturbances. Hence, there is motivation to perform a stability analysis in which a system can be proven to have an appropriately bounded response to large disturbances.

To this end, consider a broad class of nonlinear systems modeled by

$$\dot{\mathbf{x}} = \mathbf{F}(\mathbf{x}, \mathbf{u}), \tag{23.29}$$

$$\mathbf{y} = \mathbf{h}(\mathbf{x}, \mathbf{u}), \tag{23.30}$$

where $\mathbf{x} \in \Re^n$ is the state vector, $\mathbf{u} \in \Re^m$ is the input vector, and $\mathbf{y} \in \Re^p$ is the output vector. The above nonlinear model is referred to as the truth model of the underlying system. Systems of this form may be analyzed by the direct method of Lyapunov [10] and [11]. However, there are difficulties associated with applying the direct method of Lyapunov including the determination of a valid Lyapunov function candidate [20]. Analytical means are often impractical if not impossible so a numerical approach is necessary. Through the use of polytopic modeling and linear matrix inequalities the search for possible Lyapunov function candidates can be automated.

Defining local models of the form:

$$\dot{\mathbf{x}} = \mathbf{A}\mathbf{x} + \mathbf{B}\mathbf{u} + \phi_{\mathbf{x}}, \tag{23.31}$$

$$\mathbf{y} = \mathbf{C}\mathbf{x} + \mathbf{D}\mathbf{u} + \phi_{\mathbf{y}}, \tag{23.32}$$

that approximate the behavior of the truth model at a modeling point, $(\mathbf{x}_0, \mathbf{u}_0)$, of interest. The particular characteristics that the local model encapsulates varies depending on the method used in obtaining the local model. Examples include Taylor series, Teixeira-Zak [21], and the generalized Teixeira-Zak base approximations [22]. All three types of models coincide with the truth model at the operating point. In addition, the Teixeira-Zak based model and the Taylor series based model are proportional and coincide, respectively, to the first order behavior at the operating point.

One characteristic of particular interest, but not enforced by the above approximation methods, is having the right hand sides of the local model and the truth model equal one another at the equilibrium pair, (x_e, u_e), and the modeling point. Assigning coincident equilibrium pairs can be accomplished by the following procedure.

Step one: form the local model at the modeling point of interest. Step two: perform a coordinate transformation on the local model shifting the desired equilibrium pair to the origin. Step three: perform the generalized Teixeira-Zak based approximation on the shifted local model. Step four: shift the local model back to the original coordinates.

Once the local models have been obtained they are used as ingredients for the polytopic models constructed using a convex combination of local models:

$$\dot{\mathbf{x}} = \sum_{i=1}^{r} w_i(\theta) \left[\mathbf{A}_i \mathbf{x} + \mathbf{B}_i \mathbf{u} + \phi_{\mathbf{x}i} \right], \tag{23.33}$$

$$\mathbf{y} = \sum_{i=1}^{r} w_i(\theta) \left[\mathbf{C}_i \mathbf{x} + \mathbf{D}_i \mathbf{u} + \phi_{\mathbf{y}i} \right]. \tag{23.34}$$

Thus, (23.33) and (23.34) consist of r local models, defined by $\mathbf{A}_i$, $\mathbf{B}_i$, $\phi_{\mathbf{x}i}$, $\mathbf{C}_i$, $\mathbf{D}_i$, $\phi_{\mathbf{y}i}$, combined by weighting functions, W_i. The weighting functions must satisfy $0 \leq w_i(\theta) \leq 1$, $\sum_{i=1}^{r} w_i(\theta) = 1$, where θ may be a function of x or u. Polytopic models can accurately represent the nonlinear system over a wide range of operation. Although the truth model already has this property the structure of the polytopic model readily lends itself to searching for a Lyapunov function candidate.

Herein it is assumed that all of the local models have coincident equilibrium pairs and the Lyapunov function candidate is of the form:

$$V = \tilde{x}' P \tilde{x} \quad \text{where} \quad \tilde{x} = x - x_e. \tag{23.35}$$

Based upon these assumptions the following proposition can be stated.

Proposition 23.4.1 : *If there exists a common $P = P' > 0$ such that,*

$$A_i'P + PA_i < 0, \; i = 1, \ldots, r, \tag{23.36}$$

then the equilibrium state satisfying

$$A_i x_e + B_i u_e + \phi_i = 0 \; i = 1, \ldots, r \tag{23.37}$$

is globally uniformly asymptotically stable (GUAS) in the sense of Lyapunov (ISL) [22]. The search for a matrix P can be automated by setting up the system of linear matrix inequalities of the form (23.36) and using commercially available optimization routines. If a common P is found the polytopic model is globally uniformly asymptotically stable in the sense of Lyapunov. However stability analysis of the truth model is incomplete.

To complete the stability analysis of the truth model it is necessary to find the region of attraction around the equilibrium state using the direct method of Lyapunov. The region of attraction can be approximated by the largest level set of (23.35) contained within the region defined by $\dot{V} < 0$, where

$$\dot{V} = 2\tilde{x}' P\dot{x}, \qquad (23.38)$$

and $\dot{\tilde{x}} = F(\tilde{x} + x_e, \tilde{u} + u_e)$, where $\tilde{u} = u - u_e$. A Lyapunov function candidate (23.35) is constructed using P found in the polytopic model analysis. If a region of attraction is found then the truth model is uniformly asymptotically stabile (UAS) within this region. The following simple example demonstrates the potential of this analysis technique.

Polytopic Analysis Example

Consider the second order nonlinear system depicted in Figure 23.18 and having the parameters listed in Table 23.5. The source can be viewed as the NLAM of a 3-phase rectifier connected to an infinite bus [9]. The load can be viewed as the NLAM of a converter with a tightly regulated output and a valid operating range limited by the input voltage v_1, [23].

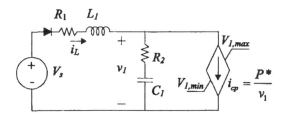

Fig. 23.18   Second-order nonlinear system.

Table 23.5   Second-Order Nonlinear System Parameters

Vs	595.49 V	$C1$	1.051 mF
$R1$	0.526 Ω	P^*	10 kW
$L1$	11.32 mH	$V1min$	550 V
$R2$	0.08305 Ω	$V1max$	650 V

The states of this system are chosen as the inductor current, i_L, and the capacitor voltage, v_c. Local models are obtained using Taylor series approximation at all combinations of $2.5A < i_L < 50A$, divided into 19 equally spaced points, and $550V < v_1 < 560V$, divided into 20 equally spaced points. The local models are then forced to have coincident equilibrium pairs using the procedure given in the local modeling section. Using linear matrix inequalities (23.36) formed using the local models a common P is found,

$$P = \begin{bmatrix} 10.16537 & 0.44366 \\ 0.44366 & 1.02148 \end{bmatrix},\tag{23.39}$$

which is symmetric and positive definite. This proves that the polytopic model is GUAS ISL.

To analyze the stability of the truth model, (23.38) is evaluated over a region of the state space surrounding the equilibrium state, see Figure 23.19. The region in which $\dot{V} > 0$ is shaded black and the operating voltage limits for the load are included as lines. The level set of V satisfying all three constraints forms an ellipse and is included along with one trajectory of the truth model. The ellipse identifies a region of uniform asymptotic stability for the truth model. The trajectory demonstrates the conservative nature of the Lyapunov based analysis.

23.4.5   Remarks on the Three Stability Analysis Methods

Each of the three stability analysis methods presented above contain serious limitations for which ADP methods may provide fundamental improvements. A discussion of these limitations and potential improvements follow. Actual stability involves the behavior of the time evolution of the system state trajectory. As discussed in the dc Stability Definition section, the determinations of the boundary between stable and unstable operation is difficult and dependent on a specific, formal definition of stability. A more useful metric would involve the calculation of a stability margin which might be a generalization of the classic Nyquist gain and phase margins.

The difficulty in experimental stability observations involves the need to perform exhaustive and potentially destructive testing; ambiguity regarding the specific onset of instability; and problems isolating the source of instability due to the dynamically interdependent nature of power systems. Compared with experimental observation, time domain simulation lessens the burden of time domain stability assessment. Unfortunately, exhaustive time domain simulation remains computationally prohibitive. Moreover, ambiguities regarding the source and onset of instability remain. ADP methods could provide fundamentally new time domain methods to detect the proximity to instability and avoid unstable modes of operation.

Frequency domain generalized immittance stability assessment provides an experimentally validated unambiguous stability assessment including stability margins but is limited to systems represented by linear networks or equivalently, small perturbations of a linear approximation of the actual system. The use of ADP methods to extend generalized immittance analysis to uncertain, stochastic and nonlinear systems could have a profound effect on the design of isolated dc power systems.

While time domain observation and immittance based analysis both offer insight into the stability of nonlinear systems Lyapunov methods identify a region of asymptotic stability about an equilibrium point. The traditional obstacles to nonlinear Lyapunov stability assessment include the search for Lyapunov function candidates and overly conservative results. Recently, a polytopic model structure has been shown which allows automation of the search for function candidates. Figure 23.19 provides an illustrative example of a stable state trajectory derived from polytopic Lyapunov stability analysis of the Navy IPS benchmark system. Although ADP methods could be applied to the generation of polytopic models, a more direct and profound opportunity involves using ADP to search for the least conservative Lyapunov function candidate. The ultimate objective involves the "holy grail" of system stability: the determination of the region of asymptotic stability for higher order nonlinear systems

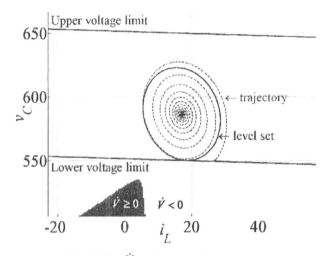

Fig. 23.19 $\dot{V}$ evaluated over the state space.

23.5   SUMMARY OF POWER SYSTEM CHALLENGES AND TOPICS

The previous sections have provided some detailed discussions of several important power systems problems illustrated by two important systems: the terrestrial and the Navy systems, respectively. In the present section, we would like to provide a comprehensive summary of what is needed in power systems applications with individual topics identified and discussed.

23.5.1   Transmission Stability Analysis

Much of the non-linear analysis of power system behavior subjected to exogenous disturbances has focused on conventional analytical methods including risk assessment and Lyapunov stability assessment. Voltage and angle stability assessments are needed to ensure dynamic reconfiguration in response to system faults and disturbances. Since the margin of stability cannot be obtained using numerical methods, an energy-based method has been proposed for angle stability. To date, real-time evaluation of the system under different loading and unknown contingencies has not been implemented.

Several researchers in the United States and abroad have done extensive research on the application of the Newton-Raphson method, which accurately shifts a set of electrical network poles and transfer function zeros to more suitable location in the complex plane to improve the harmonic voltage performance of the power system. Their work dealt with eigenvalue sensitivity coefficients, which are used for computing the element changes that are the most cost effective in computing the system Jacobian elements for the Newton's method. Realistic results of a modeled system show promising results [28].

Also work on small signal stability, control applications and harmonic analysis have been undertaken recently by researchers in determining Hopf bifurcations and minimum distances to small signal stability boundaries in the control parameter space as well as computation of reduced models of high order multivariable transfer functions [29, 30].

23.5.2   Voltage Stability

Voltage stability employs chaos and bifurcation theory to solve the differential equations and assess the stable and unstable equilibrium points (SEP) and (UEP), respectively. The technique based on load flow requires a series of load flow runs with changes in parameter space (load) and category.

Solution:
Traditional methods for solving stability problems include Proximity Indicator method, Singular Value Decomposition, Jacobian Condition Number and Energy Margin methods. To achieve a high performance method of voltage stability, the following challenges need to be addressed: (1) Increased nonlinearity of the problem space and (2) Model and parameter uncertainty. For theses challenges, adaptive control and optimization techniques are needed in solving the uncertainty problem inherent in data measurements to enhance system self-healing under any contingency.

23.5.3   Adaptive Control and Optimization

Adaptation and control strategy scheduling must consider the stochastic and dynamically interdependent attributes of power systems. An efficient optimization technique must:

- Achieve optimal scheduling subject to technical constraints,

- Adapt to perturbation of power system dynamics due over time (in state, Pareto, controls),

- Adapt to varying random noise and uncertainties,

- Adapt to changes in system structure, and, finally,

- Distinguish between observable and unobservable measurements.

The techniques available to handle these problems are based on vague methods and somewhat heuristic methods that tend to give a non-optimal solution. The system stochasticity and dynamic interdependence lead to unstable, non-feasible and non-optimal solutions.

Solution:
Using Dynamic Stochastic Optimal Power Flow (DSOPF), we need a multi-objective scheme to accommodate evolving systems dynamics in the presence of missing or corrupted measurements. State-of-the-art techniques from Operations Research, Information Technology and Computational Intelligence communities can improve existing optimization methods.

23.5.4   Reliability Assessment

One of the increasing requirements of Electric Power Systems (EPS) and Integrated Power Systems (IPS) is their ability to survive under different attacks and maintain some level of invulnerability. The assessment of indices such as Loss of Load Probability (LOLP) and Expected Unserved Energy (EUE) is aimed at the determination of what energy / load can be supplied over time for different reliability services. The cost component balances the cost of reliability improvements with the cost of service disruptions. The ultimate goal is to achieve the maximum reliability under a variety of probable contingencies.

The complexity associated with reliability improvement may significantly increase the cost of service delivery. As a result, cost effective reliability and reconfiguration assessment strategies are needed that address the following challenges: (1) Non-linearity of the power system parameters (2) Uncertainty in load demand and generation availability (3) System dynamics and (4) Stochasticity of the system parameters.

Intelligent control methods including Adaptive Critics Designs, Fuzzy Logic and Artificial Neural Networks (ANN) methods may provide adaptive stochastic solutions to the general problem formulation.

23.5.5   Economic Benefit and Control Strategy

Innovative control and optimization strategies for improving power systems efficiency and security are urgently required. Moreover, the overall nonrecurring and recurring personnel, equipment and software costs must be calculated. These infrastructure costs must be balanced by the economic and social impacts and the value of improved continuity of power delivery. Table 23.6 presents a summary of the terrestrial power system economic benefits and control strategy tradeoffs.

Cost–benefit analysis should include consideration of nonlinear system dynamics, hierarchical control law implementation, optimal algorithms and intelligent technologies to achieve the desired stability and reliability. Intelligent control solutions such as Artificial Neural Networks, Learning Algorithms, and Fuzzy Logic approaches may provide hybridized methods which extend existing tools.

To this end, we present a number of control/optimization problems as challenges to researchers wishing to explore new algorithms and determine quantifiable performance metrics. Problems associated with realistic control/optimization solutions include: (1) Incomplete system knowledge/model formulation and data, (2) Nonlinearities, (3) Noise, and (4) Un-modeled system dynamics and delays.

23.5.6   Isolated System Stability

The Navy benchmark system represents the characteristics of emerging isolated power systems including a variety of vehicle and alternative energy applications. These systems differ from conventional terrestrial systems in that they involve (1) finite intertia sources, (2) power electronic-based distribution and conversion, (3) loads and disturbancesthat are large relative to the source capacity and (4) constant power loads with potentially destabilizing system dynamics. Due to the high cost and potentially catastrophic consequences of interruption of power delivery to critical power services, the Navy benchmark problem focuses on nonlinear system stability despite stochastic model variations and disruptions. The key opportunity involves the use of ADP methods to improve power system robustness with respect to natural and hostile disruptions. One approach involves new methods to more accurately determine system stability boundaries. Improved nonlinear stochastic stability predictions would have important offline design applications, along with the potential to provide online stability margin assessments. The second approach involves the use of ADP to derive intrinsically nonlinear stabilizing controllers with improved quality and continuity of service.

Table 23.6   Power System Economic Benefit and Control Strategy

Summary of Terrestrial Reference Problems	Power System Economic Benefit and Control Strategy Objectives		
	System Requirements		New Optimization Techniques
	Existing	Future	
Voltage Stability	Static model with limited dynamics	Dynamically interdependent system with uncertainty	Voltage stability margin (VSM) constraints using ADP[1] DSOPF[2] stochastic formulation
Angle Stability	Transient model with limited dynamics	Dynamically interdependent online stochastic assessment and control	Constrained angle stability margin (ASM) using ADP DSOPF subject to contingencies
Congestion	Probability based static congestion constraints	Dynamically interdependent congestion based stochastic constraints	Dynamic margin based index for congestion subject to contingencies and available transfer capability
Reliability	Expected Unserved Energy (EUE) and Loss of Load Probability, (LOLP)	Dynamically interdependent margins subject to stochastic contingencies	Time domain continuity of service ADP DSOPF formulation
Coordinated Control	Model is either discrete or continuous	Dynamically interdependent hybrid modeling of discrete and continuous processes	Hybrid adaptive distributed control with contingency management
Unit Commitment, Optimal Power flow	Lagrange relaxation methods	Dynamically interdependent planning subject to stochastic contingencies	Dynamic and stochastic indices based on ADP and Interior Point (IP) Methods DSOPF
Cost Benefit Analysis	Net Present Value (NPV) and Annual Worth or Cost (AW/AC)	Incorporation of uncertainty into market and investment and predictions	Stochastic ADP DSOPF formulation

[1] Approximate Dynamic Programming (ADP)   [2] Dynamic Stochastic Optimal Power Flow

23.6   SUMMARY

This chapter presents a grand overview and introductory remarks for power system benchmark of both utility and Navy systems. Details of existing modeling techniques for such systems have been presented and their limitations addressed.

Conventional optimization and control system theories and tools have been used in solving the power system problems of both utility and Navy systems. However, since the computational tools developed by power system researchers are limited in some extent to dealing with uncertainties of the power system along with its multifaceted dynamics, new optimization techniques that can handle uncertainties and with enhanced capability for predicting future performances are required.

The formulation of different power system problems such as the modeling of system components with dynamics, design and analysis as well as control coordination are presented with illustrative examples using conventional techniques. Test cases of established work have been added to supplement readers understanding.

The basic idea is to extend the performance of these conventional tools by incorporating new computational intelligence approaches, which can handle foresight and uncertainty of the power grid. Approximate Dynamic Programming (ADP), a powerful tool from the computational intelligence community, is capable of handling these problems. Various power system problems are reformulated such that ADP can handle it effectively.

Bibliography

1. *Architectures for Secure and Robust Distributed Infrastructures*, University Research Initiative Web Site:
   http://element.stanford.edu/~lall/projects/architectures/.

2. J. A. Momoh, *Electric Power System Applications of Optimization*, Marcel Dekker Inc., New York, 2001.

3. A. A. Fouad and V. Vittal, *Power Systems Transient Stability Analysis Using the Transient Energy Function Method*, Prentice-Hall, Englewood Cliffs, NJ, 1992.

4. J. A. Momoh and M. El-Hawary, *Electric Systems Dynamics and Stability with Artificial Intelligence Applications*, Marcel Dekker Inc., New York, 2000.

5. S. D. Pekarek et al., A hardware power electronic-based distribution and propulsion testbed, *Proc. 6th IASTED International Multi-Conference On Power and Energy System*, Marina del Rey, CA, 2002.

6. S. D. Sudhoff, S. D. Pekarek, B. T. Kuhn, S. F. Glover, J. Sauer, and D. E. Delisle, Naval combat survivability testbeds for investigation of issues in shipboard power electronics based power and propulsion systems, *Proc. IEEE Power Engineering Society Summer Meeting*, Chicago, 2002.

7. S. D. Pekarek et al., Development of a testbed for design and evaluation of power electronic based generation and distribution system, *SAE2002 Power Systems Conference*, Coral Springs, FL, 2002.

8. S. D. Sudhoff, S. F. Glover, S. H. Zak, S. D. Pekarek, E. L. Zivi, D. E. Delisle, and D. Clayton, Stability analysis methodologies for DC power distribution systems, *Proc. Thirteenth Ship Control Systems Symposium (SCSS 2003)*, Orlando, FL, 2003.

9. P. C. Krause, O. Wasynczuk, and S. D. Sudhoff, *Analysis of Electric Machinery, 2nd Ed.*, Wiley, New York, 2002.

10. S. H. Zak, *Systems and Control*, Oxford University Press, New York, 2003.

11. H. K. Khalil, *Nonlinear Systems, 2nd Ed.*, Prentice-Hall, Upper Saddle River, NJ, 1996.

12. http://www.ESAC.info.

13. S. D. Pekarek, S. D. Sudhoff, J. D. Sauer, D. E. Delisle, and E. L. Zivi, Overview of the naval combat survivability program, *Proc. Thirteenth Ship Control Systems Symposium (SCSS 2003)*, Orlando, FL, 2003.

14. *MATLAB The Language of Technical Computing*, pp. 1760-2098, The Math-Works, Inc., Natick, MA, 2000.

15. *Advanced Continuous Simulation Language (ACSL) Reference Manual*, Aegis Simulation, Inc., Huntsville, AL, 1999.

16. S. D. Sudhoff, D. H. Schmucker, R. A. Youngs, and H. J. Hegner, Stability analysis of DC distribution systems using admittance space constraints, *Proc. Institute of Marine Engineers All Electric Ship 98*, London, 1998.

17. S. D. Sudhoff, S. F. Glover, Three dimensional stability analysis of DC power electronics based systems, *Proc. Power Electronics Specialist Conference*, pp. 101-106, Galway, Ireland, 2000 .

18. S. D. Sudhoff, S. F. Glover, P. T. Lamm, D. H. Schmucker, and D. E. Delisle, Admittance space stability analysis of power electronic systems, *IEEE Trans. Aerospace and Electronics Systems*, vol. 36. no. 3, pp. 965–973, 2000.

19. S. D. Sudhoff, S. D. Pekarek, S. F. Glover, S. H. Zak, E. Zivi, J. D. Sauer, and D. E Delisle, Stability analysis of a DC power electronics based distribution system, *SAE2002 Power Systems Conference*, Coral Springs, FL, 2002.

20. H. Lim and D. C. Hamill, Problems of computing Lyapunov exponents in power electronics, *Proc. IEEE International Symposium on Circuits and Systems*, pp. 297–301, 1999.

21. M. C. M. Teixeira and S. H. Zak, Stabilizing controller design for uncertain nonlinear systems using fuzzy models, *IEEE Trans. Fuzzy Systems*, vol. 7, no. 2, pp. 133–142, 1999.

22. S. F. Glover, S. H. Zak, S. D. Sudhoff, and E. J. Zivi, Polytopic modeling and Lyapunov stability analysis of power electronics systems, *Society of Automotive Engineers 2002 Power Systems Conference*, Coral Springs, FL, 2002.

23. S. D. Sudhoff, K. A. Corzine, S. F. Glover, H. J. Hegner, and H. N. Robey, DC link stabilized field oriented control of electric propulsion systems, *IEEE Trans. Energy Conversion*, vol. 13, no. 1, pp. 27–33, 1998.

24. A. A. Fouad and V. Vittal, Power systems transient stability analysis using the transient energy function method, in C. T. Leonides, (ed.), *Control and Dynamic Systems*, Academic Press, San Diego, CA, 1991.

25. A. J. Woods and B. F. Wollenberg, *Power Generation, Operation and Control*, Wiley, New York, 1984.

26. G. T. Heydt, *Computer Analysis Methods in Power Systems*, Macmillan, New York, 1986.

27. A. R. Bergen, *Power Systems Analysis*, Prentice-Hall, Englewood Cliffs, NJ, 1986.

28. S. Luis, N. Martins and L. T. G. Lima, A Newton-Raphson method based on eigenvalues sensitivities to improve harmonic voltage performance, *IEEE Trans. Power Delivery*, vol. 18, no. 1, 2003.

29. N. Martins, S. Gomes Jr, P. E. M. Quintao, J. C. R. Ferraz, S. L Varricchio, and A. de Castro, Some recent developments in small signal stability and control, *Power Engineering Society Winter Meeting*, vol. 2, pp. 1171–1177, 2002.

30. V. Ajjarapu, Application of bifurcation and continuation methods for the analysis of power system dynamics, *Proc. 4th IEEE Conference on Control Applications*, pp. 52–56, 1995.

Index

Printed and bound by CPI Group (UK) Ltd, Croydon, CR0 4YY

27/10/2024

14580256-0004